STATISTICS

BASIC TECHNIQUES FOR SOLVING APPLIED PROBLEMS

STATISTICS

BASIC TECHNIQUES FOR SOLVING APPLIED PROBLEMS

Stephen A. Book

California State College
Dominguez Hills

McGRAW-HILL BOOK COMPANY

New York   St. Louis   San Francisco   Auckland   Bogotá   Düsseldorf
Johannesburg   London   Madrid   Mexico   Montreal   New Delhi   Panama
Paris   São Paulo   Singapore   Sydney   Tokyo   Toronto

This book was set in Optima by Black Dot, Inc.
The editors were A. Anthony Arthur, Alice Macnow,
and Anne T. Vinnicombe;
the designer was Joan E. O'Connor;
the production supervisor was Leroy A. Young.
The drawings were done by ECL Art Associates, Inc.
Kingsport Press, Inc., was printer and binder.

STATISTICS

BASIC TECHNIQUES FOR SOLVING APPLIED PROBLEMS

1 2 3 4 5 6 7 8 9 0 K P K P 7 8 3 2 1 0 9 8 7 6

Library of Congress Cataloging in Publication Data

Book, Stephen A
Statistics: basic techniques for solving applied problems.

Includes bibliographical references and indexes.
1. Statistics.   2.   Statistics—Problems, exercises, etc.   I.   Title.
HA29.B733   519.5   76-18718
ISBN 0-07-006493-8

To our parents

CONTENTS

PREFACE

This text has as its objectives the explanation of the role of statistics in the solution of applied problems and the illustration of some basic statistical methods used in applied work. In accordance with these objectives, the discussion is applications-oriented; applied examples and exercises are the backbone of the text. All statistical methods introduced are fully explained and illustrated in terms of situations drawn from economics and the management sciences, psychology, sociology, political science and the other social sciences, and the life, medical, and agricultural sciences. The multitude of examples solved in the text, discussed by the instructor in the classroom, and worked by the student outside of class are intended to produce an understanding of the common uses of statistical reasoning in answering quantitative questions in all the above fields of knowledge.

In order to understand the applied examples and exercises, it is not necessary for the student to begin the course in possession of a formal background in mathematics. Although some experience in algebra would allow the student to progress more rapidly, the text begins at a very low level, and new mathematical techniques are introduced very slowly as needed. With the instructor providing a few minutes of appropriate (depending on the backgrounds of the students) remarks on topics in arithmetic and algebra from time to time, the student should be able to build up the skills necessary to handle each succeeding section. Naturally, in courses where students have a full high-school algebra background or even some college mathematics, it would be possible to bypass the more elementary portions of the discussion and proceed directly to the statistical methods.

Although this text focuses on applications and assumes only a very minimal mathematical background, we have included algebraic justifications of formulas and other facts. These justifications are presented in supplements to all points of the discussion where the procedures can be fully explained *without* the use of any higher mathematics, such as calculus. All of these supplements are purely optional, and they can be ignored completely without any loss to a student interested primarily in applications. Nevertheless, the presence of these justifications in the text relieves the instructor of the obligation to "prove theorems" in response to questions of "why?" from the more advanced students. The supplements illustrate that the statistical techniques studied are based on logical mathematical reasoning (not a general plan to irritate and confuse students!), and that the statistical analysis of an applied problem is a realistic and logically valid way of solving it.

The heart of the elementary statistics course consists of the first six chapters. In these chapters, the student will learn graphical and numerical ways to describe a set of data, the binomial and the normal distributions, confidence intervals, fundamentals of hypothesis testing, and linear regression and correlation. This material can be covered in one quarter or one semester. Depending on the interests of the instructor and the students, there will often also be time to discuss one or more of the special topics covered in the later chapters: curvilinear and multiple regression, probability (including the derivation of the binomial probability formula), chi-square tests of independence or goodness of fit, analysis of variance, or nonparametric statistics. In a full-year course, all this material can probably be covered.

This text has benefited greatly from the comments and suggestions of Professors James M. Edmondson of Santa Barbara City College, Walter H. McCurdy of Bradley University, Linda Mercier of the State University of New York Agricultural and Technical College at Delhi, Charles O. Minnich of Mt. San Antonio College, John S. Mowbray of Shippensburg State College, Frederic C. Parker of Jefferson Community College, and Raymond C. Sansing of Virginia Commonwealth University. To all of these professors, I wish to express my appreciation and gratitude for their detailed reading of the manuscript and their valuable suggestions for improvement.

I would also like to thank Mr. Thomas A. Cochran of the Mathematics Department of Carson High School, Carson, California, for his assistance in proofreading the manuscript and working out the solutions to all the exercises.

It was a pleasure to work with the editors, Tony Arthur, Alice Macnow, and Anne Vinnicombe of McGraw-Hill, on this project, and I would like to thank them for their expert guidance before, during, and after the writing of the manuscript.

I am grateful to Hafner Press, and to the *Biometrika* Trustees for their permission to reproduce certain statistical tables in the Appendix.

Finally, I will close by acknowledging the invaluable assistance of my wife, who read the entire manuscript and viewed it from a perspective completely different from that of all the other reviewers.

Stephen A. Book

STATISTICS

BASIC TECHNIQUES FOR SOLVING APPLIED PROBLEMS

GRAPHICAL METHODS OF ORGANIZING DATA

Statistics, an applied branch of mathematics, has as its goal the organization, analysis, and explanation of facts and figures arising out of the study of the social, behavioral, management, and natural sciences. It seeks to develop general rules of behavior of biological, economic, and psychological phenomena, for example, for the dual purposes of understanding these phenomena as they are and of controlling and improving them for society's benefit. Statistics originated almost a century ago in attempts to understand the factors influencing industrial and agricultural production, in order to better control these factors and so to increase production at reasonable cost. Today the aim of statistics is no different, although it has expanded its vista to include additional aspects of the various sciences.

As an example of one type of problem which statistics can be used to solve, consider the comparison in effectiveness between two different kinds of chemical fertilizer for use in agriculture. The researcher must account for several problems which are more complicated than the simple question: Which kind is better? These problems may be summarized as follows:

1   What do we mean by "better," and how can we measure such a quality?
2   Is the first product "better" under some circumstances, while the second is "better" in other situations?
3   If the "better" one is more expensive, is it superior enough to warrant the increased cost?

In the case of agricultural fertilizer, there are at least two components of the determination of which fertilizer is "better": a biological component and an economic component. The biological component would involve the comparative yields of, say, a certain vegetable using each kind of fertilizer. The economic component would be the relative cost of the two fertilizers. A possible measure of fertilizer effectiveness in this situation could be the number of pounds of the vegetable obtained from an acre of farmland per dollar spent on the fertilizer.

If we agree on this way of evaluating the fertilizers, we can plant 2 acres worth of the vegetable and treat 1 acre with the first kind of fertilizer and the other with the second. We can keep records on how much we spend on each fertilizer and how many pounds of vegetables we get at harvest time from each acre. Then we could say that the fertilizer whose acre had the greatest yield per dollar is the "better" fertilizer.

Some questions still remain, however. It may be that, had we chosen a different vegetable or different plots of farmland or even a different geographical area, the "better" fertilizer would have come out second best. Maybe even if we do exactly the same experiment next year, the results will be reversed. How can we be sure that the fertilizer that we have decided is "better" really is? Well, in truth, we can't be 100% sure. We can improve our degree of certainty by planting several acres of land, not just two, for our experiment and by carrying out the experiment over a period of years. How many acres and how many years? That depends on how certain we want to be that we have eliminated all random variation in arriving at our decision and that our decision is the correct one. The more acres we plant, the more certain we can be of our decision.

At this point, mathematics enters the discussion. You may recall from your study of mathematics that mathematics answers questions of how many, what percentage, etc. In carrying out a statistical analysis, the mathematics is used only for the purpose of expressing the relationships between the various numbers involved. For example, the degree of certainty in our decision between the two fertilizers will be related by a mathematical formula to the number of acres planted and the difference in the yields. If we plant a particular number of acres, we will obtain particular yields, and then using that mathematical formula, we will be able to calculate the degree of certainty in our decision.

A statistical problem begins with a question of applied interest under consideration in one of the various sciences mentioned above. However, the mathematical techniques of statistical analysis do not enter the discussion until some data bearing on the problem are accumulated. (The proper methods of accumulating data are studied in a relatively nonmathematical branch of statistics called "statistical sampling.") After the data are collected, the next step is often the organization of the data in an understandable and easily communicable format. Sometimes the data can be organized into a pictorial or graphical structure which allows the data to be perceived visually by persons not very familiar by training or experience with numerical descriptions. Visual methods, when available, are excellent vehicles for communicating statistical information quickly and clearly to those not trained in numerical thinking.

To persons accustomed to mathematical thinking,[1] however, visual methods of data description are interesting, but numerical descriptions are infinitely more useful. Numerical descriptions of data, as we will discover in future chapters of this book, communicate a wealth of detailed and precise information that belies the relative simplicity of the methods involved. It remains true nevertheless that visual descriptions of the data are very useful for purposes of communicating information. We introduce, in this chapter, some of the important techniques of pictorially organizing a set of data so as to more easily convey the information contained therein.

SECTION 1.A FREQUENCY DISTRIBUTIONS AND THEIR GRAPHS

An electric utility must anticipate the demand for electricity in the region it serves in order to be sure of being able to meet normal demands. Although it is unreasonable to expect the company to be able to meet every demand, no matter how extraordinary, the company should be able to supply the desired amount of electricity almost all the time. To get some idea of the demand for electricity, a major electric company recorded the demands upon it each day for the past 100 days. It expects to use the information obtained to help it forecast future demand, or at least to help it get an idea of maximum level of demand that it will be expected to satisfy. Table 1.1 contains the data obtained by the electric company: the demand for electric power, in millions of kilowatthours, over the past 100 days.

As can immediately be seen from Table 1.1, not much can be specified about the

[1]The objective of this book is to make you one of them!

TABLE 1.1
Demand for Electricity over 100 Days
In Millions of Kilowatthours

25	32	28	26	30	21	36	27	32	32
33	23	31	29	33	31	29	33	28	31
27	29	29	33	27	28	32	23	30	25
29	32	21	24	30	31	28	32	31	34
31	26	30	30	34	26	26	29	32	22
24	31	28	32	28	37	30	35	27	33
28	30	33	30	31	30	32	28	31	28
29	26	32	24	29	32	30	22	34	26
32	27	27	34	29	31	33	39	30	33
30	28	31	30	37	31	25	32	33	28

demand for electricity from a glance at the data. The problem is that we have an unorganized collection of numbers, and no trends indicating possible demand levels are discernible. In order to make use of the mass of available data, we must first organize it into an understandable form.

To make sense out of the numbers in Table 1.1, we divide the range of data into a reasonable number of intervals and count the number of data points falling into each interval. This procedure results in what is called a "frequency distribution" of the data, and is illustrated in Table 1.2. Because the minimum daily demand for electricity is at the level of 21 million kilowatthours over the 100-day period, while the maximum daily demand is at 39 million kilowatthours, a reasonable set of intervals to use would be 20 to 22, 23 to 25, 26 to 28, 29 to 31, 32 to 34, 35 to 37, 38 to 40. In choosing a set of intervals, care must be taken regarding the following points:

1 Every data point must fall into *exactly one* of the intervals, no more and no less.
2 The number of intervals must not be too large, for then no substantial improvement in understanding the trend of the data would be made.
3 The number of intervals must not be too small, for then variations among the data points would be somewhat obscured.
4 It is not necessary that all intervals have the same length, but extreme fluctuations in the sizes of consecutive intervals should be avoided.

Once we have agreed upon the intervals, the next step is to tally the number of data points falling in each interval. Basically, this is all there is to the task of constructing a frequency distribution. The result appears in Table 1.2. As a check on the correctness of our counting, we can sum the number of days in the far right-hand column. If our counting was correct, we should get a total of 100 days, the number of days involved in the electric company's data.

Some useful information is apparent from a glance at the frequency distribution of Table 1.2, whereas almost nothing could be learned from a glance at the original data of

TABLE 1.2
Frequency Distribution of Demand for Electricity

Interval, millions of kilowatthours	Number of Days Having Demand in the Interval	
20–22	ⵏⵏⵏⵏ	4
23–25	ⵜⵂⵍ ⵏⵏⵏ	8
26–28	ⵜⵂⵍ ⵜⵂⵍ ⵜⵂⵍ ⵜⵂⵍ ⵏⵏⵏ	23
29–31	ⵜⵂⵍ ⵜⵂⵍ ⵜⵂⵍ ⵜⵂⵍ ⵜⵂⵍ ⵜⵂⵍ ⵏⵏⵏⵏ	34
32–34	ⵜⵂⵍ ⵜⵂⵍ ⵜⵂⵍ ⵜⵂⵍ ⵜⵂⵍ ⵏ	26
35–37	ⵏⵏⵏⵏ	4
38–40	ⵏ	1
		Sum = 100

Table 1.1. For example, two things we learn from the frequency distribution are the following:

1 Approximately one-third of the time (34 days out of the 100), the demand is about 30 million kilowatthours.
2 The demand exceeds 34 million kilowatthours only about 5% of the time (the 4 days corresponding to the interval 35 to 37, and the 1 day corresponding to the interval 38 to 40).

There are two common vehicles for pictorially communicating the information in a frequency distribution. The first of these methods is the "histogram" (or bar graph), and the second is the "frequency polygon" (or line graph). Although the two types of graphs are closely related, there are important differences in the details of constructing them.

To construct a histogram, we must modify the frequency distribution in such a way that there are no gaps between intervals. For the data in Table 1.2, we close the gap between the two intervals 20 to 22 and 23 to 25 by setting the point of division at 22.5, halfway between 22 (the end of the first interval) and 23 (the beginning of the second). To maintain the relative length of the intervals, we consider that the first interval starts at 19.5 and ends at 22.5, while the second starts at 22.5 and ends at 25.5. The modified frequency distribution appears in Table 1.3. Using Table 1.3, we construct the histogram of Fig. 1.1 by drawing bars one interval wide to a height corresponding to the number of days in that interval.

To construct a frequency polygon, we have to modify the original frequency distribution (Table 1.2) in a different manner. Here we must find the midpoint of each of the intervals, for we consider each interval to be represented by its midpoint. From Table 1.2, we determine the midpoints of the intervals preceding 20 to 22 and following 38 to 40, as if those intervals really existed in the table. The frequency distribution modified in preparation for the frequency polygon appears in Table 1.4, while the frequency polygon itself appears in Fig. 1.2. The frequency polygon is constructed by connecting dots which represent the number of days that each demand level is attained, considering an interval of demand levels to be represented by its midpoint.

TABLE 1.3
Frequency Distribution of Demand for Electricity,
Modified for Construction of Histogram

Original Interval	Modified Interval Boundaries	Number of Days
20–22	19.5–22.5	4
23–25	22.5–25.5	8
26–28	25.5–28.5	23
29–31	28.5–31.5	34
32–34	31.5–34.5	26
35–37	34.5–37.5	4
38–40	37.5–40.5	1

FIGURE 1.1
Histogram of demand for electricity.

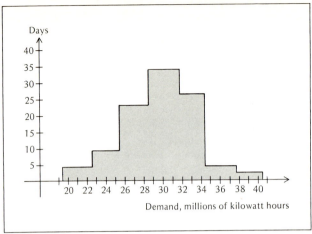

You should observe that the histogram and the frequency polygon both communicate the same information, namely, the facts contained in the frequency distribution of Table 1.2. They both illustrate that demand for electricity is rarely below 24 million kilowatthours, very often is between 26 and 34 million kilowatthours, and falls off sharply above 35 million kilowatthours. These facts were not at all evident from the original data in Table 1.1.

There are some situations in which special conditions require modified versions of the histogram or frequency polygon. For example, in a certain type of problem, it may

FIGURE 1.2
Frequency polygon of demand for electricity.

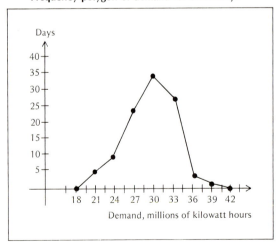

TABLE 1.4
**Frequency Distribution of Demand for Electricity,
Modified for Construction of Frequency Polygon**

Interval	Midpoint	Number of Days
(17–19)	18	0
20–22	21	4
23–25	24	8
26–28	27	23
29–31	30	34
32–34	33	26
35–37	36	4
38–40	39	1
(41–43)	42	0

be sometimes useful to use intervals of unequal length. In other cases, there may be open-ended intervals or large gaps between intervals where no data points are located. Sometimes we may need a graph which records a running cumulative total of the data points. For a discussion of these and other variations of the pictorial representations we have presented, the reader is invited to consult the indicated item in the bibliography at the end of this chapter.

EXERCISES 1.A

1 A geographer studying West Africa would like to know the size distribution of stones in the delta region where the Niger River runs into the Gulf of Guinea. He collects 50 stones and measures the diameter of each, and so obtains the following measurements, in centimeters:

4	5	9	13	8	19	9	8
17	3	6	7	10	14	20	1
6	11	2	7	12	17	2	18
3	7	14	1	8	3	12	5
6	16	8	11	4	7	6	5
4	6	10	7	12	7	6	15
5	9						

a Construct a frequency distribution of the diameters based on the intervals 1 to 4, 5 to 8, 9 to 12, 13 to 16, and 17 to 20.
b Draw a histogram which illustrates the frequency distribution of part a.
c Draw a frequency polygon which illustrates the frequency distribution of part a.

2 A popular index of common stocks traded on the New York Stock Exchange shows that current values, in dollars, of the 500 stocks listed have the following frequency distribution:

Dollar Value	No. of Stocks Having Value in Interval
0.00– 9.99	51
10.00–19.99	98
20.00–29.99	101
30.00–39.99	151
40.00–49.99	48
50.00–59.99	25
60.00–69.99	13
70.00–79.99	8
80.00–89.99	4
90.00–99.99	1

Construct a frequency polygon based on the frequency distribution of stock prices.

3 The 50 employees of a small factory have yearly earnings distributed as follows:

Earnings Level, dollars	No. of Employees at That Level
0.00– 2999.99	3
3000.00– 5999.99	5
6000.00– 8999.99	18
9000.00–11999.99	15
12000.00–14999.99	6
15000.00–17999.99	0
18000.00–20999.99	3

Draw a frequency polygon which illustrates the data.

4 One hundred housing units in a major United States city are selected at random from the official property tax list, and each is rated according to quantity and quality of plumbing, electrical, and other mechanical facilities. Each unit is assigned a score from 0 to 100, and the resulting ratings appear in the following frequency distribution:

Score	No. of Units
0.0– 19.5	30
19.5– 39.5	15
39.5– 59.5	5
59.5– 79.5	10
79.5–100.0	40

a Draw a histogram which illustrates the data.
b Construct a frequency polygon based on the data.

5 A psychologist conducted a learning experiment aimed at finding how fast monkeys were able to learn that a red light means "stop," while a green one means "go." Each time the monkey made the correct decision, he or she was rewarded with a banana, and each time the monkey made a wrong response, he or she was punished

with an electric shock. One hundred monkeys participated in the experiment, and the following data lists the number of trials necessary for each to learn the lesson:

1	6	17	11	4	18	24	34	2	26
4	2	7	3	12	5	10	1	15	3
16	5	3	8	9	13	1	20	2	14
6	2	8	4	9	5	14	1	2	10
1	7	7	6	5	10	5	9	15	3
6	25	3	8	8	1	6	13	19	10
20	2	7	12	4	9	2	7	9	18
1	6	11	3	19	5	8	3	8	10
21	30	2	5	4	7	17	23	4	9
10	1	4	3	6	16	22	28	32	5

a Set up a frequency distribution of the number of trials required to learn the lesson using the intervals 1 to 5, 6 to 10, 11 to 15, 16 to 20, 21 to 25, 26 to 30, 31 to 35.
b Construct a frequency polygon from the frequency distribution of part a.

6 The following data[1] record the forecasts made in June 1970 of the 1971 Consumer Price Index by 47 leading economists:

134.6	139.0	141.4	142.1	141.0	138.0
140.5	135.0	139.2	141.0	138.0	141.4
141.5	140.7	134.6	138.0	139.5	139.0
140.0	141.5	138.0	137.0	139.0	140.0
142.0	138.0	140.0	139.0	137.0	140.9
137.0	141.0	139.0	140.4	141.5	137.0
141.0	138.6	142.0	143.0	140.5	141.5
138.2	142.0	143.0	142.6	142.0	

a Construct a frequency distribution of the data, using five intervals.
b Draw a frequency polygon based on the frequency distribution of part a.

SECTION 1.B RELATIONSHIPS AMONG PAIRED DATA

One important property of a proposed new drug that must be thoroughly studied by medical researchers is the medication's effect on the patient's pulse rate. Without analyzing this effect, there is the possibility that the good achieved by the medication will be overshadowed by the harm done by it to other portions of the patient's system. If a researcher working on a new drug administers different doses to seven randomly selected victims (technically referred to as "subjects"), he could record their pulse rates 1 hour before and 1 hour after administration of the drug. He might obtain the data of Table 1.5 concerning the relationship between the size of the dosage and the (induced,

[1]From. J. A. Carlson, Are Price Expectations Normally Distributed?, Journal of the American Statistical Association, December 1975, p. 750.

TABLE 1.5
Effect of Dosage on Change in Pulse Rate

Subject	Dosage, milligrams	Change in Pulse Rate, beats/minute
A	1.5	8
B	2.0	10
C	2.5	14
D	1.5	10
E	3.0	15
F	2.0	11
G	2.5	12

perhaps) change in pulse rate between the two readings. Again, without some organization of the data, the information contained therein does not present itself clearly.

The method of illustration of paired data, used to visualize the relationship involved, is called a "scattergram." The data of Table 1.5 are illustrated by the scattergram of Fig. 1.3, composed of a horizontal axis on which dosage levels are marked off and a vertical axis on which pulse rate changes are indicated. Each subject is represented by a single dot indicating both the dosage and the pulse rate change levels. For example, the dot for subject B is 2.0 units to the right of the vertical axis because subject B received a dosage of 2.0 mg, and 10 units above the horizontal axis because subject B's pulse rate showed an increase of 10 beats per minute.

From the scattergram in Fig. 1.3, we can see that the general trend of the data is from lower left to upper right. This indicates that a larger dosage can be expected to result in a greater change of pulse rate than a smaller dosage. We say, in this situation, that pulse rate changes are "positively correlated" with dosage levels.

FIGURE 1.3
Scattergram of dosage versus pulse rate change.

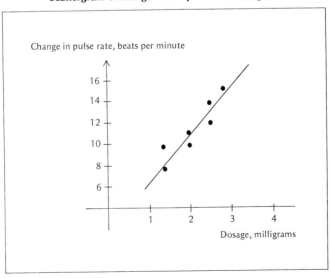

FIGURE 1.4
Scattergram of distance versus population density.

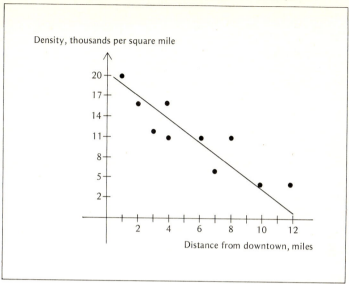

Density, thousands per square mile

The scattergram in Fig. 1.4 illustrates a set of "negatively correlated" data. That set of data resulted from a geographic study of an urban area undertaken in order to determine how the residential population density of a region within the city depends on its distance from the downtown business district. The city was divided into several zones, and the distance of each zone from the downtown area was compared with the population density of the zone. The data appear in Table 1.6. The scattergram of Fig. 1.4 portrays the trend of the data from upper left to lower right. Such a set of data is said to exhibit negative correlation because a larger distance from downtown seems to correspond to a smaller population density.

Both sets of data discussed so far have demonstrated definite directional trends.

TABLE 1.6
Data Relating Distance from Downtown
with Population Density

Zone	Distance from Downtown, miles	Population Density, thousands per square mile
#1	1	20
#2	2	15
#3	4	10
#4	3	11
#5	4	15
#6	6	10
#7	8	10
#8	10	5
#9	7	6
#10	12	5

Such trends are often called linear trends because they can be symbolized by a straight line, the "trend line," following the data in the direction of the trend. Sometimes, however, no such linear trend appears on the scattergram, as illustrated in the following examples.

Example 1.1   Sales Ability   It is an accepted fact that ability to succeed in a sales career in heavy industrial equipment is related to general mathematical ability in an unusual way. According to the prevailing theory, persons having a very low level of mathematical ability usually cannot succeed in a field dealing with equipment of such a complex and technical nature, while those having a very high level of mathematical ability often have difficulty communicating technical information to those responsible for purchasing the equipment. One major supplier wants to check on the validity of this theory.

SOLUTION   The sales headquarters of the supplier gave seven prospective applicants a general mathematics exam. Their exam scores were later compared with scores assigned to the individuals based on the sales record they had established with the company. The comparative scores are presented in Table 1.7, and the scattergram

TABLE 1.7
Data Relating Mathematical and Sales Ability

Salesperson	Math Level Based on Test of Ability	Sales Level Based on Actual Sales Record
A	0	2
B	3	9
C	1	5
D	6	2
E	2	8
F	4	8
G	5	5

FIGURE 1.5
Scattergram of math level versus sales level.

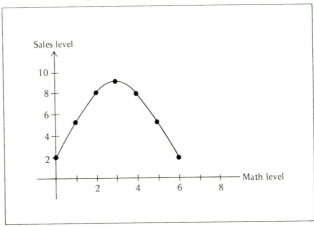

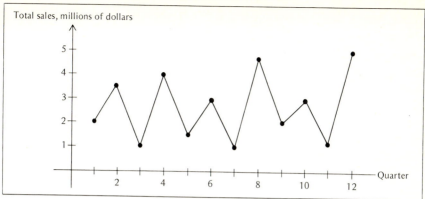

FIGURE 1.6
Scattergram of quarter versus total sales.

TABLE 1.8
Data Relating Quarter and Total Sales

Quarter	Total Sales, millions of dollars
1	2
2	3.5
3	1
4	4
5	1.5
6	3
7	1
8	4.5
9	2
10	3
11	1.5
12	5

appears in Fig. 1.5. The scattergram of Fig. 1.5, which exhibits what is called a "parabolic trend," shows clearly that sales ability is highest at the middle levels of mathematical ability and lowest for those having either very low or very high mathematical ability.

Example 1.2 Seasonal Retail Sales A retail firm whose merchandise sells well during the Christmas and Easter seasons but not as well at other times of the year might have sales totals for 12 consecutive quarters (three-month periods) as listed in Table 1.8.

SOLUTION The scattergram of the retail sales data fluctuates up and down from quarter to quarter indicating the presence of a "cyclical trend" in the original data of Table 1.8. The scattergram appears in Fig. 1.6.

EXERCISES 1.B

1 An agricultural research organization tested a particular chemical fertilizer to try to find out whether an increase in the amount of fertilizer used would lead to a corresponding increase in the food supply. They obtained the following data, based on seven plots of arable land:

Pounds of Fertilizer	Bushels of Beans
2	4
1	3
3	4
2	3
4	6
5	5
3	5

Draw a scattergram which illustrates the relationship between pounds of fertilizer and bushels of beans.

2   In a study of the efficiency of automobile engines with respect to consumption of gasoline, one particular 3990-pound car was operated at various speeds and the gasoline consumption (in miles per gallon) at each speed level was carefully measured. The following data give the results of this experiment:

Speed, miles/hour	Efficiency, miles/gallon
30	20
40	18
50	17
60	14
70	11

Construct a scattergram of speed versus efficiency on the basis of the above data.

3   A biologist conducted an experiment aimed at determining how the growth rate of a certain type of bacteria increases with the passage of time. She set up six cultures of the bacteria, studied each one's growth for a different time period, and recorded their growth rates at the conclusion of the respective time periods. The data follow:

Time Period, hours	Growth Rate, units/hour
3	10
5	13
1	8
7	13
5	11
9	15

Draw a scattergram illustrating the relationship between time period and growth rate.

4   Coronado Nautotronics of San Diego, bidding for the contract to produce the radar displays for the Navy's new fleet of patrol hydrofoils, collects the following information in an attempt to determine the cost curve for the radar displays:

Quantity Produced	Total Cost of Production Run
10	7.0
50	8.5
100	9.0
160	9.4
200	9.5
320	10.0
630	10.5
800	10.8

Draw a scattergram illustrating the cost curve of quantity produced versus total cost of production run.

5 Twelve persons selected at random are asked the following two questions by a psychologist as part of a job satisfaction study: (1) "What is your hourly pay?" (2) "How would you rate your job satisfaction on a scale of 0 (low) to 10 (high)?" The results of the survey are presented below:

Respondent	A	B	C	D	E	F	G	H	I	J	K	L
Hourly pay, dollars	8	2	6	4	4	20	10	6	6	4	11	15
Job satis-faction	6	4	5	4	3	8	7	4	1	2	9	5

Construct a scattergram of hourly pay versus job satisfaction.

6 A social worker is assigned the job of finding out whether or not a particular 6-month vocational training program is effective in increasing the income of its participants. The following data give the monthly incomes of 14 participants just about to begin training, and also 18 months later:

Income at Start of Program, hundreds of dollars	Income 18 Months Later, hundreds of dollars
2.2	3.8
0.0	4.1
3.9	4.8
6.0	6.0
1.3	4.1
3.9	5.8
2.3	0.0
0.0	0.0
1.5	0.0
4.2	4.7
1.6	4.0
2.4	4.2
3.6	4.1
3.5	0.0

Draw a scattergram which illustrates the relationship between income at the start of the program and income 18 months later.

7 An 1891 study[1] of the distribution of men and women in various occupations in England yielded the following data:

Occupational Sector	No. of Persons Employed, Thousands	
	Men	Women
Agriculture and fishing	1245	52
Building	833	8
Manufacture	2609	1530
Transport	816	10
Dealing	851	298
Industrial service	886	21
Public service and professional	563	265
Domestic service	359	1632

Construct a scattergram to illustrate the data.

SECTION 1.C VENN AND TREE DIAGRAMS

On the average, about 5% of the patients assigned to a state hospital which specializes in the treatment of respiratory diseases have tuberculosis. All persons entering this hospital are given chest x-rays, and past data indicate that 90% of those with tuberculosis have a positive x-ray (substantiating presence of the disease), while 2% of those without tuberculosis have a positive x-ray. Before the full meaning of this information will be clear to the reader or listener, it is necessary to organize the numbers in order to determine exactly what is being said.

We have a set of individuals entering a state hospital, with each individual classified according to two different, but related, criteria: (1) Does the patient have a positive x-ray? (2) Does the patient have tuberculosis? In order to facilitate our analysis, we can introduce some abbreviations; in particular, we denote by PX the set of patients having a positive x-ray and by TB the set of patients having tuberculosis. We know that the sets PX and TB are not exactly the same, because, according to the data, only 90% of those in the set TB are also in the set PX, while, on the other hand, 2% of those not in TB are in PX. Furthermore, although we know that 5% of the patients have tuberculosis, the given information does not specify the answers to the following questions:

1 How many of the patients have positive x-rays?
2 Of those having positive x-rays, how many have tuberculosis?
3 Of those not having positive x-rays, how many have tuberculosis?

[1] E. A. Wrigley (ed.), "Nineteenth-Century Society: Essays in the Use of Quantitative Methods for the Study of Social Data," Cambridge University Press, 1972, p. 246.

In symbols, these three questions can be rephrased as follows:

1 How many of the patients are in PX?
2 Of those in PX, how many are also in TB?
3 Of those not in PX, how many are in TB?

While we postpone to Chap. 8 the technical details of calculating the exact answers to the above three questions, we show here the techniques of communicating the sense of the data by means of pictorial descriptions. Two appropriate vehicles for describing the sort of data under discussion are called "Venn diagrams" and "tree diagrams." Venn diagrams convey the information given in the statement of facts about the situation, while tree diagrams yield the answers to questions 1, 2, and 3 above. If you are acquainted with elementary set theory and counting techniques, you will probably recognize some elements of both of these pictorial representations.

In the Venn diagram of Fig. 1.7, the large rectangle delineates the set of all patients entering the state hospital. We call such a set the "sample space" of the problem. It includes all objects involved in the discussion, in this case, all patients of the state hospital. In set theory, such a set is often referred to as the "universal set," but in statistics it is called the sample space. The circles represent, respectively, the sets PX and TB.

Figure 1.7 presents to the viewer two major impressions about the situation under study:

1 Among all the patients entering the state hospital, relatively few have tuberculosis and relatively few have a positive x-ray. (*Note:* The given information includes the statement that 5% of the patients have tuberculosis, so that the circle TB is a relatively small part of the sample space. We know that the circle PX is also small, because PX is composed of 90% of TB together with 2% of the patients outside of TB. Using tree diagrams, we will soon see how to find the exact size of PX.)

FIGURE 1.7
Venn diagram of x-rays versus tuberculosis.

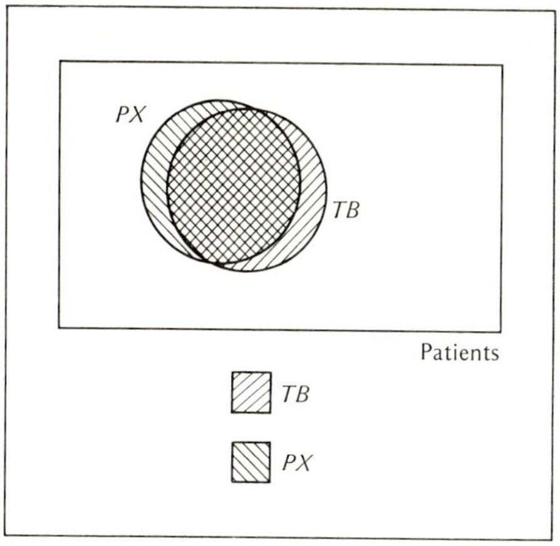

2 The great majority of patients with tuberculosis have a positive x-ray, but there are also some without tuberculosis who have a positive x-ray. (*Note:* 90% of those in TB are also in PX, while PX further includes 2% of those outside of TB.)

Although Venn diagrams have as their objective the pictorial illumination of a situation described in words, the primary objective of tree diagrams is the derivation of new facts based on the given data. More precisely, we use tree diagrams for the purpose of obtaining numerical answers to questions 1, 2, and 3 at the beginning of this section.

About the tree diagram appearing in Fig. 1.8, we make the following observations: (1) Of that 5% of patients having tuberculosis, 90% of them also have a positive x-ray. Taking 90% of 5% (by multiplying $.90 \times .05$ to get $.045 = 4.5\%$), we see that 4.5% of the patients have both tuberculosis and a positive x-ray. (2) Of that 5%, 10% do not have a positive x-ray; so multiplying yields that $.05 \times .10 = .005 = 0.5\%$ (one-half of 1%) of the original group of patients have tuberculosis but do not have a positive x-ray. (3) We similarly conclude that 1.9% of the patients have a positive x-ray but no tuberculosis, while the remaining 93.1% have no tuberculosis and no positive x-ray.

From the information conveyed by the tree diagram, we can actually determine the extent to which a positive x-ray indicates the presence of tuberculosis. More precisely, we can see that 4.5% of the patients have a positive x-ray and tuberculosis, while 1.9% have a positive x-ray but no tuberculosis. Therefore a total of $4.5\% + 1.9\% = 6.4\%$ of the patients have a positive x-ray. Of this 6.4%, only 4.5% have tuberculosis. It follows that of those patients with a positive x-ray, a fraction $4.5/6.4 = 0.703 = 70.3\%$ have tuberculosis. In summary, we can say that, although a positive x-ray is not a perfect indication of the presence of tuberculosis, about 7 out of 10 patients with a positive x-ray have tuberculosis.

Using the numbers calculated in the tree diagram of Fig. 1.8, we can improve the

FIGURE 1.8
Tree diagram of x-rays versus tuberculosis.

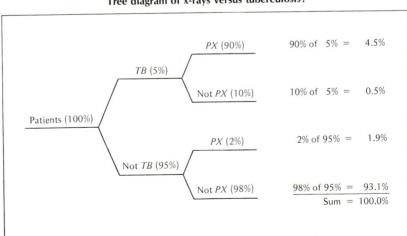

FIGURE 1.9
Sharpened Venn diagram of x-rays versus tuberculosis.

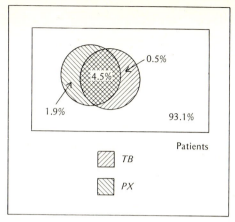

Venn diagram of Fig. 1.7 in such a way that it presents the actual percentages of patients falling in each region of the diagram. The sharpened version of the Venn diagram appears in Fig. 1.9.

EXERCISES 1.C

1 An elementary school bond referendum is defeated at the polls in a school district where 30% of the voters have children in elementary school and 70% do not. It is estimated on the basis of residential areas that 90% of those with children in school voted for the bond issue, while only 20% of those without children in school did so. To find out whether or not it would be worthwhile for the school board to send explanatory literature home with the children, a detailed analysis of the results would be useful.

   a Construct a Venn diagram classifying the voters as to how they voted and whether or not they had children in elementary school.

   b Construct a tree diagram classifying the voters.

   c Use the tree diagram of part b to determine what percentage of voters voted against the bond issue.

   d Use the tree diagram of part b to find out what percentage of those who voted against the bond issue did not have children in elementary school.

   e Sharpen the Venn diagram of part a by inserting the actual numbers determined from the tree diagram.

2 A national sociological survey indicates that 15% of all adults are college graduates, while 85% are not. Furthermore, 80% of all college graduates own their own homes, while only 30% of nongraduates are homeowners.

   a Construct a Venn diagram illustrating the relationship between college graduates and homeowners.

   b Construct a tree diagram illustrating the relationship between college graduates and homeowners in greater detail.

c Determine from the tree diagram what percentage of adults are homeowners.

d Determine from the tree diagram what percentage of homeowners are college graduates.

e Use the information obtained from the tree diagram to draw a more detailed Venn diagram of the relationship between college graduates and homeowners.

3 Eighty percent (80%) of used car and major appliance buyers are good credit risks. Seventy percent (70%) of good credit risks have a charge account at one department store at least, while 40% of bad credit risks have a charge account.

a Construct a tree diagram of the relationship between the two classifications, good credit risks and those possessing a charge account.

b From the tree diagram, find out what percentage of buyers have charge accounts.

c Determine what percentage of those with charge accounts are bad credit risks.

d Draw a Venn diagram which summarizes the information contained in the tree diagram.

4 A psychological test is designed to separate entering college freshmen into good prospects and not-so-good prospects. Among those who later performed satisfactorily during the year, 80% had passed the test. Among the students who did unsatisfactory work in their first year, only 40% had passed the test. On the whole 70% of the students tested did satisfactory work that year.

a Draw a Venn diagram of satisfactory work versus the psychological test.

b Draw a tree diagram of satisfactory work versus the psychological test.

c Use the tree diagram to find out what percentage of freshmen passed the psychological test.

d Determine from the tree diagram the percentage of those passing the test who went on to do satisfactory work.

e Sharpen the Venn diagram so that it includes the numerical information discovered in constructing the tree diagram.

SUMMARY AND DISCUSSION

Our objectives in Chap. 1 have been to introduce the role of statistical analysis in solving problems of applied value and to develop some methods of organizing and displaying statistical data. We have shown how a mass of statistical data, originally a mere collection of numbers, can be organized in tabular or graphical form to yield a comprehensive picture of the information contained within the data. The comprehensive picture of a single set of data is revealed through histograms and frequency polygons, with each type of illustration presenting a different perspective to the viewer. Numerical, rather than pictorial, methods of describing a set of data will be dealt with in Chaps. 2 and 3. Relationships between paired sets of data may be illustrated on scattergrams, which point up the general behavior of one set of data with respect to the other (called the "correlation" between the sets of data). Such relationships will be studied in more detail in Chaps. 6 and 7. Finally, if each member of a group of individuals (persons or objects) is classified according to two criteria, the associations between one criterion and the other can be studied by means of Venn and tree diagrams. Here the focus is primarily on the interrelationships between the two criteria within the group under study, rather than on an analysis of the group itself. A more comprehensive discussion of the questions posed in the section on Venn and tree diagrams will be presented in Chaps. 8 and 9.

BIBLIOGRAPHY

Statistical Sampling
Cochran, W. A.: "Sampling Techniques," 2d ed., Wiley, New York, 1963.

Deming, W. E.: "Some Theory of Sampling," Wiley, New York, 1950.

Raj, D.: "The Design of Sample Surveys," McGraw-Hill, New York, 1972.

Graphing Techniques
Lutz, R. R.: "Graphic Presentation Simplified," Funk & Wagnalls, New York, 1949.

Readings in the Applied Use of Statistics
Haber, A., et al. (eds.): "Readings in Statistics," Addison-Wesley, Reading, Mass., 1970.

Liebermann, B.: "Contemporary Problems in Statistics," Oxford University Press, New York, 1971.

Mansfield, E. (ed.): "Elementary Statistics for Business and Economics: Selected Readings," Norton, New York, 1970.

Tanur, J. M., et al. (eds.): "Statistics: A Guide to the Unknown," Holden-Day, San Francisco, 1972.

Wrigley, E. A. "Nineteenth-Century Society: Essays in the Use of Quantitative Methods for the Study of Social Data," Cambridge University Press, Cambridge, England, 1972.

Nontechnical Discussions of Statistics
Bartholemew, D. J., and E. E. Bassett: "Let's Look at the Figures," Penguin Books, Baltimore, 1971.

Campbell, S. K.: "Flaws and Fallacies in Statistical Thinking," Prentice-Hall, Englewood Cliffs, N.J., 1974.

Huff, D.: "How to Lie with Statistics," Norton, New York, 1954.

Reichmann, W. J.: "Use and Abuse of Statistics," Oxford University Press, New York, 1961.

Von Mises, R.: "Probability, Statistics, and Truth," Macmillan, New York, 1939.

How to Read Technical Statistical Analyses
Huck, S. W., W. H. Cormier, and W. G. Bounds, Jr.: "Reading Statistics and Research," Harper & Row, New York, 1974.

SUPPLEMENTARY EXERCISES

1 The following data[1] give the price per earnings (P/E) ratios of the top 60 companies in stock-market performance in 1975 as analyzed by *Forbes* magazine:

3	7	14	25	22	17
28	15	20	4	6	4
14	32	5	14	14	7
16	4	3	4	9	8
14	14	12	14	7	8
8	5	19	35	32	12
21	11	13	8	6	16
10	5	12	12	4	5
33	7	11	5	7	14
7	7	17	27	19	5

  a Construct a frequency distribution of the P/E ratios.
  b Draw a frequency polygon which illustrates the distribution of part a.

2 As part of an experiment to study the growth rate of a new variety of plant, a botanist planted 100 germinated seeds and measured the heights of the plants 60 days later. The 100 heights were as follows, in inches:

0.0	1.0	4.6	5.9	3.0	3.4	5.6	4.3	3.7	2.8
2.9	4.5	1.1	4.7	5.8	3.1	3.7	3.8	2.7	3.6
4.9	2.1	4.4	1.2	3.3	3.8	3.2	2.6	3.5	4.0
4.7	0.6	2.2	4.3	1.3	3.9	2.5	3.3	4.1	5.4
3.0	4.9	4.8	2.3	3.9	1.4	3.4	3.5	3.4	5.3
4.8	3.1	4.7	3.2	2.4	2.4	1.5	4.8	3.6	3.5
4.6	4.6	3.2	2.3	3.3	0.4	4.2	1.6	3.6	3.6
1.9	3.1	2.2	3.2	4.4	5.7	4.2	3.7	1.7	4.9
4.5	2.1	3.1	4.3	3.3	2.6	3.8	3.5	4.1	1.8
2.0	3.0	4.4	4.5	2.5	3.9	3.4	5.5	5.2	4.0

  a Set up a frequency distribution of the heights.
  b Construct a frequency polygon from the distribution of part a.
  c Draw a histogram to illustrate the frequency distribution of part a.

3 An economist wants to determine the daily demand equation for rolled steel in a small industrial town. She collects the following data relating the price with the quantity of rolled steel that can be sold at that price:

Tons of Rolled Steel That Can Be Sold	Price per Ton, hundreds of dollars
1.0	50.10
2.0	12.60
2.5	8.00

[1]*Forbes*, January 1, 1976, pp. 186–187.

Tons of Rolled Steel That Can Be Sold	Price per Ton, hundreds of dollars
4.0	3.20
5.0	2.00
6.3	1.25

Construct a scattergram of tons of rolled steel versus price per ton.

4 To check on the possibility that the slope of a delta region can be accurately predicted from a knowledge of the typical size of stones found there, a geographer gathered stones from various locations in a delta region of West Baffin Island. The following data compare the typical size of stones found at each location with the cotangent of the angle of slope of the delta there. (A larger cotangent corresponds to a smaller angle of slope.)

Typical Size of Stones, diameter in meters	Slope of Delta, 1% of cotangent of angle
0.5	3.0
3.0	0.5
1.0	2.0
2.5	1.0
2.0	1.5

Construct a scattergram illustrating the relationship, according to the above data, between size of stones and angle of slope.

5 In a study of the advertising budgets of small businesses, a consultant to the Small Business Administration collects the following data relating the size of a business' advertising budget with that business' total sales volume:

Advertising Budget, hundreds of dollars	Total Sales Volume, thousands of dollars
6	50
4	60
10	100
1	30
7	60
5	60
7	40

Construct a scattergram of advertising budget versus total sales volume.

6 About 6% of hospital patients exhibiting a particular combination of symptoms actually have cancer. Three research scientists propose a test for cancer which they claim has the following detection rates: If a patient has cancer, the test will say so in 95% of the cases; if the patient does not have cancer, the test will say *that* in 95% of the cases.

a Construct a Venn diagram of cancer versus test result.

b Construct a tree diagram illustrating the relationship between cancer and the proposed test for it.

c Use the tree diagram to find out what percentage of patients are judged by the test to have cancer.

d Use the tree diagram to find out how many of those judged by the test to have cancer actually have it.

e Sharpen the Venn diagram to include the information conveyed by the tree diagram.

7 A "stop-smoking" clinic advertises that, in a test region last August, 80% of those who tried and were able to stop smoking had participated in its program, while only 30% of those who tried and failed to stop smoking had participated. Research by the Consumer Fraud Division supports these assertions but also reveals that, of all smokers in the region who tried to "kick the habit" last August, 90% had failed, while only 10% succeeded.

a Draw a Venn diagram of stopping smoking versus participating in the clinic's program.

b Draw a tree diagram illustrating the relationship between the ability to stop smoking and the clinic's program.

c Determine from the information contained in the tree diagram what percentage of those who tried to "kick the habit" had participated in the clinic's program.

d Determine from the tree diagram what percentage of the clinic's customers actually were able to stop smoking.

e Construct a Venn diagram which contains the information contained in the tree diagram.

8 The noxious oxides of nitrogen comprise 20% by weight of all pollutants in the air in a certain metropolitan area. Automobile exhaust accounts for 70% of those noxious oxides, but only 10% of all other pollutants in the air.

a Construct a tree diagram of the relationship between pollutants in the air and automobile exhaust in the metropolitan area under study.

b From the tree diagram, find what percentage of pollutants in the air are accounted for by automobile exhaust.

c Use the tree diagram to find how much of the pollution contributed by automobile exhaust can be classified as noxious oxides of nitrogen.

d Draw a Venn diagram which provides a pictorial illustration of the information contained in the tree diagram.

NUMERICAL DESCRIPTIONS
OF DATA

As we observed from the material presented in Chap. 1, an unorganized set of data is merely a collection of numbers, seemingly listed in random fashion and conveying no clear information. To discover the message inherent in the data, it is necessary to develop an understanding of patterns and trends existing among the individual data points. Several pictorial or graphical methods of so describing the data were presented in Chap. 1. Our goal in this chapter is to develop numerical measures of a set of data which will convey an understanding of important patterns among the data.

SECTION 2.A AVERAGES

Consider a small business having 10 persons on its payroll, whose weekly salaries are listed in Table 2.1. The salaries vary from one employee to another because of job classification, experience, and other relevant, or perhaps irrelevant, factors. One question that may be of interest is, What is the average weekly salary of all those on the payroll?

TABLE 2.1
Weekly Salaries

Person	Weekly Salary, dollars
A	400
B	200
C	100
D	200
E	2500
F	200
G	300
H	600
I	200
J	300
Total payroll = 5000	

The usual method of computing the "average" salary is to divide the total number of dollars paid out weekly by the total number of persons on the payroll. Let's do it. The total number of dollars paid out is 5000, as shown at the bottom of Table 2.1, while there are 10 persons on the payroll; so this method of computing the average yields an average weekly salary of 5000/10 = 500 dollars. So far so good.

The next questions that arise are: What does the number 500 signify? What information does it communicate about the data? Why did we compute the average salary by such a procedure, and what, in fact, are we talking about when we mention the average salary? Consider the following three proposals for the meaning of the word average:

1 The average salary is the middle salary: Half of those on the payroll earn at or less than the average salary, while the other half earn at or more than the average salary.
2 The average salary is the most common salary: More people earn that salary than any other single dollar amount.
3 The average salary is the dollar amount lying halfway between the highest salary and lowest salary.

While it would be difficult to think up another possible interpretation of the word average, it is a fact that the average we have computed, namely 500 dollars, satisfies none of the above three criteria. It is something entirely different from all of them. To see this, let's calculate the specific numbers which play the roles defined by the above three descriptions of the word "average."

First, the middle salary, which is technically referred to as the "median," can be determined by listing the data points from lowest to highest and finding a number located exactly in the middle. From Table 2.2, it is apparent that the median of this set of data is the number 250, for persons C, D, F, B and I earn at or less than the 250-dollar level, while persons G, J, A, H, and E earn at or more than the 250-dollar level. (If there is an odd number of data points, the median will be the exact middle number, but if

TABLE 2.2
Weekly Salaries in
Numerical Order

Person	Weekly Salary, dollars
C	100
D	200
F	200
B	200
I	200
G	300
J	300
A	400
H	600
E	2500

there is an even number of data points, the median is usually taken to be the number halfway between the two middle numbers. In this case, the two middle numbers are 200 and 300, and so we consider 250 to be the median.) The average that we computed earlier, namely 500, is certainly not the median, as we now know that the median salary is 250 dollars. From Table 2.2, we can see further that 8 out of the 10 persons on the payroll are earning less than the so-called average salary of 500 dollars.[1]

Next, the most common salary, which is technically called the "mode," is determined by listing each possible salary level and then noting the number of persons receiving that salary. From Table 2.3, we see that the "modal" salary is 200 dollars because four persons have salaries at that level, far more than the number of persons at any other salary level. It is interesting to observe that, for this set of data at least, the modal salary of 200 dollars, the median salary of 250 dollars, and what we have been calling the average salary of 500 dollars are all different numbers.

[1]This observation has the somewhat entertaining sociological interpretation that "nearly everybody is below average."

TABLE 2.3
Persons Grouped by Salary Level

Salary Level, dollars	No. of Persons at That Salary Level
100	1 (C)
200	4 (D, F, B, I)
300	2 (G, J)
400	1 (A)
600	1 (H)
2500	1 (E)
Sum = 10	

Finally, the dollar amount lying halfway between the highest salary (2500 dollars) and the lowest salary (100 dollars) is 1300 dollars, not the 500 dollars we have calculated as the average. The amount of 1300 dollars, technically labeled the "midrange," differs from all the measures of average discussed previously, for this particular set of data.

What, then, is the meaning of the average which is calculated by summing all data points and dividing that sum by the number of data points involved? Well, the precise significance of that process is not easy to explain in a few words, and so we will postpone the details until the next section. We will note, however, that the technical name of the average computed in this manner is the "mean." In particular, 500 dollars is the mean salary of all persons on the payroll of the small business under study. In Table 2.4, we summarize the various types of averages discussed in our attempt to determine the average salary of persons on the payroll listed in Table 2.1. It is to be noted that, in this situation, all the averages differ in numerical value.

In view of the fact that the word "average" may mean many different things, and often different things to different people, how do we know which type of average is appropriate in any particular situation? As a rule, this decision is not based on mathematical considerations but instead on criteria existing within the applied field of study itself. For example, geographers talk about the mean annual temperature or mean annual rainfall in a particular geographical area; economists and sociologists are interested in the median family income of a certain community; shoe and clothing outlets have to make sure they stock the modal sizes (for there may be relatively few individuals having exactly the mean, median, and midrange sizes, and possibly no individuals at all with these sizes, as a comparison of Tables 2.1 and 2.4 will illustrate); and investors speak about the daily midrange of stock-market prices.

As an example of a set of data exhibiting quite a different relationship between the various types of averages, consider the exam scores obtained by six junior high school students on their geography exam. This set of data appearing in Table 2.5 has mean 70 (420 divided by 6), median 70 (because in numerical order the data are 40, 50, 70, 70, 90, 100), mode 70 (two students have this score, but each of the other scores occurs only once), and midrange 70 (halfway between 40 and 100). These averages are summarized in Table 2.6, which points out the fact that it is possible for all of the four averages to be the same. From Table 2.4, we have seen earlier that it is possible for all the averages to be different. The moral of the story is that the word "average" may be

TABLE 2.4
Averages of Weekly Salaries

Type of Average	Numerical Value of Average
Mean	500
Median	250
Mode	200
Midrange	1300

TABLE 2.5
Geography Exam Scores

Student	Exam Score
#1	90
#2	40
#3	50
#4	70
#5	100
#6	70
Sum = 420	

TABLE 2.6
Averages of Geography Exam Scores

Type of Average	Numerical Value of Average
Mean	70
Median	70
Mode	70
Midrange	70

interpreted in many different ways, and so to be sure of communicating the proper information, it is necessary to specify the particular type of average being used.

EXERCISES 2.A

1 The following data are the prices, in cents per pound, of imported English cheese in 12 metropolitan areas of the United States and Canada:

100	105 $^c/_\#$
110	110
130	100
120	90
110	115
90	104

  a Calculate the mean price.
  b Determine the median price.
  c Find the modal price.
  d Find the midrange of the prices.

2 The data listed below represent levels of impurity, according to a standard measurement system, in each of 16 lots of tea leaves:

5	6
4	5
25	7
9	10
11	6
7	8
12	5
20	4

  a Calculate the mean impurity level.
  b Determine the median impurity level.
  c Find the modal impurity level.
  d Determine the midrange of the impurity levels.

3 As part of a study aimed at finding out what a typical auto body repair job costs, an insurance adjusters' organization collects the following data on repair estimates,in dollars, for 15 damaged cars:

220	200
460	380
630	590
300	280
120	130
540	500
730	650
930	

    a What is the mean repair estimate?
    b What is the median repair estimate?
    c Is there a modal repair estimate?
    d What is the midrange of the estimates?

4 Find a set of three data points whose mean is larger than their median.

5 Find a set of three data points whose mean is smaller than their median.

6 Find a set of five data points having the same mean and median but having a midrange different from the mean and median.

7 Just as the median divides a set of data into two parts, a lower half and an upper half, we can define quartiles (which divide the data into four equal parts), deciles (ten equal parts), and percentiles (100 equal parts). For example, the 7th decile divides the lower 70% of the data from the upper 30%, while the 3rd quartile divides the lower 75% from the upper 25%. Using the data of Exercise 1.A.5, find
    a The median.
    b The 1st quartile.
    c The 3rd decile.
    d The 57th percentile.

8 From the data of Exercise 1.A.1, determine
    a The median.
    b The 3rd quartile.
    c The 9th decile.
    d The 38th percentile.

SECTION 2.B THE MEAN

In the previous section, we have noted that the mean (calculated by summing the data points and then dividing the sum by how many of them there are) is the most vague in meaning of the various types of averages. We pointed out, in particular, that it is not clear exactly what information is conveyed by the mean, even though it is the most commonly used measure of average. The present section has as its purpose the explanation of the significance of the mean as a measure of average.

Let us look again at the data on weekly salaries appearing in Table 2.1. If we imagine a seesaw in a children's playground marked off in units of measurement and we place bowling balls at locations corresponding to the data points, we obtain a graphical description of the facet of the data presented in Table 2.3. The picture of the seesaw appears in Fig. 2.1, with the mean value of 500 dollars clearly marked. In the picture, we see one ball at location 100, four at 200, two at 300, one at 400, one at 600, and one at 2500, in accordance with Table 2.3. What we will now show is that the seesaw will balance only if its fulcrum (balancing point) is placed at the mean value, namely, 500 in this situation.

One of the principles of the seesaw asserts that a 25-pound child will be able to balance a 50-pound child if the smaller child sits twice as far away from the center as does the larger child. In a similar spirit, we see that four bowling balls at a distance 300 from the center will balance one bowling ball at a distance 1200 from the center. The reader could draw a diagram of this, or could actually test it out (using bricks, instead of bowling balls) at a local playground. In the seesaw of Fig. 2.1, the balls at 400 and 600 balance each other around the center position of 500, while the one ball at 2500 balances the remaining seven balls. The reason for this is that the single ball at 2500 exerts a force of 2000 units on the right side of the seesaw because it is located at a distance of 2000 from the center position of 500, while on the left side of the seesaw we have forces of

One ball at 100 (*400* from the center) = 400 units
Four balls at 200 (*300* from the center) = 1200 units
Two balls at 300 (*200* from the center) = 400 units

totaling 2000 units of force on the left. The equivalence of forces on the left and right sides of the 500 location causes the balancing of the seesaw at that point.

It is in this sense, the sense of the balancing point of the data, that the mean can be considered a measure of average. The balancing property of the mean, therefore, defines how the mean describes a set of data. When we are informed that the mean of a particular set of data is 500, we should visualize in our minds the data strung out along a seesaw which is balanced at 500.

For some types of data, nevertheless, the mean is not an appropriate measure of

FIGURE 2.1
Balancing property of the mean weekly salary.

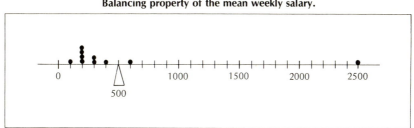

average because variations within the set of data are so extreme as to make the balancing point irrelevant or nonexistent. In the case of stock-market prices exhibiting wild fluctuations, for example, the median is generally used to measure average price per share.[1] As we shall show later, however, the mean is generally more useful than the other measures of average because it can be easily used to develop very accurate descriptions of sets of data.

Let's try to organize this balancing property into a useful technique for describing data. If we subtract the mean value of 500 from each of the data points, for example subtract 500 from 300, we get a number which tells us the distance of the data point from the mean and also tells us on which side of the mean the data point is located. Using as an example the data point 300, we calculate $300 - 500 = -200$. The negative sign indicates that 300 is to the left of the mean on the seesaw, while the 200 indicates that 300 is at a distance of 200 units away from the mean. Similarly, $600 - 500 = 100$ means that the point 600 is 100 units to the right of the mean.

We call the difference between a data point and the mean the "deviation of the data point from the mean." The deviation will be negative if the data point falls to the left of the mean (on the seesaw) and positive if the data point lies to the right of the mean. The balancing property means that the positive deviations exactly cancel the negative deviations, so that the sum of all deviations is zero. In Table 2.7, we show how the deviations sum to zero in the case of the weekly salary data of Table 2.1. The sum of the deviations is zero because the negative deviations and the positive deviations both have a magnitude of 2100.

The deviations of the various data points from the mean are very important in themselves, for they can be used to measure whether the data points are all bunched together near the mean or whether some are dispersed far away from the mean. We will deal with this question in detail in the next section. Here let us remark only that the

[1]See, for example, the article by Eugene F. Fama listed in the bibliography.

TABLE 2.7
Sum of Deviations of Weekly Salary Data

Person	Weekly Salary, dollars	Salary − Mean =	Deviation of Salary
A	400	400 − 500 =	−100
B	200	200 − 500 =	−300
C	100	100 − 500 =	−400
D	200	200 − 500 =	−300
E	2500	2500 − 500 =	2000
F	200	200 − 500 =	−300
G	300	300 − 500 =	−200
H	600	600 − 500 =	100
I	200	200 − 500 =	−300
J	300	300 − 500 =	−200
		Sum of deviations = 0	

computational procedure in Table 2.7 can be used as a check on our calculation of the mean. If the sum of the deviations turns out to be something other than zero, we have probably made an arithmetic error in our calculation of the mean.

In order to establish general procedures for analyzing a set of data, it is necessary to develop new techniques of labeling a set of data points. One convenient way of labeling a set of data points is to consider the data as a set of x's labeled as follows:

$$x_1 = \text{1st data point}$$
$$x_2 = \text{2nd data point}$$
$$x_3 = \text{3rd data point}$$
$$\cdots\cdots\cdots\cdots\cdots\cdots\cdots\cdots\cdots\cdots\cdots\cdots$$
$$x_k = k\text{th data point (pronounced ''kayth'')}$$
$$x_n = n\text{th data point (pronounced ''enth'')}$$

We use the symbol x_k to denote a representative member of the set of data points. We will agree that x_n will be the last of our data points, and this means that we are dealing with n data points, so that

$$n = \text{number of data points in our set}$$

In the case of weekly salary data of Table 2.1, we have ten data points so that $n = 10$. We label the points of Table 2.1 as follows:

$$x_1 = 400$$
$$x_2 = 200$$
$$x_3 = 100$$
$$x_4 = 200$$
$$x_5 = 2500$$
$$x_6 = 200$$
$$x_7 = 300$$
$$x_8 = 600$$
$$x_9 = 200$$
$$x_{10} = 300$$

To avoid the necessity of writing the words "sum of the data points" over and over again, we should introduce some symbols for that expression. The symbol used in statistics to signify the word sum is the capital letter Σ (pronounced "sigma") of the Greek alphabet. To symbolize the sum of the data points, we write

$$\sum_{k=1}^{n} x_k$$

which is pronounced "the sum of the x_k's from $k = 1$ to $k = n$." The meaning of that collection of symbols is therefore

$$\sum_{k=1}^{n} x_k = x_1 + x_2 + x_3 + \cdots + x_n$$

where the $k = 1$ below the Σ and the n above the Σ mean that the sum starts with the first data point x_1 and proceeds all the way to the last data point x_n, encompassing all data points in between. For the weekly salary data,

$$\sum_{k=1}^{10} x_k = x_1 + x_2 + x_3 + x_4 + x_5 + x_6 + x_7 + x_8 + x_9 + x_{10}$$
$$= 400 + 200 + 100 + 200 + 2500 + 200 + 300 + 600 + 200 + 300$$
$$= 5000$$

Using the statistical shorthand introduced so far, we can write a compact expression for the mean. In situations where the data represent a complete "population" (as do the weekly salaries of *all* 10 persons employed in the small business under study), we use the lowercase Greek letter μ (pronounced "mew") as a symbol for the mean. The expression

$$\text{The mean} = \frac{\text{the sum of all the data points}}{\text{the number of data points}}$$

can then be translated into our new shorthand as

for population   *for sample*

$$\mu = \frac{\sum_{k=1}^{n} x_k}{n} \approx \overline{x}$$

With the shorthand formula, we can express the mean of the weekly salary data as

$$\mu = \frac{\sum_{k=1}^{n} x_k}{n} = \frac{5000}{10} = 500$$

In terms of our system of labeling data points, we can construct a table, analogous to Table 2.7, which illustrates the balancing property by pointing out that the deviations always have a sum of zero. This symbolic calculation appears in Table 2.8, while a picture of a seesaw illustrating the balancing property appears in Fig. 2.2.

By using only algebraic manipulation of the symbols involved, it can be shown that

FIGURE 2.2
Balancing property of the mean.

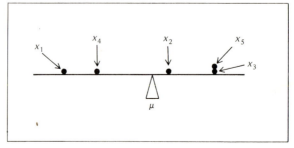

TABLE 2.8
Sum of Deviations from the Mean

	Data Points x	Deviations $x - \mu$
	x_1	$x_1 - \mu$
	x_2	$x_2 - \mu$
	x_3	$x_3 - \mu$
	$\vdots$	$\vdots$
	x_n	$x_n - \mu$
Sums	$\displaystyle\sum_{k=1}^{n} x_k$	0

the deviations add to zero for any set of data points. It is therefore a mathematical fact (more technically, a "theorem") that

$$\sum_{k=1}^{n} (x_k - \mu) = 0$$

for any possible set of data points; that is, the sum of the deviations is zero. This is the expression of the balancing property of the mean in the statistical shorthand language we have developed. For the benefit of those of you who are familiar with algebra and who would like to see the algebraic justification of the balancing property, we present the computational details in Supplement 2.1.

SUPPLEMENT 2.1   The Balancing Property of the Mean

In this supplement, we show that $\displaystyle\sum_{k=1}^{n} (x_k - \mu) = 0$ for any set of data points. We write that

$$\sum_{k=1}^{n} (x_k - \mu) = (x_1 - \mu) + (x_2 - \mu) + \cdots + (x_n - \mu)$$
$$= x_1 - \mu + x_2 - \mu + \cdots + x_n - \mu$$
$$= x_1 + x_2 + \cdots + x_n - \mu - \mu - \cdots - \mu$$
$$= \sum_{k=1}^{n} x_k - n\mu$$

because there are n μ's involved in this expression, one μ for each of the n data points. Remembering that the mean can be symbolized algebraically as

$$\mu = \frac{\displaystyle\sum_{k=1}^{n} x_k}{n}$$

we see that

$$n\mu = n \frac{\sum\limits_{k=1}^{n} x_k}{n} = \sum\limits_{k=1}^{n} x_k$$

by cancellation of the n's. Therefore

$$\sum_{k=1}^{n} (x_k - \mu) = \sum_{k=1}^{n} x_k - n\mu = \sum_{k=1}^{n} x_k - \sum_{k=1}^{n} x_k = 0$$

which is what we set out to prove.

EXERCISES 2.B

1 Draw a "seesaw" representation of the cheese price data of Exercise 2.A.1, and show that the deviations from the mean really do sum to zero.

2 Construct a "seesaw" representation of the impurity level data of Exercise 2.A.2, and make the calculations which verify that the deviations from the mean have a sum of zero.

3 Put the auto repair estimates of Exercise 2.A.3 on a "seesaw," and show that the deviations from the mean sum to zero.

4 Make up a set of any nine numbers you can think of, and verify that the deviations from the mean really sum to zero.

5 Using the data of Exercise 2.A.1:
a Show that the deviations from the *median do not* sum to zero.
b Show that the deviations from the *mode do not* sum to zero.
c Show that the deviations from the *midrange do not* sum to zero.

6 Using the data of Exercise 2.A.2:
a Show that the deviations from the *median do not* sum to zero.
b Show that the deviations from the *mode do not* sum to zero.
c Show that the deviations from the *midrange do not* sum to zero.

7 Using the data of Exercise 2.A.3, show that the deviations from the median do not sum to zero.

8 Using the data of Exercise 2.A.6, show that the deviations from the median *do* sum to zero. Can you explain why this occurs?

9 If $x_1 = 5$, $x_2 = 8$, $x_3 = 7$, and $x_4 = 19$, calculate $\sum\limits_{k=1}^{4} x_k$.

10 If

$x_1 = 17$	$x_5 = -8$
$x_2 = 9$	$x_6 = -9$
$x_3 = -6$	$x_7 = -5$
$x_4 = 14$	$x_8 = 3$

calculate $\sum\limits_{k=1}^{8} x_k$.

11 Calculate the mean of the diameters listed in the data of Exercise 1.A.1.

12 Calculate the mean learning time from the data of Exercise 1.A.5.

13 Calculate the mean forecast of the Consumer Price Index using the data of Exercise 1.A.6.

14 The following data[1] record the number of working days lost through labor disputes in England during a 16-year period:

Year	No. of Worker Days Lost, thousands
1950	1389
1951	1694
1952	1792
1953	2184
1954	2457
1955	3781
1956	2083
1957	8412
1958	3462
1959	5270
1960	3024
1961	3046
1962	5798
1963	1755
1964	2277
1965	2925

Calculate the mean yearly number of worker days lost due to labor disputes during the 16-year period.

SECTION 2.C THE STANDARD DEVIATION

If a geographer compares two regions of a continent on the same day in regard to their recorded temperatures at various times during the day, he might come up with the data presented in the top part of Table 2.9. A check of the bottom part of Table 2.9 or an application of the methods introduced in our earlier discussion of averages shows that *both* sets of temperatures have mean, median, mode, and midrange *all* equal to 50°F. No matter what you consider the appropriate measure of average here, each set of data has an average of 50°F. Because there is such substantial agreement on the averages of the temperature data, it would be important to know exactly how descriptive of a set of data the average is. In particular, when we are informed that the average (as measured by one of the methods we discussed earlier) of a certain set of data is 50, how are we to interpret that information? How are we to visualize or reconstruct in our mind the original set of data from the knowledge that the average is 50?

[1] B. R. Mitchell and H. G. Jones, "Second Abstract of British Historical Statistics," p. 51, Cambridge University Press, 1971.

TABLE 2.9
Comparison of Regional Temperatures

Hour	East Coast Temperatures, °F	West Coast Temperatures, °F
3 A.M.	40	10
6 A.M.	50	50
9 A.M.	50	50
12 P.M.	60	80
3 P.M.	50	90
6 P.M.	50	20
Mean	50	50
Median	50	50
Mode	50	50
Midrange	50	50

As the pictorial representation in Fig. 2.3 demonstrates, there can be considerable differences between data sets which have the same averages. Just because two sets of data both have an average of 50°F, this does not mean that they exhibit the same overall characteristics.

How then do the two sets of temperature data differ? As Fig. 2.3 shows, they differ primarily by how much the data points vary from the common average of 50°F. The East Coast temperatures are heavily concentrated near their average of 50°F, while the West Coast temperatures vary considerably from their average of 50°F. In particular, the East Coast temperatures remain between 40 and 60°F throughout the day, while West Coast temperatures range from a low of 10 to a high of 90°F. As this shows, the extent of variation from the average is an important element in the description of a set of data, but knowledge of the averages alone gives no information about the variation.

It would be very useful, therefore, to have a numerical measure of variation of the individual points within a set of data. We could use such a measure of variation each time we use a measure of average, for the purpose of obtaining a more accurate description of the general pattern of the data.

Consider the following proposal for a measure of variation of a set of data: the

FIGURE 2.3
"Seesaw" representations of temperature data.

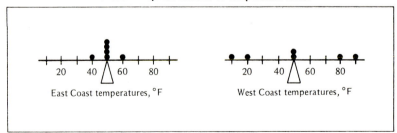

East Coast temperatures, °F   West Coast temperatures, °F

difference between the largest data point and the smallest, technically called the "range" of the data.

The computation of the range is fairly simple because it involves only the two extreme data points. For the regional temperature data appearing in Table 2.9, we can see that the range of East Coast temperatures is $60 - 40 = 20°F$, while the range of West Coast temperatures is $90 - 10 = 80°F$. A comparison of the ranges indicates that hourly West Coast temperatures are more variable than hourly East Coast temperatures, substantiating the impression given in the "seesaws" of Fig. 2.3.

As a second proposed measure of variation, we can look at the average deviation of points from their mean, technically referred to as the "mean absolute deviation." This is somewhat more difficult to compute because it involves the use of a table of all deviations, like that in Table 2.7. The preliminary computations leading to the mean absolute deviation of each set of temperature data can be found in Table 2.10. In the portion of Table 2.10 dealing with the East Coast temperatures, the column headed x lists the actual temperature data, while the column headed $x - \mu$ lists the deviation of each data point from the mean value of 50. In view of the balancing property of the mean as illustrated in Table 2.8, we know that the column headed $x - \mu$ must sum to 0. But what we are interested in for the purpose of measuring variation is the absolute magnitude of each deviation, not whether they are positive or negative. The absolute magnitudes (technically called the "absolute values") of the deviations are listed in the column headed $|x - \mu|$.[1] The mean absolute deviation is then the sum of the $|x - \mu|$ column divided by the number of data points. This procedure yields the average deviation of the data points from their mean. A set of data with its data points far from their mean, on the

---

[1] By the "absolute value" of a number, we mean the magnitude of the number, ignoring the + or − sign. For example, both $|5| = 5$ and $|-5| = 5$. The upshot of this is that, whether a number is positive or negative (or zero), its absolute value cannot be negative.

TABLE 2.10
Computations Leading to the Mean Absolute
Deviations of Temperature Data

	East Coast Temperatures x			West Coast Temperatures y		
	x	$x - \mu$	$|x - \mu|$	y	$y - \mu$	$|y - \mu|$
	40	−10	10	10	−40	40
	50	0	0	50	0	0
	50	0	0	50	0	0
	60	10	10	80	30	30
	50	0	0	90	40	40
	50	0	0	20	−30	30
Sums	300	0	20	300	0	140

$n = 6$
$\mu = \frac{300}{6} = 50$

$n = 6$
$\mu = \frac{300}{6} = 50$

average, will have a larger mean absolute deviation than a set having its points closer, on the average, to their mean.

We abbreviate the mean absolute deviation by the symbol MAD. In the mathematical shorthand introduced earlier in this chapter, we can write that

$$\sigma = 1.25 \text{ MAD}$$

$$\text{MAD} = \frac{\text{sum of absolute deviations}}{\text{number of data points}} = \frac{\sum_{k=1}^{n} |x_k - \mu|}{n}$$

The East Coast temperatures have MAD $= \frac{20}{6} = 3.33$, using the fact in Table 2.10 that the sum of the absolute deviations (the column headed by $|x - \mu|$) is 20. On the other hand, the West Coast temperatures have a MAD $= \frac{140}{6} = 23.33$, again indicating strongly that hourly West Coast temperatures are more variable than hourly East Coast temperatures.

Now that we have thoroughly discussed the use and computation of the range and the mean absolute deviation as possible measures of the variation of a set of data points, it is the author's sad duty to report that neither of these measures is widely used, except in very special cases involving extremely large variations of the data.[1] Both have been preempted in importance by a third measure of variation, technically referred to as the "standard deviation." The standard deviation, which is somewhat more complicated to calculate than either of the measures already discussed, is also more useful and more efficient than either of them.

The standard deviation is the square root of the mean of the squared deviations. (In some applied contexts, it is referred to as the "root-mean-square deviation" or the "rms deviation.") We usually denote the standard deviation by the Greek letter σ (lowercase "sigma"), and in mathematical symbolism, we express it as follows:

$$\sigma = \sqrt{\frac{\sum_{k=1}^{n} (x_k - \mu)^2}{n}}$$

To calculate σ in accordance with the above formula, we proceed as follows:

1 Calculate the deviations $x_k - \mu$ as in Table 2.10.
2 Square each deviation; i.e., multiply it by itself.
3 Sum the squared deviations.
4 Divide the sum by n, the number of data points.
5 Take the square root[2] of the resulting number.

The standard deviation, like the mean before it, has the unfortunate characteristic that its meaning is not immediately clear from its definition. In general, we can say that the standard deviation is the most widely used of the measures of variation because it

---

[1] See, for example, the article by William F. Sharpe, cited in the bibliography.
[2] The square root may be explained as follows: the square of a number, say 3, is that number multiplied by itself, namely $3 \times 3 = 9$. The square root is the reverse of this, that is $\sqrt{9} = 3$ because $3 \times 3 = 9$. In short, for positive numbers a and b, $\sqrt{a} = b$ if and only if $b^2 = a$.

TABLE 2.11
Computations Leading to the Standard Deviations of Temperature Data

	East Coast Temperatures x			West Coast Temperatures y		
	x	$x - \mu$	$(x - \mu)^2$	y	$y - \mu$	$(y - \mu)^2$
	40	−10	100	10	−40	1600
	50	0	0	50	0	0
	50	0	0	50	0	0
	60	10	100	80	30	900
	50	0	0	90	40	1600
	50	0	0	20	−30	900
Sums	300	0	200	300	0	5000

$$n = 6 \qquad\qquad n = 6$$
$$\mu = \tfrac{300}{6} = 50 \qquad\qquad \mu = \tfrac{300}{6} = 50$$

conveys the most precise and useful information about variation of the data from the mean. The sort of information it conveys will be discussed in detail in the next section. Here we concentrate on how to calculate it.

The most direct method of calculating the standard deviation is by means of a table of preliminary computations analogous to Table 2.10. In accordance with the five-step procedure discussed above, we would record the square of each deviation rather than its absolute value. A deviation of −10 would have a square of (−10) × (−10) = 100, if we recall that the product of two negative numbers is a positive number. Similarly, a deviation of 40 has a square of (40) × (40) = 1600. We construct Table 2.11 with these principles in mind. From Table 2.11, we see that the sum of squared deviations of the East Coast temperatures is

$$\sum_{k=1}^{n} (x_k - \mu)^2 = 100 + 0 + 0 + 100 + 0 + 0 = 200$$

adding up the column labeled $(x - \mu)^2$, while the sum of squared deviations of West Coast temperatures is, by adding up the column labeled $(y - \mu)^2$,

$$\sum_{k=1}^{n} (y_k - \mu)^2 = 1600 + 0 + 0 + 900 + 1600 + 900 = 5000$$

It follows that, for East Coast temperatures, the standard deviation is

$$\sigma = \sqrt{\frac{\sum_{k=1}^{n} (x_k - \mu)^2}{n}} = \sqrt{\frac{200}{6}} = \sqrt{33.3} = 5.77$$

while West Coast temperatures have standard deviation

$$\sigma = \sqrt{\frac{\sum_{k=1}^{n} (y_k - \mu)^2}{n}} = \sqrt{\frac{5000}{6}} = \sqrt{833} = 28.86$$

As is clear from the pictures in Fig. 2.3, the West Coast temperatures are more variable than the East, and this fact is borne out by a comparison of the standard deviations. The West Coast temperatures' standard deviation of 28.86 exceeds by a considerable margin the East Coast's value of 5.77.

Square Roots

Before proceeding further in our study of the standard deviation let us pause for a moment to talk about the square roots involved in the calculation. There are basically three ways of finding the square root of a number: (1) using an electronic calculator which has a square-root button; (2) carrying out the paper-and-pencil calculation by hand using the long-division-type method sometimes taught in elementary algebra courses; and (3) using the square-root tables in the back of this book. Method 1 is obviously the most efficient and most accurate. It is used by those persons who handle statistics in their daily work. However, not every calculator has a square-root button, and so method 1 will not be practical for everyone who uses this text. The hand calculation of method 2 is somewhat more complicated and time-consuming than ordinary long division. Therefore, we also consider this method impractical, although it is a fascinating experience to work out the square root by yourself. If you are interested in the details of the hand calculation method you are invited to ask your local mathematics expert, namely your instructor, for an explanation of it. This brings us to method 3. Using Table A.1 of the Appendix, we can work out the square root of any number having three significant digits. Because Table A.1 contains only the numbers 1.00 through 9.99 explicitly, it is necessary to express the number whose square root we want in terms of one of those numbers. To illustrate this procedure, we will work out the details of finding $\sqrt{33.3}$ and $\sqrt{833}$. We have

$$\sqrt{33.3} = \sqrt{3.33 \times 10} = 5.77$$

looking up $n = 3.33$ and using the fourth column of the table, the column headed $\sqrt{n \times 10}$. Next,

$$\sqrt{833} = \sqrt{8.33 \times 100} = \sqrt{8.33} \times \sqrt{100}$$

$$= 2.886 \times 10 = 28.86$$

looking up $n = 8.33$, finding $\sqrt{n} = 2.886$, and realizing that (you probably have this memorized) $\sqrt{100} = 10$. Additional examples throughout this text will continue to illustrate the procedure of finding square roots, using Table A.1.

In some situations in statistics, it is more efficient to calculate the standard deviation by an alternative procedure. The alternative procedure, euphemistically called "the shortcut method" of computing the standard deviation, does in fact simplify the computation when the mean μ does not come out "even" (namely, as a whole number exactly). For example, consider the data of Table 2.12 showing the weekly sales volume of a small auto parts outlet over a 7-week period. The mean sales volume is

TABLE 2.12
Weekly Sales of Auto Parts

Week	Sales Volume, thousands of dollars
#1	7
#2	5
#3	7
#4	8
#5	4
#6	9
#7	5
Sum = 45	

$$\mu = \frac{\sum\limits_{k=1}^{n} x_k}{n} = \frac{45}{7} = 6.43$$

which does not come out even. In Table 2.13, we illustrate the procedures for calculating the standard deviation by both the usual method and the shortcut method. You should observe that the shortcut method does indeed save some steps in this case. The shortcut formula does not require the mean to be computed first, but rather it proceeds directly to the standard deviation. The formula involves only the sum of the data points and the sum of their squares:

$$\sigma = \sqrt{\frac{n \sum\limits_{k=1}^{n} x_k^2 - \left[\sum\limits_{k=1}^{n} x_k\right]^2}{n^2}}$$

TABLE 2.13
Calculation of the Standard Deviation of Auto Parts Sales Data
Usual Method versus Shortcut Method

	Usual Method			Shortcut Method	
	x	$x - \mu$	$(x - \mu)^2$	x	x^2
	7	0.57	0.3249	7	49
	5	−1.43	2.0449	5	25
	7	0.57	0.3249	7	49
	8	1.57	2.4649	8	64
	4	−2.43	5.9049	4	16
	9	2.57	6.6049	9	81
	5	−1.43	2.0449	5	25
Sums	45	−0.01	19.7143	45	309
	$n = 7$			$n = 7$	
	$\mu = \frac{45}{7} = 6.43$				

The equivalence of the usual and the shortcut formulas for the standard deviation is demonstrated algebraically in Supplement 2.2. Both formulas give exactly the same answer in every case, except for possible differences due to rounding off decimal places. Notice that some round-off error has already occurred in the computations preparatory to the usual method. Such an error is indicated by the fact that the deviations sum to -0.01 instead of exactly 0 as the balancing property requires. The error cannot be avoided because, as a practical matter, we have to round off μ to 6.43 from its more precise value of 6.428571429. Therefore we might, in some cases, wind up with a slightly inaccurate value of σ as well. The shortcut method, however, maintains 100% accuracy in the computation table, and is therefore the preferred method of computing σ when μ does not come out "even." Furthermore, it should be noted that the shortcut formula is easier to use in computing the standard deviation by means of a calculator, for basically all that are needed are the sum of the data points and the sum of their squares. We now calculate σ both ways from the information in Table 2.13:

Usual method: $\sigma = \sqrt{\dfrac{\Sigma(x - \mu)^2}{n}} = \sqrt{\dfrac{19.7143}{7}} = \sqrt{2.82} = 1.68$ where we used

the abbreviation $\Sigma(x - \mu)^2$ in place of the more cumbersome $\displaystyle\sum_{k=1}^{n}(x_k - \mu)^2$.

Shortcut method: $\sigma = \sqrt{\dfrac{n\Sigma x^2 - (\Sigma x)^2}{n^2}} = \sqrt{\dfrac{7(309) - (45)^2}{(7)^2}}$

$$= \sqrt{\dfrac{2163 - 2025}{49}} = \sqrt{\dfrac{138}{49}} = \sqrt{2.82} = 1.68$$

where we have written Σx^2 and Σx in place of the more cumbersome $\displaystyle\sum_{k=1}^{n} x_k^2$ and $\displaystyle\sum_{k=1}^{n} x_k$.

You should observe that both methods yield the same value of σ, namely 1.68. It is also important to observe that Σx^2 and $(\Sigma x)^2$ are not one and the same. To get Σx^2, we sum the squares; while to get $(\Sigma x)^2$, we square the sum. Here $\Sigma x^2 = 309$, and $(\Sigma x)^2 = (45)^2 = 2025$. They're quite different.

While the shortcut method is therefore to be preferred when the mean does not come out "even," the usual method works slightly better when the mean does come out "even." To demonstrate this, we calculate the standard deviations of the temperature data of Table 2.9 by the shortcut method, and we compare the procedure with our work in connection with Table 2.11. The calculations for the shortcut method can be found in Table 2.14. From the computations in Table 2.14, we recalculate the standard deviation of the East Coast temperatures as follows:

$$\sigma = \sqrt{\dfrac{n\Sigma x^2 - (\Sigma x)^2}{n^2}} = \sqrt{\dfrac{6(15,200) - (300)^2}{(6)^2}} = \sqrt{\dfrac{91,200 - 90,000}{36}}$$

$$= \sqrt{\dfrac{1200}{36}} = \sqrt{33.3} = 5.77$$

and calculating the standard deviation of the West Coast temperatures by the shortcut

TABLE 2.14
Computations Leading to the Standard Deviation of Temperature Data
Shortcut Method

	East Coast Temperatures x		West Coast Temperatures y	
	x	x²	y	y²
	40	1600	10	100
	50	2500	50	2500
	50	2500	50	2500
	60	3600	80	6400
	50	2500	90	8100
	50	2500	20	400
Sums	300	15,200	300	20,000
	n = 6		n = 6	

method, we have

$$\sigma = \sqrt{\frac{n\Sigma y^2 - (\Sigma y)^2}{n^2}} = \sqrt{\frac{6(20,000) - (300)^2}{(6)^2}} = \sqrt{\frac{120,000 - 90,000}{36}}$$

$$= \sqrt{\frac{30,000}{36}} = \sqrt{833} = 28.86$$

the same values as obtained earlier, of course.

$\sigma = 1.25$ MAD

SUPPLEMENT 2.2   The Shortcut Formula for the Standard Deviation

In this supplement, we do the algebra necessary to demonstrate that the short-cut formula yields exactly the same value for σ as that given by the usual method. We first show that

$$\sum_{k=1}^{n} (x_k - \mu)^2 = \frac{n\sum_{k=1}^{n} x_k^2 - \left(\sum_{k=1}^{n} x_k\right)^2}{n}$$

The above equation holds because

$$\sum_{k=1}^{n} (x_k - \mu)^2 = \sum_{k=1}^{n} (x_k^2 - 2\mu x_k + \mu^2)$$

$$= (x_1^2 - 2\mu x_1 + \mu^2) + (x_2^2 - 2\mu x_2 + \mu^2)$$
$$+ \cdots + (x_n^2 - 2\mu x_n + \mu^2)$$

$$= (x_1^2 + x_2^2 + \cdots + x_n^2) - 2\mu(x_1 + x_2 + \cdots + x_n)$$
$$+ (\mu^2 + \mu^2 + \cdots + \mu^2)$$

$$= \sum_{k=1}^{n} x_k^2 - 2\mu \sum_{k=1}^{n} x_k + n\mu^2$$

Now recalling that $\mu = \left(\sum_{k=1}^{n} x_k \right)/n$, we insert this formula for μ into the expression above and obtain

$$\sum_{k=1}^{n} (x_k - \mu)^2 = \sum_{k=1}^{n} x_k^2 - 2 \frac{\sum_{k=1}^{n} x_k}{n} \sum_{k=1}^{n} x_k + n \left(\frac{\sum_{k=1}^{n} x_k}{n} \right)^2$$

$$= \sum_{k=1}^{n} x_k^2 - 2 \frac{\left(\sum_{k=1}^{n} x_k \right)^2}{n} + \frac{\left(\sum_{k=1}^{n} x_k \right)^2}{n}$$

$$= \sum_{k=1}^{n} x_k^2 - \frac{\left(\sum_{k=1}^{n} x_k \right)^2}{n} = \frac{n \sum_{k=1}^{n} x_k^2 - \left(\sum_{k=1}^{n} x_k \right)^2}{n}$$

It follows, then, that

$$\sigma = \sqrt{\frac{\sum_{k=1}^{n} (x_k - \mu)^2}{n}} = \sqrt{\frac{n \sum_{k=1}^{n} x_k^2 - \left(\sum_{k=1}^{n} x_k \right)^2}{n^2}}$$

which justifies the use of the shortcut formula. Incidentally, another useful fact can be extracted from this discussion as follows: Because each $(x_k - \mu)^2$ is a nonnegative (i.e., positive or zero) number, we can be sure that

$$n \sum_{k=1}^{n} (x_k - \mu)^2 \geq 0$$

But

$$n \sum_{k=1}^{n} (x_k - \mu)^2 = n \sum_{k=1}^{n} x_k^2 - \left(\sum_{k=1}^{n} x_k \right)^2$$

as we have shown by the above algebraic calculations, so that

$$n \sum_{k=1}^{n} x_k^2 - \left(\sum_{k=1}^{n} x_k \right)^2 \geq 0$$

This fact is called the "Schwarz inequality," and it guarantees that we will never have to find the square root of a negative number in the shortcut formula. Because we know, therefore, that the numerator under the square root sign in the shortcut formula can never be negative, it is a sure sign of an arithmetic error when $n\Sigma x^2$ turns out to be smaller than $(\Sigma x)^2$.

EXERCISES 2.C

1   From the data on prices of English cheese appearing in Exercise 2.A.1, calculate
    a The standard deviation of the prices.
    b The mean absolute deviation.
    c The range of the prices.
    d The standard deviation by using the shortcut formula.

2 Using the data points given in Exercise 2.A.2, find

    a The standard deviation of the impurity levels by computations based on the original formula.

    b The standard deviation by computations based on the shortcut formula.

    c The mean absolute deviation of the impurity levels.

    d The range of the impurity levels.

3 On the basis of the data of Exercise 2.A.3, calculate

    a The standard deviation of the repair estimates using the original formula for the standard deviation.

    b The standard deviation using the shortcut formula.

    c The mean absolute deviation.

    d The range.

4 Explain why the standard deviation of a set of data points will always be smaller than its range.

5 Calculate the standard deviation of the diameters listed in the data of Exercise 1.A.1.

6 Calculate the standard deviation of the learning times from the data of Exercise 1.A.5.

7 Calculate the standard deviation of the forecasts in Exercise 1.A.6.

8 Calculate the standard deviation of the yearly number of worker days lost using the data of Exercise 2.B.14.

SECTION 2.D CHEBYSHEV'S THEOREM

The mean and the standard deviation stand out among all other possible measures of average and variation, respectively. Despite their relative complexity in comparison with the other suggested measures, they are used by people in applied work and theoretical research to a degree unmatched by any of the other measures. Why is this so? What information do the mean and standard deviation convey about the data that makes them so universally valuable? The answers to these questions are given by Chebyshev's theorem.

    When we want to communicate information about the data, we ideally want our listener to be able to reconstruct the frequency polygon of the data after we finish talking. If we present the information, but the listener does not get a good idea of the nature of the data, then we have wasted our time making the presentation. Suppose you hear that Fosbert Realty takes a mean time of 14 days to sell the houses it lists with a standard deviation of 4 days. Now you know that the waiting times in days for houses, listed with Fosbert, to be sold have mean $\mu = 14$ and standard deviation $\sigma = 4$. How well do you understand the data? Do you understand the data well enough to know, for example, how many of Fosbert's houses sell within 25 days or how many stay on the market longer than 30 days? As it turns out, Chebyshev's theorem provides useful answers to both these questions. In short, Chebyshev's theorem is the primary link between the mean and standard deviation, on the one hand, and the nature of the data itself, on the other.

Before proceeding with our work, it is therefore necessary to develop an understanding of how Chebyshev's theorem turns the mean and standard deviation into useful information about the data. Chebyshev's theorem can be stated as follows:

If $x_1, x_2, \ldots, x_n$ is a set of data points having mean μ and standard deviation σ, then the proportion p of data lying farther from the mean than k standard deviations cannot exceed $1/k^2$.

In Supplement 2.3, we provide an algebraic explanation of why Chebyshev's theorem is true. Here we concentrate only on explaining what it says and how to use it for the purpose of discovering useful information about the data. In Fig. 2.4, there appears an illustration of what Chebyshev's theorem means. Forming a length of k standard deviations, namely $k\sigma$, on both sides of the mean μ, Chebyshev's theorem says that not more than a proportion $1/k^2$ of the data points can possibly fall outside this interval. This implies two equivalent assertions:

1 Not more than a proportion $1/k^2$ of the data points can have numerical values smaller than $\mu - k\sigma$ or greater than $\mu + k\sigma$.
2 At least a proportion $1 - 1/k^2$ of the data points must have values between $\mu - k\sigma$ and $\mu + k\sigma$.

The second assertion is a logical consequence of the first because if at most a fraction $1/k^2$ of the data falls outside the interval, then at least the remainder $1 - 1/k^2$ must fall inside. For example, considering an interval of three standard deviations about the mean, we see that $k = 3$, and we know that no more than $1/k^2 = \frac{1}{9}$ of the data can fall outside the interval. This automatically implies that at least $1 - 1/k^2 = 1 - \frac{1}{9} = \frac{8}{9}$ must be falling inside. This situation is illustrated graphically in Fig. 2.5.

Let us take another look at the questions involving Fosbert Realty, where the relevant data have mean $\mu = 14$ and standard deviation $\sigma = 4$. The question of how many houses remain on the market longer than 30 days can be transformed into the

FIGURE 2.4
Chebyshev's theorem.

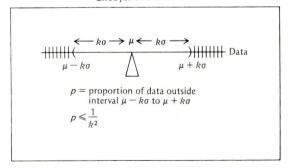

FIGURE 2.5
Chebyshev's theorem with $k = 3$.

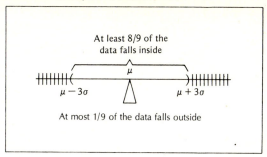

picture of Fig. 2.6. We are interested in the proportion of data points falling above 30. But 30 lies at distance of $30 - 14 = 16$ from the mean $\mu = 14$, and so we are interested in the proportion of data points falling farther from the mean than a distance of 16. Because $\sigma = 4$, the distance 16 represents a distance of $k = 16/\sigma = \frac{16}{4} = 4$ standard deviations. By Chebyshev's theorem, the proportion of data points farther from the mean than $k = 4$ standard deviations cannot exceed $1/k^2 = 1/4^2 = \frac{1}{16} = 0.0625 = 6.25\%$. Therefore, we know that no more than 6.25% of the Fosbert Realty houses remain unsold after 30 days on the listing.

The second question involving the Fosbert Realty data asks: What proportion of Fosbert's houses sell within 25 days? In Fig. 2.7, we illustrate this question. The mean $\mu = 14$, and so the relevant number 25 lies at a distance $25 - 14 = 11$ from the mean. Since $\sigma = 4$, the distance 11 can be expressed as $k = \frac{11}{4} = 2.75$ standard deviations. Chebyshev's theorem then asserts that no more than $1/k^2 = 1/(2.75)^2 = 1/7.5625 = 0.132 = 13.2\%$ of the data points can lie farther from the mean than 2.75 standard deviations. This means that no more than 13.2% of the data points can fall outside the interval from 3 to 25. In particular, no more than 13.2% of Fosbert Realty's houses are sold on days outside the interval 3 to 25 days after listing. Therefore, at least 86.8% ($= 100\% - 13.2\%$) of the houses are sold between 3 and 25 days after listing. We can therefore be sure that at least 86.8% are sold within 25 days of listing.

FIGURE 2.6
Fosbert realty data: proportion of houses on market longer than 30 days.

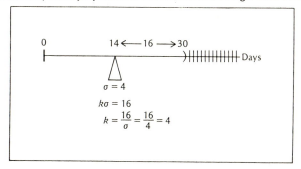

FIGURE 2.7
Fosbert realty data: proportion of houses selling within 25 days.

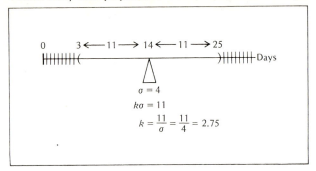

It is possible that more than 86.8% are sold within 25 days because the 86.8% figure does not include any houses sold within 3 days. Houses in the latter category are included among the 13.2% lying outside the interval. Unfortunately, Chebyshev's theorem does not provide any breakdown of the percentages falling above 25 and below 3, but only percentages falling outside and inside various intervals.

We can summarize the sort of information available from Chebyshev's theorem in a table such as Table 2.15, which is based on Fig. 2.4. Using Table 2.15 together with a knowledge of only the mean and the standard deviation, we can estimate, for any set of data points, the proportions of data falling in and out of various intervals about the mean. You should observe from Table 2.15 that, no matter what the set of data, no more than 1% of the data points can fall farther away from the mean than a distance of 10 standard deviations, and no more than 4% of the data points farther away than 5 standard deviations. It automatically follows that at least 96% of the data must fall

TABLE 2.15
Information Given by Chebyshev's Theorem

Distance from Mean to End of Interval in Standard Deviations k	Maximum Possible Percentage of Data Outside Interval $1/k^2$	At Least This Much Data Falls Within Interval $1 - 1/k^2$
1	100.00%	0.00%
2	25.00%	75.00%
3	11.11%	88.89%
4	6.25%	93.75%
5	4.00%	96.00%
6	2.78%	97.22%
7	2.04%	97.96%
8	1.56%	98.44%
9	1.23%	98.77%
10	1.00%	99.00%

within 5 standard deviations, and at least 99% within 10 standard deviations. Analogous statements can be made for any number k of standard deviations, regardless of the shape of the frequency polygon of the data.

We close this section by giving two examples of the application of Chebyshev's theorem to problems of interest in various fields.

Example 2.1   Job Aptitude Test   There are 80 new jobs opening up at an airplane manufacturing plant, but 1100 applicants show up for the 80 positions. To select the best 80 from among the applicants, the prospective employer gives a combination physical and written aptitude test, which covers mechanical skill, manual dexterity, and mathematical ability. The mean grade on this test turns out to be 175, and the scores have standard deviation 10. Can a person who scored 215 count on getting the job?

SOLUTION   From the way the question is formulated, what we need to find out is the number of persons scoring above 215. This information would be useful because if fewer than 80 applicants scored above 215, then a person who scored 215 would be among the top 80 and would therefore get one of the open positions. From Fig. 2.8, we see that the distance $k\sigma$ from the mean of 175 to the end 215 of the interval is $215 - 175 = 40$. But $k\sigma = 40$ means that $k = 40/\sigma = \frac{40}{10} = 4$ and $1/k^2 = 1/4^2 = \frac{1}{16} = .0625 = 6.25\%$. According to Chebyshev's theorem, this means that no more than 6.25% of the applicants could have obtained scores differing from 175 by more than 40 points. That is, if we look at all those who scored below 135 and all those who scored above 215, then both those together will account for no more than 6.25% of all applicants. (Conversely, at least 93.75% of the applicants must have scored between 135 and 215.) In particular, the most we can say is that no more than 6.25% could have scored above 215, in view of the fact that Chebyshev's theorem does not allow us to say how much of the 6.25% falls below 135 and how much above 215. Therefore, because .0625 × 1100 = 68.75, or 6.25% of 1100 equals 68.75, we can be sure that no more than 68 of the applicants could have scored higher than 215. (Even though 68.75 is rounded to 69, there cannot be 69 applicants scoring above 215, because no more than 68.75 could have done so.) Since there are 80 positions, therefore, a person scoring 215 indeed gets one of the jobs, for no more than 68 people could have scored higher.

FIGURE 2.8
Aptitude test scores.

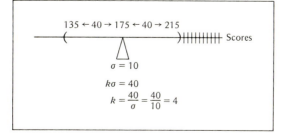

FIGURE 2.9
Cereal box contents (in ounces).

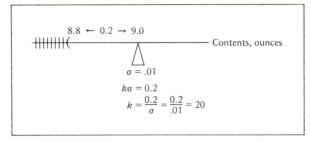

Example 2.2   Contents of Cereal Boxes   The law regulating producers of 9-ounce boxes of breakfast cereal asserts that no more than .5% (one-half of 1%) of the output of any brand of cereal (in boxes labeled as containing 9 ounces) are allowed to contain less than 8.8 ounces. One company markets a 9-ounce box of Corn Puffies, which is packed by machine in such a way that the boxes have mean weight 9.0 ounces, with a standard deviation of .01 ounce. Is the company in violation of the law?

SOLUTION   To test whether the law is being obeyed, it is necessary to find out what proportion of 9-ounce boxes of Corn Puffies contain less than 8.8 ounces. If this proportion turns out to be below .005 (one-half of 1%), then the company is conducting a legal operation. If not, some further investigation might be required. In Fig. 2.9, we are interested in the proportion of data falling in the shaded region, i.e., the proportion of data less than 8.8. Here the distance $k\sigma$ from the mean of 9.0 to the end of the interval at 8.8 is 0.2. But $k\sigma = .2$ means that $k = .2/\sigma = .2/.01 = 20$ and $1/k^2 = 1/(20)^2 = \frac{1}{400} = .0025$. Therefore by Chebyshev's theorem, no more than .0025 (one-quarter of 1%) of the Corn Puffies boxes will contain less than 8.8 ounces. Since a proportion up to .005 is allowed, the Corn Puffies company is well within legal standards because its proportion of underfilled boxes cannot exceed .0025.

SUPPLEMENT 2.3   Algebraic Justification of Chebyshev's Theorem

If, as in Fig. 2.4, we denote by p the proportion of data falling farther from the mean than k standard deviations, Chebyshev's theorem asserts that p cannot exceed $1/k^2$. In symbols, this assertion can be expressed as $p \leq 1/k^2$, where the symbol $\leq$ is read "less than or equal to," a slight expansion of the equals sign. Because a proportion p of the n data points falls outside the interval $\mu - k\sigma$ to $\mu + k\sigma$, a total of pn data points fall outside. For example, if $p = \frac{1}{10}$ and there were 60 data points in all, then the number of data points falling outside would be $pn = (\frac{1}{10})(60) = 6$. Now because

$$\sigma = \sqrt{\frac{\sum_{k=1}^{n} (x_k - \mu)^2}{n}}$$

we know that
$$\sigma^2 = \frac{\sum\limits_{k=1}^{n} (x_k - \mu)^2}{n}$$

and so
$$n\sigma^2 = \sum\limits_{k=1}^{n} (x_k - \mu)^2$$

But $\sum\limits_{k=1}^{n} (x_k - \mu)^2$ is the sum of *all* the squared deviations; if we consider the sum of squared deviations for *only* those points falling outside the interval, we will get a smaller number. This smaller number will be denoted by

$$\sum\limits_{\substack{\text{points} \\ \text{outside}}} (x_k - \mu)^2$$

For points falling outside the interval x_k differs from μ by at least a distance $k\sigma$, and so $(x_k - \mu)^2 \geq (k\sigma)^2 = k^2\sigma^2$. Since there are np points falling outside the interval, the summation just above consists of np terms, each of which is larger than $k^2\sigma^2$. Therefore

$$\sum\limits_{\substack{\text{points} \\ \text{outside}}} (x_k - \mu)^2 \geq npk^2\sigma^2$$

But $n\sigma^2 = \sum\limits_{k=1}^{n} (x_k - \mu)^2 \geq \sum\limits_{\substack{\text{points} \\ \text{outside}}} (x_k - \mu)^2$ as we have already observed so that,

combining the above two lines, we can conclude that

$$n\sigma^2 \geq npk^2\sigma^2$$

Dividing both sides of this inequality by $n\sigma^2$, we get

$$1 \geq pk^2$$

which is the same as $pk^2 \leq 1$. Dividing through by k^2, we wind up with

$$p \leq \frac{1}{k^2}$$

which is precisely the assertion that the proportion of points falling outside the interval cannot exceed $1/k^2$.

EXERCISES 2.D

1 Even among cars produced on the same assembly line, rates of emission of environmental pollutants differ. In particular, the revolutionary new Peccarie (which gets 52 miles/gallon at freeway speeds) emits an average of 1.2 milligrams of pollutants per mile with a standard deviation of 0.04 milligram. New EPA regulations prohibit the sale of the entire production run of an automobile brand if more than 2% of the cars in the run emit pollutants in excess of 1.5 milligrams/mile. Can the Peccarie be legally sold in the United States?

2 State guidelines on the number of units taken by college students during their undergraduate careers will soon specify, in the interests of reducing costs, that no student ought to be permitted to exceed 220 quarter units of work or equivalent. (186 units are required for graduation.) If graduating students at Bolsa Chica State average 203 units with a standard deviation of 4 units, at least what proportion of students already satisfy the new guidelines?

3 An electrical firm manufactures a 100-watt light bulb which, according to specifications printed on the package, has a mean life of 800 hours with a standard deviation of 40 hours. At most, what percentage of the bulbs fail to last even 700 hours?

4 Since the 55-mile/hour speed limit on freeways has been in effect, observers estimate that vehicles on the freeways travel with mean speed 55 miles/hour and standard deviation 2 miles/hour. If these estimates are accurate, at most what percentage of the vehicles travel at speeds in excess of 65 miles/hour?

5 The 435 congressional districts in the United States have a mean population of 480,000 with a standard deviation of 30,000. At most how many districts contain more than 600,000 residents?

6 Mean annual rainfall in a certain geographical region is 52.75 inches with a standard deviation of 6.50 inches. At most, what percentage of years have rainfall exceeding 70 inches?

7 A manufacturing company has a contract calling for several thousand units of steel pipe having diameter between 4.56 and 4.62 inches. If a unit of pipe has diameter outside this range, it will not fit into the intended construction position and so must be wasted. Only those units having diameter within the range are acceptable to the customer. Initial output of pipe from the assembly line projects a mean diameter of 4.59 inches, with a standard deviation of .005 inch. At least what percentage of the output will be acceptable to the customer if this trend continues?

8 A psychologist believes that it takes an average of 70 hours with a standard deviation of 10 hours to train a monkey to distinguish the colors red, yellow, and blue. If his assumption is correct, would it be reasonable that, out of a group of 360 monkeys, there were 80 monkeys able to learn the distinction in less than 40 hours of training?

9 The stock exchange is open 250 days each year. In a study of the fluctuation of the closing price of the stock of MGI, Inc., a computer readout indicated that the mean closing price for last year was 22.375 with a standard deviation of 5.000. At most, how many days did MGI stock close below 10?

10 After studying the incomes of families in a large urban area, a sociologist concludes that the mean income is 9000 dollars with a standard deviation of 2000 dollars. At most what proportion of families have incomes higher than 25,000 dollars?

SECTION 2.E STANDARD SCORES (z SCORES)

There is a certain graduate school which requires all prospective students to take either the Graduate Record Exam (GRE) or the Montana Reasoning Test (MRT) as part of the

TABLE 2.16
Test Scores Submitted by Applicants

Applicant	Test Taken	Score
A	GRE	700
B	GRE	450
C	MRT	70
D	GRE	600
E	MRT	110

application procedure. The school presumably uses the results of these tests in ranking all its applicants for the purpose of deciding which ones are to be admitted to advanced degree programs. Unfortunately, for comparison purposes, the GRE and the MRT are graded on different scales. The totality of GRE scores have mean 500 with a standard deviation of 100, while the MRT average 60 with a standard deviation of 20. Table 2.16 shows the test scores submitted by five applicants, three of whom have taken the GRE, while the other two have taken the MRT. How is the school to compare and rank the scores, some of which are based on one type of scale and the rest on another? For example, how do we know which is the better score: a 700 on the GRE or a 110 on the MRT?

The way to solve the problem is to adjust, if possible, all the scores so that they will all be on the same scale. Then it will be easy to see which scores are higher than others, relatively speaking. Such an adjustment procedure is called "standardization" of scores, and the resulting adjusted scores are called "standard scores" or "z scores." To distinguish between the z scores and the original scores, the latter are often referred to as "raw scores." The standardization procedure is another highly useful application involving the mean and the standard deviation, and further contributes to the desirability of these measures of average and variation.

Standard scores can be best described as follows:

If $x_1, x_2, \ldots, x_n$ are a set of data having mean μ and standard deviation σ, then the standard scores (or z scores) of the set of data are the numbers $z_1, z_2, \ldots, z_n$ where

$$z_1 = \frac{x_1 - \mu}{\sigma}$$

$$z_2 = \frac{x_2 - \mu}{\sigma}$$

$$\cdots\cdots\cdots\cdots$$

$$z_n = \frac{x_n - \mu}{\sigma}$$

Notice that to find the z score of a data point x, we subtract the mean of its group from x and divide the resulting difference by the standard deviation of its group. More compactly, we say

$$z = \frac{x - \mu}{\sigma}$$

Using this formula for z scores, let's find the z score for applicant A of Table 2.16. Applicant A scored 700 on the GRE, an exam which has mean 500 and standard deviation 100. Therefore the z score is

$$z_A = \frac{700 - 500}{100} = \frac{200}{100} = 2$$

On the other hand, let's take a look at applicant C. This individual scored 70 on the MRT, a test on which the mean score is 60, with a standard deviation of 20. Therefore, applicant C has z score

$$z_C = \frac{70 - 60}{20} = \frac{10}{20} = 0.5$$

Now the z scores of A and C are 2 and 0.5, respectively. The formula

$$z = \frac{x - \mu}{\sigma}$$

really specifies the location of the point x in terms of the number of standard deviations it lies above the mean of its group. Therefore, applicant A scored 2 (two) standard deviations above the mean of her group, while applicant C scored only .5 (one-half) standard deviation above the mean of his group. Relatively speaking, then, applicant A did somewhat better than applicant C.

To compute all the z scores, we set up a table of calculations like Table 2.17, recalling that GRE scores have a nationwide mean of 500 and a standard deviation of 100, while MRT scores have a nationwide mean of 60 and a standard deviation of 20. Having standardized all the scores, we can now easily rank all the applicants. Applicant E had the highest score, 2.5 standard deviations above the mean of the MRT group, while applicant B got the lowest score, .5 standard deviation *below* the GRE group. The ranking of the applicants appears in Table 2.18.

Now that we have discussed the process of standardizing sets of data which are measured according to different scales, it would be useful to know the exact way in which the data are "standardized." In short, what are the standards for standardization? Well, standardization of a set of data does the following thing: It transforms the original

TABLE 2.17
Calculation of Standard Scores (z Scores)

Applicant	Test	Raw Score x	Mean of Group μ	Standard Deviation σ	$x - \mu$	z Score $\dfrac{x - \mu}{\sigma}$
A	GRE	700	500	100	200	2.0
B	GRE	450	500	100	−50	−.5
C	MRT	70	60	20	10	.5
D	GRE	600	500	100	100	1.0
E	MRT	110	60	20	50	2.5

TABLE 2.18
Applicants Ranked by z Scores

Rank	Applicant	z Score
1	E	2.5
2	A	2.0
3	D	1.0
4	C	.5
5	B	−.5

FIGURE 2.10
The process of standardization.

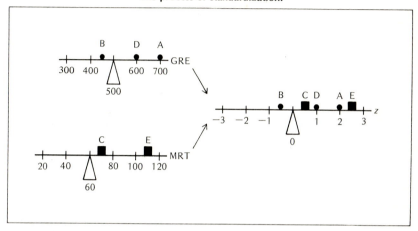

(raw) data into a set of numbers (called z scores) which have mean 0 and standard deviation 1. In fact, no matter what the original values of μ and σ were, the z scores always have mean 0 and standard deviation 1. Once two or more sets of data are standardized, they can be graphed on the same scale and compared. This graphing process is illustrated in Fig. 2.10. Details of the algebraic justification of the fact that a set of z scores, no matter what its origin, has mean 0 and standard deviation 1 appear in Supplement 2.4.

SUPPLEMENT 2.4 Algebraic Verification of Standardization

In this supplement, our objective is to show that, no matter what numbers constitute a set of data, the corresponding set of z scores always has mean 0 and standard deviation 1. If the original data points are $x_1, x_2, \ldots, x_n$, then the z scores are the numbers $z_1, z_2, \ldots, z_n$, where each z_k can be expressed as

$$z_k = \frac{x_k - \mu}{\sigma}$$

where
$$\mu = \frac{\sum_{k=1}^{n} x_k}{n} \qquad \text{and} \qquad \sigma = \sqrt{\frac{\sum_{k=1}^{n} (x_k - \mu)^2}{n}}$$

By μ_z and σ_z we denote the mean and standard deviation of the z scores, respectively. We now proceed to the algebraic calculation that $\mu_z = 0$ and $\sigma_z = 1$ for every possible set of data.

$$\mu_z = \frac{\sum_{k=1}^{n} z_k}{n} = \frac{1}{n} \sum_{k=1}^{n} z_k = \frac{1}{n}(z_1 + z_2 + \cdots + z_n)$$

$$= \frac{1}{n}\left(\frac{x_1 - \mu}{\sigma} = \frac{x_2 - \mu}{\sigma} + \cdots + \frac{x_n - \mu}{\sigma}\right)$$

$$= \frac{1}{n}\frac{(x_1 - \mu) + (x_2 - \mu) + \cdots + (x_n - \mu)}{\sigma} = \frac{1}{n\sigma}\sum_{k=1}^{n}(x_k - \mu)$$

$$= \frac{0}{n\sigma} = 0$$

in view of the balancing property of the mean as justified in Supplement 2.1. For the standard deviation,

$$\sigma_z = \sqrt{\frac{\sum_{k=1}^{n}(z_k - \mu_z)^2}{n}} = \sqrt{\frac{\sum_{k=1}^{n} z_k^2}{n}}$$

because $\mu_z = 0$, as we have just shown. Continuing,

$$\sigma_z = \sqrt{\frac{1}{n}\sum_{k=1}^{n} z_k^2} = \sqrt{\frac{1}{n}\sum_{k=1}^{n}\left(\frac{x_k - \mu}{\sigma}\right)^2}$$

$$= \sqrt{\frac{1}{n}\left[\left(\frac{x_1 - \mu}{\sigma}\right)^2 + \left(\frac{x_2 - \mu}{\sigma}\right)^2 + \cdots + \left(\frac{x_n - \mu}{\sigma}\right)^2\right]}$$

$$= \sqrt{\frac{1}{n}\left[\frac{(x_1 - \mu)^2 + (x_2 - \mu)^2 + \cdots + (x_n - \mu)^2}{\sigma^2}\right]}$$

$$= \sqrt{\frac{1}{n\sigma^2}\sum_{k=1}^{n}(x_k - \mu)^2} = \sqrt{\frac{1}{\sigma^2}\frac{1}{n}\sum_{k=1}^{n}(x_k - \mu)^2}$$

$$= \sqrt{\frac{1}{\sigma^2}}\sqrt{\frac{1}{n}\sum_{k=1}^{n}(x_k - \mu)^2} = \left(\frac{1}{\sigma}\right)(\sigma) = 1$$

in view of the formula for σ. This completes the algebraic justification of the standardization process.

EXERCISES 2.E

1 Six applicants are applying for a single opening in medical technology at a research clinic working on skin cancer. Each applicant is given a 6-hour test divided into three parts: ability to concentrate, manual dexterity, and knowledge of chemistry. Their grades on each part are as follows:

Applicant	Ability to Concentrate	Manual Dexterity	Knowledge of Chemistry
A	10	9	5
B	9	9	6
C	5	2	0
D	5	9	10
E	1	0	6
F	0	1	9

In order to make fair comparisons, the grades are first standardized within each competency area, and the applicant with the highest total of z scores gets the job.

a Standardize the test scorés of each applicant on the "ability to concentrate" section.

b Standardize all the "manual dexterity" scores.

c Standardize the "knowledge of chemistry" scores.

d Which applicant gets the job?

2 One college student took three exams on the same day. Her scores, as well as the mean and standard deviation of all the scores of her group, are presented in the following table:

Subject Matter	Student's Score	Group Mean	Group Standard Deviation
Economics	72	70.8	0.6
Mathematics	61	51.0	6.0
Psychology	50	30.0	12.0

On which exam did the student do best in comparison with the rest of her group?

3 Four applicants for flight training, two from Portland, Oregon, and two from Tampa, Florida, have submitted their scores on different flight aptitude tests. The two from Oregon had taken the Great Open Skies Flight Aptitude Test (GOSFAT), while the pair from Florida had taken the Gulf and Southern Flight Aptitude Test (GASFAT). Scores on the GOSFAT have mean 50 with standard deviation 10, while those on the GASFAT have mean 120 and standard deviation 15. The applicants and their scores are listed below:

Applicant	Test Taken	Score
#1	GOSFAT	60
#2	GOSFAT	45
#3	GASFAT	129
#4	GASFAT	115

$$\frac{60-50}{10} = 1$$
$$\frac{45-50}{10} = -.5$$
$$\frac{129-120}{15} = +.6$$
$$\frac{115-120}{15} = -.3$$

Standardize each score and rank the applicants.

4 Make up a set of five data points and standardize them. Then verify by calculation that the z scores have mean 0 and standard deviation 1.

SUMMARY AND DISCUSSION

We have devoted Chap. 2 to the development of numerical characteristics of a set of data and to an analysis of their usefulness in describing the data. We have mentioned four types of averages, the mean, the median, the mode, and the midrange, and three measures of dispersion, the standard deviation, the mean absolute deviation, and the range, and we have carefully noted the distinctions between them. Through Chebyshev's theorem, the mean and the standard deviation have been seen to be the most useful of the measures of average and dispersion, respectively. We have observed that when we apply Chebyshev's theorem to the mean and standard deviation of a set of data points, we can often answer very specific questions about the situation described by the set of data. Finally, we have discussed standardization of sets of data with the aim of being able to compare widely differing sets of data by placing all of them on the same scale. The calculation of standard scores, or z scores, illustrates another important use of the mean and standard deviation. The material of Chap. 2 provides the framework for all the discussions of data to follow in the remainder of the text. Throughout the book, we will be referring to the basic facts about the mean and standard deviation that were introduced in this chapter.

BIBLIOGRAPHY

Review of Useful Mathematical Background
Auslander, L., et al.: "Mathematics Through Statistics," Williams & Wilkins, Baltimore, 1973.

Braverman, H.: "Reviewing Statistics," Robin Hill, New York, 1971.

More on Numerical Descriptions
Chisson, B. S.: Interpretation of the Kurtosis Statistic, The American Statistician, October, 1970, pp. 19–22.

Cureton, E. C.: Letter to the Editor, The American Statistician, December 1971, p. 61.

Darlington, R. B.: Is Kurtosis Really "Peakedness"? The American Statistician, April 1970, pp. 19–22.

Fama, E. F.: The Behavior of Stock Market Prices, The Journal of Business, 1965, pp. 34–105.

Hildebrand, D. K.: Kurtosis Measures Bimodality? The American Statistician, February 1971, pp. 42–43.

Sharpe, W. F.: Mean-Absolute-Deviation Characteristic Lines for Securities and Portfolios, Management Science, October 1971, pp. 1–13.

SUPPLEMENTARY EXERCISES

1 The following data points from the files of a major life insurance company record the number of consecutive weeks that medical disability compensation was paid to 14 heart-attack victims:

9	20
21	27
14	13
23	10
8	32
16	18
7	20

a Calculate the mean number of weeks that compensation was paid.
b Find the median number of weeks that compensation was paid.
c Find the mode.
d Find the midrange.
e Calculate the standard deviation of the number of weeks of paid compensation.
f Calculate the mean absolute deviation of the number of weeks of paid compensation.
g Find the range of the number of weeks of paid compensation.

2 A local bank having 12 branch offices is considering adopting a completely computerized payroll system. As part of an analysis of the operating costs of the system, the costs of running one aspect of the system are carefully observed for the 12 branches with the following resulting data on costs per hour of operating the system:

24	20
14	24
14	14
14	20
12	14
7	12

a Calculate the mean operating cost for the 12-branch bank.
b Find the median operating cost.
c Find the modal operating cost.
d Find the midrange of the operating costs.
e Calculate the standard deviation of the 12 operating costs.
f Calculate the mean absolute deviation of the costs.
g Find the range of the operating costs.

3 In a biological study of the uniformity of growth of cherry tomato plants, the following data give the numbers of tomatoes harvested from 20 test plants after 90 days of growth:

3	8	11	17
22	24	19	14
9	1	17	18
12	9	12	25
22	17	13	14

a Calculate the mean harvest.
b Draw a seesaw diagram of the data points, and indicate the location of the mean.
c Find the median harvest.

d Calculate the standard deviation.

4 The United States Smelting, Refining, and Mining Company had, until 5 years ago, a mill at Midvale, Utah, which produced a mean of 26,000 tons of lead concentrate per year over the preceding 36 years. The standard deviation of the annual production was 3500 tons. At most, in how many of those 36 years could production have fallen below 15,500 tons?

5 An accountant commutes by automobile from his suburban home to his office in a downtown corporate headquarters. The daily trip takes, on the average, 24 minutes with a standard deviation of 4 minutes. If he were to leave his home regularly at 8:20 A.M., at least how often would he manage to make it to the office in time for the staff summary meeting at 9:00 A.M.?

6 An automobile battery carrying a 24-month guarantee actually has a mean lifetime of 30 months, with a standard deviation of 2 months.
a At most, how many of these batteries fail before 24 months have passed?
b The manufacturer is considering reducing the guarantee period in order to cut back on the expense of honoring the guarantee. At least how many of the batteries last longer than 20 months?

7 A local hospital conducts its fund-raising drives by first sending a request for a contribution to 1000 local business executives who have contributed generously in the past. If these business executives gave 50 dollars, on the average, with a standard deviation of 10 dollars, at least how many contributed more than $20?

8 One local supermarket carries 20,000 different food items. With an eye toward both publicity and cost cutting, it announces that it will refuse to carry any item which has risen in price by more than 25% during the previous year. If food prices rose by an average of 12% one year with a standard deviation of 4%, at least how many of the original 20,000 items were still on the shelves the next year?

9 An urban planner's data indicate that on Los Angeles area freeways between 4 and 6 P.M. weekdays there are 400,000 vehicles. These vehicles, furthermore, spend a mean time of 45 minutes on the freeways with a standard deviation of 15 minutes. At most, how many vehicles spend 90 minutes or more on the freeways?

10 Over the past several years, the daily prime rate of interest at major New York banks has averaged 6.2% with a standard deviation of 0.4%. At most, what proportion of the time has it exceeded 11.8%?

11 Several years ago, a group of urban planners constructed a model of a "planned" ideal city, of which one goal was that no more than 2% of the residents would have to commute longer than 20 minutes from home to work. When the actual city was completed recently, the inhabitants' mean travel time from home to work turned out to be 12 minutes with a standard deviation of 1 minute. Was the planners' goal achieved?

12 Insurance claims arrive at Buzzardbait National Life & Casualty's home office at a mean rate of 140 per week with a standard deviation of 12. At most, what proportion of weeks does the number of claims exceed 200?

13 Cars now on the road nationwide get 14 miles/gallon of gasoline, on the average, with a standard deviation of 3 miles/gallon. A plan now under study proposes to encourage engine efficiency by rebating all federal gasoline taxes to owners of those cars ranking among the top 10% in efficiency. If your car gets 25 miles to the gallon, would you qualify for a rebate?

14 The mean age of those eligible to vote in the United States is 40 years with a standard deviation of 10 years. At most what percentage of eligible voters are in the age group 18 to 21?

15 In a college course on the construction and standardization of psychological tests, three exams are given. All three exams count equally toward the student's final grade, and nothing else is taken into consideration. The following table gives the mean and standard deviation of the scores of the entire class on each of the exams:

	First Exam	Second Exam	Third Exam
Class mean	65	74	71
Class standard deviation	4	6	12

One of the students, Xerxes by name, got a 57 on the first exam, a 76 on the second exam, and a 96 on the third, while another student, Yennie, obtained a 73 on the first, an 86 on the second, and a 70 on the third. Yennie was awarded a grade of B for the course, while Xerxes received only a C. Xerxes protests to the instructor, of course, because his average on the three exams is exactly the same as Yennie's, so he feels that he deserves the same grade. Unfortunately for Xerxes, however, the instructor standardized each exam score with respect to the overall class perform- ance on that exam before computing the student's grade.
a Calculate each of Xerxes' standardized scores, and compute the average of the three.
b Standardize each of Yennie's exam scores, and average them.
c Explain why Yennie got a B and Xerxes a C.

16 Construct a set of eight data points having mean 12 and standard deviation 5.
a Standardize the data points.
b Verify by calculation that the z scores have mean 0 and standard deviation 1.

THE BASIC
PROBABILITY DISTRIBUTIONS

In the previous two chapters, several techniques for describing a set of data have been introduced in an attempt to organize the data into an understandable format. Of the various techniques discussed, probably the most useful for visualizing and communicating the behavior of a single set of data are the histogram, the frequency polygon, the mean, and the standard deviation. We have seen how to construct the histogram and frequency polygon and how to calculate the mean and standard deviation directly from the numerical values of the data points. Finally, we have developed procedures such as Chebyshev's theorem to analyze more deeply the set of data and so further improve our comprehension of it.

Our work so far, however, has dealt only with the organization of a single set of data. All the data sets that have been discussed in the text and the exercises generally have very different pictorial representations and numerical descriptions. It seems that there is no way of telling in advance how the frequency polygon will look or what the mean and standard deviation will be. Yet, as it turns out, there are often discernible patterns common to sets of data derived from many different sources. In this chapter, our objective will be to further our understanding of statistical data by studying these general patterns of behavior of sets of data. If we can fit a particular data set with which we are working into one of these general patterns, then we will know more about its overall characteristics; we will be able to say that it has all the properties of the general pattern to which it belongs, in addition to those special properties that can be discovered using the methods of the previous two chapters. These general patterns of data behavior are called "probability distributions."

SECTION 3.A PROBABILITIES AND PROPORTIONS

By the probability of an event, we usually mean the proportion of time that the event can be expected to occur in the long run. For a simple example, consider the tossing of a coin and the observation of which of the two sides, heads or tails, is facing upward when the coin lands. If the coin is tossed 10,000 times and falls heads on 3000 of these tosses, we are probably justified in asserting that the probability of heads for this particular coin is 3/10. By a "fair" coin, we mean a coin whose probability of heads is 1/2.

To give an example of how probabilities are determined from a set of data, let's take a look at the subject matter of Table 1.2, which we reproduce in a briefer format as Table 3.1. We can see from Table 3.1 that on 23 of the 100 days involved in the study the demand for electricity was between 26 and 28 million kilowatthours, inclusive. We can rephrase this fact in probability language as follows: The study indicates that the probability of a daily demand for electricity between 26 and 28 million kilowatthours is 23/100 or .23. Using symbols, we can write

$$P(26 \leq D \leq 28) = .23$$

where $P(\ )$ means "probability of," and D stands for "daily demand for electricity in millions of kilowatthours." Technically speaking, D is called a "random variable." Random variables are variable quantities which take several possible values, determined according to a probability distribution. The notation $26 \leq D \leq 28$ signifies that

TABLE 3.1
Frequency Distribution of
Daily Demand for Electricity

Demand, millions of kilowatthours	Days
20–22	4
23–25	8
26–28	23
29–31	34
32–34	26
35–37	4
38–40	1
Sum = 100	

D is at least as large as 26 ($26 \leq D$) but no larger than 28 ($D \leq 28$); i.e., D is between 26 and 28, inclusive. Similarly, we can write the following remaining probabilities:

$$P(20 \leq D \leq 22) = .04$$
$$P(23 \leq D \leq 25) = .08$$
$$P(29 \leq D \leq 31) = .34$$
$$P(32 \leq D \leq 34) = .26$$
$$P(35 \leq D \leq 37) = .04$$
$$P(38 \leq D \leq 40) = .01$$

The seven probabilities calculated above comprise what is called the "probability distribution" of the daily demand for electricity in millions of kilowatthours, or, more simply, the probability distribution of D. In Table 3.2, we formally display the probability distribution of D.

As we have already shown in Chap. 1, the probability distribution of D can be illustrated graphically by the histogram of Fig. 1.1 or the frequency polygon of Fig. 1.2. All that is necessary is to divide the number of days on the vertical scale by 100 (because a total of 100 is represented by the data) to obtain the proportions. As can be seen from Table 3.2, the probability distribution of a quantity is merely a listing of the possible values for the quantity together with the probabilities with which the quantity takes on those values.

In Table 3.3, we present the probability distribution for the toss of the coin discussed earlier in this chapter. It is to be recalled that the coin in question fell heads 3000 times in 10,000 tosses. (The coin fell tails, of course, the remaining 7000 times.)

A probability distribution is said to be "uniform" if all the possible outcomes have the same probability. A simple example of a uniform distribution would be the probability distribution of a "fair" coin. Such a coin could fall heads about half the time and tails the other half. We could then write

$$P(\text{heads}) = \frac{1}{2}$$

TABLE 3.2
Probability Distribution of Daily
Demand for Electricity D

a, millions of kilowatthours	b, millions of kilowatthours	$P(a \leq D \leq b)$
20	22	.04
23	25	.08
26	28	.23
29	31	.34
32	34	.26
35	37	.04
38	40	.01
		Sum $= 1.00$

TABLE 3.3
Probability Distribution of
Outcomes of Coin Toss
For a Coin That Fell Heads
3000 Times in 10,000 Tosses

Outcome	Probability
Heads	3/10
Tails	7/10
	Sum = 1

$$P(\text{tails}) = \frac{1}{2}$$

for a fair coin. If a quantity is uniformly distributed and has n possible outcomes, it is automatic that each of the n possible outcomes has probability $1/n$. For example, if 16 evenly matched golfers play in a tournament, then each has probability $1/16$ of winning. This estimate of probabilities comes from the observation that if the tournament were repeated many, many times, each golfer would win a proportion of 1 out of every 16 times, on the average. This is what we mean by the uniform distribution.

In the remaining sections of this chapter, we will study in detail what are perhaps the two most useful probability distributions encountered in elementary statistical investigations: the binomial distribution and the normal distribution. Probabilities and proportions are discussed further from another point of view in Chap. 8.

EXERCISES 3.A

1 From the data of Exercise 1.A.1, construct a probability distribution of the diameters of stones found in the delta region under study.

2 Using the data of Exercise 1.A.5, set up a probability distribution of the learning times.

3 Construct a probability distribution of the housing ratings which are given by the frequency distribution of Exercise 1.A.4.

4 Based on the Consumer Price Index forecasts of Exercise 1.A.6, set up a probability distribution of the forecasts.

SECTION 3.B THE BINOMIAL DISTRIBUTION

The repeated tossing of a coin provides perhaps the simplest example of a process which generates binomially distributed data. There are only two possible outcomes, heads and tails, with $P(\text{heads}) = 1/2$ and $P(\text{tails}) = 1/2$ if the coin is fair. Furthermore, each toss of the coin is independent of all previous tosses.[1] These are the characteristic properties of what is called a binomial experiment: there are only two possible outcomes ("binomial," from Greek and Latin roots, means "having two names"), their probabilities remain constant throughout the experiment, although they do not necessarily have to equal 1/2, and each trial of the experiment is independent of all previous trials. Many examples of applied interest can be organized as binomial experiments, and therefore any resulting data taken in such a situation will have the binomial distribution. Consider the following few cases:

1 **A marketing survey** Several shoppers near the dairy case of a supermarket are asked the question: Would you try a carton of chocolate flavored buttermilk if such a product were developed? The two possible outcomes are "yes" and "no." Presumably each shopper's opinion would not depend on that of any other shopper if the questioning were properly conducted. The probability of primary concern involved here is $p = P(\text{yes})$, the proportion of shoppers who would be willing to try the new product.

2 **A political poll** Registered voters are asked whether or not they favor repeal of a certain tax law. The two possible outcomes are "for repeal" and "against repeal." If the poll is properly conducted, each voter's response should be independent of that of all other voters. The relevant probability is $p = P(\text{for repeal})$, the proportion of voters favoring repeal.

3 **A medical test** Certain patients are given blood tests for anemia. The two possible outcomes are "is anemic" and "is not anemic." We can be reasonably sure that whether or not a particular patient is anemic does not depend on the condition of any other patient. The probability here is $p = P(\text{is anemic})$, the proportion of patients showing anemia.

4 **A psychological cure** Some patients are given a new drug whose job is to cure mental depression. The two possible outcomes are "cured" and "not cured." Each patient's reaction to the drug ordinarily does not depend on the reaction of other patients. The probability $p = P(\text{cured})$ is the proportion of patients successfully treated by the new drug.

[1]This statement about coin tossing has generated substantial philosophical controversy over the years, even though it can be proved by a simple experiment. The allegation has been made that, if a fair coin falls heads 10 times in a row, it is very likely to fall tails on the 11th toss. Much money has been lost by gambling on this untrue assertion. Even though the "law of averages" implies that a coin will fall heads about half the time and tails the other half, this does not place a moral obligation on the individual coin to pay back those 10 tails it owes. In fact, during the 11th toss, the coin itself does not even remember that it has fallen heads on the last 10 consecutive tosses. All it knows while it is spinning in the air is that, because of its structure, it is equally likely to fall heads or tails on any given toss, including the 11th. The reader is invited to check this out by conducting a coin-tossing experiment. Take a handful of, say, 500 pennies and toss them in the air in a closed room having no holes in the floor. About half the coins will fall tails. Remove them, and toss the remaining ones a second time. Again, half will have fallen tails. Remove them and proceed. Each time only half the coins will fall tails, even though every coin tossed has fallen heads on all previous tosses.

5 An agricultural experiment Seeds are planted in soil to which a proposed new chemical fertilizer, to aid in seed germination, has been added. The two possible outcomes are "germinated" and "did not germinate." The germination of each individual seed is not affected by the germination of any other. The probability of interest is $p = P$(germinated), the proportion of seeds that germinate with the aid of the new fertilizer.

6 An actuarial problem Several men aged 45 each apply for 10-year term life insurance policies. The two possible outcomes are "died before age 55" and "lived to age 55." Whether or not one particular man dies before he reaches the age of 55 does not seem to be dependent on what happens to any other policyholder. The relevant probability is $p = P$(death before age 55), the proportion of 45-year-old men who die before they reach age 55.

All binomial experiments can be completely characterized by two numbers. The first is p, the probability of one of the outcomes; the other outcome must then have probability $1 - p$, for if P(heads) $= 1/2$, then P(tails) $= 1/2$, and if P(yes) $= 4/5$, then P(no) $= 1/5$. (The reason for this is that the probabilities must add up to 1.00 in order to represent 100% of the occurrences.) The second is n, the number of times the experiment is repeated. All properties and mathematical formulas involved in the study of binomial experiments can be expressed in terms of the numbers n and p. Numbers that are characteristic of a situation are called the "parameters" of the problem. For this reason, much of statistics is concerned with the study of parameters, because if we can figure out what the parameters of a situation are, then we can determine everything else from them.[1]

To investigate the workings of a binomial experiment from the statistical point of view and to follow the action of the parameters, let's take a look at some of the above examples in more detail.

Example 3.1 Coin Tossing Suppose we take three fair coins, toss them, and count the number of heads we have obtained. Here the parameters are $n = 3$ and $p = 1/2$, where $p = P$(heads). We can denote by H the total number of heads obtained with the three coins. The quantity H is a random variable taking the possible values 0, 1, 2, and 3, because with three coins, we can get none, one, two, or three heads. (There is no way to get four or more heads with three coins.) We would like to compute the following probabilities:

$P(H = 0)$ = probability of getting no heads with three coins
$P(H = 1)$ = probability of getting one head with three coins
$P(H = 2)$ = probability of getting two heads with three coins
$P(H = 3)$ = probability of getting three heads with three coins

---

[1]However, some types of questions do not easily lend themselves to the analysis of parameters. To answer these questions, we need an entirely different class of procedures, called "nonparametric" statistics, a brief introduction to which appears in Chap. 11.

SOLUTION   As described in Sec. 3.A, the probabilities above comprise the "probability distribution" of the number of heads obtained in the toss of three coins. In this case, we say that the number of heads has the binomial distribution with parameters $n = 3$ and $p = 1/2$.

Now, how do we compute these probabilities? In Table 3.4, we list all possible outcomes of the toss of the three coins. As can be seen from Table 3.4, there are eight possible ways the three coins, as a group, can turn up. Because $p = 1/2$, we know that, for each of the three coins, heads and tails are equally likely. Therefore outcome 1 is just as likely to occur as outcome 8, and, in fact, all eight outcomes are equally likely to occur. This means that we can expect each outcome to occur about 1/8 or 12.5% of the time.

We are not directly interested in these outcomes, however, but in the number of heads obtained in each case. From Table 3.4, we can see that we obtain three heads from outcome 1; two heads from outcomes 2, 3, and 4; and one head from outcomes 5, 6, and 7; and no heads from outcome 8. Therefore, we can expect to get three heads whenever outcome 1 occurs, namely 12.5% of the time. We express this in mathematical symbols by writing $P(H = 3) = .125$. We can expect to get two heads 12.5% of the time when outcome 2 occurs, 12.5% of the time when outcome 3 occurs, and also 12.5% of the time when outcome 4 occurs. Therefore, we would expect to obtain two heads a total of 12.5% + 12.5% + 12.5% = 37.5% of the time. In mathematical symbols, we write $P(H = 2) = .375$. By similar reasoning, it can be shown that $P(H = 1) = .375$ and $P(H = 0) = .125$. By analogy with Table 3.2, we collect all these results and display them in the probability distribution of Table 3.5. In Fig. 3.1, we present the histogram of the probability distribution of Table 3.5.

In our calculation of the probability distribution of H, we relied very heavily on the fact that all eight outcomes of the coin tossing were equally likely to occur. This situation was reflected in the value of the parameter p, which was equal to 1/2. In the case $p \neq 1/2$, we no longer will have equally likely outcomes. When $p > 1/2$, we are more likely to get a head than a tail, and when $p < 1/2$, we are more likely to get a tail. In such cases, the calculation of the probability distribution is somewhat more difficult. Although there is a formula for calculating the various probabilities involved, to under-

TABLE 3.4
Outcomes of the Toss of Three Coins

Outcome	First Coin	Second Coin	Third Coin
#1	Heads	Heads	Heads
#2	Heads	Heads	Tails
#3	Heads	Tails	Heads
#4	Tails	Heads	Heads
#5	Tails	Heads	Tails
#6	Tails	Tails	Heads
#7	Heads	Tails	Tails
#8	Tails	Tails	Tails

FIGURE 3.1
Histogram of the binomial distribution with parameters $n = 3$, $p = \frac{1}{2}$.

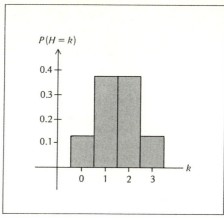

stand it fully requires the knowledge of some concepts which we have not yet explained. These concepts will be discussed in Chap. 8, where probability theory is developed in some detail. We will therefore have to postpone to Sec. 8.E the full explanation of how to calculate these binomial probabilities directly. So that we will be able to study binomially distributed data in the meantime, Table A.2 of the Appendix presents a comprehensive analysis of the binomial distribution. By use of that table, we can find out all relevant probabilities for binomial distributions having various numerical values for n and p.

How to Use the Table of the Binomial Distribution

As an example of how to use the table, let's recompute the material in Table 3.5, using Table A.2 instead of the facts about coin tossing.

The first fact we must note about Table A.2 is that the numbers there are values of $P(H \leq k)$ instead of $P(H = k)$. With this in mind, we reproduce in Table 3.6 the portion of

TABLE 3.5
Binomial Distribution with Parameters $n = 3$, $p = 1/2$
Number of Heads, H

k	$P(H = k)$
0	.125
1	.375
2	.375
3	.125
Sum $= 1.000$	

TABLE 3.6
Table A.2 for $n = 3, p = .50$

n	k	$p = .50$
3	0	.1250
	1	.5000
	2	.8750
	3	1.0000

Table A.2 dealing with $n = 3$ and $p = 1/2$ ($p = .50$ in the language of Table A.2). We then proceed to explain how to derive Table 3.5 from the information of Table 3.6. As has been just mentioned, the numbers in the right-hand column of Table 3.6 are probabilities of the type $P(H \leq k)$. This means that

$$.1250 = P(H \leq 0) = P(H = 0)$$
$$.5000 = P(H \leq 1) = P(H = 0) + P(H = 1)$$
$$.8750 = P(H \leq 2) = P(H = 0) + P(H = 1) + P(H = 2)$$
$$1.0000 = P(H \leq 3) = P(H = 0) + P(H = 1) + P(H = 2) + P(H = 3)$$

From the above information, we would like to compute the individual probabilities $P(H = 0)$, $P(H = 1)$, $P(H = 2)$, and $P(H = 3)$. First of all, since H can never be negative, $P(H \leq 0)$ is the same as $P(H = 0)$, so that

$$P(H = 0) = .1250 = .125$$

Then, because $P(H = 0) + P(H = 1) = .5000$, it follows that

$$P(H = 1) = .5000 - P(H = 0) = P(H \leq 1) - P(H \leq 0)$$
$$= .5000 - .1250 = .3750 = .375$$

Furthermore, because $P(H = 0) + P(H = 1) + P(H = 2) = .8750$, we have that

$$P(H = 2) = .8750 - [P(H = 0) + P(H = 1)]$$
$$= P(H \leq 2) - P(H \leq 1)$$
$$= .8750 - .5000 = .3750 = .375$$

And, as $P(H = 0) + P(H = 1) + P(H = 2) + P(H = 3) = 1.0000$, we get

$$P(H = 3) = 1.0000 - [P(H = 0) + P(H = 1) + P(H = 2)]$$
$$= P(H \leq 3) - P(H \leq 2)$$
$$= 1.0000 - .8750 = .1250 = .125$$

We can summarize the above computations as follows:

$$P(H = 0) = P(H \leq 0) = .125$$
$$P(H = 1) = P(H \leq 1) - P(H \leq 0) = .375$$
$$P(H = 2) = P(H \leq 2) - P(H \leq 1) = .375$$
$$P(H = 3) = P(H \leq 3) - P(H \leq 2) = .125$$

It should now be observed that these are exactly the numbers appearing in the probability distribution of Table 3.5.

As we continue to work with binomially distributed data in future portions of this book, we will refer to Table A.2 whenever numerical calculations are required. As another example of the use of Table A.2, consider the following.

Example 3.2   A Political Poll   Suppose it is really true that 60% of the voters are for repeal of a certain local tax law. If we were to select a random sample of seven voters, what would be the probability that a majority of the sample would be against repeal, thus making it appear that the antirepeal forces are in the lead?

SOLUTION   Here we are dealing with a binomial experiment with $n = 7$. This can be viewed as analogous to tossing seven coins (representing the seven voters), each marked "for repeal" on one side and "against repeal" on the other. The fact that 60% of the voters are for repeal can be translated into coin tossing language as the statement that $p = P(\text{for repeal}) = .60$. Therefore, if we abbreviate by F the number of voters that are *for* repeal, we see that F has the binomial distribution with parameters $n = 7$ and $p = .60$. Equivalently, we say that F is a binomially distributed random variable with parameters $n = 7$ and $p = .60$. In Table 3.7, we reproduce the relevant portion of the table of binomial distribution, Table A.2 of the Appendix. We have therefore completed the first part of the solution, which is to decide which portion of Table A.2 applies to the problem at hand.

For the second part of the solution, we must decide how to make use of the information in Table A.2 in order to answer the question. The question here asks what the probability is that a majority of voters in the sample is against repeal. To say that a majority of the seven voters is against repeal means that four or more are against repeal. This is equivalent to saying that three or fewer are for repeal, or that $F \leq 3$. It follows that

$$P \text{ (a majority of the sample is against repeal)} = P(3 \text{ or fewer are for repeal})$$
$$= P(F \leq 3)$$

in view of the fact that F represents the number of voters that are for repeal. Now, as we have pointed out earlier, Table A.2 (and so Table 3.7) contain exactly probabilities of

TABLE 3.7
Table A.2 for $n = 7, p = .60$

n	k	$p = .60$
7	0	.0016
	1	.0188
	2	.0963
	3	.2898
	4	.5801
	5	.8414
	6	.9720
	7	1.0000

the form $P(F \leq k)$. Therefore, in Table 3.7, to the right of 3 (of the k column), we find $P(F \leq 3) = .2898$. Therefore, the probability is 28.98%, or almost 29%, that the majority of our sample will be against repeal, even though 60% of the entire voting population are for repeal.

In Table 3.8 and Fig. 3.2, respectively, we record for illustrative purposes the probability distribution and the histogram of F. The calculations used to derive the information from Table 3.7 are also shown.

As the above example shows, there is more than one chance in four that a poll of seven voters would mislead the pollster into believing that one side was in the lead, even when the other side was really ahead 60% to 40%. The problem here is that seven voters do not constitute a sample large enough to inspire real confidence in the results. Suppose, for purposes of comparison, we had taken instead a sample of 19 voters instead of 7. Then, to assert that a majority is against repeal means that 9 or fewer are for repeal, namely $F \leq 9$. In Table A.2, for $n = 19$ and $p = .60$, we read that $P(F \leq 9) = .1861 = 18.61\%$. Therefore, working with 19 voters instead of 7, we would be able to reduce our chances of being seriously misled from 28.98% down to 18.61%. Our chances of error would naturally be even smaller were we to take a sample of 50 or 100 or more voters. Although Table A.2 contains the binomial probabilities only for values of n up to $n = 20$, we will learn in the next two sections how to use Table A.3 for calculating binomial probabilities having the parameter n larger than 20.

The next example illustrates another way to use the material of Table A.2.

Example 3.3   Effectiveness of a New Drug   Suppose a new drug aimed at alleviating mental depression appears on the market. Prior to its licensing for public use, an intensive testing program has established that the drug is 80% effective; this means that it can be expected to alleviate mental depression in about 80% of the patients for whom it is prescribed. If we put 15 patients on the new drug, what are the chances that 12 or more of them will benefit from it?

TABLE 3.8
Binomial Distribution with Parameters $n = 7, p = .60$
Number of Voters for Repeal F

k	$P(F \leq k) - P(F \leq k - 1) =$		$P(F = k)$
0	.0016 − .0000	=	.0016
1	.0188 − .0016	=	.0172
2	.0963 − .0188	=	.0775
3	.2898 − .0963	=	.1935
4	.5801 − .2898	=	.2903
5	.8414 − .5801	=	.2613
6	.9720 − .8414	=	.1306
7	1.0000 − .9720	—	.0280
		Sum =	1.0000

FIGURE 3.2
Histogram of the binomial distribution with parameters $n = 7$, $p = .60$.

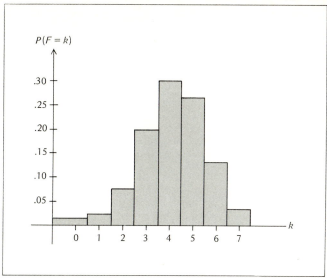

SOLUTION   What we have here is a set of binomial data with parameters $n = 15$ and $p = .80$. If we denote by B the number of patients who benefit from the drug, we are interested in $P(B \geq 12)$, which is the same as $P(B > 11)$. Table A.2, as it is constructed, works better with $P(B > 11)$ than it does with $P(B \geq 12)$. In fact, the table contains neither of these items, but it does show that $P(B \leq 11) = .3518$ for $n = 15$, $p = .80$, and $k = 11$. This means that the chances are 35.18% that 11 or fewer patients benefit. It follows automatically that the chances are $100\% - 35.18\% = 64.82\%$ that more than 11 patients benefit. We can compactify this calculation by writing

$$P(B > 11) = 1 - P(B \geq 11) = 1 - .3518$$
$$= .6482$$
$$= 64.82\%$$

Therefore, if the drug is 80% effective, the probability is 64.82% that 12 or more of our 15 patients will benefit from it.

In Table 3.9 and Fig. 3.3, respectively, we derive the probability distribution and graph the histogram of a binomial random variable having parameters $n = 15$ and $p = .80$.

Mean and Standard Deviation of the Binomial Distribution

Before going on to the next section, we should call attention to a few more facts about the binomial distribution that we will find useful in our future work with statistics. These

TABLE 3.9
Binomial Distribution with Parameters $n = 15, p = .80$
Number of Patients Who Benefit B

n	k	$P(B \leq k)$	$P(B \leq k) - P(B \leq k - 1) =$		$P(B = k)$
15	0	.0000*	.0000 − .0000	=	.0000*
	1	.0000*	.0000 − .0000	=	.0000*
	2	.0000*	.0000 − .0000	=	.0000*
	3	.0000*	.0000 − .0000	=	.0000*
	4	.0000*	.0000 − .0000	=	.0000*
	5	.0001	.0001 − .0000	=	.0001
	6	.0008	.0008 − .0001	=	.0007
	7	.0042	.0042 − .0008	=	.0034
	8	.0181	.0181 − .0042	=	.0139
	9	.0611	.0611 − .0181	=	.0430
	10	.1642	.1642 − .0611	=	.1031
	11	.3518	.3518 − .1642	=	.1876
	12	.6020	.6020 − .3518	=	.2502
	13	.8329	.8329 − .6020	=	.2309
	14	.9648	.9648 − .8329	=	.1319
	15	1.0000	1.0000 − .9648	=	.0352

Sum = 1.0000

*A probability of .0000 signifies less than 1 chance in 20,000.

facts involve the mean and standard deviation of a binomial distribution which, as the material of Chap. 2 has shown, are important in understanding the behavior of random variables having that distribution.

To introduce the concept of the mean of a probability distribution, let's take a look

FIGURE 3.3
Histogram of the binomial distribution with parameters $n = 15, p = .80$.

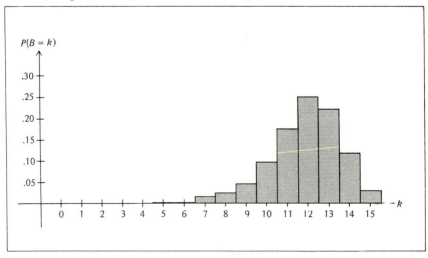

at some coin tossing examples. Suppose we take a fair coin, having $p = P(\text{heads}) = .50$, and toss it 10 times. We would expect to get 5 heads in the 10 tosses. In fact, if several people got together, and each tossed a fair coin 10 times, the number of heads tossed by each would average out to something near 5. It would therefore be reasonable to agree that the mean number of heads obtained in 10 tosses of a fair coin is 5.

Now, what would the situation be if we were to toss another coin which was not fair but instead had probability of heads $p = .80$? Then we would expect to obtain 8 heads in our 10 tosses, and so it would be reasonable to say that the mean number of heads in 10 tosses of the other coin is 8. For larger values of p, therefore, the mean number of heads would also be larger.

Suppose, on the other hand, we were to toss each coin 50 times instead of 10? Well, with a fair coin we would expect to get 25 heads, but with the coin having $p = .80$ we would expect 80% heads, or 40 heads out of the 50 tosses.

In general, suppose we take a coin having probability p of heads, and suppose we toss it n times. How many heads can we expect to get? We would expect that a proportion p of the n tosses would turn up heads. Therefore the number of heads expected would be $p \times n = np$ heads. Notice that this is exactly the procedure we followed in the specific cases above. In 10 tosses ($n = 10$) of a fair ($p = .50$) coin, we would expect $np = 10 \times .50 = 5$ heads, but in 50 tosses ($n = 50$) of a coin having $p = .80$, we would expect $np = 50 \times .80 = 40$ heads. We summarize these calculations and some more like it in Table 3.10.

As Examples 3.1, 3.2, and 3.3 made clear, coin tossing is merely one special case of a large class of situations in which the data follow binomial distributions. Furthermore, as we have seen, whenever a random variable has a binomial distribution, all relevant probabilities can be determined from Table A.2 of the Appendix, which is classified according to the parameters n and p of the random variable under study. Therefore, mathematically speaking, all binomially distributed random variables having the same numerical values of n and p as parameters are, in some sense, the same because they take the same numerical values with the same probabilities. By similar reasoning, we can see that all these random variables having the same n and p will also have the same

TABLE 3.10
The Mean Number of Heads in Coin Tossing

Number of Tosses n	Probability of Heads p	Expected Number of Heads np
10	.30	3
10	.50	5
10	.80	8
50	.30	15
50	.50	25
50	.80	40
200	.30	60
200	.50	100
200	.80	160

TABLE 3.11
Calculation of μ and σ for Some Binomial Distributions

n	p	$\mu = np$	$1 - p$	$np(1 - p)$	$\sigma = \sqrt{np(1 - p)}$
10	.30	3	.70	2.1	1.45
10	.50	5	.50	2.5	1.58
10	.80	8	.20	1.6	1.26
50	.30	15	.70	10.5	3.24
50	.50	25	.50	12.5	3.54
50	.80	40	.20	8.0	2.83

mean, namely np, regardless of whether they are the mean number of heads tossed, the mean number of voters for repeal, the mean number of patients benefiting, or the mean of any other binomially distributed random variable having parameters n and p.

We are therefore justified in using the formula

● $$\mu = np \qquad (3.1)$$

to calculate the mean of any set of binomially distributed data having parameters n and p. To work out the standard deviation of a set of binomial data is somewhat more complicated. An algebraic derivation using formula (8.13) of Sec. 8.E for the binomial distribution (from which the probabilities appearing in Table A.2 were computed) would show that we should use the formula

● $$\sigma = \sqrt{np(1 - p)} \qquad (3.2)$$

to calculate the standard deviation of a set of binomial data having parameters n and p. We will have occasion to use these formulas for μ and σ in applied contexts from time to time as we proceed through this course. In Table 3.11, we present some sample calculations of μ and σ.

EXERCISES 3.B

1 Suppose we take four fair coins, toss them, and count the number of heads obtained, denoting the resulting number of heads by the random variable H.
   a What are the parameters of the binomial random variable H?
   b What are the possible values of H?
   c Construct the probability distribution of H.

2 An executive in charge of new product development at a local dairy feels that about 10% of the population would be willing to try a carton of chocolate-flavored buttermilk, while the other 90% would not be at all interested. To support her feeling with some observable facts, she conducts a survey at a local supermarket, asking each of 15 persons whether or not they would be willing to try a carton. She denotes by Y the random variable indicating the number of persons responding "yes" to the question.
   a What are the parameters of the random variable Y?
   b What are the possible values of Y?

c Calculate the mean of Y.
d Calculate the standard deviation of Y.
e Construct the probability distribution of Y.

3 Blood tests for anemia are given to 12 patients manifesting certain special symptoms. Usually, 20% of patients having those symptoms are judged by the test to have some form of anemia. We denote by A the random variable indicating the number of persons of the 12 tested who are judged to have some form of anemia.
  a State the parameters of the random variable A.
  b List the possible values of A.
  c Calculate the mean of the random variable A.
  d Calculate the standard deviation of the random variable A.
  e Set up the probability distribution of A.

4 The manufacturer of a new chemical fertilizer guarantees that with the aid of his fertilizer 80% of the seeds planted will germinate. To check on his assertion, we plant 17 seeds as directed and denote by G the number that germinate.
  a What should the parameters of G be?
  b What are the possible values of G?
  c What is the probability that fewer than 10 of our 17 seeds will germinate?
  d Construct the probability distribution of G.

5 Twenty (20) men aged 45 apply for a 10-year term life insurance policy. In order to know how much to charge for the policy, the issuing company needs to be able to estimate how many of the applicants will die before age 55, in which case the company will have to pay off on the policy. Actuarial studies indicate that of all 45-year old men in the applicants' occupational grouping 5% will die before they reach age 55.
  a Find the parameters of the random variable D which denotes the number of applicants who die before they reach age 55.
  b What is the probability that none of the 20 applicants will die before age 55, making it a very profitable deal for the insurance company?
  c Find the probability that exactly 1 of the 20 will die during the 10-year term.
  d What are the chances that more than 5 of the 20 applicants will die before age 55?
  e What is the probability that fewer than 10 of the applicants will die during the specified period?
  f Calculate the probability that more than 5 but fewer than 10 applicants will not make it to 55.

6 After reading a news reporter's allegation that 65% of the membership of the U.S. House of Representatives favors a certain controversial change in the banking laws, a staff assistant to the major lobbying group opposing the change conducts a quick "straw poll" of 14 randomly selected House members.
  a If the reporter's story is correct, what are the chances that fewer than half of the 14 members surveyed would favor the change?
  b What is the probability that exactly half of those surveyed favor the change?
  c What would be the probability that more than 12 of the 14 members polled would favor the change?
  d What are the chances that more than half but fewer than three-fourths of those questioned would favor the change?

SECTION 3.C THE NORMAL DISTRIBUTION

There are two fundamental sources from which normally distributed data arise: repetition of the same physical process over and over again (such as the machine filling of 8-ounce cans of tomato sauce) and artificial standardization of test scores (such as scores on the Graduate Record Examinations, the Admission Test for Graduate Study in Business, etc.). A set of data can often be recognized as having an approximately normal distribution if its frequency polygon is a "bell-shaped" curve, as illustrated in Fig. 3.4. The word "normal" comes from historical usage, and is not meant to imply that other distributions are "abnormal" in any sense. As examples of the occurrence of normally distributed sets of data in situations of applied interest, consider the following:

1   **Contents of cereal boxes:**   It is a well-known fact that Corn Puffies are packed by machine into 9-ounce boxes. The great majority of boxes contain amounts very close to 9 ounces, a little more or a little less, but every once in a while a box shows up having contents of quite a bit more or quite a bit less. Contents in ounces of boxes of Corn Puffies are normally distributed with mean $\mu = 9$ and standard deviation $\sigma = .01$. The fact that the standard deviation is so small, relatively speaking, means that virtually all the boxes have almost exactly 9 ounces of cereal, while very, very, very few differ appreciably from it. The intense concentration of data points near 9 results in the tall and thin bell-shaped frequency polygon of Fig. 3.5.

2   **Heights and weights of persons:**   People living in a certain area, whether the specified area is a political subdivision, such as a city, state, or nation, or a geographical subdivision, such as a valley, plain, delta region, have normally distributed heights and weights. Other descriptive measures such as shirt or shoe sizes are also normally distributed. A particular group of people may have mean height, $\mu = 5.5$ feet, with a standard deviation, $\sigma = .4$ feet. The degree of concentration near the mean would therefore be not as extreme as in the case of the Corn Puffies data of the previous example which had a smaller σ. It follows logically that the lessened degree of concentration near the mean would result in a flatter bell-shaped curve. The normal curve of heights appears in Fig. 3.6.

3   **Scores on standardized tests:**   Scores of all students taking the verbal portion of the Graduate Record Examinations (GRE) are artificially standardized so as to have a

FIGURE 3.4
A "bell-shaped" curve.

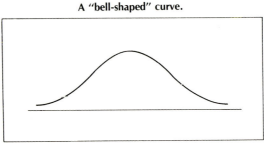

FIGURE 3.5
Frequency polygon of normally distributed contents of Corn Puffies boxes.

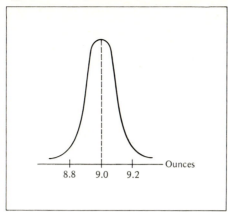

normal distribution with mean $\mu = 500$ and standard deviation $\sigma = 100$. The purpose of such standardization is to allow persons who use the results of such tests to be better able to interpret the results in terms of the relative performances of several candidates. The frequency polygon of the GRE scores is graphed in Fig. 3.7.

Each set of normally distributed data can be completely characterized by two numbers, its mean μ and its standard deviation σ. As explained in Chap. 2, μ specifies the location of the mean while σ measures the extent to which the data points are concentrated near the mean. The numbers μ and σ, since they completely describe the data, are the parameters of the normal distribution.

In order to study the details of the normal distribution, let's analyze the six examples that follow, each of which looks at the data from a different point of view.

Example 3.4   The Graduate Record Examination   As we know, verbal scores on the GRE are normally distributed with mean $\mu = 500$ and standard deviation $\sigma = 100$. If a student scored 643, what percentage of those who took the test ranked higher than she did?

FIGURE 3.6
Normal curve of people's heights.

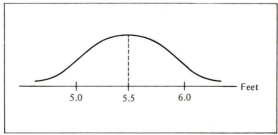

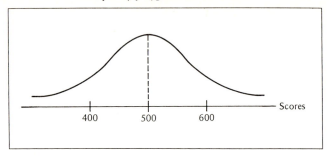

SOLUTION   The graph of the normal distribution involved in this example is shown in Fig. 3.8. The problem is to calculate the proportion of data falling in the shaded region to the right of the number 643. There is a formula for calculating the proportions of a set of normally distributed data that fall in various intervals. Unfortunately, however, to be able to read this formula out loud, let alone to know how to use it, requires a considerable knowledge of calculus. Fortunately, we have in Table A.3 of the Appendix a table of the various probabilities connected with a set of normally distributed data. Before we go any further, therefore, it is necessary to discuss how to use Table A.3. It should also be pointed out here that there are some other types of probability distributions whose frequency polygons appear bell-shaped, but these cannot be the normal distribution unless their probabilities are consistent with those of Table A.3.

How to Use the Table of the Normal Distribution

First of all, as the commentary preceding Table A.3 indicates, it is a table of the "standard" normal distribution. What does this mean? If you recall the discussion of standard scores, or z scores, in Sec. 2.E, you should note that we can calculate the z scores of any set of data and that these z scores have a probability distribution with mean $\mu = 0$ and standard deviation $\sigma = 1$. The operation of calculating the z scores of a set of data is called "standardizing" the data. If we calculate the z scores of a normally distributed set of data (namely, if we standardize a set of normally distributed data), we will obtain a set of normally distributed data having mean $\mu = 0$ and standard deviation $\sigma = 1$. The resulting normal distribution having the parameters $\mu = 0$ and $\sigma = 1$ is called the standard normal distribution, so that the z scores of any normal distribution have the standard normal distribution.

It would be grossly impractical to print tables of normal distributions for *every possible combination* of the numbers μ and σ, as this would require an extremely large number of pages, infinitely many, to be sure. Fortunately, however, in problems dealing with *any* normal distribution, it is sufficient to use the table of the standard normal distribution, merely by working the z scores instead of the original data points (raw scores). To see how this is done, let's return to the GRE example.

What we want to know is the proportion of scores falling above 643. This propor-

FIGURE 3.8
Proportion of GRE scores above 643.

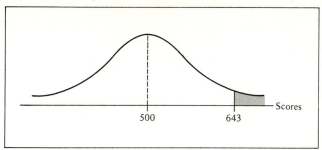

tion will be exactly the same as the proportion of z scores falling above the z score corresponding to 643. Now, as you recall from Sec. 2.E, the z score corresponding to the number x is given by the formula

$$z = \frac{x - \mu}{\sigma}$$

Therefore the z score corresponding to 643 is

$$z = \frac{643 - 500}{100} = \frac{143}{100} = 1.43$$

because $\mu = 500$ and $\sigma = 100$ in this situation.

The original and standardized normal distributions involved are illustrated in Fig. 3.9. As the pictures indicate, we can determine the proportion of candidates scoring above 643 on the verbal portion of the GRE merely by calculating the proportion of a normal distribution which falls *above* the z score 1.43. Table A.3 is basically a list of z scores, together with the proportion of a set of standard normal data falling *below* the z score. Looking for the z score $z = 1.43$ in Table A.3, we find next to it the proportion

FIGURE 3.9
Original and standardized GRE scores.

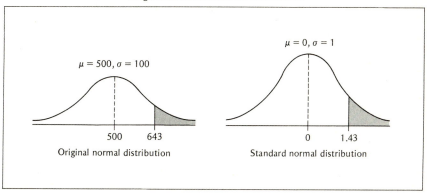

TABLE 3.12
A Small Portion of
Table A.3

z	Proportion
1.43	.9236

.9236, which indicates that 92.36% of a standard normal distribution lies *below* the z score z = 1.43. We reproduce the relevant portion of Table A.3 in Table 3.12. It follows, then, that 100% − 92.36% = 7.64% of the distribution must lie *above* z = 1.43. Returning to the GRE data, we can conclude that 7.64% of all candidates taking the GRE score above 643. A student scoring 643 can therefore be sure of easily qualifying for the top 10% of all students taking the GRE, since only 7.64% could have scored higher.

Example 3.5   The Graduate Record Examination (Another View)   Now that we know that only 7.64% of students score above 643, it would be useful and interesting to know what score will qualify a student for the top 5% of all candidates.

SOLUTION   As it turns out, we also can answer this question by using Table A.3. A graph illustrating the situation appears in Fig. 3.10, with the letter x denoting the (unknown) test score which separates the lower 95% of candidates from the upper 5%. Because $\mu = 500$ and $\sigma = 100$, the z score of the unknown x can be written as

$$z = \frac{x - 500}{100}$$

Because of this algebraic relationship between x and z, we will be able to compute x if we know the numerical value of z. As the picture of the standard normal distribution of Fig. 3.10 shows, 5%, or .0500, of the z scores will be above z, since 5% of the

FIGURE 3.10
Original and standardized GRE scores.

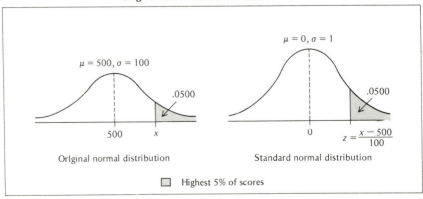

TABLE 3.13
A Small Portion of
Table A.3

z	Proportion
1.64	.9495
1.65	.9505

original scores are above x. This implies that 95%, or .9500, of the z scores will be below z. From Table A.3, we can find the particular z which has 95% of the z scores below it. All we have to do is to find the *proportion* in Table A.3 that is closest to .9500, and then we read off the corresponding value of z.

The relevant portion of Table A.3 is reproduced in Table 3.13. Unfortunately, our table is not sufficiently detailed as to include the number .9500 exactly.[1] As Table 3.13 shows, there are two proportions, .9495 and .9505, which are equally close to .9500. In a case like this, it is customary to choose the z score that is closer to 0; so let us say that the z score in question is z = 1.64. For purposes of comparison, the graph of the standard normal distribution illustrating the position of z = 1.64 appears in Fig. 3.11, together with the standard normal distribution of Fig. 3.10. From the pictures in Fig. 3.11, we note the following two facts:

1 5% of a standard normal distribution lies to the right of

$$z = \frac{x - 500}{100}$$

2 5% of a standard normal distribution lies to the right of

$$z = 1.64$$

A comparison of the two pictures in Fig. 3.11 then shows that

$$\frac{x - 500}{100} = 1.64$$

Successive algebraic operations then yield

$$x - 500 = (1.64)(100) = 164$$

and so $$x = 164 + 500 = 664$$

We have therefore found the numerical value (664) of

$$x = \text{the test score which separates the lower}$$
$$95\% \text{ of candidates from the upper } 5\%$$

[1]Those students who go on to use statistics in their professional work will have available more detailed tables. The tables in this book are merely textbook examples to illustrate how to use such tables—in some cases, they will not be precise enough for professional use.

FIGURE 3.11
A comparison of standard normal distributions.

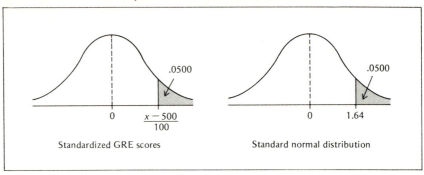

It follows that a score above 664 on the verbal portion of the GRE will put a student in the top 5% of all persons taking the test.

Example 3.6   Suit Sizes   Men living in a certain urban area have mean suit size 38 with a standard deviation of 2.5. Furthermore, their suit sizes are normally distributed. If a small ready-to-wear clothing store plans to begin business with an initial stock of 500 suits, how many of these should be of sizes between 35 and 40?

SOLUTION   First of all, we should remark that men's *actual* suit sizes are *continuous*, while ready-to-wear suits are manufactured only in *discrete* sizes. For example, an individual man may have size 35.8430067, but he will have to buy a size 36 suit, even though it will not fit him exactly. The normal distribution is a continuous distribution because Table A.3 admits all numbers as possible values of z. For example, to compute the proportion of men who are fitted with size 36 suits, we would find the proportion of men having actual physical sizes between 35.5 and 36.5. We are, in fact, computing the probability that a normal random variable with mean 38 and standard deviation 2.5 will have a value between 35.5 and 36.5. The binomial distribution, on the other hand, is a discrete[1] distribution, because only certain numbers are admitted as possible values. For example, if we toss a coin three times, we can get only 0, 1, 2, or 3 heads. There is no way of getting 2.5 or 2.6768 as a possible value of a binomial random variable.

With the above facts in mind, we see that we need to compute the proportion of men having actual suit sizes between 34.5 and 40.5. If we do this, we will know what proportion of men will need suits labeled between 35 and 40. The diagrams appear in Fig. 3.12, where the z score of 34.5 is

$$z(34.5) = \frac{34.5 - \mu}{\sigma} = \frac{34.5 - 38}{2.5} = \frac{-3.5}{2.5} = -1.40$$

---

[1]The spelling of the word "discrete," as opposed to "discreet," does not necessarily imply that the binomial distribution is indiscreet. It merely implies that it is not indiscrete, namely it is not continuous.

FIGURE 3.12
Original and standardized suit sizes.

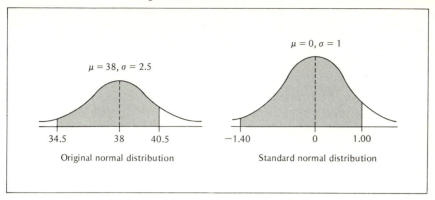

and the z score of 40.5 is

$$z(40.5) = \frac{40.5 - \mu}{\sigma} = \frac{40.5 - 38}{2.5} = \frac{2.5}{2.5} = 1.00$$

From Table A.3, or from the portion of it appearing in Table 3.14, we see that 8.08% of the z scores lie below −1.40 while 84.13% lie below 1.00. Of the 84.13% that lie below 1.00, then, 8.08% also lie below −1.40, and so the difference 84.13% − 8.08% = 76.05% falls between −1.40 and 1.00. Returning to the original normal distribution of suit sizes, we conclude that 76.05% of men have sizes between 34.5 and 40.5 and would therefore be fitted with suits labeled 35 to 40. Of the initial stock of 500 suits, then, the store should order 76.05% or 380 suits of sizes between 35 and 40.

Example 3.7   Child Development   Experimental results in child development show that ages at which children learn to walk are normally distributed with mean 11.2 months and standard deviation 0.4 month. One theory classifies as "early walkers" those children aged among the youngest 2% according to the time they learn to walk. If a child is to qualify as an early walker, by what age must he or she learn to walk?

SOLUTION   If we denote by x the (unknown) age which qualifies a child as an early walker, then the definition of early walker implies that 2% of the normally distributed

TABLE 3.14
A Small Portion of
Table A.3

z	Proportion
−1.40	.0808
1.00	.8413

TABLE 3.15
A Small Portion of
Table A.3

z	Proportion
-2.06	.0197
-2.05	.0202

data falls below (i.e., to the left of) x. This means that 2% of a standard normal distribution lies to the left of

$$z = \frac{x - \mu}{\sigma} = \frac{x - 11.2}{.4}$$

From Table A.3, or the remnant of it appearing in Table 3.15, we see that the z score of x must be -2.05 (because .0202 is closer to .0200 than is .0197). Therefore

$$\frac{x - 11.2}{.4} = -2.05$$

and some algebraic transformations yield that, successively,

$$x - 11.2 = (-2.05)(.4) = -.82$$

and $\qquad x = -.82 + 11.2 = 10.38$

It follows that children who learn to walk at age 10.38 months or earlier are classified as early walkers. The diagrams of the relevant normal distributions appear in Fig. 3.13.

Example 3.8 Penguin Behavior A biologist participating in a research project in Antarctica is studying the behavior of penguins during the long period of time that they

FIGURE 3.13
Original and standardized ages of learning to walk (months).

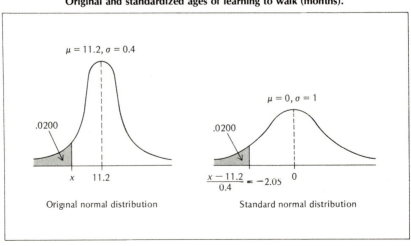

FIGURE 3.14
Original and standardized times penguins remain in water.

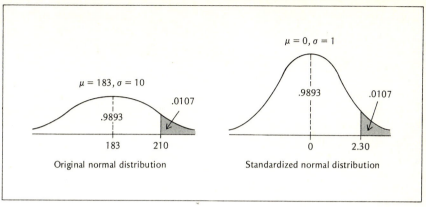

TABLE 3.16
A Small Portion of
Table A.3

z	Proportion
2.30	.9893

live entirely in the water. On the average, a certain breed of penguin remains in the water 183 days with a standard deviation of 10 days. Furthermore, the number of days such a penguin remains in the water is a normally distributed random variable. What percentage of the penguins remain in the water longer than 7 months (210 days)?

SOLUTION   The question asks for the probability that a normally distributed random variable with mean $\mu = 183$ and standard deviation $\sigma = 10$ exceeds $x = 210$. A diagram of the situation is presented in Fig. 3.14. A transformation to z scores gives the z score of $x = 210$ as

$$z = \frac{x - \mu}{\sigma} = \frac{210 - 183}{10} = \frac{23}{10} = 2.30$$

From the portion of Table A.3 that appears in Table 3.16, we see that a standard normal random variable is less than or equal to 2.30 about 98.93% of the time. It follows that such a variable exceeds 2.30 about 1.07% of the time, and we can therefore conclude that only 1.07% of the penguins remain in the water longer than 210 days.

EXERCISES 3.C

1   The length of time it takes grade-school pupils to complete a standard nationwide arithmetic progress test is a normally distributed random variable having mean 58 minutes and standard deviation 9.5 minutes. The educational psychologist who

designed the exam wants it to be completed by 90% of the pupils who take it within a specified time allowed for the test. How much time should be allowed?

2 Records of the Wetspot Washing Machine Company indicate that the length of time their washing machines operate without requiring repairs is normally distributed with mean 4.3 years and standard deviation 1.6 years. The company repairs free any machine which fails to work properly within one year after purchase. What percentage of Wetspot machines require these free repairs?

3 The mathematics portion of the nationally administered Graduate Record Examination is graded in such a way that the scores are normally distributed with mean 500 and standard deviation 100.
   a What proportion of applicants score 682 or higher?
   b What proportion score between 340 and 682?

4 Men's shoe sizes nationwide are normally distributed with mean 10.5 and standard deviation 1.2. What proportion of men have shoe sizes between 8.25 and 12.25?

5 Lamps used in residential area street lighting are constructed to have a mean lifetime of 400 days with a standard deviation of 30 days. Furthermore, their lifetimes are normally distributed.
   a What percentage of such lamps last longer than one year (365 days)?
   b What percentage last between 375 and 425 days?
   c What percentage last longer than 480 days?

6 Rainfall for the month of March in a geographic area in South America is normally distributed with mean 3.6 inches and standard deviation 1.0 inch. What proportion of years have total rainfall in March exceeding 2.0 inches but not exceeding 3.0 inches?

7 A cafeteria vending machine dispenses 6 ounces of coffee per cup, on the average, in such a way that the amount dispensed per cup is a normally distributed random variable. How fine should the machine be "tuned"; namely, to what level should the standard deviation be set so that 99% of the cups are filled with at least 5.9 ounces of liquid?

8 One psychology instructor makes sure that every set of examination scores in her course is normally distributed. Then she assigns A grades to the top 15% of the scores, while the bottom 15% are awarded F's. A particular set of exam scores turns out to have a mean of 60 and a standard deviation of 16. What score is the dividing line between A and B, and what score between D and F?

9 A well-known law asserts that no more than .5% of a company's cereal boxes labeled as containing 9 ounces of cereal may contain less than 8.8 ounces. A company planning to produce a 9-ounce box of Rice Puffies purchased a newly devised box-filling machine which fills cereal boxes in such a way that the contents of an individual box are a normally distributed random variable having *standard deviation* of 0.05 ounce. The machine has a dial with which the company can set the *mean* contents to any desired level. What is the lowest value to which the company can set the mean and still remain within the law?

10 The New York Stock Exchange is open 250 days each year. Closing prices of shares of Brownline Copper Co. (BCC) averaged $21\frac{1}{2}$ in the past year with a standard

deviation of 4.3. Furthermore, BCC stock closed above $32\frac{1}{4}$ on 28 different days during the year. Were closing prices of BCC shares normally distributed?

11 A small auto parts company markets a 12-volt battery which, according to statistical studies, lasts an average of 1200 days with a standard deviation of 50 days. The lifetimes of the batteries were also shown to be approximately normally distributed. The company would like to put a guarantee on its product so that no more than 10% of the batteries will fail before the guarantee runs out. For how many days should the battery be guaranteed?

12 The growing season, planting to harvesting, of a certain variety of tomato is approximately normally distributed with a mean of 100 days and a standard deviation of 10 days. What percentage of such tomatoes can be harvested within 75 days of planting?

SECTION 3.D THE NORMAL APPROXIMATION TO THE BINOMIAL DISTRIBUTION

In this section our goal is to clear up a question involving the binomial distribution that was brought up near the end of Sec. 3.B. Recall that Table A.2, of probabilities of the binomial distribution, contains information only for situations when the parameter n has a value of 20 or less. Because of this restriction we have been unable up until now to calculate, for example, the probability of tossing 40 or fewer heads in 100 tosses of a fair coin, because $n = 100$ for this problem. Fortunately, there is a result from advanced mathematics called the central limit theorem (to be discussed in greater detail in Sec. 4.B) which gives the surprising information that the normal distribution can be used to find that probability.

What the central limit theorem says in regard to this problem is that, when $np(1 - p)$ is greater than 5 in magnitude, a binomial random variable with parameters n and p can be considered normally distributed with parameters $\mu = np$ and $\sigma = \sqrt{np(1 - p)}$. In this form, the central limit theorem is referred to as "the normal approximation to the binomial distribution." In 100 tosses of a fair coin, the parameters of the binomial random variable H (the number of heads obtained in the 100 tosses) are $n = 100$ and $p = .50$. Therefore $np(1 - p) = 100(.50)(1 - .50) = 50(.50) = 25 > 5$, so that the "normal approximation to the binomial distribution" applies to this problem. Therefore H can be considered as a normal random variable with parameters $\mu = np = 100(.50) = 50$ and $\sigma = \sqrt{np(1 - p)} = \sqrt{100(.50)(1 - .50)} = \sqrt{25} = 5$. Furthermore, since H is a discrete random variable, as discussed in Example 3.6, we actually should calculate the probability that $H < 40.5$, namely, that we toss fewer than 40.5 heads, if we are looking for the probability of tossing 40 or fewer heads in 100 tosses of the coin. The bell-shaped curves for the coin-tossing situation are presented in Fig. 3.15. The z score of 40.5 is

$$z = \frac{40.5 - \mu}{\sigma} = \frac{40.5 - 50}{5} = \frac{-9.5}{5} = -1.90$$

and Table A.3 indicates that the proportion of a standard normal distribution falling

FIGURE 3.15
Original and standardized number of heads tossed.

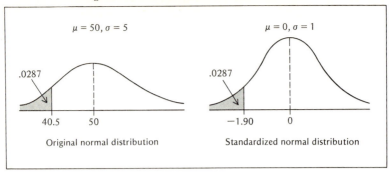

below −1.90 is .0287. Therefore, the probability of tossing 40 or fewer heads in 100 tosses of a fair coin is only 2.87%, or 287 chances out of 10,000.

We present the following additional example of the use of the normal approximation to the binomial distribution.

Example 3.9 A Political Poll   (This example is a modified version of Example 3.2.) Suppose it is really true that 60% of the voters are for repeal of a certain local tax law. If we were to select a random sample of 70 voters, what is the probability that a majority of the sample (36 or more voters) would be against repeal, thus making it appear that the antirepeal forces are in the lead?

SOLUTION   Here we are dealing with a binomial random variable F, the number of voters that are *for* repeal, having parameters $n = 70$ and $p = .60$. The normal approximation is valid because $np(1 - p) = 70(.60)(1 - .60) = 42(.40) = 16.8$ which is greater than 5. The probability that a majority of the sample of 70 is against repeal is the same as the probability that F is fewer than 34.5, and this latter probability is the same as the proportion of a standard normal distribution that falls below

$$z = \frac{34.5 - \mu}{\sigma}$$

where

$$\mu = np = 70(.60) = 42$$

and

$$\sigma = \sqrt{np(1 - p)} = \sqrt{16.8} = 4.1$$

We then have

$$z = \frac{34.5 - 42}{4.1} = \frac{-7.5}{4.1} = -1.83$$

From Table A.3, we find out that the probability that a standard normal random variable falls below −1.83 is .0336 or 3.36%. Therefore, if we take a sample of 70 voters, our chances of being misled into believing that the antirepeal forces are in the lead are only 3.36%.

Referring back to Example 3.2 of Sec. 3.B, we note that our chances of being so misled were 28.98% when we used a sample of 7 voters and 18.61% when we used a sample of 19 voters. As we have just shown, the probability of being misled drops considerably, to 3.36%, when a sample of 70 voters is used. This demonstrates convincingly that by increasing the sample size we can obtain a substantial decrease in our probability of error.

EXERCISES 3.D

1 As part of a coin-tossing experiment to check on the laws of probability (presumably to find out whether they should be repealed), a fair coin is tossed several times, and the number of heads obtained is carefully counted and denoted by H.
   a If the coin is tossed 4 times, find the probability that H, the number of heads, will be at least 2 and no larger than 3.
   b If the coin is tossed 40 times, find the probability that H will be at least 20 and no larger than 30.
   c If the coin is tossed 400 times, find the probability that H will be at least 200 and no larger than 300.
   d If the coin is tossed 4000 times, find the probability that H will be at least 2000 and no larger than 3000.

2 The manufacturer of a new chemical fertilizer guarantees that with the aid of her fertilizer, 80% of the seeds planted will germinate. If we believe the manufacturer's assertion:
   a With what probability can we expect more than 13 of 17 seeds planted to germinate?
   b With what probability can we expect more than 130 of 170 seeds planted to germinate?
   c With what probability can we expect more than 1300 of 1700 seeds planted to germinate?
   d With what probability can we expect more than 13,000 of 17,000 seeds planted to germinate?

3 Actuarial studies indicate that of all 45-year old men in a certain occupational grouping 5% will die before they reach age 55. As part of the process of deciding whether or not to offer a 10-year term group life insurance policy to all 45-year old men in that occupational grouping, an insurance company must know the answers to the following questions:
   a If 20 men apply for the policy, what are the chances that more than 2 of them will die before reaching age 55?
   b If 200 men apply, what are the chances that more than 20 of them will die before age 55?
   c If 2000 men apply, what are the chances that more than 200 of them will die before age 55?
   d If 20,000 men apply, what are the chances that more than 2000 of them will die before age 55?

4 A news reporter alleges that 65% of the membership of the 50 state legislatures

favors a certain controversial change in the banking laws. If the reporter's story is correct:

a Find the probability that no more than half of 10 randomly selected legislators support the change.

b Find the probability that no more than half of 100 randomly selected members actually support the change.

c Find the probability that no more than half of 400 randomly selected members actually support the change.

5 A new drug aimed at alleviating mental depression seems to be 80% effective, according to an intensive prelicensing testing program.

a If the new drug is prescribed for 15 patients, with what probability can we expect fewer than 11 to benefit from it?

b If the drug is prescribed for 150 patients, how likely is it that fewer than 110 will benefit?

c Of 1500 patients taking the drug, what are the chances that fewer than 1100 will benefit from it?

d With 15,000 patients being given the new drug, what is the probability that fewer than 11,000 will benefit from it?

SECTION 3.E PERCENTAGE POINTS OF THE NORMAL DISTRIBUTION

In many problems of applied interest that we will encounter in the rest of the book, we will be working with various quantities which have the standard normal distribution. One of the questions we will be interested in is where these quantities are located on the normal scale, relative to the bulk of the distribution. To be better able to specify the location on the normal distribution of such quantities, we will introduce here a system of labeling the distribution.

Using the Greek letter α (pronounced "alpha") to represent a proportion of data, we will use the symbol $z_{\alpha/2}$ to signify the z value of a standard normal distribution which has to its right a fraction $\alpha/2$ of the data. The meaning of $z_{\alpha/2}$ is illustrated in Fig. 3.16. Because $z_{\alpha/2}$ has a proportion $\alpha/2$ of the data to its right, it must have the remainder, a

FIGURE 3.16
The meaning of $z_{\alpha/2}$.

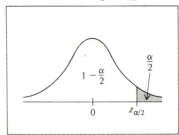

FIGURE 3.17
The meaning of z_α.

TABLE 3.17
A Small Portion of
Table A.3

z	Proportion
1.96	.9750

TABLE 3.18
A Small Portion of
Table A.3

z	Proportion
1.64	.9495
1.65	.9505

proportion $1 - \alpha/2$, to its left. For example, taking $\alpha = .05$, we have $\alpha/2 = .025$. Here $z_{\alpha/2}$ is $z_{.025}$, the z score which has .025 or 2.5% of the data to its right and therefore .975 or 97.5% to its left. We can find the numerical value of $z_{.025}$ by using Table A.3 of the Appendix because $z_{.025}$ is the z score corresponding to the proportion .9750. The portion of Table A.3, the part reproduced in Table 3.17, shows that $z_{.025} = 1.96$.

Suppose that for $\alpha = .05$ we needed to know z_α, the z score which has a proportion α of the data to its right. As illustrated in Fig. 3.17, this means that a fraction $1 - \alpha$ of a set of standard normal data lies to the left of z_α. When $\alpha = .05$, z_α is $z_{.05}$, which has 5% of the data to its right and $1 - \alpha = 1 - .05 = .95 = 95\%$ to its left. From the portion of Table A.3 which is reproduced in Table 3.18, we see that the exact value of $z_{.05}$ lies somewhere between 1.64 and 1.65. As we have agreed to choose the z score that is closer to 0 in cases like this, we shall set $z_{.05} = 1.64$.

In Table 3.19, we list z_α and $z_{\alpha/2}$ for several selected values of α. The number z_α corresponds to the numbers $1 - \alpha$ of Table A.3, while the $z_{\alpha/2}$'s correspond to the $1 - \alpha/2$'s. In Fig. 3.18, we illustrate the position of various percentage points of the normal distribution. We will make extensive use of percentage points throughout the rest of the book, beginning with the next chapter.

TABLE 3.19
Some Selected Percentage Points of
the Normal Distribution

α	$1 - \alpha$	z_α	$\alpha/2$	$1 - \alpha/2$	$z_{\alpha/2}$
.20	.8000	.84	.10	.9000	1.28
.10	.9000	1.28	.05	.9500	1.64
.05	.9500	1.64	.025	.9750	1.96
.025	.9750	1.96	.0125	.9875	2.24
.02	.9800	2.05	.01	.9900	2.33
.01	.9900	2.33	.005	.9950	2.57
.005	.9950	2.57	.0025	.9975	2.81

FIGURE 3.18
Some percentage points of the normal distribution.

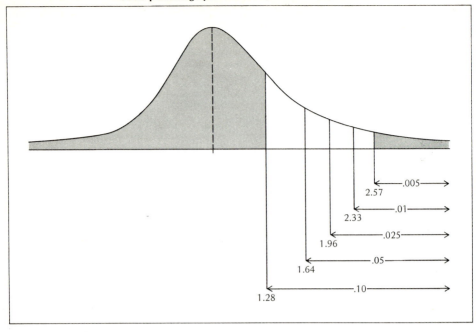

EXERCISES 3.E

1 For each of the following numerical values of α, find z_α:

a $\alpha = .30$. **e** $\alpha = .015$.
b $\alpha = .15$. **f** $\alpha = .008$.
c $\alpha = .075$. **g** $\alpha = .002$.
d $\alpha = .03$. **h** $\alpha = .001$.

2 For each of the following numerical values of α, find $z_{\alpha/2}$:

a $\alpha = .30$. **e** $\alpha = .015$.
b $\alpha = .15$. **f** $\alpha = .008$.
c $\alpha = .075$. **g** $\alpha = .002$.
d $\alpha = .03$. **h** $\alpha = .001$.

3 If $z_\alpha = 1.41$, find α.

4 If $z_\alpha = 2.93$, find α.

5 If $z_{\alpha/2} = 2.19$, find α.

6 If $z_{\alpha/2} = 1.20$, find α.

SUMMARY AND DISCUSSION

In Chap. 3, we have continued our development of "descriptive statistics" by studying general patterns of data called probability distributions. We have concentrated especially on the binomial and normal distributions, the two most common types of distributions which describe behavior of data arising out of applications in the social, natural, and managerial sciences. The techniques developed in this chapter will be utilized throughout the remainder of the book as we enter the field of "statistical inference," the drawing of conclusions (inferences) from a study of patterns inherent in a set of collected data. Here we have seen how to recognize some kinds of data which have the binomial distribution and other kinds which have the normal distribution. Future chapters will be devoted to special methods of analyzing binomial, normal, and other patterns of data. In Chap. 9, we will even introduce a formal procedure for determining whether or not a particular set of data is binomial, normal, or neither of these. The notion of a probability distribution is a fundamental idea in the organized study of statistics, for we learn from it that sets of data are not merely haphazard collections of numbers but are, in fact, based on general rules of behavior.

BIBLIOGRAPHY

Additional Types of Probability Distributions and Their Properties
Derman, C., L. Gleseer, and I. Olkin: "A Guide to Probability Theory and Its Application," pp. 247–518, Holt, Rinehart & Winston, New York, 1973.

Officer, R. R.: The Distribution of Stock Returns, *Journal of the American Statistical Association*, December 1972, pp. 807–812.

Tables of Probability Distributions
Fama, E., and R. Roll: Some Properties of Symmetric Stable Distributions, *Journal of the American Statistical Association*, September 1968, pp. 817–836.

Harter, H. L., and D. B. Owen: "Selected Tables in Mathematical Statistics," vol. I (1970), vol. II (1974), vol. III (1976), American Mathematical Society, Providence, R.I.

Random Variables as a Foundation for Statistics
Chapman, D. G., and R. A. Schaufele: "Elementary Probability Models and Statistical Inference," pp. 10–207, Xerox, Waltham, Mass., 1970.

SUPPLEMENTARY EXERCISES

1 A company psychologist estimates that 30% of the employees of a large corporation are regular smokers of cigarettes. If her estimate is correct:
   a What is the probability that of 10 randomly selected employees more than 4 are smokers?
   b What is the probability that of 100 randomly selected employees more than 40 are smokers?

2 One public health physician claims that 60% of those with lung disorders are heavy smokers of cigarettes. If his assertion is correct:
   a Find the probability that of 20 such patients recently admitted to a hospital fewer than half are heavy smokers.

b Find the probability that of 200 such patients recently admitted fewer than half are heavy smokers.

3 The latest figures indicate that 15% of all licensed drivers in one major metropolitan area have a drinking problem. If, on one particular day, 170 drivers apply for the normal three-year renewal of their licenses, what is the probability that more than 25% of them are problem drinkers?

4 One urban affairs sociologist feels that 55% of the adult residents of a particular major city have been victimized by a criminal at one time or another.
  a If his figures are correct, find the probability that of a random sample of 18 residents fewer than 5 have ever been victimized.
  b Under the sociologist's assumptions, what would be the probability that fewer than 500 of 1800 residents have ever been victimized?

5 An accountant reports to her superiors that 35% of all accounts which are currently overdue will eventually require legal action to force payment. If her estimate is correct:
  a What is the probability that of 12 currently overdue accounts fewer than 3 will actually require legal action?
  b Find the probability that of 120 overdue accounts fewer than 30 will require legal action.

6 A psychologist specializing in marriage counseling claims that his program of "conscientious communication" can prevent divorce in 80% of the cases he works on.
  a If his claims are true, what is the probability that of 20 couples currently involved in his program more than 18 will stay together?
  b What is the probability that of 200 cases in his files fewer than 150 were able to avoid divorce?

7 The time it takes a college student to complete a standard psychological test is a normally distributed random variable with mean 80 minutes and standard deviation 10 minutes.
  a How much time should be allowed for the test so that 90% of the students will be able to finish it?
  b How much time should be allowed for the test so that 95% of the students will be able to finish it?
  c How much time should be allowed for the test so that 99% of the students will be able to finish it?

8 A proposed test for the presence of diabetes assigns to each patient tested a number indicating that person's level of sugar in the urine. Patients with diabetes have levels that are normally distributed with a mean of 56 and a standard deviation of 3, while nondiabetic patients have levels that are normally distributed with a mean of 46 and a standard deviation of 10. The proposed test is intended to be primarily a preliminary indication of the presence of the disease, and so those patients having levels above 50.5 are scheduled for a more detailed battery of tests, while those with levels below 50.5 are considered to be nondiabetic.
  a What percentage of diabetic patients are incorrectly judged by the proposed test to be nondiabetic?

b What percentage of nondiabetic patients are unnecessarily scheduled for the more detailed battery of tests?

9 A particular job requires very relaxed individuals. There is available a psychological test of aptitude for this job on which more relaxed persons score higher than less relaxed ones. The scores on this test are standardized so as to be normally distributed with mean 70 and standard deviation 10. What score is required to rank among the top 5% of all those taking the test?

10 The main plant of General Fabricators is lighted by several thousand light bulbs whose length of life is normally distributed with a mean of 2000 hours and a standard deviation of 100 hours. Instead of replacing each bulb as it burns out (individual attention requires greater expense overall), the plant replaces all bulbs at the same time on a regular schedule. The management would like to set up the replacement schedule in such a way that no more than 1% of the bulbs will have burned out before being replaced. After how many hours of use should all the bulbs be replaced in order to achieve this goal?

11 A cafeteria vending machine dispenses coffee into 6-ounce cups in such a way that the actual amount dispensed into any particular cup is a normally distributed random variable with a standard deviation of .07 ounce. The vending machine has a dial in its interior with which the company can set the mean amount dispensed to any desired level. At what level should the mean be set so that a 6-ounce cup will overflow only 2% of the time?

12 A major nut company markets a bag of peanuts labeled as containing 10 ounces of nuts. The bags are packed by machine in such a way that they can have any desired mean weight, but their standard deviation will always be 0.1 ounce. Furthermore the weights will always be normally distributed.
a To what mean weight should the machine be set so that not more than .25% of the bags contain fewer than 9.8 ounces?
b If the mean is set at the level decided upon in part a, what percentage of the bags will contain more than 10.3 ounces?

13 An economist, studying the fluctuation of vegetable prices in small grocery stores in rural areas, concludes that the weekly average of prices per pound of ten basic vegetables has mean 30 cents and standard deviation 7.8 cents.
a If the economist does not know the probability distribution of the vegetable prices, what should his answer be to the question: At most what proportion of the time does the weekly average exceed 48 cents per pound?
b If the economist tests the data (using methods discussed in Sec. 8.C) and finds the data to be normally distributed, how should he answer the question?
c Explain why the two answers are different, and explain the relationship between them.

14 A psychologist working for the National Safety Council is assigned the task of measuring the reaction times of experienced interstate drivers when they are confronted with an animal crossing the highway in front of them at night. She discovers that the reaction times are approximately normally distributed with mean 2.1 seconds and standard deviation .7 second. Earlier studies of animals crossing highways have found that, if the driver does not react before 3.6 seconds have passed, the animal will be hit. What proportion of the time is the animal hit?

15 One mathematics instructor at a large university always gives A's to 10% of the students in his large introductory course, B's to 20%, C's to 40%, D's to 20%, and F's to the remaining 10%. On one final exam, the scores were normally distributed with mean 70 and standard deviation 10. What ranges of scores would qualify students for A's, B's, C's, D's, and F's?

16 The "Poisson distribution" is another probability distribution which is often used in applied work.[1] It arises in situations such as the following: If a certain species of insect populates a forest clearing in such a way that there are 3.6 specimens per square foot on the average, what are the chances of finding two or fewer specimens in a particular 1 square foot area under study? If we denote by X the random variable representing the actual number of specimens found in the 1 square foot area, then X has the Poisson distribution with parameter $\lambda = 3.6$. The parameter λ (pronounced "lambda") represented the *average* number of specimens per square foot, while the random variable X represents the *actual* number of specimens found in a particular 1 square foot area under study. Here we would like to find $P(X \leq 2)$, the probability that two or fewer specimens are found. Table A.10 of the Appendix contains the Poisson probabilities. Looking in the row corresponding to $\lambda = 3.6$ and the column corresponding to $N = 2$, we see that $P(X \leq 2) = .303$. Therefore the probability is $.303 = 30.3\%$ that a particular 1 square foot area will contain two or fewer specimens.

a Find the probability that a particular 1 square foot area will contain seven or fewer specimens.

b What are the chances that it contains more than nine specimens?

c Determine the probability that it contains fewer than five specimens.

17 A sociologist observes from criminology data that bank robberies in one major city occur at the rate of 2.4 per day on the average.

a Find the probability that more than six bank robberies will occur tomorrow.

b What are the chances that no bank robberies will occur next Tuesday?

18 Fire insurance claims come in to her office at the rate of four per hour on the average, according to an insurance adjuster in a heavily populated region of one southern state.

a Find the probability that in the next hour, fewer than two claims will come in.

b What is the probability that during the next hour, eight or more claims will come in?

19 A supermarket manager notes that 1.6 customers, on the average, arrive at the check-out stands each minute. In order to arrange the hiring and scheduling of checkers, he needs to know the answers to questions like the following:

a What is the probability that more than five customers will arrive at the stands during the next minute?

b What are the chances that the next minute will pass with no customers having arrived at the check-out stands?

c What is the probability that two or more customers will arrive during the next minute?

[1]There is nothing fishy about this distribution.

ESTIMATION BY
CONFIDENCE INTERVALS

What we have been discussing so far in this course is generally referred to as "descriptive statistics." Descriptive statistics is that aspect of statistics whose primary objective is to describe, graphically and numerically, the characteristic behavior of sets of data. However, in most cases of applied interest, the data we have available will be only a small fraction of the totality of existing information on the subject, and so we cannot be certain that the data we have will be a true representation of the existing situation. For example, suppose a distributor of citrus fruit wants to know the mean weight of California grapefruits for the purpose of estimating shipping costs. The way to calculate this would, of course, be to weigh each California grapefruit, add all the weights, and divide by the total number of grapefruits. To do this would be very difficult, if it were at all even possible. The best that can be done is to select a "random sample" of a relatively few grapefruits (a few compared to the total number of California grapefruits in existence), and calculate the mean weight of all the grapefruits in the sample. One of the most important questions in statistics then surfaces: What is the relationship between this "sample mean" and the item we are looking for, the "true mean" weight of all California grapefruits? To answer this question requires the development of a second aspect of statistics, the aspect called "statistical inference." In statistical inference, our goal is to determine what conclusions about a situation may be validly and logically drawn ("inferred") on the basis of a random sample of statistical data. Except for some additional material on probability calculations in Chap. 8, the remainder of this book is primarily concerned with statistical inference.

SECTION 4.A THE PROBLEM OF ESTIMATING THE TRUE MEAN

Let's denote the true mean weight of all California grapefruits by the Greek letter μ. How can we calculate μ? Well, we need to weigh all California grapefruits and then calculate the mean of all the weights so obtained. As a practical matter, however, it is virtually impossible to do this. Some of the reasons why the exact value of μ cannot be directly calculated are: (1) It would involve too much work and expense to remove every single mature California grapefruit from its tree and weigh it. (2) It would create an instantaneous glut on the grapefruit market exceeding the capacity of all cold storage facilities, thus wiping out a substantial portion of the citrus industry. (3) By the time the results were announced, they would be invalid because many of the grapefruits used in the study would have been eaten while new ones (of possible different weights) would have matured on the trees. (4) The results, even if obtainable, would be valid for only one season anyway, and it would be ridiculous to repeat this procedure every year. What we need therefore is an indirect method of getting a reasonably accurate measurement of μ, a method that is quite a bit simpler than the procedure just described. It is the objective of statistical inference to come up with such a method of determining the numerical value of μ.

Because it is not possible to weigh every possible grapefruit, how many of them should we weigh? Well, it seems reasonable that the more grapefruits we weigh, the closer our average of those weighed will be to the true mean weight (if such a thing indeed exists) of all California grapefruits. Suppose, for example, we take a sample of n randomly selected California grapefruits. By "randomly selected," we mean chosen in such a way that every grapefruit in the state has an equal chance of winding up in the sample. From our random sample of n grapefruits, we compute the following numbers:

x_1 = weight of the 1st grapefruit of the sample
x_2 = weight of the 2nd grapefruit of the sample
. . . .
x_k = weight of the kth grapefruit of the sample
. . . .
x_n = weight of the nth grapefruit of the sample

Using these n numbers, which are the weights of the n grapefruits in our random sample, we can compute the sample mean:

$$\bar{x} = \frac{\sum_{k=1}^{n} x_k}{n} = \frac{x_1 + x_2 + \cdots + x_n}{n} \tag{4.1}$$

Here the sample mean, the mean of the numbers in the sample, is denoted by the symbol $\bar{x}$ (pronounced "x bar"). By way of contrast, μ is often referred to as the "population mean" to signify that it is the mean of an entire population, not merely of a sample.

Now, in regard to the sample mean, the following two questions arise: (1) Does $\bar{x} = \mu$; that is, does the sample mean of the n randomly selected grapefruits equal the true mean of all California grapefruits? (2) If $\bar{x} \neq \mu$, how close does $\bar{x}$ come to μ; that is, how close is the sample mean to the true mean? Before we attempt to discuss in detail the answers to these questions, we should remind ourselves that we do not know what the numerical value of μ is; μ is an unknown quantity, and we are trying to find out what its numerical value is.

Let's describe, if we can, the relationship between the sample mean $\bar{x}$ and the true mean μ. Does $\bar{x} = \mu$? If we choose n grapefruits at random and weigh them, will their mean weight be the same as the mean weight of all California grapefruits? In answer to this question, we have to say "probably not," for it would be highly fortuitous if we were to pick n grapefruits (10, 500, 2000, or whatever number n may be) whose mean turned out to be exactly the same as the overall mean of all California grapefruits. In fact, suppose I pick n grapefruits, you pick n grapefruits, and six other people we know each pick n grapefruits. Then each of us will have his or her personal $\bar{x}$. More likely than not, these eight $\bar{x}$'s not only will differ from μ, but they will also differ from one another. All eight of us will each come up with a different value for $\bar{x}$. Of course, then, it is nonsensical to expect $\bar{x}$ to be the same as μ.

Now that we have agreed that $\bar{x}$ is probably not the same as μ, it would be useful to know how close it actually is to μ. The surprising answer to this question is that we really have no idea how close $\bar{x}$ is to μ. Our difficulty here is rooted in the fact that we do not really know the numerical value of μ; therefore, no matter what the value of $\bar{x}$ is, we cannot tell how close it is to μ.

To summarize our answers to questions 1 and 2 posed above, we can say: (1) $\bar{x}$ is, in all probability, not the same as μ, but we really can't be sure, because we don't know what the value of μ is. (2) We don't even know how close $\bar{x}$ is to μ, for we don't know what μ itself is, and so there is no way of telling how close any number is to it.

Well, the chances for an accurate estimate of μ seem sort of bad, don't they? It looks as though we have gone just about as far as we can go. At this point, fortunately, we have available some major theoretical results from advanced mathematics. It is the job of professional mathematicians to figure out ways of getting us out of situations like this. Unfortunately, though, these results provide only a partial solution to the problem. They allow us to use $\bar{x}$ to estimate μ only in the following two situations: (1) when our random sample consists of a large number of data points, say more than 30; or (2) when the original population under study is normally distributed. If neither of these requirements is satisfied, namely, if we are working with a nonnormal population and have fewer than 30 data points available, the methods to be introduced in this chapter don't work.[1]

In the next two sections, we will show how to estimate the true mean μ in the two situations where the above conditions are satisfied.

[1]Discussion of some substitute procedures can be found in the references listed in the bibliography.

EXERCISES 4.A

1 Give at least one reason why each of the following true means cannot be calculated exactly:
   a The true mean protein content of eggs produced in Denmark.
   b The true mean height of American adults.
   c The true mean cost of weekly groceries for a family of four.
   d The true mean number of persons per household in the United States.
   e The true mean number of pupils in second-grade classrooms in the Chicago public school system.
   f The true mean ozone level in a cubic meter of air over Los Angeles County today.
   g The true mean rainfall in the state of Louisiana last week.

2 Explain why each of the following true means can be calculated exactly:
   a The true mean paid attendance per game at the home games of the New York Mets last season.
   b The true mean grade-point average of all students attending the University of Arizona last term.
   c The true mean number of votes cast per congressional district during the 1976 general election.
   d The true mean number of empty seats on all nonstop commercial flights between Boston and Chicago last month.

SECTION 4.B CONFIDENCE INTERVALS BASED ON LARGE SAMPLES

Our objective in this section is to develop the procedures for estimating the true mean μ from a random sample consisting of a large number of data points. We refer to such random samples, which as a rule of thumb contain more than 30 data points, as "large samples." (By "small samples," we mean sets of data consisting of 30 or fewer data points.)

We noted earlier that, when we choose a sample of n data points and calculate their sample mean $\bar{x}$, we really don't know how close $\bar{x}$ is to the true mean μ. It seems reasonable, however, that the more data points we have in our sample, the closer $\bar{x}$ will be to μ. For example, if I tried to estimate the true mean weight of California grapefruits by randomly selecting 25 grapefruits and averaging their weights, while you tried to do the job with a sample of 200 grapefruits, we'd have to agree that your $\bar{x}$ would probably be closer to μ than would my $\bar{x}$. The formal statement of this principle is a mathematical theorem called the "law of large numbers," or, more popularly, the "law of averages." A fairly elementary wording of the law of large numbers reads as follows:

The law of large numbers: If a population has the true mean μ, and $\bar{x}$ is the sample mean of a random sample of n members of the population, then, as n grows larger and larger (as we take more and more data points for our sample), the sample mean $\bar{x}$ becomes closer and closer to the true mean μ.

Because of the law of large numbers, we know that we can improve our estimate of

μ by using random samples of larger and larger numbers of data points. Although we know our estimate $\bar{x}$ will be getting closer and closer to μ, we unfortunately still do not know exactly how far we can expect our $\bar{x}$ to be from μ. And this, of course, is the key to finding out something about μ.

A second mathematical theorem, called the "central limit theorem," goes a long way toward breaking this impasse. Roughly speaking, it asserts that for large values of n the set of all possible sample means of random samples consisting of n data points is a set of normally distributed data. More precisely, we can state the central limit theorem as follows:

The central limit theorem: If a population has the true mean μ and the true standard deviation σ, then the probability distribution of the set of all possible sample means of n members of the population becomes closer and closer to the normal distribution with mean μ and standard deviation $\sigma/\sqrt{n}$ as n grows larger and larger.

For all practical purposes, if n is larger than 30, namely, if we are working with a set of more than 30 data points, then the distribution of all possible sample means is close enough to the normal for us to consider it to *be* the normal distribution. For large samples, then, we consider the numbers $\bar{x}$ to be normally distributed with mean μ and standard deviation $\sigma/\sqrt{n}$. This situation is illustrated in Fig. 4.1.

In any particular applied problem under study, consequently, it is necessary to use Table A.3 of the Appendix, the table of the normal distribution. Table A.3, as you will recall, is actually a table of the *z scores* of the normal distribution, and it is therefore more useful to work with z scores of the possible sample means than with the original sample means themselves. In view of the fact that the $\bar{x}$'s are normally distributed with mean μ and standard deviation $\sigma/\sqrt{n}$, we know that their z scores, which are numbers of the form

$$z = \frac{\bar{x} - \mu}{\sigma/\sqrt{n}}$$

have the standard normal distribution. A graph of the distribution of the z scores of

FIGURE 4.1
Distribution of sample means for large samples.

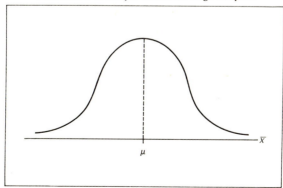

FIGURE 4.2
Distribution of z scores of sample means for large samples.

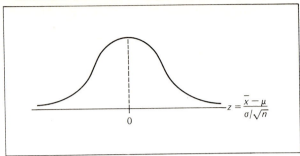

the possible sample means, a standard normal distribution, appears in Fig. 4.2. It should be recalled from Sec. 2.E that any set of z scores, in particular those of a standard normal distribution, has mean 0 and standard deviation 1.

To summarize what we have said so far, the set of z scores of all possible sample means of size n (where n is a number greater than 30) has the standard normal distribution. In terms of the percentage points of the normal distribution, this means that a fraction $1 - \alpha$ or a percentage $(1 - \alpha) \times 100\%$ of the z scores falls between the numbers $-z_{\alpha/2}$ and $z_{\alpha/2}$. For example, taking $\alpha = .05$ so that $1 - \alpha = 1 - .05 = .95$ and $(1 - \alpha) \times 100\% = (.95) \times 100\% = 95\%$, we see that 95% of the z scores fall between -1.96 and 1.96. This situation is illustrated in Fig. 4.3. In probability language, we can express this fact by saying that the probability is 95% that for any particular sample mean $\bar{x}$ its z score

$$z = \frac{\bar{x} - \mu}{\sigma/\sqrt{n}}$$

will fall between -1.96 and 1.96.

Figure 4.4 provides another view of how the central limit theorem works. That

FIGURE 4.3
The middle 95% of z scores of sample means (large samples).

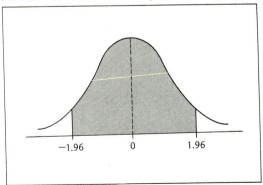

graph shows that 10% of the z scores fall above 1.28. We can therefore say that 10% of the sample means result in a z score above 1.28, so that there is a 10% chance that your z score (or mine or anybody else's, for that matter) will turn out to be larger than 1.28.

Unfortunately, although we can be fairly sure that 10% of the z scores will be larger than 1.28, we can never tell whether a particular z score is among that upper 10%. Why is this so? Well, the reason goes back to the fact that we do not know what μ is, and therefore we can never calculate the value of the z score

$$z = \frac{\bar{x} - \mu}{\sigma/\sqrt{n}}$$

In fact, not only do we not know what μ is, but we also do not generally know what σ is. Therefore, we cannot substitute enough numbers into the formula for z in order to calculate it. However, even though we cannot tell whether a particular z score is, for example, between -1.96 and 1.96, we do know that 95% of them are. And we can use this information to help us in estimating the numerical value of the true mean μ.

We first express in mathematical symbolism the statement that 95% of the z scores fall between -1.96 and 1.96. In algebraic language,[1] we can be 95% sure that

$$-1.96 < z < 1.96$$

Replacing z by its formula, we see that we can be 95% sure that

$$-1.96 < \frac{\bar{x} - \mu}{\sigma/\sqrt{n}} < 1.96$$

Multiplying across the above inequality (i.e., an equation with $<$ or $>$ instead of $=$) by the denominator $\sigma/\sqrt{n}$ gives us 95% certainty that

$$-1.96 \frac{\sigma}{\sqrt{n}} < \bar{x} - \mu < 1.96 \frac{\sigma}{\sqrt{n}}$$

[1]The notation $-1.96 < z < 1.96$ represents a combination of the two statements $-1.96 < z$ (z is greater than -1.96) and $z < 1.96$ (z is less than 1.96).

FIGURE 4.4
The upper 10% of z scores of sample means (large samples).

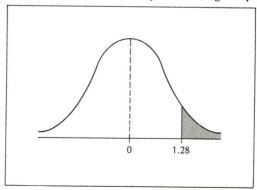

FIGURE 4.5
A fact about symmetric intervals.

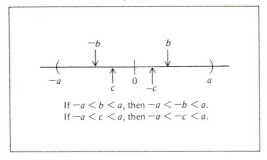

If $-a < b < a$, then $-a < -b < a$.
If $-a < c < a$, then $-a < -c < a$.

Now, referring to Fig. 4.5, which shows that whenever a number is between $-a$ and a, its negative is also between $-a$ and a, we can be 95% sure that

$$-1.96\frac{\sigma}{\sqrt{n}} < -(\bar{x} - \mu) < 1.96\frac{\sigma}{\sqrt{n}}$$

To simplify the term $-(\bar{x} - \mu)$ in the middle of the above inequality, we can remove the parentheses algebraically as follows:

$$-(\bar{x} - \mu) = -\bar{x} + \mu = \mu - \bar{x}$$

so that we can now be 95% sure that

$$-1.96\frac{\sigma}{\sqrt{n}} < \mu - \bar{x} < 1.96\frac{\sigma}{\sqrt{n}}$$

Finally, if we add $\bar{x}$ to all three sides of the inequality, we see that we can be 95% sure that

$$\bar{x} - 1.96\frac{\sigma}{\sqrt{n}} < \mu < \bar{x} + 1.96\frac{\sigma}{\sqrt{n}}$$

For the last page or so, we have been working only with the manipulation of the mathematical symbols. Perhaps now is a good time to go back to the English language and reflect upon what the resulting statement really means. A direct translation of the last mathematical statement just above would read as follows: We can be 95% sure that the true mean μ falls between $\bar{x} - 1.96(\sigma/\sqrt{n})$ and $\bar{x} + 1.96(\sigma/\sqrt{n})$, where

$\bar{x}$ = the sample mean based on our particular random
sample of n members of the population

σ = the true standard deviation of the population

n = the numbers of data points in our sample

Let's return to the problem introduced in Sec. 4.A of finding the true mean weight of California grapefruits. Suppose we try to estimate the true mean μ by selecting a sample of $n = 64$ grapefruits. We weigh each of these grapefruits, and they turn out to have a mean weight of $\bar{x} = 11.36$ ounces. What we now want to know is how close we can expect our sample of 11.36 to be to the true mean μ. Well, by the development above, we can be 95% sure that

$$11.36 - 1.96\frac{\sigma}{\sqrt{64}} < \mu < 11.36 + 1.96\frac{\sigma}{\sqrt{64}}$$

Taking note of the fact that $\sqrt{64} = 8$ and that $1.96/8 = 0.245$, it follows that we can be 95% sure that the true mean weight μ of California grapefruits falls between $11.36 - 0.245\sigma$ ounces and $11.36 + 0.245\sigma$ ounces, where σ is the true standard deviation of the weights of all California grapefruits. Until we find out the numerical value of σ, we therefore will not know how good our estimate of μ is. Unfortunately, the same difficulties that prevented us from directly measuring μ also prevent us from directly measuring σ.

It turns out that for large values of n (i.e., $n > 30$) there is a way of estimating σ to a degree of accuracy sufficient for completing the problem at hand. What we do is to replace σ in the formula we have been using by a number s, called the "sample standard deviation," which is an estimate of the true standard deviation σ calculated from the n data points of the random sample. The formula for the sample standard deviation s is a slight variation of the formula for σ introduced in Sec. 2.C. It is as follows:

$$s = \sqrt{\frac{\sum_{k=1}^{n}(x_k - \bar{x})^2}{n-1}}. \tag{4.2}$$

Comparing this latter formula with the one for σ, we note two changes: (1) μ is replaced by $\bar{x}$. (2) In the denominator, n is replaced by $n - 1$. The change from μ to $\bar{x}$ is necessary because we are dealing with a random sample of which we know (or can calculate) the sample mean, but we have no access to the true mean μ. The replacement of n by $n - 1$ is less easy to understand. Basically, the idea here is that if we calculate s according to the formula presented above, the mean of all possible values of s obtainable from samples of size n will be σ, in the same way that the mean of all values of $\bar{x}$ is μ. If we use n, however, instead of $n - 1$ in the denominator of s, the possible values of s will average out to something slightly different from σ. The justification of this fact requires the use of some advanced mathematical principles, the details of which can be found on pages 112 to 113 of the Chapman-Schaufele book listed in the bibliography of Chap. 3.

As an equivalent shortcut formula for the sample standard deviation, we have

$$s = \sqrt{\frac{n\sum_{k=1}^{n}x_k^2 - \left(\sum_{k=1}^{n}x_k\right)^2}{n(n-1)}} \tag{4.3}$$

As we mentioned earlier in connection with the shortcut formula for the population standard deviation, this formula is the one more adaptable for use with a pocket calculator.

We will see in the next section how to compute the numerical value of s (it will turn out to be virtually the same as the calculation of σ in Chap. 2), but let us now return to the grapefruit example. Suppose the 64 grapefruits in our random sample turned out to have sample standard deviation $s = 1.76$ ounces. Replacing σ in our most satisfactory result so far, the assertion of 95% certainty that

$$11.36 - .245\sigma < \mu < 11.36 + .245\sigma$$

by $s = 1.76$ implies that we can be 95% sure that

$$11.36 - (.245)(1.76) < \mu < 11.36 + (.245)(1.76)$$
$$11.36 - .43 < \mu < 11.36 + .43$$
$$10.93 < \mu < 11.79$$

This, now, is the type of statement we have been seeking.

Our conclusion about the mean weight of California grapefruits can be expressed as follows: On the basis of a random sample of 64 California grapefruits, we can be 95% sure that the true mean weight of all California grapefruits is somewhere between 10.93 and 11.79 ounces.

The interval 10.93 to 11.79 is said to be a "95% confidence interval" for μ. It allows us to come up with not only an estimate of μ but also a measure of how accurate we can expect our estimate to be. A picture of this confidence interval appears in Fig. 4.6.

The confidence interval whose development we have just completed is only one of many, many confidence intervals that may be used to estimate μ. In fact, it is not *the* confidence interval for μ, but it is a 95% confidence interval for μ based on a random sample of size 64. If, for example, we wanted to be 98% sure of having an interval in which μ fell, we would have to modify our present interval somewhat, by replacing the 1.96 in the formula (and in Fig. 4.3) by $z_{.01} = 2.33$. Based on the same sample of 64 data points, we could be 98% sure that

$$11.36 - (2.33)\left(\frac{1.76}{\sqrt{64}}\right) < \mu < 11.36 + (2.33)\left(\frac{1.76}{\sqrt{64}}\right)$$
$$11.36 - .51 < \mu < 11.36 + .51$$
$$10.85 < \mu < 11.87$$

namely, that the mean weight of California grapefruits lies somewhere between 10.85 and 11.87 ounces.

FIGURE 4.6

A 95% confidence interval for the mean weight of California grapefruits.

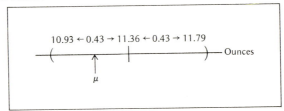

Notice that the 98% confidence interval for μ is wider than the 95% interval based on the same data. It is logical that this be so, because if we want to be more certain of having an interval that contains μ, we are going to have to give ourselves a little more room by widening our interval. Similarly, if we were to narrow our interval, there might be only an 80% chance of its containing μ. We summarize this discussion in Table 4.1, which demonstrates the procedure of obtaining confidence intervals of various degrees of confidence. For purposes of comparison the various intervals are graphed in Fig. 4.7. The calculations therein are based on the fact (discussed in detail in Sec. 3.E) that the middle $(1 - \alpha)100\%$ of the standard normal distribution lies between the numbers $-z_{\alpha/2}$ and $z_{\alpha/2}$. This fact results in the replacement of the 1.96 in our discussion by $z_{\alpha/2}$. We can therefore be $(1 - \alpha)100\%$ sure that

●
$$\bar{x} - z_{\alpha/2}\,\frac{s}{\sqrt{n}} < \mu < \bar{x} + z_{\alpha/2}\,\frac{s}{\sqrt{n}} \qquad (4.4)$$

based on a random sample of n data points having sample mean $\bar{x}$ and sample standard deviation s. Equation (4.4) above is the general formula for computing confidence intervals for means based on large samples.

When we say that we are 95% sure that the true mean μ falls between $11.36 - .43$ and $11.36 + .43$, we are saying that we are 95% sure that μ is within a distance .43 of 11.36. This way of expressing the situation can be seen in Fig. 4.6, where attention is drawn to the fact that 10.93 is located a distance .43 to the left of 11.36, while 11.79 is .43 to the right. Therefore, if we were to use the sample mean of 11.36 ounces as an estimate of the true mean μ, we could be 95% sure of being in error by no more than .43 ounce. If we look at the general situation this way, we can define the "error of estimation"

●
$$E = z_{\alpha/2}\,\frac{s}{\sqrt{n}} \qquad (4.5)$$

FIGURE 4.7
Graphical representations of various confidence intervals for the true mean weight of California grapefruits.

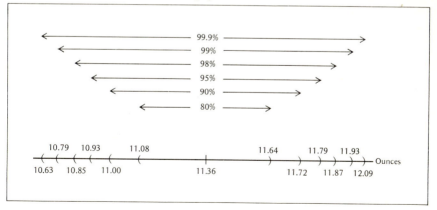

in the following terms: We can be $(1 - \alpha)100\%$ sure that the maximum possible error in using $\bar{x}$ as an estimate of μ cannot exceed E. In terms of error of estimation, then, we can recast the subject matter of Table 4.1 into the form used in Table 4.2. The two procedures, confidence interval and error of estimation, are merely ways of viewing the same information from different vantage points. The salient fact that is especially apparent from Table 4.2 is the parallel behavior of the degree of confidence and the error of estimation. As the degree of confidence increases, so does the error of estimation. As we have pointed out earlier in the discussion, this is entirely logical: The larger our error of estimation, the more confident we can be of having included the true mean μ within the induced bounds.

As typical examples of the sort of questions that can be answered using the procedures developed so far in this chapter, we present the three that follow.

Example 4.1   Automobile Usage   A random sample of 100 private automobiles registered in the province of Ontario put a mean yearly mileage of 6500 on their odometers with a standard deviation of 1400 miles. These facts were brought out as part of a statistical study of automobile usage commissioned by a local insurance agency. Find a 99% confidence interval for the true mean yearly mileage of all private cars in the province.

SOLUTION   For a 99% confidence interval, we have that

$$(1 - \alpha)100\% = 99\%$$

Dividing by 100%, we get

$$1 - \alpha = .99$$

so that

$$-\alpha = .99 \quad 1 = \quad .01$$

and finally

$$\alpha = .01$$

It follows that

$$\frac{\alpha}{2} = .005$$

TABLE 4.1
Various Confidence Intervals for the True Mean Weight of California Grapefruits
Based on a Sample of $n = 64$ Grapefruits Having Sample Mean
$\bar{x} = 11.36$ and Sample Standard Deviation $s = 1.76$

Degree of Confidence $(1 - \alpha)100\%$	α	$\alpha/2$	$z_{\alpha/2}$	$z_{\alpha/2}\dfrac{s}{\sqrt{n}}$	Confidence Interval $\bar{x} - z_{\alpha/2}\dfrac{s}{\sqrt{n}} < \mu < \bar{x} + z_{\alpha/2}\dfrac{s}{\sqrt{n}}$
80%	.20	.10	1.28	.28	$11.08 < \mu < 11.64$
90%	.10	.05	1.64	.36	$11.00 < \mu < 11.72$
95%	.05	.025	1.96	.43	$10.93 < \mu < 11.79$
98%	.02	.01	2.33	.51	$10.85 < \mu < 11.87$
99%	.01	.005	2.57	.57	$10.79 < \mu < 11.93$
99.9%	.001	.0005	3.3	.73	$10.63 < \mu < 12.09$

TABLE 4.2
**Maximum Possible Errors of Estimation of the True Mean Weight
of California Grapefruits**
Based on a Sample of $n = 64$ Grapefruits Having Sample
Mean $x = 11.36$ and Sample Standard Deviation $s = 1.76$

Degree of Confidence $(1 - \alpha)100\%$	α	$\alpha/2$	$z_{\alpha/2}$	Maximum Possible Error of Estimation $E = z_{\alpha/2}s/\sqrt{n}$
80%	.20	.10	1.28	.28
90%	.10	.05	1.64	.36
95%	.05	.025	1.96	.43
98%	.02	.01	2.33	.51
99%	.01	.005	2.57	.57
99.9%	.001	.0005	3.3	.73

from which we conclude that

$$z_{\alpha/2} = z_{.005} = 2.57$$

The data give us the information that

$$n = 100$$
$$\bar{x} = 6500$$
$$s = 1400$$

Applying the formula (4.4), we know that

$$\bar{x} - z_{\alpha/2}\frac{s}{\sqrt{n}} < \mu < \bar{x} + z_{\alpha/2}\frac{s}{\sqrt{n}}$$

where μ is the true mean yearly mileage of all private cars in the province. Inserting the actual numerical values of the quantities involved yields

$$6500 - (2.57)\frac{1400}{\sqrt{100}} < \mu < 6500 + (2.57)\frac{1400}{\sqrt{100}}$$

$$6500 - \frac{(2.57)(1400)}{10} < \mu < 6500 + \frac{(2.57)(1400)}{10}$$

$$6500 - 359.8 < \mu < 6500 + 359.8$$

$$6140.2 < \mu < 6859.8$$

Therefore we can be 99% sure that the true mean yearly mileage of all private cars in the province of Ontario lies between 6140.2 and 6859.8 miles.

Example 4.2 Soft-Drink Machines Owners of a chain of coin-operated beverage machines advertise that their machines dispense, on the average, 7 ounces of soft drink per cup. A consumer testing organization, which is interested in whether a particular machine is meeting its specifications, wants to run a study after which they can be 95% sure of having the true mean amount dispensed by the machine correctly estimated to within an error of .01 ounce (one-hundredth of an ounce). If a preliminary analysis indicates that the amounts dispensed have standard deviation .12 ounce (twelve-hundredths of an ounce), how many sample cupfuls does the organization need to measure in order to achieve the desired level of accuracy?

SOLUTION   The unknown here is n, the number of data points required for the analysis, in this case the number of sample cupfuls needed. Ordinarily, n is known, along with the standard deviation and α, and it is desired to compute the error. Here, however, the desired error level, $E = .01$, is given, and we want to know instead the number n of data points required. We use the error formula (4.5)

$$E = z_{\alpha/2} \frac{s}{\sqrt{n}}$$

As we have just mentioned, the level of error is

$$E = .01$$

The degree of confidence is 95% so that $\alpha = .05$ and $\alpha/2 = .025$. Therefore

$$z_{\alpha/2} = z_{.025} = 1.96$$

Finally, in place of the sample standard deviation (which is unavailable, for we have not yet taken the actual sample), we use the preliminary estimate that

$$s = .12$$

Inserting these numbers into their proper places in the error formula (4.5), we find that

$$.01 = (1.96) \frac{.12}{\sqrt{n}}$$

Successive steps of algebraic operations yield

$$(.01) \sqrt{n} = (1.96)(.12)$$
$$\sqrt{n} = \frac{(1.96)(.12)}{.01}$$
$$\sqrt{n} = 23.52$$

It follows that $n = (23.52)^2 = 553.19$. To achieve the desired level of accuracy, then, would require at least 553.19 data points. Therefore, we would need 554 sample cupfuls for the survey. (Notice that even though 553.19 rounds down to 553, the latter number would leave us just short of the desired accuracy. That 554th data point puts us over the top.)

Example 4.3   Reading Scores   The administrator of an elementary school district would like to estimate the mean reading level as measured by the standard Zeta-Epsilon Basic Educational Skills Test (ZEBEST) of all pupils under her direction. She chooses a random sample of 60 pupils from the district and gives the ZEBEST to each. The pupils obtain a mean score of 550 points with a standard deviation of 90. How sure can the administrator be that she has the true mean for the entire district estimated correctly to within 20 points?

SOLUTION   Here α is the unknown item, for that is related to the degree of confidence, $(1 - \alpha)100\%$. From the results of the study, we know that

$$n = 60$$
$$\bar{x} = 550$$
$$s = 90$$

($\bar{x}$ turns out to be unneeded for the solution of this problem.) The desired level of error is $E = 20$. Substituting into the equation,

$$E = z_{\alpha/2} \frac{s}{\sqrt{n}}$$

we obtain the relationship

$$20 = z_{\alpha/2} \frac{90}{\sqrt{60}}$$

A little algebra yields the equation

$$z_{\alpha/2} = \frac{(20)\sqrt{60}}{90} = 1.72$$

because $\sqrt{60} = \sqrt{6 \times 10} = 7.746$. From Fig. 4.8 and Table A.3 of the Appendix, we see that

$$1 - \alpha = .4573 + .4573 = .9146$$

or $\quad 1 - \alpha = 1 - 2\left(\frac{\alpha}{2}\right) = 1 - 2(.0427) = 1 - .0854 = .9146$

Therefore the administrator can be $(1 - \alpha)100\% = (.9146)100\% = 91.46\%$ sure that she has the true district mean score estimated correctly to within 20 points.

FIGURE 4.8
Finding α when $z_{\alpha/2} = 1.72$.

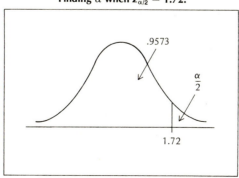

1 For 64 randomly selected months in the past, the Oryx Veterinary Supply Company (OVS) has sold an average of 16,224.72 dollars worth of goods each month with a standard deviation of 205.00 dollars. Find a 90% confidence interval for the true mean monthly sales level of OVS.

2 A biologist studying the water life of the Gulf of California has found 36 examples of a certain breed of fish. The 36 fish have a mean length 11 inches with a standard deviation of 1.5 inches. What can she present as a 95% confidence interval for the true mean length of all fish of that breed living in the Gulf of California?

3 An auto parts dealer plans on advertising the average lifetime of the auto batteries he sells. To find out what that average lifetime is, he gives away 100 randomly selected batteries to 100 randomly selected customers with the proviso that the customer use the battery continuously with no maintenance, except for adding water as needed and cleaning the terminals, until it fails. The lifetimes, in years, of the 100 batteries turn out to have a mean 2.0 years with a standard deviation of .156 year.
  a Determine a 90% confidence interval for the true mean lifetime of the dealer's batteries.
  b Determine a 95% confidence interval for the true mean lifetime.
  c Determine a 99% confidence interval for the true mean lifetime.

4 A group of fifth-grade children who have participated in a special remedial reading program are tested for reading improvement. A random sample of 81 children had a mean increase of 18.4 points with a standard deviation of 6.9. Find a 97% confidence interval for the mean increase in reading scores for the entire group.

5 A statewide real estate brokerage company would like to be able to forecast the true mean price of a house with an accuracy within 1500 dollars, for such information will allow the company to develop accurate estimates of its projected earnings. A survey of 100 recently sold houses shows a mean price of 33,400 dollars with a standard deviation of 7100 dollars. How certain can the company be that the true mean is within 1500 dollars of its estimate of 33,400 dollars?

6 An agricultural zoologist wants to be 95% sure of having the true mean gestation period of a new variant of dairy cow estimated correctly to within 1.3 days. It seems reasonable to use a standard deviation of 5 days for the variant under study. On how many cows are data needed in order to achieve the desired level of accuracy?

7 The Eastingfield Electric Company, which produces light bulbs, is required by law to estimate statistically the true mean lifetime of its bulbs. In particular, it must be 95% sure of having the true mean estimated accurately to within 25 hours. If the lifetimes have standard deviation 200 hours, how many bulbs must be tested in order to achieve the desired accuracy in estimating the true mean?

8 Find a 90% confidence interval for the true mean forecast of the Consumer Price Index based on the data of Exercise 1.A.6.

SECTION 4.C CONFIDENCE INTERVALS BASED ON SMALL SAMPLES

In the previous section, we have shown how to estimate the true mean of a population if we have a random sample of more than 30 data points. Sometimes, however, it is not

possible or economically feasible to use a large sample. For example, a steel mill may want to estimate temperatures, pressures, etc., in various parts of its furnaces and may not have as many as 30 furnaces available for testing. A political scientist may want to study the voting patterns of a particular occupational group, but there may not be data available for 30 elections in which this group has participated. In situations such as these, where sufficient data are unobtainable, the central limit theorem of Sec. 4.B simply does not apply. We therefore have no guarantee that the z scores of the sample means

$$z = \frac{\bar{x} - \mu}{\sigma/\sqrt{n}}$$

are normally distributed. Use of the normal distribution in establishing confidence intervals then cannot be logically justified, and so a new approach to the problem is needed.

As we pointed out at the end of Sec. 4.A, we can handle the case of small samples (30 or fewer data points) if we have reason to believe that the data points came from a normally distributed population. This will be the case, for example, in situations satisfying those criteria for normally distributed data presented at the beginning of Sec. 3.C. Just as the central limit theorem revealed (in the large-sample case) that the numbers

$$z = \frac{x - \mu}{\sigma/\sqrt{n}}$$

are approximately normally distributed, mathematical calculations show that if the overall population from which a random sample is selected has the normal distribution, then, for all samples of size n (no matter how large or small), the numbers

$$t = \frac{\bar{x} - \mu}{s/\sqrt{n}}$$

have the t distribution of Table A.4 of the Appendix, where

$\bar{x}$ = sample mean
s = sample standard deviation
n = number of data points in the sample
μ = true mean of the population

The use of Table A.4 is valid only when the overall population has the normal distribution, because Table A.4, the table of percentage points of the t distribution, is derived by mathematical calculations from Table A.3, the table of the normal distribution.

Even though the numbers

$$t = \frac{\bar{x} - \mu}{s/\sqrt{n}}$$

have the t distribution for all values of n, both large and small, Table A.4 is commonly used only for small samples of $n \leq 30$. The reason for this is that for large values of n the t distribution is very close to the normal distribution because of the increasing influence of the central limit theorem. Common statistical practice therefore calls for the use of Table A.4 when $n \leq 30$ and Table A.3 when $n > 30$.

Beyond the switch from Table A.3 to Table A.4, there is no recognizable difference in one's calculation of confidence intervals. In formula (4.4),

$$\bar{x} - z_{\alpha/2} \frac{s}{\sqrt{n}} < \mu < \bar{x} + z_{\alpha/2} \frac{s}{\sqrt{n}}$$

we simply replace $z_{\alpha/2}$ by $t_{\alpha/2}(n-1)$, the appropriate number selected from Table A.4. Here the number $n - 1$ is called the "degrees of freedom" and appears in the column headed "df" in the table. It is through the degrees of freedom that the amount of data points makes its influence felt on the t distribution. The expression itself is a traditional one, based on a geometric interpretation of the standard deviation. We find the proper t value at the intersection of the row corresponding to the number $n - 1$ (one less than the number of data points) and the column headed $t_{\alpha/2}$ (for the relevant value of α). The formula for confidence intervals based on small samples is therefore

- $$\bar{x} - t_{\alpha/2}(n-1) \frac{s}{\sqrt{n}} < \mu < \bar{x} + t_{\alpha/2}(n-1) \frac{s}{\sqrt{n}} \qquad (4.6)$$

The following two examples illustrate the use of this formula, particularly how to choose the correct value of $t_{\alpha/2}(n-1)$.

Example 4.4   Company Cars   A corporation that maintains a large fleet of company cars for the use of its sales staff needs to determine the average number of miles driven monthly per salesperson. A random sample of 16 monthly car-use records was examined, yielding the information that the 16 salespersons under study had driven a mean of 2250 miles with a standard deviation of 420 miles. Find a 95% confidence interval for the true mean monthly number of miles per salesperson for the entire sales staff.

SOLUTION   We can summarize the information contained in the sample by recording that

$$n = 16$$
$$\bar{x} = 2250$$
$$s = 420$$

Because $n = 16 < 30$, use of the percentage points of the t distribution, instead of those of the normal distribution, is required. Since we want a 95% confidence interval, we know that $\alpha = .05$ so that $\alpha/2 = .025$. As $n = 16$, we know that df $= n - 1 = 16 - 1 = 15$, and so the number we need from Table A.4 is $t_{\alpha/2}(n-1) = t_{.025}(15)$. We reproduce the relevant portion of Table A.4 in Table 4.3, which shows that $t_{.025}(15) = 2.131$. Now we have in our possession all the components of formula (4.6), the formula for confidence intervals based on small samples, where μ stands for the true mean monthly number of miles per salesperson for the entire sales staff. Therefore, inserting the numer-

TABLE 4.3
A Portion of
Table A.4

df	$t_{.025}$	df
15	2.131	15

ical values for $n, \bar{x}, s$, and $t_{\alpha/2}(n - 1)$, we see that we can be 95% sure that

$$2250 - (2.131)\frac{420}{\sqrt{16}} < \mu < 2250 + (2.131)\frac{420}{\sqrt{16}}$$

$$2250 - \frac{(2.131)(420)}{4} < \mu < 2250 + \frac{(2.131)(420)}{4}$$

$$2250 - 223.755 < \mu < 2250 + 223.755$$

$$2026.245 < \mu < 2473.755$$

Therefore, we can be 95% sure that the true mean monthly number of miles driven per salesman for the entire sales staff is somewhere between 2026. 245 and 2473.755 miles.

Example 4.5   Manufacture of Antibiotics   Because pharmaceutical companies manufacture antibiotics in large vats for later bottling in smaller units, it is necessary to test potency levels at several locations in the vats before bottling. Although it is impossible to achieve complete uniformity in potency levels of the bottled product, such testing is required in order to obtain the mean potency level of the entire batch. In one such test, the readings at 12 randomly selected locations in the vat were

8.9   9.0   9.1   8.9   9.1   9.0
9.0   9.0   8.9   8.8   9.1   9.2

Find a 98% confidence interval for the true mean potency reading for the entire batch.

SOLUTION   In this problem, we know immediately that $n = 12$, but we will have to compute $\bar{x}$ and s from the data. We proceed to make the calculations by the methods introduced in Chap. 2. The basic table of calculations appears in Table 4.4 and shows that

$$\bar{x} = 9.0$$
$$s = 0.113$$

Because we are looking for a 98% confidence interval, we are using $\alpha = .02$ so that $\alpha/2 = .01$. For

$$n = 12$$

we know that df $= n - 1 = 12 - 1 = 11$, and so we obtain from Table A.4 the relevant

TABLE 4.4
Calculation of $\bar{x}$ and s for the
Antibiotic Potency Data

x	$x - \bar{x}$	$(x - \bar{x})^2$
8.9	−.1	.01
9.0	.0	.00
9.1	.1	.01
8.9	−.1	.01
9.1	.1	.01
9.0	.0	.00
9.0	.0	.00
9.0	.0	.00
8.9	−.1	.01
8.8	−.2	.04
9.1	.1	.01
9.2	.2	.04
108.0	.0	.14

$$\bar{x} = \frac{108.0}{12} = 9.0$$

$$s = \sqrt{\frac{.14}{11}} = \sqrt{.0127} = .113$$

value of $t_{\alpha/2}(n - 1)$, namely,

$$t_{.01}(11) = 2.718$$

as shown in Table 4.5. We insert these numbers into the formula for confidence intervals based on small samples, where μ is the true mean potency level of the entire batch. Again, we can be $(1 - \alpha)100\%$ sure that

$$\bar{x} - t_{\alpha/2}(n - 1)\frac{s}{\sqrt{n}} < \mu < \bar{x} + t_{\alpha/2}(n - 1)\frac{s}{\sqrt{n}}$$

On the basis of the data collected, we can then be 98% sure that

$$9.0 - (2.718)\frac{.113}{\sqrt{12}} < \mu < 9.0 + (2.718)\frac{.113}{\sqrt{12}}$$

$$9.0 - \frac{(2.718)(.113)}{3.464} < \mu < 9.0 + \frac{(2.718)(.113)}{3.464}$$

TABLE 4.5
A Portion of
Table A.4

df	$t_{.01}$	df
11	2.718	11

$$9.0 - 0.089 < \mu < 9.0 + .089$$
$$8.911 < \mu < 9.089$$

We can therefore be 98% sure that the true mean potency level of the entire batch of antibiotic in the vat falls between 8.911 and 9.089.

EXERCISES 4.C

1 For 18 randomly selected months in the past, the Oryx Veterinary Supply Company (OVS) has sold an average of 16,224.72 dollars worth of goods each month with a standard deviation of 205.00 dollars. Find a 90% confidence interval for the true mean monthly sales level of OVS.

2 The Desert Breeze Casualty Insurance Company of Trona wants to estimate the mean number of auto accidents its clients cause per month. They have collected the following data over 12 randomly selected months in the recent past:

Month	No. of Accidents
May 1974	24
July 1974	16
October 1974	20
November 1974	18
January 1975	22
March 1975	8
June 1975	12
July 1975	28
September 1975	20
December 1975	14
April 1976	10
October 1976	12

Find an 80% confidence interval for the true mean number of auto accidents caused per month by the company's clients.

3 An auto parts dealer plans on advertising the average lifetime of the auto batteries he sells. To find out what that average lifetime is, he gives away 10 randomly selected batteries to 10 randomly selected customers with the proviso that the customer use the battery continuously with no maintenance, except for adding water as needed and cleaning the terminals, until it fails. The lifetimes, in years, of the 10 batteries turn out to be as follows:

$$2.2 \quad 1.9 \quad 2.0 \quad 2.1 \quad 1.8 \quad 2.1 \quad 2.1 \quad 2.1 \quad 1.7 \quad 2.0$$

a Determine a 90% confidence interval for the true mean lifetime of the dealer's batteries.

b Determine a 95% confidence interval for the true mean lifetime.

c Determine a 99% confidence interval for the true mean lifetime.

4 A random sample of 25 cigarettes of the Puffer's Heaven brand had a mean nicotine content of 22 milligrams with a standard deviation of 4 milligrams. Find

an 80% confidence interval for the true mean nicotine content of Puffer's Heaven cigarettes.

5 Find a 90% confidence interval for the mean yearly number of worker days lost due to labor disputes based on the data of Exercise 2.B.14.

SECTION 4.D ESTIMATING STANDARD DEVIATIONS

The problem of estimating the true standard deviation σ of a population and the basic principles underlying the idea of confidence intervals for standard deviations are essentially the same as those for the corresponding questions involving means. It is therefore not necessary to redevelop the ideas here, and we will concentrate instead on the technical aspects of estimating the true population standard deviation σ.

   If we have a random sample of n data points, $x_1, x_2, \ldots, x_n$, we have been and will be using formula (4.2), namely,

$$s = \sqrt{\frac{\sum_{k=1}^{n} (x_k - \bar{x})^2}{n - 1}}$$

to provide an estimate of σ. The number s, as we have already pointed out, is called the sample standard deviation. Its most important property is that the squared sample standard deviations s^2 of all possible samples of size n will average out to σ^2. In other words, σ^2 is the mean of all possible values of s^2, based on samples of size n, so that we can reasonably expect our value of s to be fairly close to the true standard deviation σ.

   Exactly how close s will be to σ can never be known, of course, because we can never know σ exactly (for reasons discussed in Sec. 4.A in regard to μ). However, in view of some results from advanced mathematics, it turns out that the quantity

$$\chi^2 = \frac{(n - 1) s^2}{\sigma^2}$$

where χ is the Greek letter lowercase chi (pronounced "kye"), has the chi-square distribution of Table A.6, with $n - 1$ degrees of freedom, if the data come from a population that is normally distributed. The frequency polygon of the chi-square distribution, together with a pair of complementary percentage points $\chi^2_{1-\alpha/2}$ and $\chi^2_{\alpha/2}$, is presented in Fig. 4.9. As the picture shows, the numbers $\chi^2_{1-\alpha/2}$ and $\chi^2_{\alpha/2}$ are chosen in such a way that $(1 - \alpha)100\%$ of the numbers

$$\chi^2 = \frac{(n - 1)s^2}{\sigma^2}$$

fall between them. We can use this information, together with the table of percentage points of the chi-square distribution (Table A.6), in order to calculate confidence intervals for the true population standard deviation σ.

FIGURE 4.9
The frequency polygon of the chi-square distribution.

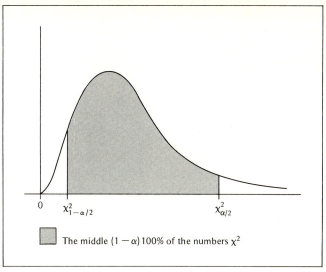

The middle $(1 - \alpha)\,100\%$ of the numbers χ^2

Figure 4.9 shows that we can be $(1 - \alpha)100\%$ sure that

$$\chi^2_{1-\alpha/2}(n - 1) < \frac{(n - 1)s^2}{\sigma^2} < \chi^2_{\alpha/2}(n - 1)$$

where the $n - 1$ in parentheses stands for the number of degrees of freedom, df. If we apply some basic algebraic techniques to the above inequality, we can derive the formula giving confidence intervals for σ. First of all, if a and b are positive numbers such that $a < b$, then it is a fact of arithmetic that $1/b < 1/a$. One example of this is the fact that $\frac{1}{10} < \frac{1}{3}$, since $3 < 10$. Therefore, we obtain from the inequality above that, with probability $1 - \alpha$,

$$\frac{1}{\chi^2_{\alpha/2}(n - 1)} < \frac{\sigma^2}{(n - 1)s^2} < \frac{1}{\chi^2_{1-\alpha/2}(n - 1)}$$

It follows next that we can be $(1 - \alpha)100\%$ sure that

$$\frac{(n - 1)s^2}{\chi^2_{\alpha/2}(n - 1)} < \sigma^2 < \frac{(n - 1)s^2}{\chi^2_{1-\alpha/2}(n - 1)}$$

Finally for the $(1 - \alpha)100\%$ confidence interval for σ, we can be $(1 - \alpha)100\%$ sure that, for small samples,

●

$$\sqrt{\frac{(n - 1)s^2}{\chi^2_{\alpha/2}(n - 1)}} < \sigma < \sqrt{\frac{(n - 1)s^2}{\chi^2_{1-\alpha/2}(n - 1)}} \qquad (4.7)$$

As an application of formula (4.7), we present the following example.

Example 4.6 Manufacture of Antibiotics   In the manufacture of antibiotics in large vats, it is important to know the variability in potency of the product from location to location in the vat. If the variability is too great, the antibiotic could be virtually useless because medical personnel would not be able to judge its effectiveness in any given situation. The standard deviation provides one way of measuring this variability. If potency readings at 12 randomly selected locations in the vat were

$$8.9 \quad 9.0 \quad 9.1 \quad 8.9 \quad 9.1 \quad 9.0$$
$$9.0 \quad 9.0 \quad 8.9 \quad 8.8 \quad 9.1 \quad 9.2$$

find a 90% confidence interval for the true standard deviation in potency of the entire batch.

SOLUTION   The sample standard deviation of this set of data has been calculated in Table 4.4 to be $s = 0.113$. For the 90% confidence interval, we have $\alpha = .10$, and so $\alpha/2 = .05$. Here $n = 12$, so that $n - 1 = 11$. The chi-square values appearing in the above formula are $\chi_{.05}^2(11) = 19.675$ and $\chi_{.95}^2(11) = 4.575$, as illustrated in Table 4.6 which is a selected portion of Table A.5. Inserting all these numbers into the formula above, we see that we can be 90% sure that

$$\sqrt{\frac{(12-1)(.113)^2}{19.675}} < \sigma < \sqrt{\frac{(12-1)(.113)^2}{4.575}}$$

$$\sqrt{\frac{(11)(.012769)}{19.675}} < \sigma < \sqrt{\frac{(11)(.012769)}{4.575}}$$

$$\sqrt{.00714} < \sigma < \sqrt{.0307}$$

$$.0845 < \sigma < .1752$$

Therefore we can be 90% sure that the true standard deviation of the potency level of the entire batch lies somewhere between .0845 and .1752.

As can be noticed from a peek at Table A.5 of the Appendix, chi-square values are commonly used only for df's of 30 or below. Formula (4.7) can therefore be viewed as the formula for confidence intervals based on small samples. For large samples, an analog of the central limit theorem of Sec. 4.B takes over and asserts that the distribution of s is approximately normal with mean σ and standard deviation $\sigma/\sqrt{2n}$. Therefore, we can be $(1 - \alpha)100\%$ sure that

$$-z_{\alpha/2} < \frac{s - \sigma}{\sigma/\sqrt{2n}} < z_{\alpha/2}$$

TABLE 4.6
A Small Portion of
Table A.5

df	$\chi_{.95}^2$	$\chi_{.05}^2$
11	4.575	19.675

Some algebraic transformations show that the above inequality can be successively written as

$$-z_{\alpha/2}\frac{\sigma}{\sqrt{2n}} < s - \sigma < z_{\alpha/2}\frac{\sigma}{\sqrt{2n}}$$

and then

$$\frac{z_{\alpha/2}}{\sqrt{2n}} < \frac{s}{\sigma} - 1 < \frac{z_{\alpha/2}}{\sqrt{2n}}$$

dividing by σ, and then

$$1 - \frac{z_{\alpha/2}}{\sqrt{2n}} < \frac{s}{\sigma} < 1 + \frac{z_{\alpha/2}}{\sqrt{2n}}$$

adding 1 to all three sides, and then

$$\frac{1}{1 + \dfrac{z_{\alpha/2}}{\sqrt{2n}}} < \frac{\sigma}{s} < \frac{1}{1 - \dfrac{z_{\alpha/2}}{\sqrt{2n}}}$$

inverting all fractions. Finally, we can be $(1 - \alpha)100\%$ sure that for large samples

● $$\frac{s}{1 + \dfrac{z_{\alpha/2}}{\sqrt{2n}}} < \sigma < \frac{s}{1 - \dfrac{z_{\alpha/2}}{\sqrt{2n}}}$$ (4.8)

We have the following example to illustrate the use of this formula.

Example 4.7   California Grapefruits   A distributor of citrus fruits judges his stock not only by magnitude of weight but also by uniformity of weight. The standard deviation is, as we know, an efficient way of measuring uniformity: The smaller the standard deviation, the less variation there is from the mean, and so the more uniform the weights are. If the distributor selects a random sample of 64 California grapefruits, and these turn out to have (sample) mean 11.36 ounces and (sample) standard deviation 1.76 ounces, find a 95% confidence interval for the true standard deviation of all California grapefruits.

SOLUTION   We use formula (4.8) because we need a large-sample confidence interval based on $n = 64$ data points. For the 95% confidence interval, we use $\alpha = .05$ so that $z_{\alpha/2} = 1.96$. Further, we have $s = 1.76$. The sample mean $\bar{x}$ plays no role in the estimation of the standard deviation. We can therefore be 95% sure that

$$\frac{1.76}{1 + \dfrac{1.96}{\sqrt{128}}} < \sigma < \frac{1.76}{1 - \dfrac{1.96}{\sqrt{128}}}$$

$$\frac{1.76}{1 + \dfrac{1.96}{11.3}} < \sigma < \frac{1.76}{1 - \dfrac{1.96}{11.3}}$$

$$\frac{1.76}{1.173} < \sigma < \frac{1.76}{0.827}$$

Finally, we can be 95% sure that $1.50 < \sigma < 2.13$, namely, that the true standard deviation of California grapefruits lies between 1.50 and 2.13 ounces.

EXERCISES 4.D

1 Using the data of Exercise 4.C.2,
  a Find a 90% confidence interval for the true standard deviation of the number of auto accidents caused per month by the company's clients.
  b Find a 95% confidence interval for the true standard deviation.
  c Find a 99% confidence interval for the true standard deviation.

2 From the information presented in Exercise 4.C.4, determine a 90% confidence interval for the true standard deviation of the nicotine content of all Puffer's Heaven cigarettes.

3 Over a period of 50 months, the Desert Sands Casualty Insurance Company of Stovepipe Wells conducted an actuarial study which showed that its clients caused an average of 17 auto accidents per month (as a group, of course, not individually) with a standard deviation of 6 accidents.
  a Find an 80% confidence interval for the true standard deviation of the number of auto accidents caused per month by the company's clients.
  b Find a 90% confidence interval for the true standard deviation.
  c Find a 95% confidence interval for the true standard deviation.

4 Using the data of Exercise 4.C.1, find a 90% confidence interval for the true standard deviation of monthly sales of OVS.

5 Calculate a 95% confidence interval for the standard deviation of Consumer Price Index forecasts based on the data of Exercise 1.A.6.

6 Find a 90% confidence interval for the standard deviation of the yearly number of worker days lost because of labor disputes based on the data of Exercise 2.B.14.

SECTION 4.E ESTIMATING PROPORTIONS

In addition to needing estimates of true means or standard deviations, we often have to estimate proportions. The objective of a political poll, for example, is to estimate the proportion of registered voters favoring one or another candidate or position. The objective of a test of effectiveness of a new drug is to estimate the proportion of cases satisfactorily treated by the drug. And the objective of scholastic examinations is to estimate the proportion of subject matter learned by each student.

When we are estimating a proportion, we collect a set of binomially distributed data having parameters n (the number of data points) and p (the true proportion we want to estimate). We can therefore use the facts about the binomial distribution, recorded in the last paragraph of Sec. 3.B, that

$$\mu = np$$

and
$$\sigma = \sqrt{np(1 - p)}$$

We can insert these expressions into formula (4.4) for large sample confidence intervals. We denote by $\hat{p}$("p-hat") the sample proportion based on the data, for example, the proportion of sampled voters who favor the position in question, the proportion of test cases satisfactorily treated by the new drug, or the proportion of examination questions correctly answered by the student. Then in the formula (4.4):

1 μ, the true mean, is replaced by np.
2 $\bar{x}$, the sample mean, is replaced by $n\hat{p}$.
3 $s/\sqrt{n}$, the standard deviation appearing in the central limit theorem, is replaced by $\sqrt{n\hat{p}(1 - \hat{p})}$, its sample analog in the normal approximation to the binomial distribution discussed at the end of Sec. 3.C.

We can therefore be $(1 - \alpha)100\%$ sure that

$$n\hat{p} - z_{\alpha/2}\sqrt{n\hat{p}(1 - \hat{p})} < np < n\hat{p} + z_{\alpha/2}\sqrt{n\hat{p}(1 - \hat{p})}$$

The above formula is valid for $n > 20$, essentially the domain of validity of the normal approximation to the binomial distribution, which we discussed in Sec. 3.D.

Upon applying some algebraic operations to the above formula, we derive a $(1 - \alpha)100\%$ confidence interval for the true proportion p. We divide the entire expression through by n, and we obtain the statement of our $(1 - \alpha)100\%$ certainty that

● $$\hat{p} - z_{\alpha/2}\sqrt{\frac{\hat{p}(1 - \hat{p})}{n}} < p < \hat{p} + z_{\alpha/2}\sqrt{\frac{\hat{p}(1 - \hat{p})}{n}}$$ (4.9)

Formula (4.9) gives us large sample $(n > 20)$ confidence intervals for proportions. Let's now proceed to an example of its use.

Example 4.8 A Political Poll (This example continues the framework of Examples 3.2 and 3.9.) Of a sample of 700 voters, 390 favor repeal of a certain local tax law. Find a 95% confidence interval for the true proportion of all voters favoring repeal.

SOLUTION Here we have $\alpha = .05$, so that $z_{\alpha/2} = 1.96$. We have $n = 700$, and we see that the sample proportion favoring repeal is

$$\hat{p} = \tfrac{390}{700} = .557$$

Therefore, denoting by p the true proportion favoring repeal, we can be 95% sure, in view of formula (4.9), that

$$.557 - (1.96)\sqrt{\frac{.557(1 - .557)}{700}} < p < .557 + (1.96)\sqrt{\frac{(.557)(1 - .557)}{700}}$$

$$.557 - (1.96) \sqrt{\frac{(.557)(.443)}{700}} < p < .557 + (1.96) \sqrt{\frac{(.557)(.443)}{700}}$$

$$.557 - (1.96) \sqrt{.000353} < p < .557 + 1.96 \sqrt{.000353}$$

$$.557 - (1.96)(.0188) < p < .557 + (1.96)(.0188)$$

$$.557 - .037 < p < .557 + .037$$

We conclude that on the basis of the 700 voters sampled we can be 95% sure that the true proportion favoring repeal is somewhere between .520 and .594, namely, between 52 and 59.4%. (In particular, we are over 95% sure that a majority of voters favors repeal.)

We can also use the confidence interval formula for proportions in order to determine how many data points are needed to achieve a desired level of accuracy. By analogy with the technique developed in connection with Example 4.2, we start with the error formula which asserts that if

$$E = z_{\alpha/2} \sqrt{\frac{\hat{p}(1 - \hat{p})}{n}}$$

then $\hat{p} - E < p < \hat{p} + E$. We can solve the error formula for n, the unknown number of data points, by the following sequence of steps:

$$\sqrt{n} = \frac{z_{\alpha/2} \sqrt{\hat{p}(1 - \hat{p})}}{E}$$

so that

$$n = \left[\frac{z_{\alpha/2} \sqrt{\hat{p}(1 - \hat{p})}}{E} \right]^2$$

Now, we have not yet taken the data because we are trying to figure out how many data points to use. Therefore, we do not know what the value of $\hat{p}$ will be, and so we really cannot use the above formula for n. However, it is a fact of arithmetic that $\hat{p}(1 - \hat{p})$ will *always* be less than .25 no matter what number (between 0 and 1, of course) $\hat{p}$ turns out to be.[1] Therefore $\sqrt{\hat{p}(1 - \hat{p})}$ will always be less than $\sqrt{.25} = 0.5$, so that, no matter what the eventual value of $\hat{p}$ turns out to be, it will always be sufficient to use

●
$$n = \left[\frac{(z_{\alpha/2})(.5)}{E} \right]^2 \qquad (4.10)$$

data points. This value of n will be sufficient to achieve the desired level of accuracy because it will always be a larger number than

$$\left[\frac{z_{\alpha/2} \sqrt{\hat{p}(1 - \hat{p})}}{E} \right]^2$$

the smallest acceptable value of n, since .5 will always be larger than $\sqrt{\hat{p}(1 - \hat{p})}$. Let's look at an example.

[1]Try a few numerical values of $\hat{p}$ to confirm this fact.

Example 4.9 Effectiveness of a New Drug (This example is related to Example 3.3.) A new drug for alleviating mental depression is proposed, and we want to estimate its effectiveness. If we want to be 98% sure of having the true proportion of patients helped by the drug correctly estimated to within .03, how many patients do we need for our study?

SOLUTION Here $\alpha = .02$, so that $z_{\alpha/2} = z_{.01} = 2.33$. Furthermore, the desired level of accuracy is given by $E = .03$. Using formula (4.10) we see that a sufficient number of data points for our study would be

$$n = \left[\frac{(z_{\alpha/2})(.5)}{E} \right]^2 = \left[\frac{(2.33)(.5)}{.03} \right]^2 = (38.83)^2$$
$$= 1508$$

Therefore a sample of 1508 patients will be sufficient to guarantee the level of accuracy desired in our study of the drug effectiveness. On the basis of a sample of that size we can be 98% sure of having the true proportion of patients helped correctly estimated to within .03.

As an aside to the conclusion of the above example, it is interesting to note that the Gallup Poll, an organization which undertakes to predict the outcome of elections, uses a sample of 1500 voters in its surveys.

EXERCISES 4.E

1 The manufacturer of a new chemical fertilizer would like to know the proportion of seeds planted in soil containing the fertilizer that will germinate.
   a On the basis of an experiment in which 130 seeds germinated out of 170 that were planted, find a 95% confidence interval for the true proportion of such seeds that will germinate in soil containing the fertilizer.
   b If the manufacturer wants to be 95% sure of having the true proportion correctly estimated to within .02, how many seeds are required for the experiment?
   2/100
2 An insurance company is planning to offer a 10-year term group life insurance policy to all 45-year old men in a certain occupational grouping. In order to determine a reasonable premium, the company conducts an actuarial study of all men of that age in that occupational grouping.
   a If a survey of 2000 such men in the past showed that 150 died before reaching age 55, construct a 99% confidence interval for the true proportion of such men who die before age 55.
   b Records on how many men are required in order to be 99% sure of having the true percentage correctly estimated to within 1%?
3 A news reporter would like to predict the percentage of members of Congress who favor a certain controversial change in the banking laws. If a poll of 120 members yields 72 who support the change, find a 90% confidence interval for the true percentage of members favoring the change.
4 A company psychologist would like to estimate the proportion of regular smokers of

cigarettes among the company's employees. Research turns up 45 smokers in a random sample of 195 employees.

a Find an 80% confidence interval for the true proportion of regular smokers among the company's employees.

b Find a 98% confidence interval for the true proportion.

5 A sociologist studying the prevalence of crime in one major city asks 1800 randomly selected residents whether or not they have ever been victimized by a criminal. Of those surveyed, 700 responded in the affirmative.

a Find a 90% confidence interval for the true percentage of city residents who have been victimized.

b Find a 95% confidence interval for that percentage.

c Find a 99% confidence interval for the percentage.

d How many residents must the sociologist poll in order to be 95% sure of having the true percentage estimated correctly to within 1.5%?

SUMMARY AND DISCUSSION

This chapter has been the first chapter of the portion of this book dealing with statistical inference, the science of drawing conclusions from information contained in a sample of data. The major problem of statistical inference is the attempt to determine to what extent a sample can be considered to be representative of the underlying population. Using two spectacular mathematical results, the law of large numbers and the central limit theorem, we have developed the method of confidence intervals to solve that problem. We have presented confidence interval formulas which show the extent to which the sample mean represents the true population mean, the sample standard deviation represents the true population standard deviation, and the sample proportion represents the true population proportion.

BIBLIOGRAPHY

Variations on the Traditional Approach to Confidence Intervals

Burstein, H.: Finite Population Correction for Binomial Confidence Limits, *Journal of the American Statistical Association*, March 1975, pp. 67–69.

Crisman, R.: Shortest Confidence Interval for the Standard Deviation, *Journal of Undergraduate Mathematics*, September 1975, pp. 57–62.

Guenther, W. C.: Shortest Confidence Intervals, *The American Statistician*, February 1969, pp. 22–25.

Guenther, W. C.: Unbiased Confidence Intervals, *The American Statistician*, February 1971, pp. 51–53.

Machel, R. E., and J. Rosenblatt: Confidence Intervals Based on a Single Observation, *Proceedings of the IEEE*, August 1966, pp. 1087–1088.

Tables of the Proper Sample Size

Burstein, H.: "Sample-Size Tables for Quality Control and Auditing," Redgrave Information Resources, Westport, Conn. 1973.

SUPPLEMENTARY EXERCISES

1 Proposed SEC regulations state that an auditor using random sampling must be 99% sure of having one class of companies' mean amount of accounts receivable estimated correctly to within 5 dollars. If experience has shown that accounts receivable of this type of company generally have standard deviation of sixty dollars, how many accounts must the auditor include in his random sample in order to achieve the specified accuracy?

2 In connection with a study on voting patterns, a political scientist would like to choose a random sample of voters in order to estimate the mean yearly income level for a certain precinct. It is known from earlier studies of this kind that such incomes have standard deviation 3000 dollars in a given precinct. If the political scientist wants to be 90% sure of having the true mean income estimated correctly to within 100 dollars, how many voters does he need for the random sample?

3 The research branch of a major automobile manufacturer has been working on an engine that is supposed to be able to get more miles per gallon of gasoline. In 9 trial runs during a preliminary testing period, the engine was used on a 100-mile track, and the amount of gasoline (in gallons) needed to cover the route was recorded each time, as follows:

   2.5   3.0   2.0   2.4   2.9   2.6   2.7   2.1   2.3

   a Find a 90% confidence interval for the engine's true mean consumption of gasoline per 100 miles.
   b Find a 90% confidence interval for the engine's true standard deviation of consumption of gasoline per 100 miles.
   c If it is reasonable to use .35 as an estimate of the standard deviation, how many trial runs would be needed to be able to estimate the true mean consumption to within .1 gallon per 100 miles with 98% confidence?
   d Using .35 as an estimate of the standard deviation, how sure can we be of having the true mean consumption correctly estimated to within .1 gallon if we base our estimate on a sample of 40 trial runs?

4 A political science researcher studying the loss of seats in the U.S. House of Representatives to the party in power at off-year elections found that in 36 such elections the party in power lost an average of 11.5 seats with a standard deviation of 5. With what confidence can we assert that the true mean loss to the party in power is somewhere between 9 and 14 seats?

5 A sociological survey involving 400 randomly selected families in one section of a large city showed that the average family surveyed had 2.37 children with a standard deviation of 1.80 children. Find a 95% confidence interval for the true mean number of children per family in that section of the city.

6 As part of a study of highway safety, a random sample of 25 automobiles from the freeways of Los Angeles were tested for their stopping ability after application of the brakes. The 25 cars required a mean distance of 148 feet to stop with a standard deviation of 21 feet. Find a 90% confidence interval for the true mean stopping distance of all cars on Los Angeles freeways.

7 In connection with a biological study of the uniformity of growth of cherry tomato plants, 60 plants had produced an average of 14.8 pounds of tomatoes with a standard deviation of 6.5 pounds after 100 days of growth.
  a Find a 90% confidence interval for the true standard deviation of cherry tomato yield, in pounds.
  b Find a 98% confidence interval for the standard deviation of the yield.

8 As part of an analysis of the relationship between smoking and lung disease, a public health physician conducts a survey of hospital patients with lung disorders with the goal of finding out what percentage of such patients are heavy smokers.
  a Of 200 such patients, the physician determines that 120 are heavy smokers. Find a 95% confidence interval for the true percentage of lung patients that are heavy smokers.
  b If an expanded survey reveals that 1200 out of 2000 patients are heavy smokers, what would be a 95% confidence interval for the true percentage?
  c If the physician wants to be 95% sure of having the true percentage estimated correctly to within 1%, how many patients must he include in his survey?

9 Of 300 licensed drivers selected at random from Department of Motor Vehicles records, it is learned that 45 are problem drinkers. Construct a 90% confidence interval for the true proportion of licensed drivers that are problem drinkers.

10 In an election in which Able is running against Baker, a survey of 1500 voters shows that 780 intend to vote for Able.
  a Find an 80% confidence interval for the true proportion of the vote going to Able.
  b Find a 90% confidence interval for the true proportion of the vote going to Able.
  c With what probability can we assert that Able will win a majority of the votes?

11 To estimate the proportion of a company's accounts that have defaulted over the years, an accountant chooses a random sample of 2000 accounts and discovers that 60 have defaulted.
  a Find a 95% confidence interval for the true proportion of defaulting accounts.
  b If it were necessary to be 99.9% sure of having the true proportion estimated to within .01, how large a random sample would be needed to guarantee that level of accuracy?

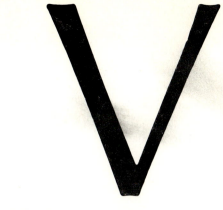

TESTING STATISTICAL HYPOTHESES

One of the fundamental principles of the scientific method for increasing our knowledge of the world around us is the formulation of a hypothesis.[1] Today this and other principles of the scientific method have just as much validity for the social and management sciences as they do for the natural sciences. When we formulate a hypothesis, we define very clearly the problem under discussion and we provide the framework for gathering the relevant facts. An analysis of the accumulated facts (or data) then leads us to accept or reject the hypothesis.

For example, an educational psychologist might want to know whether pupils who have participated in the Head Start program at an early age do better in sixth-grade mathematics than those from similar backgrounds who have not participated. One way she could set up the problem would be to define the hypothesis

> H : Head Start participants do no better in sixth-grade
> mathematics than those from similar backgrounds
> who have not participated

[1]Some individuals use the terminology "an hypothesis" by analogy with "an honor," "an historian," "an horse."

and the alternative

A : Head Start participants do better than those who have not participated

The next step would be to choose a random sample of sixth-grade pupils who have participated and a random sample of those who have not and to give all pupils selected an examination in sixth-grade mathematics. A comparison of the examination scores of the two groups would then lead the psychologist to reject or accept the hypothesis H. By rejecting the hypothesis H in favor of the alternative A, she would be concluding that the Head Start program tends to improve mathematics ability at the sixth-grade level; by failing to reject H, she would be concluding that it does not.

It should be noticed that the hypothesis H is formulated in a negative sort of way. Rather than asserting that Head Start pupils do better than others, it states that they actually do not do better. Such a formulation of the hypothesis H has the effect of placing the "burden of proof" of its effectiveness on the Head Start program itself. In order for us to decide that Head Start is really effective, the data must move us to make the decision to reject H. Unless the test scores of Head Start graduates are substantially above those of nonparticipants, we will merely accept H. If, however, the Head Start scores are significantly higher, we then take action to reject H and to conclude that Head Start participants really do better. Because the logic of the situation often requires H to be structured in this negative way, we usually refer to H as the "null hypothesis."

Our objective in this chapter is to show how to formulate a hypothesis in an applied situation and how to use statistical methods of data analysis for the purpose of deciding whether to reject or accept that hypothesis.

SECTION 5.A THE STATISTICAL DECISION-MAKING PROCESS

Suppose that an individual complaining of stomach pains comes to a doctor's office. Mentally, the doctor immediately formulates the hypothesis

H : The person has appendicitis

and the alternative

A : The person has a stomachache

The doctor's job is to decide between these two possibilities. We can run some medical tests in this and any other such situation, for example, blood tests, urine samples, x-rays, etc. After all the data are assembled, the individual in question fits into one of the following categories:

1 He almost certainly has appendicitis.
2 He probably has appendicitis.
3 The evidence is inconclusive.
4 He probably does not have appendicitis.
5 He almost certainly does not have appendicitis.

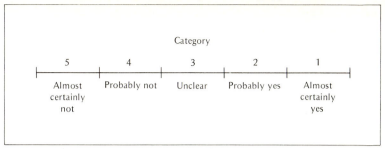

FIGURE 5.1
Categories of individuals in regard to having appendicitis.

Figure 5.1 illustrates the situation. There is no such thing as absolute certainty in a situation like this until it is almost too late—emergency appendectomies are not very unusual operations.

After we arrive at our decision concerning the medical state of the individual in question, we may have made one of two types of errors. These are formally designated "Type I" and "Type II" and they are as follows:

Type I Error   We reject H when it is really true (in this case, we decide that the individual has a stomachache when he in fact has appendicitis).

Type II Error   We accept H when it is really false (in this case, we decide that the individual has appendicitis when he in fact has a stomachache).

(Of course, we might possibly arrive at the correct decision: We could reject H when it's really false or accept it when it's really true.)

There are two questions we must settle in connection with the Type I and Type II errors: (1) Which of the two errors is the worse? (2) How often are we willing to make the worse of the two errors?

Regarding the appendicitis problem, we would probably agree that the Type I error is the worse. If we make a Type I error, we are risking the life of the patient, whereas if we make a Type II error, we are merely wasting his money, the insurance company's money, and the medical staff's time.

How often would we be willing to make a Type I error? Well, I don't know about you, but if I were the doctor, I'd prefer never to make a Type I error. What decision strategy can we follow that would guarantee our never making a Type I error? If you think about it, you will see that the only way we can be 100% sure of never making a Type I error would be to always decide that the person has appendicitis and then to operate accordingly. This way, we can be sure of never rejecting H when it's really true because we will never reject H—never when it's true and never when it's false.

Imagine the situation for a moment. The individual comes to our office saying, "My tummy hurts," and before he finishes the sentence, we're rushing him off to the appendectomy ward. Although we are guaranteed never to make a Type I error, you'll have to agree that this diagnostic technique is somewhat impractical. Why? Primarily because we will be doing appendectomies on many, many people who do not have appendicitis. More precisely, in our zeal to avoid Type I errors, we will be committing Type II errors whenever it is possible to do so. If we consider the fact that probably fewer than one-tenth of one percent of persons with stomach pains actually have appendicitis, it's truly ridiculous to perform an appendectomy on everyone who walks through the door.

Well, now we're back where we started. We agree that we cannot operate on everyone, and therefore we're going to have to prescribe Pepto-Bismol for some of our patients. Yet, every time we let a person slip through with a bottle of Pepto-Bismol instead of an appendectomy, we're risking a Type I error. Even if we let only one person through without an appendectomy, there is some chance that the person in question actually has appendicitis. What do we do? How many people do we let by? How many do we operate on?

The problem revolves around the difficulty of deciding on an acceptable level of Type I errors. How often are we willing to make a Type I error; i.e., how many persons who really have appendicitis (unknown to us at the time) are we willing to send home with a bottle of Pepto-Bismol?

Of course, there is no real answer to this question. This is a subjective decision that must take many factors into account. If we denote by α (pronounced "alpha") our probability of making a Type I error and by β (pronounced "baytah") our probability of making a Type II error, there is an inverse relationship between α and β. If α is very low, we would not be making a Type I error too often. Therefore we would not be rejecting H (and consequently asserting A to be true), unless it were fairly certain that H was false. It would therefore be necessary to accept H in some questionable cases, and so our probability of a Type II error would be high. So a low value of α would almost automatically lead to a high value of β. The reverse is also true: A high value of α means that we are rejecting H in questionable cases, so that we accept H only when it is probably true, and β is therefore low. It is therefore necessary for us to weigh the relative seriousness of the two errors, not merely to decide separately on an acceptable level of Type I errors. As we have seen in the appendicitis example above, the insistence on a low incidence of Type I errors almost certainly leads to an unacceptably high level of Type II errors. On the other hand, we can't really accept too high a level of Type I errors either.

As far as the mathematics of the situation is involved, our discussion so far shows that we cannot guarantee that both

$$\alpha = P(\text{Type I error})$$

and

$$\beta = P(\text{Type II error})$$

will be within acceptably low limits. If we insist on a low α, we may be forced into accepting a high β. If we insist on a low β, we may be forced into accepting a high α. To make the best of a bad situation, we use the following procedure for handling statistical hypotheses:

1 We decide which of the two errors, Type I or Type II, is the worse.
2 If we think the Type I error is the worse, we choose a testing procedure which makes α acceptably low, and we agree to be satisfied with whatever the value of β turns out to be.

3 If we think the Type II error is the worse, we choose a testing procedure which makes α high, and then we can expect β to be correspondingly low.

Despite the general rule of the inverse behavior of α and β, it unfortunately turns out that there is usually not a simple mathematical relationship between α and β. This means that in most problems where we choose the level of α (whether it be high or low), we cannot say exactly what the value of β is. This is too bad, but not very much can be done about it for the computation of the exact value of β in many applied situations is a very difficult mathematical problem and depends on the relationships among many unknown quantities. The only thing we can really be sure of is that as α increases from low to high, β will decrease from high to low. For this reason, it is advisable whenever possible to construct H and A in such a way that the Type I error will be the worse, and then our ability to choose α will give us some control over that error.

Our choice of α can therefore be viewed as a subjective measure, on our part, of the relative seriousness of the two types of errors. If we think the Type I error is much more serious in a particular applied context, we should choose α to be low. If we think the Type II error is very serious, we should choose α to be high. For this reason, α is said to be the "significance level of the test." When a researcher or manager chooses a particular significance level α as part of the analysis of a problem, he is morally (although not legally) obligated to state at the outset what level of α is being used. Any decision that is made on the basis of a statistical analysis must include a statement of the numerical value of α, for only in this way can a reader of the report know how the writer viewed the relative seriousness of the two errors. Sometimes, the reader may not have the same subjective viewpoint as the writer and may therefore disagree strenuously with the writer's choice of α. In some borderline cases, a change in the numerical value of α may even result in a change in the decision. (Some practical examples of this will be illustrated in the remaining sections of this chapter.) Therefore, the choice of α could be the determining factor in making the decision, and so it must be reported as an integral part of the research study.

You have probably heard or read somewhere the allegation, "anything can be proved using statistics." The greater part of whatever truth there may be to this allegation arises from disagreements between opposing parties in regard to the appropriate level of significance to be used in analyzing the data, not from disagreements over the numerical calculations based on the data. As a prime example of this sort of situation, we can look at the recent history of disputes between "environmentalists" and "economic growth advocates" over offshore oil drilling in locations such as California's Santa Barbara Channel. The hypothesis to be tested is

> H : Offshore oil drilling *does not* cause significant
> environmental damage

and the alternative is

> A : Offshore oil drilling *does* cause significant
> environmental damage

In this situation, the significance level would be

$$\alpha = P(\text{Type I error}) = P(\text{rejecting } H \text{ when it's really true})$$
$$= \text{probability of asserting that offshore oil}$$
$$\text{drilling does cause significant environmental}$$
$$\text{damage } when, \text{ in fact, it } does \text{ } not$$

while

$$\beta = P(\text{Type II Error}) = P(\text{accepting } H \text{ when it's really false})$$
$$= \text{probability of asserting that offshore oil drilling } does \text{ } not$$
$$\text{cause significant environmental damage when, in fact, it}$$
$$does$$

From the environmentalists' point of view, the Type II error would be the worse, for the environmentalist believes that the possibility of significant damage to the environment outweighs any economic advantage to be gained from offshore oil drilling. In the opinion of advocates of economic growth, the Type I error is the worse of the two because such an error would tend to stagnate economic growth, leading to loss of jobs and a lowered standard of living, even if there were no real evidence that significant environmental damage would occur. It is not that environmentalists are necessarily out to wreck the economy, nor are economic growth advocates deliberately out to destroy the environment. What is really the case is that, if an error (Type I or Type II) is to be made, the environmentalist would rather err on the side of the environment (while risking some economic dislocation), and the economic growth advocate would rather err on the side of economic well-being (while risking some environmental problems). Environmentalists would then base a statistical analysis on a high level of significance α, because they would insist that $\beta = P(\text{Type II error})$ be held as low as possible. Growth advocates, on the other hand, would favor a low level of significance, for they would feel that $\alpha = P(\text{Type I error})$ itself should be low. Disputes between environmentalists and economic growth advocates, then, do not usually center on the statistical information gathered on the question at hand, but only on the level of significance α appropriate to the problem. We summarize this discussion in Table 5.1.

One more comment before leaving this section: What do we mean by "low" α, and what do we mean by "high" α? Theoretically speaking, α can be any number

TABLE 5.1
Choice of Level of Significance α
Environmentalists versus Economic Growth Advocates
H : Drilling Does Not Cause Environmental Damage
A : Drilling Does Cause Environmental Damage

		Significance Level	
Primary Concern	Worse Error	α	β
Environment	Type II	High	Low
Economy	Type I	Low	High

between 0 and 1 (0 to 100%) because it represents the proportion of time we are willing to make a Type I error in order to avoid a Type II error. As a practical matter, however, we must not set α too low, for this will result in too many Type II errors, and we must not, of course, set α too high, for that would result in too many Type I errors. Because the choice of α is subjective, a decision to be made by the researcher or manager on the scene, it would be best to go back to the professional literature on the subject to find out what the experts in the various fields are using. An examination of the professional publications listed in the bibliography indicates that a "high" level of α usually means that α is between .02 and .10, while a "low" level of α is taken to be between .005 and .02. If the Type I error is the worse, the typical value of α is .01, but if the Type II error is the worse, α is typically taken to be .05.

The remainder of this chapter will consist of exemplary problems in applied contexts involving the testing of hypotheses about population means, standard deviations, and proportions. Additional types of statistical hypotheses commonly formulated and tested in connection with applied work will be discussed in portions of Chaps. 6 and 9, as well as the entirety of Chaps. 10 and 11.

EXERCISES 5.A

1 Many statistical studies have been conducted in an attempt to find out whether or not smoking really causes cancer. To put the burden of proof on those who feel that it does, we can formulate the hypothesis

$$H : \text{Smoking does not cause cancer}$$

and the alternative

$$A : \text{Smoking causes cancer}$$

a The cigarette industry would be unfairly hurt by a definitive decision that smoking causes cancer if the opposite were, in fact, true. To protect the cigarette industry, should α be chosen high or low?

b The American Cancer Society wants to eliminate all possible causes of cancer, even if their connection with cancer is not conclusively proved beyond all doubt. Should the Society argue for a high or low value for α?

2 A jury is supposed to convict an individual on trial for a string of murders and assaults if the individual is guilty beyond a "reasonable" doubt. The word "reasonable" is related to the numerical value of α in testing the hypothesis

$$H : \text{The individual is innocent}$$

against the alternative

$$A : \text{The individual is guilty}$$

a To avoid convicting an innocent person, should the jury choose a high value for α or a low value?

b To avoid releasing a guilty person to possibly continue the wave of attacks, should the jury select a high value for α or a low value?

3 A business executive has been assigned the task of deciding whether or not to invest

several thousand dollars in a new business venture. He has to choose between the hypothesis

$$H : \text{The venture will succeed}$$

and the alternative

$$A : \text{The venture will fail}$$

a If there is a shortage of investment capital, so that only the safest investments should be undertaken, should the executive use a high value or a low value of α?

b If investment capital is plentiful, allowing the funding of some possibly risky ventures, should α be chosen as high or low?

SECTION 5.B ONE-SAMPLE TESTS FOR THE MEAN OF A POPULATION

Because of the expense involved in manufacturing Nitro-Plus, a new chemical fertilizer, the fertilizer must produce an average yield of more than 20,000 pounds of tomatoes per acre within a specified growing period in order to be economically feasible. To check on whether or not this level of yield can reasonably be expected from the product, an independent agricultural testing organization chooses 40 acres of farmland at random in various geographical regions, fertilizes each with Nitro-Plus, and proceeds to plant tomatoes intensively. At the conclusion of the specified growing period, yields are collected and analyzed. The results show that the 40 acres had a mean yield of 20,400 pounds, and the yield per acre varied widely, having standard deviation 1200 pounds. On the basis of the test results involving 40 randomly selected acres, can we conclude that the *true* mean yield really exceeds 20,000 pounds per acre?

In order to put the burden of proof on the fertilizer, we denote the true mean yield by the symbol μ, and we define the hypothesis to be

$$H : \mu = 20,000 \text{ (the true mean yield fails to exceed 20,000)}$$

and the alternative

$$A : \mu > 20,000 \text{ (the true mean yield exceeds 20,000)}$$

What is a reasonable significance level for this test? Well, in this case, the Type I error (rejecting H when it's really true) would be to conclude that the fertilizer produced the desired level of yield when, in fact, it did not. The Type II error, on the other hand, would be to say that Nitro-Plus does not meet the specifications when, in fact, it really does. We might be able to agree that the Type I is the worse error from the farmer's viewpoint for it would eventually lead to financial failure of the operation, whereas the Type II would only prevent us from taking advantage of an effective new fertilizer. Suppose, therefore, we set the significance level low, say $\alpha = .01$.

Now, how do we use the data to decide whether to accept or reject the hypothesis H? If we recall the statement of the central limit theorem that appears in Sec. 4.B, we know that the collection of all possible sample means $\bar{x}$ of size $n = 40$ comprises a set of normally distributed data having mean μ and standard deviation $s/\sqrt{n}$. More precisely, we know that for all possible samples composed of $n = 40$ data points, the numbers

$$z = \frac{\bar{x} - \mu}{s/\sqrt{n}}$$

have the standard normal distribution. Assuming that $\mu = 20,000$ (namely, that H is really true), under what circumstances would we be misled into rejecting H anyway? We would tend to reject H if the sample mean $\bar{x}$ came out to be substantially larger than $\mu = 20,000$.

What do we mean by "substantially larger"? Well, this is the point at which the significance level of the test enters the picture. As we have noted that a reasonable level of significance here would be $\alpha = .01$, this means that we want to reject H only 1% of the time when it, H, is really true. Because we would tend to reject H if $\bar{x}$ turned out to be much larger than $\mu = 20,000$, the rejection of H would be indicated by a large value of

$$z = \frac{\bar{x} - 20,000}{s/\sqrt{n}}$$

because z would be large when $\bar{x}$ is much larger than 20,000. (z is actually a standardized measure of the difference between $\bar{x}$ and 20,000.) Now, when H is really true, there is a 1% chance that z would exceed $z_{.01} = 2.33$, if we recall the discussion of percentage points of the normal distribution in Sec. 3.E. Therefore, if we agree to reject H if

$$z = \frac{\bar{x} - 20,000}{s/\sqrt{n}}$$

turns out to be larger than $z_{.01} = 2.33$, then we will have only a 1% chance of rejecting the hypothesis H that $\mu = 20,000$ when it is really true. This means that our test of whether or not the true mean yield per acre of tomatoes using Nitro-Plus fertilizer really exceeds 20,000 pounds will have significance level $\alpha = .01$.

The set of values of z which leads us to reject H is called the "rejection region" of the test. In our example involving tomato yield using Nitro-Plus fertilizer, we would reject H if z came out to be larger than $z_{.01} = 2.33$. Therefore, the rejection region of this test is the region where $z > 2.33$. A diagram with the rejection region shaded appears in Fig. 5.2. In view of our discussion, we can formalize the hypothesis testing mechanism for this problem as follows: for

$$H : \mu = 20,000$$
$$A : \mu > 20,000$$

compute
$$z = \frac{\bar{x} - 20,000}{s/\sqrt{n}}$$

and reject H at level $\alpha = .01$ if

$$z > z_{.01} = 2.33$$

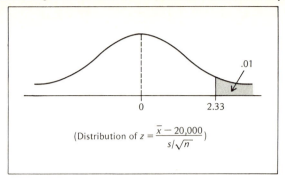

FIGURE 5.2
Rejection region for H when $\alpha = .01$ (tomatoes-fertilizer example).

$$(\text{Distribution of } z = \frac{\bar{x} - 20,000}{s/\sqrt{n}})$$

To complete this problem and to make our decision regarding the economic feasibility of the Nitro-Plus fertilizer, it remains only to calculate z and to decide whether or not it exceeds 2.33.

From the data accumulated by the independent testing organization, a study of $n = 40$ randomly selected acres gave a sample mean yield of $\bar{x} = 20,400$ with a sample standard deviation of $s = 1200$. Inserting these numbers into the formula for z, we find that

$$z = \frac{\bar{x} - 20,000}{s/\sqrt{n}} = \frac{20,400 - 20,000}{1200/\sqrt{40}} = \frac{400}{1200/6.325} = \frac{(400)(6.325)}{1200}$$
$$= 2.11$$

As we agreed to reject H if $z > 2.33$ and z actually turned out to be 2.11 which is less than 2.33, we cannot reject H. Therefore, at significance level $\alpha = .01$, we cannot conclude that the true mean yield exceeds 20,000 pounds per acre. Our experimental yields have not been large enough to provide convincing proof of the economic feasibility of Nitro-Plus.

Suppose we had looked at the problem from a different point of view. Suppose the economic goal of running a profitable operation had been outweighed by the necessity of producing a larger food supply as soon as possible. In such a situation, we would probably be less concerned with the precise level of mean yield and more concerned merely with the ability of Nitro-Plus to produce a large yield. In testing

$$H : \mu = 20,000$$
against
$$A : \mu > 20,000$$

we would therefore be likely to consider the Type II error (accepting H when it's really false) to be the worse, for it would lead us into rejecting the use of Nitro-Plus in some cases when the fertilizer would be truly effective. In our desire to avoid a Type II error, we would then be willing to ease up somewhat on the probability of a Type I error, perhaps by raising the significance level from $\alpha = .01$ to $\alpha = .05$. At significance level

$\alpha = .05$, our rejection rule would be to reject H in favor of A if

$$z > z_{.05} = 1.64$$

where
$$z = \frac{\bar{x} - 20,000}{s/\sqrt{n}}$$

as before. The rejection region is illustrated in the diagram of Fig. 5.3.

In viewing the problem from this different point of view, the only thing that has changed is the level of significance α. The data, of course, remain the same; our opinion of the problem will have no effect on the data, for they record only how many pounds of tomatoes are actually produced using the Nitro-Plus fertilizer. Therefore we still have $n = 40$, $\bar{x} = 20,400$, and $s = 1200$, so that

$$z = \frac{\bar{x} - 20,000}{s/\sqrt{n}} = \frac{20,400 - 20,000}{1200/\sqrt{40}} = 2.11$$

as before. Now, however, we have agreed to reject H if $z > 1.64$. Because $z = 2.11$ has turned out to be larger than 1.64, we would therefore make the decision to reject the hypothesis

$$H : \mu = 20,000 \text{ (the true mean yield fails to exceed 20,000)}$$

in favor of the alternative

$$A : \mu > 20,000 \text{ (the true mean yield exceeds 20,000)}$$

In the circumstances of the decline in importance of the precise economic break-even point and the parallel rise in importance of the food output itself, our conclusion would then be that Nitro-Plus increases tomato output to an extent sufficient to justify its use. At significance level $\alpha = .05$, our experimental results indicate that Nitro-Plus will be able to produce yields in excess of 20,000 pounds per acre. A tabular comparison between $\alpha = .01$ and $\alpha = .05$ is presented in Table 5.2.

FIGURE 5.3
Rejection region for H when $\alpha = .05$ (tomatoes-fertilizer example).

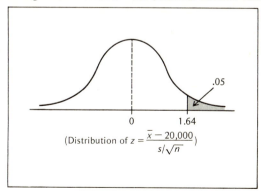

$$\left(\text{Distribution of } z = \frac{\bar{x} - 20,000}{s/\sqrt{n}}\right)$$

TABLE 5.2
Decision Making in the Tomatoes-Fertilizer Example
$H : \mu = 20,000; A : \mu > 20,000$

Significance Level	Rejection Rule	Calculated Value of z	Decision
$\alpha = .01$	$z > 2.33$	$z = 2.11$	Accept H
$\alpha = .05$	$z > 1.64$	$z = 2.11$	Reject H

The statistical test we have just conducted in our analysis of the tomatoes-fertilizer example is called the "one-sample z test." It is a one-sample test because only one set of data is used in the calculations, and it is a z test because the availability of more than 30 data points (here $n = 40$) allows us to apply the central limit theorem for the purpose of using percentage points of the standard normal distribution. One-sample tests for the mean of a population are characterized by the use of three distinct means: The true mean μ of the population, the sample mean $\bar{x}$ based on the data, and finally the "hypothesized mean" which we shall denote by the symbol μ_0 ("mew-zero"). In the tomatoes-fertilizer example, the hypothesized mean is $\mu_0 = 20,000$, for the hypothesis reads

$$H : \mu = 20,000$$

The statement that $H : \mu = 20,000$ does not really say that the true mean μ is 20,000, but merely that we are hypothesizing that it is. We therefore refer to 20,000 as the "hypothesized mean" and we denote it by μ_0. The sample mean is $\bar{x} = 20,400$, and the numerical value of the true mean μ is, as always, unknown to us.

The format of the one-sample z test is often as it was in the tomatoes-fertilizer example. In particular, we test the hypothesis

$$H : \mu = \mu_0$$

against the alternative

$$A : \mu > \mu_0$$

We compute

$$z = \frac{\bar{x} - \mu_0}{s/\sqrt{n}} \tag{5.1}$$

and we reject H at level α if $z > z_\alpha$. Sometimes, however, the alternative is formulated differently because of the nature of the question being asked. The following example provides an illustration.

Example 5.1 Calculator Battery Lifetimes One manufacturer of pocket calculators advertises that its battery pack allows the calculator to operate continuously for 22 hours on the average without recharging. To check on the validity of the manufacturer's claim, a prospective corporate customer tested a sample of 50 calculators, and these turned out to have mean 21.8 hours with a standard deviation of 0.9 hour. If a significance level of $\alpha = .10$ seems to be appropriate, is there sufficient evidence to indicate that the true mean duration of continuous operation (without recharging) is actually less than the 22 hours claimed by the manufacturer?

SOLUTION   Denoting by μ the true mean duration of continuous operation (without recharging) of the battery pack, we must test the hypothesis

$$H : \mu = 22$$
$$A : \mu < 22$$

at the level of significance $\alpha = .10$. Because we have a single sample of data and $n = 50 > 30$ data points, the appropriate method of analysis is again the one-sample z test. We compute, as before,

$$z = \frac{\bar{x} - 22}{s/\sqrt{n}}$$

and we must determine which values of z will lead us to reject H in favor of A. We will be inclined to reject H if $\bar{x}$ turns out to be considerably smaller than 22, relatively speaking, because if our sample mean comes out quite a bit lower than 22, we would have a hard time convincing ourselves that the true mean was really 22. Yet, as $\alpha = .10$ here, we would be making a Type I error that 10% of the time z actually falls below $-z_{.10} = -1.28$, even when μ is really 22. Therefore, if we decide to reject

$$H : \mu = 22$$

in favor of

$$A : \mu < 22$$

if z turns out to be less than (even more negative than) $-z_{.10} = -1.28$, our test would have significance level $\alpha = .10$, indicating a 10% chance of a Type I error. We therefore agree to reject H at level $\alpha = .10$ in this situation if $z < -1.28$. A diagram of the rejection region is presented in Fig. 5.4.

FIGURE 5.4
Rejection region for H when $\alpha = .10$ (calculator battery example).

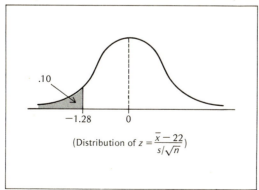

It now remains only to calculate the numerical value of z and to determine whether or not it is more negative than -1.28. From the data, we see that

$$n = 50$$
$$\bar{x} = 21.8$$
$$s = .9$$

We therefore calculate z as follows:

$$z = \frac{\bar{x} - 22}{s/\sqrt{n}} = \frac{21.8 - 22}{.9/\sqrt{50}} = \frac{-.2}{.9/7.07} = \frac{(-.2)(7.07)}{.9} = -1.57$$

We have agreed to reject H if z turns out to be less than (more negative than) -1.28, and now we see that z which equals -1.57 actually has turned out to be less than -1.28. Therefore, as $z = -1.57 < -1.28$, we must reject

$$H : \mu = 22$$

in favor of

$$A : \mu < 22$$

As level $\alpha = .10$, our conclusion is that the mean duration of continuous operation (without recharging) of the battery pack is actually less than the 22 hours claimed by the manufacturer.

The two situations studied before, namely, the tomatoes-fertilizer example and the calculator battery example, called for the use of the one-sample z test. As we have already noted, we need a one-sample test in problems where a single set of data is compared with a predetermined number, in the above two cases the hypothesized mean. The z in the one-sample z test refers to the fact that more than 30 data points are available, so that we can base our analysis of the problem on the z scores of the normal distribution. What happens in a one-sample problem where fewer than 30 data points are available? A brief glance at the discussion near the beginning of Sec. 4.C should provide the answer. In situations where the underlying population (from which the sample of data points is randomly selected) is normally distributed, we can replace the rejection regions based on the z scores of Appendix Table A.3 by rejection regions based on the t distribution of Table A.4. In cases where the underlying population cannot reasonably be considered to have the normal distribution, the use of Table A.4 cannot ordinarily be justified and an entirely different approach is required. Details of this different approach, "nonparametric statistics," are postponed to Chap. 11. Here we shall study how to handle one-sample problems involving fewer than 30 data points by applying the one-sample t test.

The following two examples illustrate the use of the one-sample t test.

Example 5.2 Body Temperatures A medical doctor who is also an amateur anthropologist is interested in finding out whether the average body temperature of Alaskan Eskimos is significantly lower than the usual American average, which is 98.6°F. Eight Eskimos selected at random from the State of Alaska census lists had the following recorded body temperatures in degrees:

$$98.5 \quad 98.1 \quad 98.6 \quad 98.7 \quad 98.4 \quad 98.9 \quad 98.0 \quad 98.4$$

At significance level $\alpha = .05$, do the results of the study support the assertion that Eskimos really have lower body temperatures than Americans who are natives of warmer climates?

SOLUTION   Because $n = 8 < 30$, the solution of this problem requires the use of the one-sample t test. If we denote by μ the true mean body temperature of Alaskan Eskimos, we want to test the hypothesis

$$H : \mu = 98.6 \text{ (Eskimos have the same average body}$$
$$\text{temperature as other Americans)}$$

against the alternative

$$A : \mu < 98.6 \text{ (Eskimos have lower body temperature)}$$

at the level of significance $\alpha = .05$. If we were using the z test, we would reject H if $z < -z_{.05} = -1.64$. However, as we are using the t test, we replace $z_{.05}$ by $t_{.05}(n - 1)$ where $n - 1$ is the degrees of freedom required in using Table A.4. Therefore, as $n = 8$, we have df $= n - 1 = 8 - 1 = 7$. We compute

●
$$t = \frac{\bar{x} - \mu_0}{s/\sqrt{n}} \tag{5.2}$$

in which $\mu_0 = 98.6$ here, and we reject H in favor of A if $t < -t_{.05}(7) = -1.895$. We illustrate the relevant portion of Table A.4 in Table 5.3 and the rejection region in Fig. 5.5.

It remains, then, only to compute the value of

$$t = \frac{\bar{x} - 98.6}{s/\sqrt{n}}$$

and to check whether or not $t < -1.895$. We can insert the value $n = 8$ immediately, but the numerical values of $\bar{x}$ and s will have to be computed directly from the data. As in Chaps. 2 and 4, we will set up a table of calculations to do the job. The calculations, which can be found in Table 5.4, show that $\bar{x} = 98.45$ and $s = 0.298$. Substitution of

TABLE 5.3
A Small Portion
of Table A.4

df	$t_{.05}$
7	1.895

FIGURE 5.5
Rejection region for *H* when $\alpha = .05$ (body temperature example).

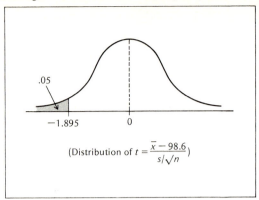

(Distribution of $t = \dfrac{\bar{x} - 98.6}{s/\sqrt{n}}$)

these numbers into the formula for *t* yields

$$t = \frac{\bar{x} - 98.6}{s/\sqrt{n}} = \frac{98.45 - 98.6}{.298/\sqrt{8}} = \frac{-.15}{.298/2.83} = \frac{(-.15)(2.83)}{.298} = -1.42$$

Now, recall that we agreed to reject *H* in favor of *A* if *t* turned out to be less than -1.895. Well, as it turns out, $t = -1.42$ which is *not* less than -1.895, because it is less negative and, therefore, larger. Therefore, since $t = -1.42 > -1.895$, we *cannot* reject

TABLE 5.4
Calculations Leading to the Mean and
Standard Deviation of the Body
Temperature Data

x	$x - \bar{x}$	$(x - \bar{x})^2$
98.5	.05	.0025
98.1	−.35	.1225
98.6	.15	.0225
98.7	.25	.0625
98.4	−.05	.0025
98.9	.45	.2025
98.0	−.45	.2025
98.4	−.05	.0025
Sums 787.6	.00	.6200

$n = 8$

$$\bar{x} = \frac{\Sigma x}{n} = \frac{787.6}{8} = 98.45$$

$$s = \sqrt{\frac{\Sigma(x - \bar{x})^2}{n - 1}} = \sqrt{\frac{.62}{7}} = \sqrt{.0886} = .298$$

H in favor of *A*. We conclude at level $\alpha = .05$ that the data *do not support* the assertion that Eskimos have lower body temperatures, on the average, than Americans of warmer climates.

Example 5.3 Depth Perception A psychologist, conducting a study of the average person's ability to judge distances, sets up a test of depth perception in which randomly selected individuals attempt to estimate the distance between two markers. The markers were actually 2.5 feet apart, while the 10 participants in the study gave the following estimates:

$$2.1 \quad 1.8 \quad 2.3 \quad 2.3 \quad 2.6 \quad 2.5 \quad 2.3 \quad 2.5 \quad 2.1 \quad 2.5$$

At level $\alpha = .05$, do the results of the study indicate that persons have difficulty in accurately estimating the correct distance?

SOLUTION   A one-sample *t* test is required for the solution of this problem in view of the fact that we have fewer than 30 data points available, namely 10. We use μ to represent the true mean estimate of the distance between the markers by the entire population in question, which is probably all adults resident in some geographical location. We then test the hypothesis

$$H : \mu = 2.5 \text{ (on the average, persons can accurately}$$
$$\text{estimate the distance between the markers)}$$

against the alternative

$$A : \mu \neq 2.5 \text{ (on the average, persons have difficulty}$$
$$\text{in accurately estimating the distance)}$$

Under what circumstances would we feel compelled to reject *H* in favor of *A*? Well, as $\bar{x}$ is an approximation of μ, we would be inclined to reject *H* if the number

$$t = \frac{\bar{x} - 2.5}{s/\sqrt{n}}$$

turned out to be either too large or too small because both these extremes would tend to indicate that μ is relatively far from 2.5.

For a significance level of $\alpha = .05$, we are willing to make a Type I error 5% of the time; this means that a rejection region of the sort pictured in Figs. 5.2 to 5.5 must have a shaded region representing a probability of .05. In the present example, however, the rejection region would have to appear in two parts, one part at each extreme of the graph of the *t* distribution. It would make sense to organize the rejection region in such a way that the far right part and the far left part each account for 2.5% of the possible values of *t*. The alternative $A : \mu \neq 2.5$ does not allow us to favor one extreme over the other. We would therefore be using the 2.5% percentage point of the *t* distribution, the number $t_{.025}(n - 1) = t_{.025}(9) = 2.262$, to yield our points of division between the rejection and acceptance regions. A small portion of Table A.4 is reproduced in Table 5.5 to point up the selection of $t_{.025}(9)$. In testing the hypothesis

$$H : \mu = 2.5$$

TABLE 5.5
A Small Portion of
Table A.4

df	$t_{.05}$	$t_{.025}$
9	1.833	2.262

against

$$A : \mu \neq 2.5$$

we would therefore reject H at level $\alpha = .05$ if either t were more negative than $-t_{.025}(9)$ $= -2.262$ or if t were greater than $t_{.025}(9) = 2.262$. In other words, we would compute

$$t = \frac{\bar{x} - 2.5}{x/\sqrt{n}}$$

and we would reject H in favor of A if $t < -2.262$ or if $t > 2.262$. We can express the two-part rejection rule more compactly in terms of the absolute value of t: We reject H if $|t| > 2.262$. The rejection region is illustrated in the diagram of Fig. 5.6.

Now that we have agreed to reject $H : \mu = 2.5$ in favor of $A : \mu \neq 2.5$ if $|t| > 2.262$, it remains only to compute the numerical value of

$$t = \frac{\bar{x} - 2.5}{s/\sqrt{n}}$$

and to observe whether or not its absolute value exceeds 2.262. The preliminary calculations appearing in Table 5.6 are aimed at obtaining the numerical values of $\bar{x}$ and s directly from the psychologist's 10 data points. From those calculations, we discover that $\bar{x} = 2.3$ and $s = 0.245$. It should be observed that we have used the short-cut formula for computing the standard deviation, namely, formula (4.3) which was introduced in Sec. 4.B. There is no reason why it is necessary to use the short-cut formula in

FIGURE 5.6
Rejection region for H when $\alpha = .05$ (depth perception example).

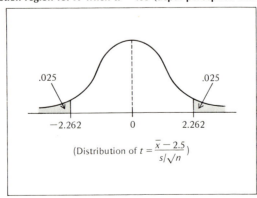

TABLE 5.6
Calculations Leading to the Mean and
Standard Deviations of the
Depth Perception Data
Shortcut Formula

x	x^2
2.1	4.41
1.8	3.24
2.3	5.29
2.3	5.29
2.6	6.76
2.5	6.25
2.3	5.29
2.5	6.25
2.1	4.41
2.5	6.25
23.0	53.44

$n = 10$

$$\bar{x} = \frac{\Sigma x}{n} = \frac{23.0}{10} = 2.3$$

$$s = \sqrt{\frac{n \Sigma x^2 - (\Sigma x)^2}{n(n-1)}} = \sqrt{\frac{10(53.44) - (23.0)^2}{(10)(9)}}$$

$$= \sqrt{\frac{534.4 - 529}{90}} = \sqrt{\frac{5.4}{90}} = \sqrt{.06} = .245$$

this example and no reason why it is to be preferred to the direct method used in the calculations of Table 5.4. We merely use it to provide an illustration of the options available to the student. Those persons having access to a calculator with memory might be able to simultaneously retain running totals of both the x and x^2 columns without writing down every single entry in the table. Calculation of s will then be speeded up considerably, and for this reason the short-cut formula for s is sometimes referred to as the "calculator formula for s."

Having accumulated the information that $n = 10, \bar{x} = 2.3$, and $s = 0.245$, the next step is to insert these numbers into the formula for t. Carrying out this procedure, we get

$$t = \frac{\bar{x} - 2.5}{s/\sqrt{n}} = \frac{2.3 - 2.5}{.245/\sqrt{10}} = \frac{-.2}{.245/3.162} = \frac{(-.2)(3.162)}{.245} = -2.58$$

We agreed to reject H if $|t| > 2.262$. Because $t = -2.58$, it follows that $|t| = 2.58$ which indeed exceeds 2.262, and we should therefore reject H. We conclude at level $\alpha = .05$ that persons generally have difficulty in accurately measuring distance, because our analysis of the psychologist's study reveals that, for markers placed 2.5 feet apart, the average person will tend to estimate the distance as being significantly different from 2.5.

TABLE 5.7
Rejection Rules in the Presence of Various
Alternatives for Tests Involving
Population Means

$$H : \mu = \mu_0$$

Alternative	Reject H in Favor of A at Significance Level α if:
$A : \mu > \mu_0$	$z > z_\alpha$
$A : \mu < \mu_0$	$z < -z_\alpha$
$A : \mu \neq \mu_0$	$|z| > z_{\alpha/2}$

NOTE: Although the above information is given in the language of the one-sample z test, it is necessary merely to replace the z_α and $z_{\alpha/2}$ by $t_\alpha(n-1)$ and $t_{\alpha/2}(n-1)$ in order to obtain the rejection rules for the one-sample t test.

The depth perception example has provided an illustration of what is called a "two-tailed t test" (not to be confused with the "two-taled tee test" used in resolving disputes over golfing scores). A two-tailed test is a statistical test having a two-part rejection region of the sort pictured in Fig. 5.6. Such a rejection region arises in tests of population means when the alternative is of the form

$$A : \mu \neq \mu_0$$

rather than $A : \mu > \mu_0$ or $A : \mu < \mu_0$. For a two-tailed test having significance level α, we reject H if $|t| > t_{\alpha/2}(n-1)$, using $\alpha/2$ instead of α to locate the rejection region. For quick reference in dealing with situations like those discussed in the present section and also those of the next two sections, we can collect the various possible alternatives together with the rejection rules corresponding to them. This listing of possible alternatives appears in Table 5.7, and use of that table will make it unnecessary to draw diagrams such as Figs. 5.2 to 5.6 for each and every problem that comes up. Table 5.7 incorporates the ideas that went into the determination of each of the rejection regions pictured above. We shall have occasion to refer to Table 5.7 during the next two sections of the present chapter and also when we study some sections of Chap. 11.

EXERCISES 5.B

1 In the 1950s, the average number of children per family in the United States was 2.35. One sociologist is interested in finding out whether the number of children per family has declined from that level during the present decade. He selects a random sample of 81 families from around the nation and discovers that they have 2.17 children on the average with a standard deviation of 0.9 children. Do the data support at level $\alpha = .05$ the contention that families of the 1970s have on the average fewer children than families of the 1950s?

2 An automobile manufacturer claims that his cars use an average of 4.50 gallons of gasoline for each 100 miles. A consumer organization tests 36 of the cars and finds that a mean of 4.65 gallons with a standard deviation of 0.36 gallon is used for each 100 miles by the cars tested. At significance level $\alpha = .01$, do the data provide evidence that the car really uses more than 4.50 gallons on the average per 100 miles?

3 A machine that puts out plastic piping is, according to contractual specifications, supposed to be producing pipe of diameter 4 inches. Piping that is either too large or too small will not meet the construction requirements. As part of a routine quality control test, a random sample of 50 pieces were carefully measured, and their diameters had mean 4.05 inches and standard deviation 0.12 inch. At level $\alpha = .05$, can we conclude that the true mean diameter differs significantly from 4 inches?

4 It is the policy of one school district to hire credentialed reading specialists whenever the true mean reading score of the district's sixth-graders falls below 40, as measured by a particular standard test. If a random sample of 25 pupils had mean reading score 38 with a standard deviation of 3, is the hiring of reading specialists justified
a At significance level $\alpha = .05$?
b At significance level $\alpha = .005$?

5 A tobacco processor certified that his cigarettes contained no more than 21 milligrams of nicotine on the average. A random sample of 25 cigarettes was analyzed by an independent medical researcher and found to have a mean of 22.6 milligrams of nicotine with a standard deviation of 3 milligrams. At level $\alpha = .05$, do the data provide enough evidence to assert that the processor's certification was fallacious?

6 A small factory produces timing devices under subcontract to a major aerospace company. Ideally, the devices have a mean timing advance of 0.00 second per week. (A *negative* time advance is called a time lag.) A sample of six devices, tested as they came off the assembly line, had the following advances:

Serial no. of device	00693A	01014F	01983B	04441W	15672P	18449G
Time advance as tested	0.1	0.2	−0.1	0.2	−0.1	0.0

At significance level $\alpha = .10$, do the data provide evidence that the true mean advance is different from 0.00, so that the production process requires some adjustment?

7 If a new, slightly cheaper process for mining copper is adopted, it will not be economical if it yields fewer than 50 tons of ore per day in an experimental mine. The results of a 5-day trial period are summarized by the following daily production figures (in tons):

$$50 \quad 47 \quad 53 \quad 51 \quad 52$$

At significance level $\alpha = .01$, do the results indicate that the mean production does not fall below 50 tons per day, and so the new process is economical?

8 A prestigious financial publication asserted in late May of 1970 that leading economists thought that the 1971 Consumer Price Index would fall below 140. Is this assertion supported at significance level $\alpha = .10$ by the data of Exercise 1.A.6?

SECTION 5.C TWO-SAMPLE TESTS FOR THE DIFFERENCE OF MEANS

A local board of education, with a view toward eventually cutting costs, conducts a comparison test of the length of time that light bulbs last before requiring replacement. A sample of 40 randomly selected Universal Electric bulbs, the brand currently used, has a mean lifetime of 1110 hours with a standard deviation of 20 hours. A cheaper brand (Light, Ltd.) which is under consideration as a replacement is also tested, and a random sample of 60 bulbs turns out to have a mean lifetime of 1081 hours with a standard deviation of 18 hours. The Board of Education would like to use the cheaper bulbs unless the results of the comparison test indicate strongly that the cheaper ones are significantly worse.

In answering this question, it is apparent that we are dealing with two distinct samples of data—the 40 Universal Electric bulbs and the 60 Light, Ltd. bulbs. We are comparing these two samples against each other rather than against a hypothesized mean, as was done in the previous section. Because more than 30 data points are available, the appropriate statistical test to use is the two-sample z test. If we set

$$\mu_U = \text{true mean lifetime of Universal Electric bulbs}$$

and $$\mu_L = \text{true mean lifetime of Light, Ltd. bulbs}$$

then the Board of Education wants to test the hypothesis

$$H : \mu_L = \mu_U \text{ (there is no real difference between the cheaper and the more expensive bulbs)}$$

versus

$$A : \mu_L < \mu_U \text{ (the cheaper bulbs are really worse in length of lifetime)}$$

The formula for the two-sample z test for the difference of means is

$$z = \frac{\bar{x}_L - \bar{x}_U}{\sqrt{\dfrac{s_L^2}{n_L} + \dfrac{s_U^2}{n_U}}} \tag{5.3}$$

where
$n_L = $ number of Light, Ltd. bulbs in the sample
$n_U = $ number of Universal Electric bulbs in the sample
$\bar{x}_L = $ sample mean of Light, Ltd. bulbs
$\bar{x}_U = $ sample mean of Universal Electric bulbs

s_L = sample standard deviation of Light, Ltd. bulbs
s_U = sample standard deviation of Universal Electric bulbs

(Here the denominator of z, called the "pooled" standard deviation of the two samples of data, plays the role of $s/\sqrt{n} = \sqrt{s^2/n}$.) By analogy with the second in the list of alternatives presented in Table 5.7, we should reject $H : \mu_L = \mu_U$ in favor of $A : \mu_L < \mu_U$ at significance level α if $z < -z_\alpha$.

What would be an appropriate level of significance for this problem? Well, because the Light, Ltd. bulbs are cheaper, the Board would prefer to use that brand unless it were convincingly demonstrated that they are really worse. This means that they would not like to reject H unless it were really false. Therefore, a test procedure with a very low probability of making a Type I error would be quite appealing to the board. Since the significance level α is the probability of a Type I error, α should be chosen to be very small, say $\alpha = .005$. For this value of α, we would have

$$z = z_{.005} = 2.57$$

It follows that we should reject H in favor of A at level $\alpha = .005$, concluding thereby that the cheaper bulbs are really worse than the more expensive ones, if

$$z < -2.57$$

We are now ready to proceed to the computation of the numerical value of z.
From the data, we know that

$$n_L = 60$$
$$n_U = 40$$
$$\bar{x}_L = 1081$$
$$\bar{x}_U = 1110$$
$$s_L = 18$$
$$s_U = 20$$

Inserting these numbers in their proper places in the formula for z, we get

$$z = \frac{\bar{x}_L - \bar{x}_U}{\sqrt{\dfrac{s_L^2}{n_L} + \dfrac{s_U^2}{n_U}}} = \frac{1081 - 1110}{\sqrt{\dfrac{(18)^2}{60} + \dfrac{(20)^2}{40}}} = \frac{-29}{\sqrt{\dfrac{324}{60} + \dfrac{400}{40}}} = \frac{-29}{\sqrt{5.4 + 10.0}}$$

$$= \frac{-29}{\sqrt{15.4}} = \frac{-29}{3.29} = -7.39$$

We agreed to reject H if $z < -2.57$. As it turned out, $z = -7.39 < -2.57$, and so we must reject H. Our conclusion is that, at level $\alpha = .005$, the cheaper Light, Ltd. bulbs are really worse than the Universal Electric bulbs in the sense that they last significantly fewer hours. On the basis of the information gained from this study, we would advise the Board to continue using Universal Electric light bulbs.

We complete this section by presenting two examples of the two-sample t test, the version of the two-sample test which is used when only small samples of data are available.

Example 5.4 Pest Control A state department of public health, faced with a serious infestation of houseflies, decides to conduct a pilot test of effectiveness between two methods of insect control, the organic method and the chemical method. The organic method, which consists of saturating a community with nonpoisonous spiders, is applied in one neighborhood, while the chemical method, using a combination of poisonous sprays and bait, is applied in a second neighborhood comparable to the first. Useful data are collected on six houses in the first neighborhood and on nine in the second during a 48-hour test period. The data are obtained by counting the number of houseflies visually observed making nuisances of themselves during the test period. In the first neighborhood, the one treated by the organic method, the six houses checked had, respectively, 41, 20, 19, 36, 38, and 26 houseflies roaming free. In the second neighborhood, the one treated by the chemical method, the nine houses checked had, respectively, 9, 26, 16, 10, 31, 28, 35, 15, and 10 houseflies roaming free. At significance level $\alpha = .01$, do the test results indicate that the chemical method is more effective than the organic method in reducing the number of houseflies roaming free throughout the neighborhood?

SOLUTION   We must use the two-sample t test because of the relatively small number of data points available, six in one sample and nine in the other. We define the following symbols for the true means relevant to the problem:

$$\mu_o = \text{true mean number of flies successfully evading}$$
$$\text{the organic method of insect control}$$

$$\mu_c = \text{true mean number of flies successfully evading}$$
$$\text{the chemical method of insect control}$$

We then want to test the hypothesis

$$H : \mu_o = \mu_c \text{ (there is no difference in effectiveness}$$
$$\text{between the organic and chemical}$$
$$\text{methods)}$$

$$A : \mu_o > \mu_c \text{ (the chemical method is more effective)}$$

The number of degrees of freedom (df) appropriate to the two-sample t test is

$$df = n_o + n_c - 2$$

where n_o = number of data points in the first sample (organic)
                 n_c = number of data points in the second sample (chemical)

In this problem $n_o = 6$ and $n_c = 9$, so that

$$df = n_o + n_c - 2 = 6 + 9 - 2 = 13$$

Because our alternative is $A : \mu_o > \mu_c$, a glance at Table 5.7 on page 152 shows that we should reject H in favor of A if

$$t > t_\alpha(n_o + n_c - 2) = t_{.01}(13) = 2.650$$

in view of the fact that our significance level here is $\alpha = .01$.

The only task remaining in the solution of this problem is the computation of the numerical value of t, which we have to compare with $t_{.01}(13) = 2.650$. The value of t in the two-sample t test for the difference of means is calculated using the formula

$$t = \frac{\bar{x}_o - \bar{x}_c}{\sqrt{\frac{\Sigma(x_o - \bar{x}_o)^2 + \Sigma(x_c - \bar{x}_c)^2}{n_o + n_c - 2}\left(\frac{1}{n_o} + \frac{1}{n_c}\right)}} \quad (5.4)$$

where
$\bar{x}_o$ = sample mean number of houseflies successfully evading organic method of control

$\bar{x}_c$ = sample mean number of houseflies successfully evading chemical method of control

The quantities $\Sigma(x_o - \bar{x}_o)^2$ and $\Sigma(x_c - \bar{x}_c)^2$ are the sums of squared deviations from the mean, analogous to the number computed by summing the third column of Table 5.4, for example. We calculate the components of the formula for t by constructing two tables like that of Table 5.4, one such table for each of the two samples of data. The calculations are presented in Table 5.8. From the information presented in Table 5.8, we see that the components of the formula for t are

$$n_o = 6$$
$$n_c = 9$$
$$\bar{x}_o = 30$$
$$\bar{x}_c = 20$$
$$\Sigma(\bar{x}_o - \bar{x}_o)^2 = 458$$
$$\Sigma(\bar{x}_c - \bar{x}_c)^2 = 808$$

TABLE 5.8
Calculations Leading to the Two-Sample t Statistic
Pest Control Data

Organic Method of Control			Chemical Method of Control		
x_o	$x_o - \bar{x}_o$	$(x_o - \bar{x}_o)^2$	x_c	$x_c - \bar{x}_c$	$(x_c - \bar{x}_c)^2$
41	11	121	9	−11	121
20	−10	100	26	6	36
19	−11	121	16	−4	16
36	6	36	10	−10	100
38	8	64	31	11	121
26	−4	16	28	8	64
180	0	458	35	15	225
			15	−5	25
			10	−10	100
			180	0	808

$n_o = 6$

$\bar{x}_o = \frac{180}{6} = 30$

$\Sigma(x_o - \bar{x}_o)^2 = 458$

$n_c = 9$

$\bar{x}_c = \frac{180}{9} = 20$

$\Sigma(x_c - \bar{x}_c)^2 = 808$

Inserting these numbers into their proper places in the formula for t, we obtain

$$t = \frac{\bar{x}_O - \bar{x}_C}{\sqrt{\dfrac{\Sigma(x_O - \bar{x}_O)^2 + \Sigma(x_C - \bar{x}_C)^2}{n_O + n_C - 2}\left(\dfrac{1}{n_O} + \dfrac{1}{n_C}\right)}} = \frac{30 - 20}{\sqrt{\dfrac{458 + 808}{6 + 9 - 2}\left(\dfrac{1}{6} + \dfrac{1}{9}\right)}}$$

$$= \frac{10}{\sqrt{\left(\dfrac{1266}{13}\right)\left(\dfrac{15}{54}\right)}} = \frac{10}{\sqrt{\dfrac{18,990}{702}}} = \frac{10}{\sqrt{27.05}} = \frac{10}{5.2} = 1.923$$

We agreed to reject

$$H : \mu_O = \mu_C$$

in favor of

$$A : \mu_O > \mu_C$$

if t turned out to be larger than $t_{.01}(13) = 2.650$. However, $t = 1.923 < 2.650$, so that we cannot reject H. At significance level $\alpha = .01$, we therefore conclude that the data do not provide evidence sufficient to assure us that the chemical method of pest control is more effective than the organic method.

Example 5.5 **Snake Sizes**   One rare species of desert snake can be found in both Clark County, Nevada, and San Bernardino County, California. A herpetologist wants to compare the sizes of the species in the two regions. She traps and measures seven snakes in Nevada and eight in California. The seven Nevada snakes average 3.4 feet in length with a standard deviation of 1.0 foot, while the eight California snakes have a mean length of 3.1 feet with a standard deviation of .8 foot. At significance level $\alpha = .05$, does the herpetologist have evidence sufficient to conclude that the Nevada and California snakes really differ in length?

SOLUTION   The problem calls for a two-sample t test of the hypothesis

$$H : \mu_N = \mu_C$$

against the alternative

$$A : \mu_N \neq \mu_C$$

where
$$\mu_N = \text{true mean length of Nevada snakes}$$
$$\mu_C = \text{true mean length of California snakes}$$

We have used the alternative $A : \mu_N \neq \mu_C$ because the herpetologist wanted to know only whether or not snake sizes differed in the two geographical locations. She was not interested in finding out which of the two areas had the larger snakes, but only whether the sizes were different in the two regions.

   The number of degrees of freedom for this problem is df $= n_N + n_C - 2 = 7 + 8 - 2 = 13$, and Table 5.7 requires us to reject H at level $\alpha = .05$ if

$$|t| > t_{\alpha/2}(\text{df}) = t_{.025}(13) = 2.160$$

because our alternative is $A : \mu_N \neq \mu_C$ in this case. It now remains only to compute the value of

$$t = \frac{\bar{x}_N - \bar{x}_C}{\sqrt{\dfrac{\Sigma(x_N - \bar{x}_N)^2 + \Sigma(x_C - \bar{x}_C)^2}{n_N + n_C - 2}\left(\dfrac{1}{n_N} + \dfrac{1}{n_C}\right)}}$$

and to check to see whether or not $|t| > 2.160$.

From the information given in the statement of the problem, we know the following quantities:

$$n_N = 7$$
$$n_C = 8$$
$$\bar{x}_N = 3.4$$
$$\bar{x}_C = 3.1$$
$$s_N = 1.0$$
$$s_C = 0.8$$

Inserting as many of these numbers into the formula for t as will fit, we obtain that

$$t = \frac{3.4 - 3.1}{\sqrt{\dfrac{\Sigma(x_N - \bar{x}_N)^2 + \Sigma(x_C - \bar{x}_C)^2}{7 + 8 - 2}\left(\dfrac{1}{7} + \dfrac{1}{8}\right)}}$$

Unfortunately, we cannot complete the calculation of t because we were not given the numerical values of $\Sigma(x_N - \bar{x}_N)^2$ and $\Sigma(x_C - \bar{x}_C)^2$. How are we to find out what these numbers are?

Well, how did we manage to do it in the last example? In that example, we had access to all the original data points; so we merely computed the sums of squared deviations from the mean directly. This computation was carried out explicitly in Table 5.8. This time we unfortunately do not know the original data points, but instead we know only how many there were and what their sample mean and standard deviation were. What are we going to do?

The quantities we need to know, namely, $\Sigma(x_N - \bar{x}_N)^2$ and $\Sigma(x_C - \bar{x}_C)^2$, are mathematically related to the sample standard deviations by the formulas

$$s_N = \sqrt{\frac{\Sigma(x_N - \bar{x}_N)^2}{n_N - 1}}$$

and

$$s_C = \sqrt{\frac{\Sigma(x_C - \bar{x}_C)^2}{x_C - 1}}$$

for those formulas would be used to compute s_N and s_C if we did know or were able to compute $\Sigma(x_N - \bar{x}_N)^2$ and $\Sigma(x_C - \bar{x}_C)^2$. Now, however, the shoe is on the other foot: We know s_N and s_C, but we want to know $\Sigma(x_N - \bar{x}_N)^2$ and $\Sigma(x_C - \bar{x}_C)^2$. Therefore, we develop a formula for calculating $\Sigma(x_N - \bar{x}_N)^2$ and $\Sigma(x_C - \bar{x}_C)^2$ based on our knowledge of the values of s_N and s_C.

In the above formulas for s_N and s_C, we can square both sides, obtaining thereby

$$s_N^2 = \left[\sqrt{\frac{\Sigma(x_N - \bar{x}_N)^2}{n_N - 1}}\right]^2 = \frac{\Sigma(x_N - \bar{x}_N)^2}{n_N - 1}$$

and

$$s_C^2 = \left[\sqrt{\frac{\Sigma(x_C - \bar{x}_C)^2}{n_C - 1}}\right]^2 = \frac{\Sigma(x_C - \bar{x}_C)^2}{n_C - 1}$$

It follows algebraically from the above results that

$$\Sigma(x_N - \bar{x}_N)^2 = (n_N - 1)s_N^2$$

and

● $$\Sigma(x_C - \bar{x}_C)^2 = (n_C - 1)s_C^2 \tag{5.5}$$

Using these expressions, we can indirectly calculate the numerical values of $\Sigma(s_N - \bar{x}_N)^2$ and $\Sigma(x_C - \bar{x}_C)^2$ that we need in order to complete the computation of t. In particular, we get that

$$\Sigma(x_N - \bar{x}_N)^2 = (n_N - 1)s_N^2 = (7 - 1)(1.0)^2 = (6)(1) = 6$$

and $$\Sigma(x_C - \bar{x}_C)^2 = (n_C - 1)s_C^2 = (8 - 1)(.8)^2 = (7)(.64) = 4.48$$

Inserting these numbers into their proper places in the expression for t, we see that

$$t = \frac{3.4 - 3.1}{\sqrt{\frac{6 + 4.48}{7 + 8 - 2}\left(\frac{1}{7} = \frac{1}{8}\right)}} = \frac{0.3}{\sqrt{\left(\frac{10.48}{13}\right)\left(\frac{15}{56}\right)}} = \frac{0.3}{\sqrt{\frac{157.2}{728}}} = \frac{0.3}{\sqrt{0.216}}$$

$$= \frac{0.3}{0.465} = 0.646$$

The numerical value of t therefore turned out to be $t = 0.646$, while, as you will recall, we agreed to reject H if $|t| > 2.160$. In view of the fact that $|t| = 0.646 < 2.160$, we cannot reject H at level of significance $\alpha = .05$. We therefore conclude that the data do not reveal a significant difference between the lengths of the Nevada and California snakes.

Conditions on the Use of the Two-Sample Tests

With this example, we have completed our discussion of two-sample tests for the difference of means. Before we go on to another topic, we should point out the conditions under which we may validly use these tests for the solution of applied problems. When large samples of data ($n > 30$) are available, the central limit theorem discussed in Sec. 4.B allows us to use the percentage points of the normal distribution in making a statistical decision. When only small samples are available, we must restrict ourselves to normally distributed data as we did in Sec. 4.C. This restriction is necessary only because the table of the t distribution is derived from the table of the normal distribution for the expressed purpose of taking care of small-sample situations. In connection with the two-sample t test, two further restrictions are required: (1) It must be reasonable for us to assume that both samples of data have approximately the same true standard deviation. (2) Both samples of data must be selected independently of each other; i.e., individual members of one sample must bear no relation to individual members of the other sample.

In summary, we use a two-sample t test for the difference of means when the following four criteria are met:

1 Fewer than 30 data points are available.
2 Both sets of data come from normally distributed populations.
3 Both sets of data have the same true population standard deviation.
4 The sets of data are independent of each other.

Whenever a problem seems to call for the two-sample t test, then, the above criteria must be carefully observed so that we can be sure that our solution of the problem is based on sound mathematical principles instead of on the shifting sands of convenience. In future sections and chapters, we therefore must also study ways of checking whether or not these criteria are satisfied in any particular problem of interest. Criterion 1, of course, can be checked merely by counting the number of available data points. For criterion 2, we will study in Sec. 9.C the "chi-square test of normality," the objective of which is to decide whether or not a set of randomly selected data has come from a population that is normally distributed. In Sec. 5.E, we will show how to check criterion 3 by testing for the difference of two standard deviations. Criterion 4 can be analyzed only by thinking about the nature of the problem under discussion and the way in which the data were gathered.

What do we do in two-sample problems when one or more of these criteria are not met? If criterion 1 is not met but all the others are, we can use the two-sample z test. If criteria 2 or 3 or both are not met, we must use a nonparametric test to study the differences between the sets of data. Finally, when all criteria except 4 are satisfied, we must use the "paired-sample t test," which we will present in the next section.

EXERCISES 5.C

1 A new chemical fertilizer, Nitro-Plus, yielded 20,400 pounds of tomatoes on the average with a standard deviation of 1200 pounds on 40 randomly selected acres of farmland. On another 100 randomly selected acres, the standard organic fertilizer produced a mean yield of 19,000 pounds with a standard deviation of 1000 pounds. At significance level $\alpha = .01$, do the results of the comparison indicate that the chemical fertilizer really produces larger yields than the organic fertilizer?

2 Two methods of teaching remedial reading to 10-year-old youngsters, the "Chicago" method and the "Boston" method, were applied to several randomly selected "slow readers" divided into two groups. On a reading test given to both groups at the conclusion of the teaching program, the 60 pupils exposed to the "Chicago" method scored a mean of 20 with a standard deviation of 7, while the 50 students using the "Boston" method had an average score of 22 with a standard deviation of 8. At level $\alpha = .05$, can we say that the two methods differ significantly in effectiveness?

3   In a study of whether or not cigarette smoke inhibits the growth of mice, 15 mice born at the same time were fed exactly the same diet in order to prepare them for the experiment. All other living conditions were identical except that 7 of the mice lived in an atmosphere heavily laden with cigarette smoke while the other 8 lived in a normal air environment. A biologist participating in the study collects the following data on weight gains of the mice over a period of time:

Weight Gains of Mice, grams

Atmosphere with Cigarette Smoke	Atmosphere of Normal Air
6	11
7	8
6	10
10	9
8	9
5	6
7	8
	11

At level $\alpha = .05$, are the data sufficient to conclude that cigarette smoke significantly inhibits the growth of mice?

4   Peanuts are known to be an excellent source of protein, and they are therefore a good substitute for meat and fish when prices are high. To find out whether roasted peanuts have less protein than raw peanuts, a nutritionist selected a random sample of 20 bags of peanuts for testing. She chose 8 at random and roasted them and then measured the protein content of all 20 bags. The data follow:

Protein Content of Bags of Peanuts, grams

Raw Peanuts	Roasted Peanuts
10	8
8	6
6	5
10	10
8	7
10	9
10	6
8	5
6	
10	
4	
6	

At significance level $\alpha = .01$, decide whether or not the roasting process reduces the protein content of peanuts.

5 An economist wants to know whether Commonwealth trade arrangements result in lower prices in Canada for British imports. As an example, he obtains price data on English cheese in six major metropolitan areas in the United States and six in Canada. The data, in cents per pound, follow:

U.S. prices	100	110	130	120	110	90
Canada prices	105	110	100	90	115	105

At significance level $\alpha = .10$, can the economist conclude that English cheese sells at lower prices in Canada?

6 A housing study conducted in 1970 showed that Honolulu had an average monthly rent of 158 dollars per dwelling unit (the highest among U.S. cities) with a standard deviation of 30 dollars, based on an analysis of 12 randomly selected units. In 1975, a random sample of 16 units had a mean rental price of 180 dollars with a standard deviation of 35 dollars. Does the study contain enough information to assert at level $\alpha = .10$ that Honolulu rents increased significantly in the 5-year period?

7 School buildings can be built more economically if state regulations on the required number of windows per classroom are relaxed. In an era of increasing costs combined with an eroding financial structure, the only thing standing between this fact and its immediate implementation is a psychological theory that the lack of windows in the classroom tends to increase anxiety among pupils (presumably because there is no escape). To check out the theory, an educational psychologist prepared a test of anxiety and administered it to a class of 15 pupils who had spent a month attending classes in an experimental windowless classroom and to a class of 20 pupils (acting as controls) who had spent the same month studying the same material in an ordinary classroom. Those pupils in the windowless classroom had a mean anxiety level of 105 with a standard deviation of 25, while those in the ordinary room had a mean anxiety level of 95 with a standard deviation of 30. At level $\alpha = .01$, do the test results indicate that lack of windows in the classroom leads to a significant increase in anxiety among schoolchildren?

SECTION 5.D A TEST FOR PAIRED SAMPLES

In this section, our goal will be to introduce the "paired-sample t test" which tests the difference between two sets of data where the individual members of one sample are *directly related* ("paired") to corresponding individual members of the other sample. As you can recall from the last few paragraphs of the previous section, the two-sample t test can be used only if the individual members of one sample are independent of the individual members of the other sample. Therefore, the paired-sample t test can be viewed as a replacement for the two-sample t test when the two samples of data are not independent of each other.

We will first show why we really need the paired-sample t test, that is, why the two-sample t test cannot be applied to paired data. The best way to illustrate this is to present an actual example of a real two-sample problem which cannot be solved using the two-sample t test. By the words "cannot be solved," we *do not mean* that we are unable to compute the numerical value of t by formula (5.4), but rather we *mean instead* that the results we get by carrying out this process do not make sense. In Example 5.6 below, we have a perfectly logical problem involving the effectiveness of two different reducing diets. When we apply the two-sample t test, however, the answers we get are obviously nonsensical. Then, after we reassess the problem, we will be able to see exactly why the two-sample t test failed to give an adequate solution and exactly what steps are required in order to solve the problem correctly.

Example 5.6 Weight-Reducing Diets A person who feels the need to lose some weight is considering going on one of the two famous reducing diets, the Drink-More diet and the Eat-Less diet, but only if statistical analysis shows that one or both of these diets are effective in inducing weight loss. As it turns out, five acquaintances have tried the Drink-More diet, and another five have tried the Eat-Less diet. The weights of the 10 acquaintances, both before beginning and after completing their diet programs, are recorded in Table 5.9. (To avoid embarrassment to those participating in the programs, we have not used their real names.) The question we have to answer is this: Are either of the diets really effective at significance level $\alpha = .05$ in producing weight loss?

DISCUSSION OF THE PROBLEM   As we can discover by looking at the data of Table 5.9, the Drink-More diet is virtually ineffective in the great majority (80%) of cases studied. The Eat-Less diet, on the other hand, seems to let everybody lose approximately 10 pounds. On the surface, it therefore seems that the Eat-Less diet is effective in producing weight loss, whereas the Drink-More diet is really not effective. This distinction between the diets ought to be reflected in our statistical analysis of the problem.

ATTEMPTED SOLUTION USING THE TWO-SAMPLE t TEST   We first test the effectiveness of the Drink-More diet using the two-sample t test. If we use the symbols

$$\mu_B = \text{true mean weight of group before dieting}$$
$$\mu_A = \text{true mean weight of group after dieting}$$

we need to test the hypothesis

$$H : \mu_B = \mu_A \text{ (the diet is not effective)}$$

TABLE 5.9
Before and After Weights of Ten Dieting Acquaintances

	Drink-More Diet			Eat-Less Diet	
Acquaintance	Weight Before	Weight After	Acquaintance	Weight Before	Weight After
A	150	150	F	150	141
B	160	160	G	160	150
C	170	170	H	170	160
D	180	179	I	180	170
E	190	141	J	190	179

versus

$$A : \mu_B > \mu_A \text{ (the diet is effective)}$$

The alternative is structured in such a way as to read that the diet is effective if a significant amount of weight is lost, namely, if mean weight before dieting exceeds mean weight after dieting.

Using the two-sample t test, we agreed to reject H in favor of A at significance level $\alpha = .05$ if

$$t > t_\alpha(n_B + n_A - 2) = t_{.05}(8) = 1.860$$

because $n_B = 5$ and $n_A = 5$. Table 5.10 shows the usual calculations for the two-sample t test. After inserting the components of t in their proper places in the formula, we obtain

$$t = \frac{\bar{X}_B - \bar{X}_A}{\sqrt{\dfrac{\Sigma(x_B - \bar{X}_B)^2 + \Sigma(x_A - \bar{X}_A)^2}{n_B + n_A - 2}\left(\dfrac{1}{n_B} + \dfrac{1}{n_A}\right)}} = \frac{170 - 160}{\sqrt{\dfrac{1000 + 922}{5 + 5 - 2}\left(\dfrac{1}{5} + \dfrac{1}{5}\right)}}$$

$$= \frac{10}{\sqrt{\left(\frac{1922}{8}\right)\left(\frac{2}{5}\right)}} = \frac{10}{\sqrt{96.1}} = \frac{10}{9.803} = 1.020$$

We have agreed to reject H and so to say that the Drink-More diet is effective if t turned out to exceed 1.860. However, it turned out that $t = 1.020 < 1.860$, and so we cannot reject H. The two-sample t test therefore asserts at level $\alpha = .05$ that the Drink-More diet is not effective.

So far so good. The result we have just obtained confirmed our original suspicion that the Drink-More diet could not be counted on to give an individual a reasonable amount of weight loss. Now let's apply the same procedure to analyze the Eat-Less diet (which from direct observation of the data seems to almost guarantee each participant a weight loss of 10 pounds). We are still testing the hypothesis

$$H : \mu_B = \mu_A$$

TABLE 5.10
Calculations Leading to the Two-Sample t Statistic
Drink-More Diet Data

Weight Before			Weight After		
x_B	$x_B - \bar{X}_B$	$(x_B - \bar{X}_B)^2$	x_A	$x_A - \bar{X}_A$	$(x_A - \bar{X}_A)^2$
150	−20	400	150	−10	100
160	−10	100	160	0	0
170	0	0	170	10	100
180	10	100	179	19	361
190	20	400	141	−19	361
850	0	1000	800	0	922

$n_B = 5$ \qquad\qquad $n_A = 5$

$\bar{X}_B = \frac{850}{5} = 170$ \qquad $\bar{X}_A = \frac{800}{5} = 160$

$\Sigma(x_B - \bar{X}_B)^2 = 1000$ \qquad $\Sigma(x_A - \bar{X}_A)^2 = 922$

versus

$$A : \mu_B > \mu_A$$

and we still reject H at level $\alpha = .05$ when $t > 1.860$, because df $= n_B + n_A - 2 = 5 + 5 - 2 = 8$, as before. It therefore remains only to compute the value of t and to note whether or not it exceeds 1.860. The relevant calculations appear in Table 5.11, and they show that

$$t = \frac{\bar{x}_B - \bar{x}_A}{\sqrt{\dfrac{\Sigma(x_B - \bar{x}_B)^2 + \Sigma(x_A - \bar{x}_A)^2}{n_B + n_A - 2}\left(\dfrac{1}{n_B} + \dfrac{1}{n_A}\right)}} = \frac{170 - 160}{\sqrt{\dfrac{1000 + 922}{5 + 5 - 2}\left(\dfrac{1}{5} + \dfrac{1}{5}\right)}}$$

$$= \frac{10}{\sqrt{\left(\dfrac{1922}{8}\right)\left(\dfrac{2}{5}\right)}} = \frac{10}{\sqrt{96.1}} = \frac{10}{9.803} = 1.020$$

Again $t = 1.020 < 1.860$, and so we cannot reject H in favor of A in the Eat-Less diet. Therefore the two-sample t test gives us the same result for the Eat-Less diet as it did for the Drink-More diet, namely, that, at level $\alpha = .05$, the Eat-Less diet cannot be considered an effective way to lose weight.

ANALYSIS OF THE ATTEMPTED SOLUTION   Despite the fact that there exist important differences between the two diets, differences which are evident from a glance at the data of Table 5.9, the two-sample t test has decided at level $\alpha = .05$ that neither diet can be considered significantly effective in producing weight loss. The statistical test has apparently taken no account of the fact that although 80% of those on the Drink-More diet lost virtually no weight, everyone on the Eat-Less diet successfully lost approximately 10 pounds. Even more surprising than the fact that the two-sample t test judged both diets to be similarly noneffective is the fact that the numerical values of t turned out to be exactly the same in both instances. In fact, all calculations leading to the numerical value of t were also identical in both instances. A comparison of Tables 5.10 and 5.11 reveals that they would be carbon (or photo-) copies of each other, except for the

TABLE 5.11
Calculations Leading to the Two-Sample t Statistic
Eat-Less Diet Data

x_B	$x_B - \bar{x}_B$	$(x_B - \bar{x}_B)^2$	x_A	$x_A - \bar{x}_A$	$(x_A - \bar{x}_A)^2$
150	−20	400	141	−19	361
160	−10	100	150	−10	100
170	0	0	160	0	0
180	10	100	170	10	100
190	20	400	179	19	361
850	0	1000	800	0	922

$n_B = 5$
$\bar{x}_B = \frac{850}{5} = 170$
$\Sigma(x_B - \bar{x}_B)^2 = 1000$

$n_A = 5$
$\bar{x}_A = \frac{800}{5} = 160$
$\Sigma(x_A - \bar{x}_A)^2 = 922$

positions of the numbers 141, -19, and 361 in the Weight After portion of the tables. In Table 5.10 these numbers appear at the bottom of the Weight After columns, whereas in Table 5.11 they appear at the top. Otherwise, the calculations toward the two-sample t test are the same for both diets. To summarize what we have just said, we can assert merely that the two-sample t test is incapable of detecting the distinctions between the obviously ineffective Drink-More diet and the obviously effective Eat-Less diet.

Why did the two-sample t test fail to give a logical solution to this problem? Well, let's review the basic principles underlying the two-sample t test. We tested, in both cases, the hypothesis

$$H : \mu_B = \mu_A$$

against the alternative

$$A : \mu_B > \mu_A$$

In other words, we tested whether or not the overall average (true mean) of the group's weight before undertaking the diet was significantly greater than the overall average of the group's weight after completing the diet. In the Drink-More diet, the group weighed a total of 850 pounds (for an average of 170) before dieting and a total of 800 (for an average of 160) after dieting, as we can see from Table 5.10. Therefore, participants in the Drink-More diet lost 10 pounds, on the average. Table 5.11 shows that exactly the same *average* results have been obtained in the Eat-Less diet. Participants in that dieting program weighed an average of 170 pounds before and 160 pounds after, for an average loss of 10 pounds. It is this average loss for which the two-sample t test looks, and that is what it has found. Because the average weight loss is the same for both diets, the two diets are equally effective (in this case, equally noneffective) in the eyes of the two-sample t test.

But is it the overall average weight loss that we are primarily interested in? We can answer this question by going back to the reasoning that led us to believe that the Drink-More diet was not really effective, while the Eat-Less diet was. As demonstrated in the Drink-More data of Table 5.9, acquaintances A, B, C, and D lost virtually no weight, while acquaintance E lost enough weight for the entire group to have an average loss of 10 pounds. The Eat-Less diet of the same table, on the other hand, shows that each participant individually lost approximately 10 pounds; so the diet could reasonably be said to guarantee everyone a loss of around 10 pounds. Therefore it was the *individual* changes in weight from before to after, not the overall group averages, that influenced us to think that the Eat-Less diet was effective in producing weight loss, while the Drink-More diet was not. It is these individual differences which are not reflected at all in the calculations for the two-sample t test, for those calculations take only overall averages into account. Therefore, because the effectiveness of the diets hinges on the individual instead of the overall average weight loss, the two-sample t test is not an appropriate vehicle for solving this problem.

What we require in order to make a proper analysis of the diets is a method in which the calculations are based on individual weight losses. Such a method is the "paired-sample t test." The paired-sample t test involves not the difference between the average before weight and average after weight, but instead the differences between the individual before and after weights. We now proceed to the formal discussion of the paired-sample t test.

CORRECT SOLUTION USING THE PAIRED-SAMPLE t TEST   We first test the effectiveness of the Drink-More diet using the paired-sample t test. We define the quantity

μ_d = true mean individual difference between the
before weight and the after weight

and we want to test the hypothesis

$H : \mu_d = 0$ (the diet is not effective)

versus

$A : \mu_d > 0$ (the diet is effective)

The alternative is structured so that the diet is judged to be effective if the average individual loses a positive amount of weight, namely, if the average individual's weight before exceeds his weight after. To carry out the paired-sample t test, we would reject H in favor of A at level α if $t > t_\alpha(n - 1)$, where

● $$t = \frac{\bar{d}\sqrt{n}}{s_d} \tag{5.6}$$

in which

n = number of before-after pairs
$\bar{d}$ = sample mean of the individual differences
s_d = sample standard deviation of the individual
differences

It is to be noted that the above formula for t emphasizes throughout the *individual* differences, in contrast to the two-sample t test which involved only overall averages. It is usually more convenient, when working with the paired-sample t test, to compute s_d using the shortcut formula for the standard deviation given by

$$s_d = \sqrt{\frac{n\Sigma d^2 - (\Sigma d)^2}{n(n - 1)}}$$

as in formula (4.3) of Sec. 4.B. Here

Σd = sum of the individual differences
Σd^2 = sum of the squares of the individual differences

Taking $\alpha = .05$ as our level of significance and noting that there are $n = 5$ before-after pairs comprising the Drink-More data of Table 5.9, we agree to reject

$H : \mu_d = 0$

in favor of

$$A : \mu_d > 0$$

if $t > t_{.05}(4) = 2.132$. The computations needed for the numerical determination of t are presented in Table 5.12, and they yield that

$$t = \frac{\bar{d}\sqrt{n}}{s_d} = \frac{(10)\sqrt{5}}{21.81} = \frac{(10)(2.236)}{21.81} = 1.025$$

Therefore t has turned out to be 1.025, which is not larger than 2.132. As we have agreed to reject H in favor of A if $t > 2.132$, we cannot reject H under the circumstances. We, therefore, conclude at level $\alpha = .05$, using the paired-sample t test, that the Drink-More is not effective in producing individual weight loss.

Now that the paired-sample t test has confirmed our earlier judgment that the Drink-More diet does not produce significant individual weight losses, we await its decision on the Eat-Less diet. It was that diet which provided us with the first intimation that the analysis based on the two-sample t test was breaking down. It remains to be seen whether the paired-sample t test will be able to detect the substantial individual weight losses that we know are present in the data. We are still testing the hypothesis

$$H : \mu_d = 0$$

versus

$$A : \mu_d > 0$$

and we still will reject H at level $\alpha = .05$ if $t > 2.132$, because df $= n - 1 = 5 - 1 = 4$, as before. Table 5.13 shows the calculations used in finding the numerical value of t.

TABLE 5.12
Calculations Leading to the Paired-Sample t Statistic
Drink-More Diet Data

Acquaintance	Weight Before x_B	Weight After x_A	$d = x_B - x_A$	d^2
A	150	150	0	0
B	160	160	0	0
C	170	170	0	0
D	180	179	1	1
E	190	141	49	2401
			Sums 50	2402

$n = 5$

$\bar{d} = \dfrac{\Sigma d}{n} = \dfrac{50}{5} = 10$

$s_d = \sqrt{\dfrac{n\Sigma d^2 - (\Sigma d)^2}{n(n-1)}} = \sqrt{\dfrac{5(2402) - (50)^2}{(5)(4)}} = \sqrt{\dfrac{12{,}010 - 2500}{20}}$

$\quad = \sqrt{\tfrac{9510}{20}} = \sqrt{475.5} = 21.81$

TABLE 5.13
Calculations Leading to the Paired-Sample t Statistic
Eat-Less Diet Data

Acquaintance	Weight Before X_B	Weight After X_A	$d = X_B - X_A$	d^2
F	150	141	9	81
G	160	150	10	100
H	170	160	10	100
I	180	170	10	100
J	190	179	11	121
			Sums 50	502

$n = 5$

$$\bar{d} = \frac{\Sigma d}{n} = \frac{50}{5} = 10$$

$$s_d = \sqrt{\frac{n\Sigma d^2 - (\Sigma d)^2}{n(n-1)}} = \sqrt{\frac{5(502) - (50)^2}{(5)(4)}} = \sqrt{\frac{2510 - 2500}{20}}$$

$$= \sqrt{\tfrac{10}{20}} = \sqrt{.5} = .7071$$

Inserting the numerical values $n = 5$, $\bar{d} = 10$, and $s_d = 0.7071$ (obtained from the calculations of Table 5.13) into the formula for t, we get

$$t = \frac{\bar{d}\sqrt{n}}{s_d} = \frac{10\sqrt{5}}{.7071} = \frac{(10)(2.236)}{.7071} = 31.62$$

We have agreed to reject H if $t > 2.312$. As it turns out, $t = 31.62$, a number substantially in excess of 2.312, so that we can proceed with the greatest confidence to the rejection of H. In the eyes of the paired-sample t test, then, there is little doubt that the Eat-Less diet is effective at level $\alpha = .05$ or at any other level for that matter. The numerical value $t = 31.62$ is so overwhelming that we must conclude that, at any reasonable level α, the Eat-Less diet is truly effective in producing individual weight loss.

The conclusions based on the use of the paired-sample t test have therefore provided us with statistical justification of the effectiveness of the Eat-Less diet and the ineffectiveness of the Drink-More diet. Because of its emphasis on individual differences between related pairs of data points instead of on the overall group averages, the paired-sample t test has successfully distinguished between the fundamental characteristics of the two diets under study. The two-sample t test should be used only when we want to investigate the difference between the mean of two independent sets of data, while the paired-sample t test should be used whenever consideration of individual differences between related pairs of data points is paramount.

Example 5.7 Cloud Seeding A meteorologist participating in a project to determine to what extent, if any, mankind can influence local weather conditions has set up an experiment to test the effectiveness of the present methods of cloud seeding in the artificial production of rainfall. Two farming areas in Nebraska with similar past meteorological records, lying 150 miles apart in a north–south direction, were selected for the experiment. The Fairbury area is regularly seeded throughout the year, but the Norfolk area is left unseeded. Their monthly precipitation for the year is recorded in Table 5.14. At level $\alpha = .01$, do the results of the experiment indicate that cloud seeding significantly increases monthly precipitation?

SOLUTION   We use the paired-sample t test in preference to the two-sample t test because the columns of data are paired by month. We would not want to conclude that cloud seeding was effective merely on the basis of a large increase in rainfall for one or two months. We would want instead to establish a pattern of increased rainfall regularly throughout the year. The only way we can address ourselves to this question is to apply the paired-sample t test.

We want to test the hypothesis

$$H : \mu_d = 0 \text{ (cloud seeding does not increase precipitation)}$$

versus

$$A : \mu_d < 0 \text{ (cloud seeding increases precipitation)}$$

where
$\mu_d =$ true mean difference between precipitation in an unseeded region and precipitation in a seeded region

The alternative $A : \mu_d < 0$ has been selected (instead of $A : \mu_d > 0$) because, as the calculations in Table 5.15 show, the individual difference d is given by the subtraction

TABLE 5.14
Nebraska Precipitation Data from
Cloud-Seeding Experiment

Month	Norfolk (Unseeded), inches of precipitation	Fairbury (Seeded), inches of precipitation
Jan.	1.5	1.4
Feb.	1.4	1.4
Mar.	2.2	2.6
Apr.	2.6	2.5
May	3.9	4.8
June	4.5	4.3
July	3.7	4.0
Aug.	3.4	3.5
Sept.	4.0	3.9
Oct.	2.5	2.6
Nov.	1.9	1.7
Dec.	1.5	1.4

$$d = \text{Norfolk precipitation minus Fairbury precipitation}$$
$$= \text{unseeded precipitation minus seeded precipitation}$$

The alternative $A : \mu_d < 0$ therefore expresses the statement that seeding increases rainfall, for it means that the bulk of the d's are negative, and, consequently, Norfolk precipitation is usually less than Fairbury precipitation. Using Table 5.7 to help us determine the rejection rule, we then see that we should reject H at level α if $t < -t_\alpha(n - 1)$. Here $\alpha = .01$ and $n = 12$ (for there is one pair for each month), and so we agree to reject

$$H : \mu_d = 0$$

in favor of

$$A : \mu_d < 0$$

if $t < -t_{.01}(11) = -2.718$. All that remains now is to determine the numerical value of t, the preliminary calculations for which appear in Table 5.15 and show that $n = 12$, $\bar{d} = -.0833$, and $s_d = .3186$. Inserting these numbers into the expression for t, we have that

$$t = \frac{\bar{d}\sqrt{n}}{s_d} = \frac{(-.0833)\sqrt{12}}{.3186} = \frac{(-.0833)(3.464)}{.3186} = -.906$$

Therefore, $t = -.906 > -2.718$ (because $-.906$ is *less negative* than -2.718), and since we agreed to reject H if $t < -2.718$, the result does not permit us to reject H. At level $\alpha = .01$, we then conclude that the experiment seems to show that cloud seeding does *not* significantly increase monthly precipitation.

Before we close this section, it would be instructive to comment on the origin of the formula

$$t = \frac{\bar{d}\sqrt{n}}{s_d}$$

used in the paired-sample t test. Toward this end, let's take a look at Tables 5.6 and 5.15 with a view toward comparing them. If we compare Table 5.6 with the last two columns of Table 5.15, we see that the tables have exactly the same structure. The only differences are the numerical values of the data points and the fact that the columns of Table 5.6 are headed by x's, while the columns of Table 5.15 are headed by d's. Other than that, we are really performing the same set of calculations. Because Table 5.6 contains calculations used in connection with the one-sample t test, it seems that Table 5.15 could also be viewed in terms of the one-sample t test. That is, in fact, exactly what the paired-sample t test is—a one-sample t test, where the one sample is the set of differences (d's). To test the hypothesis $H : \mu_d = 0$ using the one-sample t test on the d's would call for the formula

$$t = \frac{\bar{x} - \mu_0}{s/\sqrt{n}} = \frac{\bar{d} - 0}{s_d/\sqrt{n}}$$

TABLE 5.15
Calculations Leading to the Paired-Sample *t* Statistic
Cloud-Seeding Data

Month	Inches of Precipitation		$d = x_u - x_s$	d^2
	Unseeded Area x_u	Seeded Area x_s		
Jan.	1.5	1.4	.1	.01
Feb.	1.4	1.4	.0	.00
Mar.	2.2	2.6	−.4	.16
Apr.	2.6	2.5	.1	.01
May	3.9	4.8	−.9	.81
June	4.5	4.3	.2	.04
July	3.7	4.0	−.3	.09
Aug.	3.4	3.5	−.1	.01
Sept.	4.0	3.9	.1	.01
Oct.	2.5	2.6	−.1	.01
Nov.	1.9	1.7	.2	.04
Dec.	1.5	1.4	.1	.01
			Sums −1.0	1.20

$n = 12$

$$\bar{d} = \frac{\Sigma d}{n} = \frac{-1.0}{12} = -.0833$$

$$s_d = \sqrt{\frac{n\Sigma d^2 - (\Sigma d)^2}{n(n-1)}} = \sqrt{\frac{12(1.20) - (-1.0)^2}{(12)(11)}} = \sqrt{\frac{14.4 - 1.0}{132}}$$

$$= \sqrt{\frac{13.4}{132}} = \sqrt{.1015} = .3186$$

But a little algebra shows that

$$t = \frac{\bar{d} - 0}{s_d/\sqrt{n}} = \frac{\bar{d}}{s_d/\sqrt{n}} = \left(\frac{\bar{d}}{1}\right)\left(\frac{\sqrt{n}}{s_d}\right) = \frac{\bar{d}\sqrt{n}}{s_d}$$

which is the formula for the paired-sample *t* statistic. Therefore, the paired-sample *t* test can be considered as a special case of a one-sample *t* test. It just happens that the one sample of data is obtained by computing the pair-by-pair differences of two original sets of data, instead of being observed directly from an experiment or survey.

EXERCISES 5.D

1 Some psychologists think that there is a statistical correlation between smoking and absenteeism. The management of a leather goods factory has under consideration a stop-smoking incentive plan, and they would therefore be interested in knowing whether employees who have stopped smoking have better absenteeism records than they had while they were smoking. Nine such employees are randomly selected, and their absenteeism records before and after they stopped smoking are compared. The data follow:

Employee	A	B	C	D	E	F	G	H	I
Days absent per year while smoking	20	30	14	6	42	19	18	12	24
Days absent per year after stopping smoking	10	20	16	5	40	15	22	10	20

At significance level $\alpha = .05$, does the comparison tend to support the theory that employees have less absenteeism after they stop smoking?

2 At various points along a valley, a team of geographers measured the angles of slope on both sides of the valley wall. Their problem was to determine whether or not there were significant differences between the slope angles of the northeast (NE) slope and the southwest (SW) slope. The data they collected are as follows:

Miles upstream	10	20	30	40	50	60	70	80
NE slope, degrees	15	24	28	31	35	34	35	36
SW slope, degrees	16	22	25	30	34	32	36	34

At level $\alpha = .05$, can the researchers conclude that the angles of slope of the northeast wall differ significantly from the angles of slope of the southwest wall?

3 Ten sets of identical twins were administered drugs intended to increase the pulse rate. The goal of the experiment was to determine whether identical twins have similar reactions to medication. The following data on increase in pulse rates was observed:

Twin Set	Pulse Rate Increase of Older Twin	Pulse Rate Increase of Younger Twin
#1	12	19
#2	14	13
#3	8	6
#4	11	18
#5	14	12
#6	12	15
#7	13	10
#8	15	18
#9	18	21
#10	17	22

At level $\alpha = .10$, decide whether or not identical twins react similarly to medication affecting the pulse rate.

4 Environmental Nucleonics of Barstow instituted a time-consuming and rather expensive occupational health and safety program, the goal of which was to reduce the number of hours of working time lost because of on-the-job accidents. The following data were collected in their six branches before and after institution of the program:

**Hours Lost per Employee
per Quarter**

Branch	Before	After
Barstow, Ca.	46	43
Hoboken, N.J.	72	66
Ensenada, B.C.	50	51
Elko, Ne.	38	37
Rawlins, Wy.	78	73
Vancouver, B.C.	66	60

At level $\alpha = .01$, decide whether or not the data indicate that the program is effective in reducing time lost because of on-the-job accidents.

5   An agricultural experiment station tested the comparative yields of two varieties of corn, a new experimental variety and a standard variety, to find out whether or not the new variety would permit greater food production on the same land. Each of seven farmers was asked to grow both varieties on similar plots of land and to give each the exact same amount of treatment, water, fertilizer, etc. The yields, in bushels, are recorded below:

Farmer	A	B	C	D	E	F	G
New variety	28.2	24.6	29.7	20.5	34.6	27.1	31.4
Standard variety	27.2	24.3	29.0	22.5	34.2	26.8	30.4

At level $\alpha = .05$, do the results of the experiment confirm the claim that the new variety is significantly more productive than the standard variety?

6   A local school system paired 16 girls in the first grade of elementary school according to psychological test scores, socioeconomic status of family, general health, and family size, as part of a study of the effectiveness of the kindergarten experience in helping a pupil in first grade. One member of each pair had attended an optional kindergarten the year before, while the other had not. Halfway through the year (first grade) the 16 girls were tested for mastery of the first-grade material taught, and they had the following scores:

Pair	#1	#2	#3	#4	#5	#6	#7	#8
With kindergarten	83	74	67	64	70	67	81	64
Without kindergarten	78	74	63	66	68	63	77	65

Does the study indicate, at level $\alpha = .05$, that the kindergarten experience seems to help first graders?

7   The following data[1] give the value of exports of the seven member nations of the West African Customs Union (in billions of CFA francs, the standard currency unit for the region) in the years 1961 and 1966:

[1]From "Surveys of African Economies," vol. 3, p. 41, International Monetary Fund, Washington, D.C., 1970.

Nation	1961	1966
Dahomey	3.7	2.6
Ivory Coast	43.6	76.7
Mali	3.5	3.2
Mauritania	.5	17.0
Niger	3.8	8.6
Senegal	30.7	36.8
Upper Volta	.9	4.0

At level $\alpha = .05$, do the data indicate that the region's exports increased significantly in the period between 1961 and 1966?

SECTION 5.E TESTS FOR THE STANDARD DEVIATION

Up to this point, we have concentrated on statistical tests of hypotheses involving the mean of a population or the difference between means of two populations. In this section, our objective will be to undertake analogous problems involving standard deviations. In particular, we will present examples of the one-sample test for the standard deviation of a population and the two-sample test for the difference between the standard deviations of two populations.

Example 5.8 Body Temperatures Recall the doctor–anthropologist of Example 5.2, who has conducted a study of body temperatures of Alaskan Eskimos. Suppose, in addition to the question discussed in the earlier example, we wanted to test the hypothesis that the body temperatures of Alaskan Eskimos had a standard deviation below .5°F (one-half degree). In view of the fact that eight individuals randomly selected from the population had recorded temperatures of

$$98.5 \quad 98.1 \quad 98.6 \quad 98.7 \quad 98.4 \quad 98.9 \quad 98.0 \quad 98.4$$

degrees, respectively, do the results of the study support at level $\alpha = .05$ the assertion that body temperatures of Eskimos have standard deviation below .5°F?

SOLUTION The procedure we follow in solving this problem is called the "one-sample chi-square test for the population standard deviation." As in the case of the small-sample confidence intervals for the population standard deviation (studied in Sec. 4.D), the use of the chi-square distribution is valid only in situations where the underlying population is normally distributed. Therefore, only when the population has the normal distribution can we apply the small-sample methods (both one-sample and two-sample) of this section. Returning now to the problem at hand, we prepare to apply the one-sample chi-square test. The general structure of this test is as follows: We test the hypothesis

$$H : \sigma = \sigma_0$$

versus

$$A : \sigma < \sigma_0$$

where
$$\sigma = \text{true population standard deviation}$$
$$\sigma_0 = \text{hypothesized standard deviation}$$

We then reject H in favor of A at significance level α if

$$\chi^2 < \chi^2_{1-\alpha}(n - 1)$$

where
$$n = \text{number of data points in the sample}$$

and

●
$$\chi^2 = \frac{(n - 1)s^2}{\sigma_0^2} = \frac{\Sigma(x - \bar{x})^2}{\sigma_0^2} \qquad (5.7)$$

We have encountered all these quantities before, in connection with confidence intervals for standard deviations. Here s is the sample standard derivation that we have used many times before. It is an algebraic fact that

$$(n - 1)s^2 = \Sigma(x - \bar{x})^2$$

because of the formula for the sample standard deviation, which asserts that

$$s = \sqrt{\frac{\Sigma(x - \bar{x})^2}{n - 1}}$$

Because $n = 8$ and $\alpha = .05$, we agree to reject

$$H : \sigma = 0.5$$

in favor of

$$A : \sigma < 0.5$$

if
$$\chi^2 < \chi_{.95}{}^2(7) = 2.167$$

as in Table A.6 of the Appendix. It, therefore, remains only to compute the numerical value of

$$\chi^2 = \frac{\Sigma(x - \bar{x})^2}{\sigma_0^2}$$

Well, we know from the formulation of the hypothesis that $\sigma_0 = .5$, and we can refer back to the calculations of Table 5.4 where it is shown that $\Sigma(x - \bar{x})^2 = .62$. It follows that

$$\chi^2 = \frac{\Sigma(x - \bar{x})^2}{\sigma_0^2} = \frac{.62}{(.5)^2} = \frac{.62}{.25} = 2.48$$

Naturally, if Table 5.4 had not already been in existence, we would have to construct one like it in order to find out what $\Sigma(x - \bar{x})^2$ is. As we have agreed to reject H in favor of A if $\chi^2 < 2.167$, we note that $\chi^2 = 2.48 > 2.167$; so we cannot reject H. We therefore conclude, at level $\alpha = .05$, that there is not enough evidence to support the assertion that the body temperatures of Eskimos have standard deviation below .5°F.

As we discussed at the end of Sec. 5.B, for tests involving means, there are three possible alternatives in one-sample tests involving the hypothesis $H : \sigma = \sigma_0$. These are $A : \sigma > \sigma_0$, $A : \sigma < \sigma_0$, and $A : \sigma \neq \sigma_0$. By analogy with the information presented in Table 5.7 (involving means), we record the rules for rejecting $H : \sigma = \sigma_0$ in favor of each of these alternatives in cases involving standard deviations. The three rejection rules are presented in Table 5.16.

Example 5.9   Calculator Battery Lifetimes   One manufacturer of pocket calculators advertises that its battery pack allows the calculator to operate continuously for 22 hours on the average without recharging. (See Example 5.1.) A prospective corporate customer, however, is often interested in the uniformity of the product as well, for it would be quite disconcerting to order a supply of 100 calculators for company use and then find that half operate continuously for 38 hours, while the other half operate continuously only for 6 hours. Such a shipment would certainly satisfy the criterion of an average operating time of 22 hours, but the tremendous variation in quality would considerably reduce the ultimate value to the user of the total package of 100 calculators. Suppose, then, that the corporate customer insists on a uniformity of operation requiring that the standard deviation of battery pack lifetimes not exceed one-half hour (.5 hour). Based on a sample of 50 calculators whose battery lifetimes turn out to have mean 21.8 hours with a standard deviation of .9 hour, can the customer conclude at level $\alpha = .01$ that the true standard deviation of battery lifetimes is in excess of the acceptable .5 hour?

SOLUTION   We note that $\sigma_0 = .5$, $s = .9$, $n = 50$, and the sample mean of 21.8 is not relevant to the problem. We are therefore testing the hypothesis

$$H : \sigma = .5 \text{ (battery pack uniformity is acceptable)}$$

against the alternative

$$A : \sigma > .5 \text{ (battery pack lifetimes vary too much)}$$

TABLE 5.16
Rejection Rules in the Presence of Various
Alternatives for One-Sample Tests of
Population Standard Deviations
$H : \sigma = \sigma_0$

Alternative	Reject H in Favor of A at Significance Level α if:
$A : \sigma > \sigma_0$	$\chi^2 > \chi_\alpha^2(n - 1)$
$A : \sigma < \sigma_0$	$\chi^2 < \chi_{1-\alpha}^2(n - 1)$
$A : \sigma \neq \sigma_0$	$\chi^2 < \chi_{1-\alpha/2}^2(n - 1)$ or $\chi^2 > \chi_{\alpha/2}^2(n - 1)$

Because $n = 50 > 30$, we must use a large-sample, instead of a small-sample, test. We compute

$$\chi^2 = \frac{(n-1)s^2}{\sigma_0^2}$$

as before, and then we calculate

● $$z = \sqrt{2\chi^2} - \sqrt{2(n-1)} \qquad (5.8)$$

which has the standard normal distribution for large values of n. In accordance with the appropriate adjustments to Table 5.16, we therefore reject

$$H : \sigma = .5$$

in favor of

$$A : \sigma > .5$$

at level α if $z > z_\alpha$. For large samples, we therefore use the "one-sample z test for the population standard deviation." Since $\alpha = .01$, we agree to reject H if

$$z > z_{.01} = 2.33$$

So it remains only to find the numerical values of χ^2 and z. From the data we know that

$$n = 50$$
$$s = .9$$

and from the formulation of the hypothesis, that

$$\sigma_0 = .5$$

Therefore,

$$\chi^2 = \frac{(n-1)s^2}{\sigma_0^2} = \frac{(49)(.9)^2}{(.5)^2} = \frac{(49)(.81)}{.25} = 158.76$$

and so $z = \sqrt{2\chi^2} - \sqrt{2(n-1)} = \sqrt{2(158.76)} - \sqrt{2(50-1)}$
$$= \sqrt{317.52} - \sqrt{98} = 17.819 - 9.899 = 7.92$$

As we have agreed to reject H if $z > 2.33$, we must reject H in favor of A because $z = 7.92 > 2.33$. We then conclude, at level $\alpha = .01$, that the true standard deviation of the battery pack lifetimes is significantly larger than .5 hour, and they are consequently too variable to meet the customer's specifications.

Because the large-sample test for the population standard deviation is a z test, all three rejection rules are based on the percentage points of the normal distribution. Table 5.7 is therefore the table of rejection rules to be used if only the μ's are replaced by σ's.

The two-sample test for the difference of standard deviations introduces a new distribution into our discussion. This new distribution is the F distribution appearing in Table A.5 of the Appendix. (As well as seeing action in the remainder of this section, the

F distribution will appear in the future throughout the entirety of Chap. 10.) The following two examples illustrate the use of the two-sample F test for the difference of standard deviations.

Example 5.10 Supermarket Prices It is known that some supermarkets give trading stamps with purchases, while others do not. By and large, most people realize that those supermarkets that give trading stamps generally post slightly higher prices in order to cover the additional cost of the stamps. On the other hand, one consumer organization thinks that stores which do not give stamps are prone to offer "loss leaders" or other price adjustment gimmicks instead of giving uniformly lower prices. In other words, the consumer organization believes that prices are more variable among supermarkets not offering trading stamps than among those that do. To test out its guess, the organization checks banana prices (per pound) at four supermarkets that give trading stamps and six that do not. The results of the survey follow:

	Prices per pound of bananas, cents					
At supermarkets that give stamps	12	18	14	16		
At supermarkets that give no stamps	10	16	16	12	8	10

At level $\alpha = .05$, do the data support the allegation that prices are more variable among supermarkets not giving trading stamps?

SOLUTION   The appropriate way to answer this question is by means of the two-sample F test. We define the quantities

$$\sigma_T = \text{true standard deviation of banana prices at}$$
$$\text{supermarkets that give trading stamps}$$
$$\sigma_N = \text{true standard deviation of banana prices at}$$
$$\text{supermarkets that do not give trading stamps}$$

We then want to test the hypothesis

$$H : \sigma_N = \sigma_T \text{ (banana prices are equally variable)}$$

against the alternative,

$$A : \sigma_N > \sigma_T \text{ (banana prices are more variable at}$$
$$\text{supermarkets not offering trading stamps)}$$

To test H against A, we compute

$$s_T = \text{sample standard deviation of banana prices at}$$
$$\text{supermarkets that give trading stamps}$$
$$s_N = \text{sample standard deviation of banana prices at}$$
$$\text{supermarkets that do not give trading stamps}$$

and we note the numerical values of

n_T = number of data points in the sample of supermarkets
that give trading stamps

n_N = number of data points in the sample of supermarkets
that do not give trading stamps

We then calculate the value of

●
$$F = \frac{s_N^2}{s_T^2}$$
(5.9)

and we reject H at significance level α if

$$F > F_\alpha(n_N - 1; n_T - 1)$$

Degrees of Freedom of the *F* Distribution

A word about the F distribution is now in order. The percentage points of the F distribution are of the form

$$F_\alpha(\text{dfn}; \text{dfd})$$

where   dfn = degrees of freedom in the numerator

dfd = degrees of freedom in the denominator

The fact that the percentage points of the F distribution are classified by two indices of degrees of freedom contrasts markedly with those of the t and chi-square distributions which are classified by only one index of degrees of freedom. In the F test for the difference of standard deviations, each index of degrees of freedom is the number of data points involved minus one. For example, dfn equals one less than the number of data points used in computing the numerator of F. In the banana price example, therefore,

$$\text{dfn} = n_N - 1 = 6 - 1 = 5$$

because six data points were used in the computation of the numerator s_N^2. By a similar reasoning process, we see that

$$\text{dfd} = n_T - 1 = 4 - 1 = 3$$

Therefore, at level $\alpha = .05$, we agree to reject

$$H : \sigma_N = \sigma_T$$

in favor of $$A : \sigma_N > \sigma_T$$

if $$F > F_{.05}(5, 3) = 9.01$$

TABLE 5:17
A Small Portion of Table A.5
Values of $F_{.05}$

	Degrees of Freedom for Numerator	
		5
Degrees of Freedom for Denominator	3	9.01

In Table 5.17, we exhibit a small portion of Table A.5, showing where to find $F_{.05}(5, 3)$.

It now remains only to calculate the value of F and to observe whether the value so calculated exceeds 9.01 or not. To calculate F, we must first calculate s_N^2 and s_T^2. It may look to you as though we must calculate s_N and s_T first and then square the resulting numbers. However, the nature of the standard deviation allows us to shortcut this procedure somewhat. Because

$$s_N = \sqrt{\frac{\Sigma(x_N - \bar{x}_N)^2}{n_N - 1}}$$

we obtain by squaring both sides of this equation that

$$s_N^2 = \frac{\Sigma(x_N - \bar{x}_N)^2}{n_N - 1}$$

Similarly,

$$s_T^2 = \frac{\Sigma(x_T - \bar{x}_T)^2}{n_T - 1}$$

We can use these formulas to calculate the numerator and denominator of the F statistic when the actual values of the data points are known to us. The calculational details involving the banana price data are presented in Table 5.18.

The calculations made in Table 5.18 show that $F = 1.68$. As we have agreed to reject H in favor of A at level $\alpha = .05$ if $F > 9.01$, our calculation that $F = 1.68 < 9.01$ shows that we cannot reject H. On the basis of the given data, therefore, we do not have enough evidence to assert, at level $\alpha = .05$, that banana prices are more variable at supermarkets which do not give trading stamps.

Rejection Rules for the F Test

Before we present a second example illustrating the application of the two-sample F test for the difference of standard deviations, let's set up a table of rejection rules analogous to those of Tables 5.7 and 5.16. It turns out that the F test has a curious characteristic in this regard: There are only two possible alternatives, instead of the three appearing in

TABLE 5.18

Calculations Leading to the Two-Sample F Statistic
Banana Price Data

Supermarkets Not Offering Stamps			Supermarkets Offering Stamps		
x_N	$x_N - \bar{x}_N$	$(x_N - \bar{x}_N)^2$	x_T	$x_T - \bar{x}_T$	$(x_T - \bar{x}_T)^2$
10	-2	4	12	-3	9
16	4	16	18	3	9
16	4	16	14	-1	1
12	0	0	16	1	1
8	-4	16	60	0	20
10	-2	4			
72	0	56			

$$\bar{x}_N = \tfrac{72}{6} = 12 \qquad\qquad \bar{x}_T = \tfrac{60}{4} = 15$$

$$s_N^2 = \frac{\Sigma(x_N - \bar{x}_N)^2}{n_N - 1} \qquad\qquad s_T^2 = \frac{\Sigma(x_T - \bar{x}_T)^2}{n_T - 1}$$

$$= \frac{56}{6 - 1} = \frac{56}{5} = 11.2 \qquad\qquad = \frac{20}{4 - 1} = \frac{20}{3} = 6.67$$

$$F = \frac{s_N^2}{s_T^2} = \frac{11.2}{6.67} = 1.68$$

Tables 5.7 and 5.16. Taking as our hypothesis

$$H : \sigma_1 = \sigma_2$$

the only alternatives that the F test, as we have organized it, knows how to handle are

$$A : \sigma_1 > \sigma_2$$

and

$$A : \sigma_1 \neq \sigma_2$$

Because of the complicated nature of the F distribution, involving as it does two indices of degrees of freedom, it turns out that we cannot use the information presented in Table A.5 to test H against the alternative

$$A : \sigma_1 < \sigma_2$$

We could test H against this alternative if we had access to an expanded version of Table A.5, twice the size of the present one. As the following discussion shows, however, it is not really necessary to go to this trouble, for a simple reformulation of the alternative

$$A : \sigma_1 < \sigma_2$$

will turn it into

$$A : \sigma_2 > \sigma_1$$

and this pattern is acceptable to the F test. In short, the F test can be used only if the alternative contains the symbol ">" or "≠." If < is present in the statement of the alternative, the populations must be interchanged so that the < is reversed. For example, in the question concerning the variability of banana prices, we wanted to test the hypothesis that σ_T and σ_N were the same, as opposed to the alternative that σ_T was smaller than σ_N, namely, that prices varied less among supermarkets offering trading stamps. If we had set up our hypothesis testing problem as testing

$$H : \sigma_T = \sigma_N$$

versus

$$A : \sigma_T < \sigma_N$$

we could not make use of Table A.5 in applying the F test. But all that is necessary to solve the problem is to reformulate the hypothesis and alternative by interchanging the two populations. After the interchange, we get

$$H : \sigma_N = \sigma_T$$

versus

$$A : \sigma_N > \sigma_T$$

and in this formulation the problem can be solved by the methods illustrated in Example 5.10. With these facts in mind, we present the rejection rules for the two-sample F test in Table 5.19. Our final example of this section provides an illustration of the alternative $A : \sigma_1 \neq \sigma_2$ and the statistical procedures used to test $H : \sigma_1 = \sigma_2$ against it.

Example 5.11 Development of Arithmetic Skills   An educational psychologist wants to check the theory that elementary school pupils of different ages vary to different extents in their development of arithmetic skills. Some experts hold that the degree of variation among third-grade pupils in arithmetic ability does not differ significantly from that among sixth-grade pupils, while others believe that there is a substantial difference in variation within the pupils of each age group. The psychologist therefore conducts the following experiment: He randomly selects a group of 20 third-grade pupils and 25 sixth-grade pupils and gives them comparable arithmetic tests appropriate to their educational level. As the psychologist is interested only in variation of scores within each group, he does not even bother to record the mean scores but only the sample standard deviations. The 20 third-grade pupils obtain a standard deviation of 7 on their scores, while the 25 sixth-grade pupils have a standard deviation of 11. At level $\alpha = .02$, do the results of the study indicate the existence of a significant difference in variability of test scores within the two groups?

SOLUTION   We apply a two-sample F test of the hypothesis

$$H : \sigma_3 = \sigma_6 \text{ (variation does not differ between groups)}$$

versus

$$A : \sigma_3 \neq \sigma_6 \text{ (variation does differ between groups)}$$

TABLE 5.19

**Rejection Rules for the Two Possible Alternatives for the
Two-Sample F Test of Population Standard Deviations**

$$H : \sigma_1 = \sigma_2$$

Alternative	Reject H in Favor of A at Significance Level α if:	F
$A : \sigma_1 > \sigma_2$	$F > F_\alpha(n_1 - 1; n_2 - 1)$	$F = \dfrac{s_1^2}{s_2^2}$
$A : \sigma_1 \neq \sigma_2$	$F > F_{\alpha/2}(n_L - 1; n_S - 1)$	$F = \dfrac{\text{larger } s^2}{\text{smaller } s^2}$

NOTES: For the alternative $A : \sigma_1 \neq \sigma_2$, the test statistic is

$$F = \frac{\text{larger } s^2}{\text{smaller } s^2}$$

where larger s^2 = larger of the two squared sample standard deviations s_1^2 and s_2^2

smaller s^2 = smaller of the two squared sample standard deviations s_1^2 and s_2^2

Therefore the degrees of freedom for the numerator and denominator are, respectively, $n_L - 1$ and $n_S - 1$, where

n_L = number of data points in the sample having the larger squared standard deviation

n_S = number of data points in the sample having the smaller squared standard deviation

where σ_3 = true standard deviation of third-grade pupils

σ_6 = true standard deviation of sixth-grade pupils

Using symbols with analogous meanings, we summarize the data as

$$n_3 = 20$$
$$s_3 = 7$$
$$n_6 = 25$$
$$s_6 = 11$$

According to the rejection rule presented in Table 5.19, we should reject H at level $\alpha = .02$ if

$$F > F_{\alpha/2}(n_L - 1, n_S - 1) = F_{.01}(n_6 - 1, n_3 - 1) = F_{.01}(24, 19) = 2.92$$

because s_6^2 is the larger s^2, and s_3^2 is the smaller. The numerical value of $F_{.01}(24, 19)$ has been read off from Table A.5, more precisely from the portion of it appearing in Table 5.20. The numerical value of F is

$$F = \frac{\text{larger } s^2}{\text{smaller } s^2} = \frac{s_6^2}{s_3^2} = \frac{11^2}{7^2} = \frac{121}{49} = 2.47$$

As we have agreed to reject H in favor of A if $F > 2.92$, the result that $F = 2.47 < 2.92$

TABLE 5.20
A Small Portion of Table A.5
Values of $F_{.01}$

	Degrees of Freedom for Numerator	
		24
Degrees of Freedom for Denominator	19	2.92

does not lead us to reject H. We therefore conclude at level $\alpha = .02$ that the data presented do not provide convincing evidence of differing degrees of variability in arithmetic skills between the two age groups.

One final note on the F test: As with the t and chi-square tests before it, the F test is valid only when it is reasonable to assume that both sets of data come from normally distributed populations. This restriction is necessary because the same technical mathematics involved in the construction of Table A.3 (the table of the standard normal distribution) is used in the construction of Table A.5 (the table of the F distribution).

EXERCISES 5.E

1   In Exercise 5.B.3, contractual specifications require a machine to produce plastic piping of diameter 4 inches. A random sample of 50 pieces coming off the assembly line had standard deviation .12 inch. At level $\alpha = .05$, can we conclude from the sample that the diameter of the piping produced has standard deviation in excess of .10 inch and is therefore too variable to be useful?

2   A manufacturer of chemicals for use in the pharmaceutical industry conducts a study of the time it takes after receipt of an order for the chemicals to reach the customer's plant. The following data give the transit times from the manufacturer to the customer of six randomly selected shipments of the past month:

Shipment number	017	009	021	058	031	002
Transit time, days	6.0	5.0	3.0	4.5	4.0	5.5

At level $\alpha = .01$, is it reasonable to assert that the transit times have standard deviation which does not exceed 1.0 day?

3   From the data of Exercise 5.B.6, dealing with the time advance of aerospace timing devices, can we say at level $\alpha = .05$ that the time advance has standard deviation larger than .05 second?

4   The following table gives the wholesale price index and the index of industrial production in the United States on July 1 of each year of a 5-year period:

Year	Wholesale Prices	Industrial Production
#1	80	110
#2	80	90
#3	90	80
#4	80	100
#5	70	120

Do the data indicate at level $\alpha = .05$ that the index of industrial production is significantly more variable than the wholesale price index?

5 In Exercise 5.C.5, an economist obtained data on the price of English cheese in various metropolitan areas of the United States and Canada. Can he conclude from the data at level $\alpha = .10$ that the prices vary to approximately the same extent in the two countries?

SECTION 5.F TESTS FOR PROPORTIONS

We wind up the present chapter on testing statistical hypotheses with a discussion of some applied situations in which proportions and percentages are the primary items of interest. As has been pointed out in Sec. 3.B and Sec. 4.E, many questions involving proportions can be better understood if they are structured according to the format of the binomial distribution. This observation retains its validity in the context of the present section as well. As we shall soon see from the examples, only the small-sample tests make explicit use of the binomial distribution of Table A.2. The large-sample tests of proportions are based on percentage points of the normal distribution, because for large numbers of data points the normal approximation to the binomial distribution comes into play. Consider the following example of a small-sample test of proportion.

Example 5.12 Income Levels   In the opinion of the research staff attached to a congressional subcommittee on the economics of urban areas, 30% of full-time employed persons in the Dallas metropolitan area have incomes below 8000 dollars per year. Conflicting testimony given the committee by a consulting sociologist claims that the true percentage in question is substantially larger than 30%. To resolve the disagreement, the committee appoints an independent researcher to study the situation. The researcher surveys a random sample of 17 persons employed full time in the Dallas metropolitan area and finds that seven have incomes below 8000 dollars per year. Does her survey support at level $\alpha = .05$ the testimony that the true percentage in question is in excess of 30%?

SOLUTION   We must choose between the hypothesis

$$H : p = .30 \text{ (committee staff opinion is correct)}$$

and the alternative

$$A : p > .30 \text{ (consulting sociologist is correct)}$$

where $\quad p = $ true proportion of full-time employed persons
in the Dallas metropolitan area having incomes
below 8000 dollars per year

Because we are using a level of significance of $\alpha = .05$, we want the probability of a Type I error, namely, the probability of rejecting H when it is really true, to be no more than 5%. Now, in testing

$$H : p = .30$$

versus

$$A : p > .30$$

we would be inclined to reject H if substantially more than 30% of the 17 persons sampled have incomes below 8000 dollars per year. How are we to determine the meaning of the phrase "substantially more"?

In the small-sample test of proportion, we reason as follows: If the hypothesis

$$H : p = .30$$

were really true, then, as we are dealing with a random sample of 17 persons, the random variable

$$B = \text{the number of persons in the sample having}$$
$$\text{incomes below 8000 dollars per year}$$

has the binomial distribution with parameters $p = .30$ and $n = 17$. (Perhaps, at this point, you should glance back to Sec. 3.B, to remind yourself of the details of the binomial distribution.) For easy reference, we reproduce in the first three columns of Table 5.21, the portion of Table A.2 concerned with the binomial distribution having parameters $p = .30$ and $n = 17$. As we have noted in the above paragraph, we would be inclined to reject H if the random variable B turned out to be very large. Again, what does "very large" mean? Well, as we have been doing throughout this whole chapter, we define "very large" by referring to the significance level α. The number represents the probability of our rejecting H when H (namely, $p = .30$) is really true. As $\alpha = .05$ for this example, we want to find the numerical value of k such that the probability is .05 that B is larger than k. If we find this numerical value of k, then we will agree to reject H in favor of A at level $\alpha = .05$ if

$$B > k$$

because our probability of rejecting H when it's really true would be .05. This value of k is called the "critical value of k." Therefore, if $B > k$, we would reject

$$H : p = .30$$

TABLE 5.21
A Small Portion of Table A.2
With Adjustment

n	k	Values of $P(B \leq k)$ $p = .30$	Values of $P(B > k) = 1 - P(B \leq k)$
17	0	.0023	.9977
	1	.0193	.9837
	2	.0774	.9226
	3	.2019	.7981
	4	.3887	.6113
	5	.5968	.4032
	6	.7752	.2248
	7	.8954	.1046
	8	.9597	.0403
	9	.9873	.0127
	10	.9968	.0032
	11	.9993	.0007
	12	.9999	.0001
	13	1.0000	.0000
	14	1.0000	.0000
	15	1.0000	.0000
	16	1.0000	.0000
	17	1.0000	.0000

in favor of

$$A : p > .30$$

and we would assert at level $\alpha = .05$ that the true percentage of persons employed full time having incomes below 8000 dollars in the Dallas metropolitan area is indeed in excess of 30%.

We look for the critical value of k by using the fourth column of Table 5.21. While the third column gives the number $P(B \leq k)$, we have adjusted these numbers (by subtracting them from 1) to record the items $P(B > k)$ in the fourth column. We seek the critical value of k, namely, the value of k for which $P(B > k) = .05$ as follows: Halfway down the fourth column of Table 5.21, we discover the facts that

$$P(B > 7) = .1046$$

and

$$P(B > 8) = .0403$$

From those facts, we see that the critical value of k for $\alpha = .05$ should be somewhere between 7 and 8. However, there are unfortunately no possible values of k between 7 and 8 since the binomial distribution is a discrete distribution, as discussed in Sec. 3.B. We therefore come up with the unpleasant conclusion that there is no rejection rule having level of significance $\alpha = .05$. The two rejection rules having level of significance just above and just below $\alpha = .05$ are:

1 Reject H at level $\alpha = .1046$ if $B > 7$.

2 Reject H at level $\alpha = .0403$ if $B > 8$.

(Recall that B is the number of persons in the sample having incomes below 8000 dollars.)

We now must choose the more appropriate of these two significance levels, and we would probably agree that it should be $\alpha = .0403$, for that is the one closest to .05. We therefore agree to reject H at level $\alpha = .0403$ if $B > 8$.

It now remains only to find the numerical value of B, and this value is evident from the data. It is clear from the data that 7 of the 17 persons studied have incomes below 8000 dollars. Therefore $B = 7$. As we have agreed to reject H in favor of A at level $\alpha = .0403$ if $B > 8$, and it turned out that $B = 7 < 8$, we cannot reject H. Our conclusion at level $\alpha = .0403$ is then that the independent researcher's survey does *not* support the assertion that the percentage of full-time employed persons having incomes below 8000 dollars significantly exceeds 30%.

As it turned out, in the above example we would have failed to reject H at level $\alpha = .1046$ also. This is due to the fact that the rejection rule told us to reject H if $B > 7$, whereas B turned out to equal 7 exactly. Therefore $B > 7$ did not occur, and so we do not reject H. Because we did not reject H at level $\alpha = .0403$ and we also did not reject it at level $\alpha = .1046$, it is reasonable to conclude that we also would not reject H at level $\alpha = .05$. This kind of argument can be advanced only because .05 is sandwiched between two significance levels, at both of which H was rejected.

In the next example, we run more quickly through the technique of carrying out the small-sample test of proportion.

Example 5.13 Reliability of a Medical Test It is a known medical fact that roughly 40% of the adult population of Illinois has been exposed, at one time or another, to a certain minor urinary tract infection. Two NIH medical researchers propose a new test whose objective is to reveal whether or not an adult has been exposed to this disease. The test will generally register negative in virtually all of those who have *not* been exposed to the disease, but the researchers suspect that it also fails to detect past exposure in a large percentage of those who have been exposed to it. To check on the reliability of the proposed new test, the researchers select a random sample of 20 members of adult population of Illinois and test each for past exposure to the disease. They will consider the test to be unreliable if substantially less than 40% of the sample is judged to have been exposed to the disease. At a significance level of $\alpha = .05$, should the test be considered unreliable if it detects past exposure to the disease in only 4 of the 20 persons sampled?

SOLUTION   If we set

$$p = \text{true proportion of adults in which the test}$$
$$\text{detects past exposure to the disease}$$

then we want to test the hypothesis

$$H : p = .40 \text{ (the test is reliable)}$$

against the alternative

$$A : p < .40 \text{ (the test is unreliable)}$$

We must therefore carry out the small-sample test of proportion, using the binomial distribution with parameters $p = .40$ and $n = 20$, as the random sample is composed of 20 persons. If the hypothesis H is really true, the random variable

$$D = \text{number of adults in which past exposure to the}$$
$$\text{disease is detected}$$

has the binomial distribution with parameters $p = .40$ and $n = 20$. We would be inclined to reject

$$H : p = .40$$

in favor of

$$A : p < .40$$

if the numerical value of D turned out to be "small." Since our probability of rejecting H is $\alpha = .05$, we would reject H if D fell below the critical value of k for which

$$P(D < k) = .05$$

From Table A.2, for $p = .40$ and $n = 20$, we find that

$$P(D \leq 3) = .0160$$

and

$$P(D \leq 4) = .0510$$

Because $D \leq 4$ means the same thing as $D < 5$, it follows that

$$P(D < 5) = .0510$$

Therefore, we have discovered that at level $\alpha = .0510$ the critical value of k is 5.
It would then be logical to agree to reject

$$H : p = .40$$

in favor of

$$A : p < .40$$

if $D < 5$, for the significance level, the probability of rejecting H when it's really true, would then be $\alpha = .0510$, as close to $\alpha = .05$ as we could expect to get in a discrete situation.

Looking at the data, we see that the test has detected past exposure in only 4 out of the 20 persons sampled. Therefore $D = 4 < 5$; so we do reject H. At level $\alpha = .0510$, we are therefore confident in asserting that the test should be considered an unreliable detector of past exposure to the disease in question.

Now we develop the one-sample and two-sample z tests for handling large-sample problems involving proportions. The following example illustrates the "one-sample z test for the proportion of a population."

Example 5.14   A Political Poll   A local journalist reports, after talking to all his friends, that 60% of the state's electorate supports the reelection of the incumbent U.S. Senator. The challenger naturally takes issue with this report and orders his political science research staff to check the accuracy of the claim. The staff polls 1500 voters statewide and finds that 784 of those polled support the incumbent. At level $\alpha = .01$, does the poll tend to refute the journalist's report?

SOLUTION   Denote by p the true proportion of voters who support the incumbent. We want to test the hypothesis

$$H : p = .60 \text{ (the true proportion supporting the incumbent is 60\%)}$$

against the alternative

$$A : p < .60 \text{ (less than 60\% support the incumbent)}$$

In accordance with the principles of the normal approximation to the binomial distribution (since $n = 1500 > 30$), we compute

●
$$z = \frac{\hat{p} - p_0}{\sqrt{\dfrac{p_0(1 - p_0)}{n}}} \qquad (5.10)$$

where p_0 = hypothesized proportion supporting the incumbent
$\hat{p}$ = sample proportion supporting the incumbent
n = number of voters involved in the poll

(It should be noted that the formula is reminiscent of the one used in the one-sample z test for the mean of population.) We then reject H at level α if

$$z < -z_\alpha$$

This rejection rule has been chosen on the basis of the three options presented in Table 5.7, upon replacing the μ's by p's.
We therefore agree to reject

$$H : p = .60$$

in favor of

$$A : p < .60$$

at level $\alpha = .01$ if

$$z < -z_{.01} = -2.33$$

It now remains only to compute the numerical value of z. From the data accumulated in the poll, we know that

$$n = 1500$$

and

$$\hat{p} = \frac{784}{1500} = .5227$$

and, on the basis of the question asked, we see that the hypothesized proportion is

$$p_0 = .60$$

It follows automatically that

$$z = \frac{\hat{p} - p_0}{\sqrt{\dfrac{p_0(1 - p_0)}{n}}} = \frac{.5227 - .60}{\sqrt{\dfrac{(.60)(.40)}{1500}}} = \frac{-.0773}{\sqrt{.00016}} = \frac{-.0773}{.01265} = -6.114$$

As we agreed to reject H if z turned out to be less than (more negative than!) -2.33, the result that $z = -6.114 < -2.33$ leads us to reject H. At level $\alpha = .01$, the challenger's poll therefore tends to refute the journalist's report that 60% of the voters support the incumbent. The polls indicate that the true percentage supporting the incumbent is significantly below 60%.

As a final example of the application of hypothesis testing to the study of proportions, the following situation calls for the "two-sample z test for the difference of proportions."

Example 5.15 Advertising Effectiveness The advertising department of a major Southern California automobile dealership, which has branches in Orange and Los Angeles Counties, wants to know whether its advertising is really effective in influencing prospective customers. In particular, the department wants to find out whether more station wagons can be sold by saturating an area with advertising urging customers to buy station wagons. To test out the effectiveness of such an advertising campaign, the dealership for one week floods Orange County with advertising, while maintaining a low profile in Los Angeles County. Over the next 2 months 120 of 400 cars it sells at the Orange County branch are station wagons, while 150 of the 600 cars it sells at the Los Angeles branch are station wagons. Do the results of the study tend to indicate that a proportionately higher number of station wagons was sold in Orange County, where the advertising was concentrated? Use a level of significance of $\alpha = .05$.

SOLUTION   We define some symbols as follows:

p_H = true proportion of station wagons sold in region of *heavy* advertising

p_N = true proportion of station wagons sold in region of *normal* advertising

We want to test the hypothesis

$$H : p_H = p_N \text{ (advertising does not affect sales}$$
$$\text{of station wagons)}$$

against the alternative

$$A : p_H > p_N \text{ (heavy advertising increases sales}$$
$$\text{of station wagons)}$$

From the data, we can determine the items

n_H = number of data points in the region of heavy
  advertising

$\hat{p}_H$ = sample proportion of station wagons sold in
  the region of heavy advertising

n_N = number of data points in the region of normal
  advertising

$\hat{p}_N$ = sample proportion of station wagons sold in
  the region of normal advertising

$\hat{p}$ = sample proportion of station wagons sold in
  both regions combined

Sales records indicate that

$$n_H = 400$$
$$\hat{p}_H = \frac{120}{400} = .30$$
$$n_N = 600$$
$$\hat{p}_N = \frac{150}{600} = 0.25$$
$$\hat{p} = \frac{120 + 150}{400 + 600} = \frac{270}{1000} = 0.27$$

In accordance with the rejection rules of Table 5.7, we calculate

●
$$z = \frac{\hat{p}_H - \hat{p}_N}{\sqrt{\hat{p}(1 - \hat{p})\left(\dfrac{1}{n_H} + \dfrac{1}{n_N}\right)}} \tag{5.11}$$

(The formula might remind you of the one for the two-sample t test for the difference of
means), and then we reject

$$H : p_H = p_N$$

in favor of

$$A : p_H > p_N$$

at level α if

$$z > z_\alpha$$

Here $\alpha = .05$; so we agree to reject H if

$$z > z_{.05} = 1.64$$

Upon inserting the numerical values of the components of the formula for z, we get

$$z = \frac{\hat{p}_H - \hat{p}_N}{\sqrt{\hat{p}(1 - \hat{p})\left(\frac{1}{n_H} + \frac{1}{n_N}\right)}} = \frac{.30 - .25}{\sqrt{(.27)(.73)\left(\frac{1}{400} + \frac{1}{600}\right)}}$$

$$= \frac{.05}{\sqrt{(.197)\left(\frac{1000}{240,000}\right)}} = \frac{.05}{\sqrt{(.197)(.004167)}} = \frac{.05}{\sqrt{.00082}}$$

$$= \frac{.05}{.02866} = 1.745$$

Therefore $z = 1.745 > 1.64$; so we must reject H in favor of A at level $\alpha = .05$. We conclude that at level $\alpha = .05$, the advertising campaign can be considered to have been effective in increasing the proportion of station wagons sold.

One final remark on the above example: If the proper level of significance had been $\alpha = .01$ in the opinion of the advertising department, the advertising campaign would have been judged noneffective. This is due to the fact that the rejection rule would have indicated the rejection of H when $z > z_{.01} = 2.33$. Since z turned out to be 1.745, H could not validly have been rejected, and we would have to say that the advertising did not seem to affect sales. We have here one more example of the importance of making an appropriate subjective choice of the significance level α, the probability of rejecting H when it's really true.

EXERCISES 5.F

1   To find out whether or not a particular coin is really "fair" (equally likely to fall heads or tails), the coin is tossed several times and the number of heads resulting is carefully noted.
    a If 4 out of 10 tosses result in heads, can we assert at level $\alpha = .05$ that the coin is fair?
    b If 40 out of 100 tosses result in heads, can we conclude at level $\alpha = .05$ that the coin is fair?

2   An executive in charge of new product development at a local dairy feels that at least 10% of the population would be willing to try a carton of chocolate-flavored buttermilk, while the other 90% would not be at all interested. Is her opinion substantiated at level $\alpha = .10$ by a survey of 50 randomly selected persons of whom four expressed interest?

3   The manufacturer of a new chemical fertilizer advertises that with the aid of his fertilizer, at least 80% of the seeds planted will germinate. Does an experiment in which 10 seeds out of 17 germinate tend to refute his claims at level $\alpha = .05$?

4 A "straw poll" of 14 randomly selected members of the 50 state legislatures reveals 7 who favor a certain controversial change in the banking laws.

   a At level $\alpha = .10$, do the results of the poll conflict with a news reporter's assertion that at least 60% of the membership supports the change?

   b At level $\alpha = .10$, do the results of the poll conflict with a banking lobbyist's assertion that no more than 40% of the membership supports the change?

5 A straw poll of 140 randomly selected legislators turns up 70 who favor a certain controversial change in the banking laws.

   a At level $\alpha = .10$, do the results of the poll conflict with a news reporter's assertion that at least 60% of the membership supports the change?

   b At level $\alpha = .10$, do the results of the poll conflict with a banking lobbyist's assertion that no more than 40% of the membership supports the change?

6 One public health physician conducts a study to determine whether persons in his geographical area of responsibility are more likely to have lung disorders if they are heavy smokers than if they are not. Of 200 heavy smokers aged 50 or over, 10% report some sort of lung problem, while of 300 persons who are not heavy smokers, only 7% suffer from lung ailments. At level $\alpha = .05$, can the physician conclude that heavy smokers are significantly more likely than other persons to contract lung disorders?

7 A survey of 50 licensed drivers living in an urban area turns up 8 who have a drinking problem, while a survey of 30 drivers in a rural area reveals that 4 are problem drinkers. Can we conclude at level $\alpha = .05$ that the proportion of problem drinkers differs significantly between urban and rural areas?

SUMMARY AND DISCUSSION

In this chapter, we have introduced the procedure of formulating a question in statistical terms and then answering it using available data. As we have seen, the first step is setting up a hypothesis and an alternative, and the second is using the data to help us choose between them. We have illustrated the technique of testing statistical hypotheses in examples involving means, standard deviations, and proportions. The general problem of statistical inference that we first encountered in our study of confidence intervals in Chap. 4, namely, the fact that we cannot be 100% sure of our decision, is still with us. It is reflected in the fact that each statistical decision is made in terms of a "level of significance," which is our probability of rejecting the hypothesis we formulated when that hypothesis is actually true. Having chosen a level of significance gives us an idea of our chances of making an incorrect choice between hypothesis and alternative. Our study of testing statistical hypotheses does not end here, but continues into Chaps. 6 and 9 to 11, as we continue to ask important questions and to answer them on the basis of a collection of statistical data.

BIBLIOGRAPHY

Significance Levels

Feinberg, W. E.: Teaching the Type I and Type II Errors: The Judicial Process, *The American Statistician*, June 1971, pp. 30–32.

Friedman, H.: Trial by Jury: Criteria for Convictions, Jury Size and Type I and Type II Errors, *The American Statistician*, April 1972, pp. 21–23.

Morrison, D. E., and R. E. Henkel (eds.): "The Significance Test Controversy—A Reader," Aldine, Chicago, 1970.

Webster, G. W., and T. A. Cleary: A Proposal for a New Editorial Policy in the Social Sciences, *The American Statistician*, April 1970, pp. 16–19.

Calculating the Type II Error

Cohen, J.: "Statistical Power Analysis for the Behavioral Sciences," Academic Press, New York, 1969.

Special Testing Situations

Gibbons, J. D., and J. W. Pratt: P-values: Interpretation and Methodology, *The American Statistician*, February 1975, pp. 20–25.

John, S.: Tables for Comparing Two Normal Variances or Two Gamma Means, *Journal of the American Statistical Association*, June 1975, pp. 344–347.

Scheffé, H.: Practical Solutions of the Behrens-Fisher Problem, *Journal of the American Statistical Association*, December 1970, pp. 1501–1508.

Professional Publications Carrying Statistical Analyses

American Journal of Political Science
American Journal of Public Health
American Journal of Psychology
American Sociological Review
The American Statistician
Bell Journal of Economics
Biometrics
Decision Sciences
Ecological Monographs
Ecology
Educational and Psychological Measurement
Journal of Accounting Research
Journal of the American Statistical Association
Journal of Educational Psychology
Management Science
Pacific Sociological Review
Psychometrika
Psychological Reports
Review of Economics and Statistics

SUPPLEMENTARY EXERCISES

1 The manufacturer of a hospital room humidifier advertises that the control dial on the device will maintain a mean room humidity of 80%, with a standard deviation of 2.0%. Its performance was carefully observed during a 1-day test at 60 different times throughout the day. The mean humidity level turned out to be 78.3 with a standard deviation of 2.9.

    a At level $\alpha = .01$, do the observed data tend to contradict the manufacturer's claim that the true mean humidity level will be 80?

    b At level $\alpha = .05$, does the 1-day test indicate that the true standard deviation is somewhat larger than the advertised value of 2.0?

2 In a routine check of drug packaging, the FDA counts the contents of 16 bottles of aspirin labeled as containing 500 tablets each. The 16 bottles have mean contents 494 aspirins with a standard deviation of 30.

    a Can the FDA assert at level $\alpha = .05$ that the bottles are underfilled?

    b Can they assert at level $\alpha = .05$ that the true standard deviation of such bottles is significantly above 15?

3 A company that markets canned carrots advertises that their 16-ounce cans actually have mean weight of 16.1 ounces with a standard deviation of .1 ounce, so that a substantial percentage of cans exceed the stated contents (16.0 ounces). To test the validity of the company's assertion, a state consumer agency selects a random sample of 11 cans and carefully weighs the contents of each of them. The 11 weights are as follows:

    16.1   16.0   16.2   15.9   16.0   16.1   16.1   15.9   16.1   16.0   15.6

    a At significance level $\alpha = .05$, can the agency assert that the true mean contents are below the 16.1 ounces claimed by the company?

    b At level $\alpha = .05$, do the data indicate that the true mean contents are below the 16.0 ounces printed on the label?

    c At level $\alpha = .05$, can the agency conclude from the data that the true standard deviation exceeds 0.1 ounce?

4 A truck farmer decides to test a new nonpolluting insecticide which also, according to the inventor, definitely reduces the loss attributable to a common pest. The farmer has a feeling that the new spray is really worse than the standard (polluting) spray; so he treats 31 acres with the new spray and 41 with the standard spray throughout the growing season. For the new insecticide, the mean yield per acre turned out to be 980 pounds, with a standard deviation of 60 pounds; while for the standard spray there was a mean yield of 1040 pounds with a standard deviation of 50 pounds.

    a At level $\alpha = .01$, do the results of the test provide evidence that the new spray is significantly worse than the standard one?

    b At level $\alpha = .05$, can we conclude that the new spray is more variable in its performance than the standard spray in the sense of having a larger yield standard deviation?

5 In order to compare the merits of two short-range ground-to-ground antitank weapons, 8 of the conventional type and 10 of the more expensive wire-guided rockets are tested. The conventional type have mean target error of 5.2 feet with a standard deviation of 1.8 feet, while the wire-guided ones have mean target error of 3.6 feet with a standard deviation of 1.5 feet. At level $\alpha = .10$, can we say that the more expensive rockets are really more accurate?

6 After the recent detonation of a 20-kiloton nuclear bomb in the atmosphere at the Lop Nor testing site in northwestern China, scientists in Japan and the United States immediately began to measure the increase in radiation levels due to the nuclear fallout. The radiation cloud, released directly into the atmosphere, was first detected by instruments in Japan 5 days after the test, and it reached eight western states of the United States a week later. The following data give the radiation amounts (in picocuries of radium per cubic meter of air above the usual natural level of 2.1) at six geographic stations in Japan and eight in the United States:

Japan	.2	.4	.5	.1	.3	.3		
United States	.2	.3	.1	.1	.2	.2	.4	.1

a At level $\alpha = .05$, decide whether nuclear fallout levels in Japan are greater than those in the United States after release of a radiation cloud into the atmosphere over China.

b At level $\alpha = .10$, can we assert that radiation levels within each of the two countries have the same degree of variability, even though the United States is much farther away from China than is Japan?

7 It is the job of a tea tester for an English tea-importing company to decide which tea leaves are most suitable for English Breakfast tea. The tea tester must decide between two lots of leaves: an inexpensive lot and an expensive lot. The tea tester and his laboratory staff come up with the following levels of impurities in eight randomly selected leaves of each lot:

Inexpensive lot	3	4	25	9	11	4	12	20
Expensive lot	6	5	7	10	6	8	5	4

The tea tester knows, of course, that he cannot validly use the t test to decide whether or not both lots of leaves have the same average impurity levels unless both populations (inexpensive and expensive impurity levels) have the same standard deviation. At significance level $\alpha = .02$, do the data indicate that the two standard deviations are the same?

8 The research department of a company that produces industrial string proposes a new technique for strengthening the string and making it able to withstand stronger forces. To find out whether the new technique really does strengthen the string, five lengths of string are produced by the standard technique and five by the new technique and the samples are compared for

their respective breaking points, in pounds of force. The breaking points are as follows:

| Standard technique | 144 | 131 | 155 | 126 | 134 |
| New technique | 139 | 154 | 132 | 143 | 147 |

a At level $\alpha = .10$, does the comparison indicate that the breaking points of both groups have the same standard deviation?

b At level $\alpha = .10$, does the comparison support the contention of the research department that the new technique produces stronger string?

9 Several insurance adjusters were concerned about the unusually high repair estimates they seemed to be getting from Fosbert's U-bet Repair Station. To test their suspicions, they brought each of eight damaged cars to Fosbert's and also to the auto repair shop of Nickle's Department Store, a concern generally regarded as reliable. They obtained the following estimates, in hundreds of dollars:

Car Number	#1	#2	#3	#4	#5	#6	#7	#8
Fosbert estimate	2.1	4.5	6.3	3.0	1.2	5.4	7.3	9.3
Nickle estimate	2.0	3.8	5.9	2.8	1.3	5.0	6.5	8.6

At level $\alpha = .01$, can the adjusters conclude from their survey that Fosbert's estimates are significantly higher than Nickle's?

10 A committee of elementary school teachers is assigned the task of comparing the "new math" with the "old math" in the ability to develop pupils' skills in arithmetic. Twenty pupils (10 from a group being taught the "new math," and 10 from a group being taught the "old math") were paired on the basis of age, athletic ability, family income, and general health; and then the committee administered the same arithmetic skills test to all 20 students. The scores follow:

Pair	A	B	C	D	E	F	G	H	I	J
New math group	100	50	10	60	80	60	80	40	20	20
Old math group	90	60	20	60	70	70	90	50	10	40

Do the test scores indicate at level $\alpha = .05$ that the committee should report that the "new math" is less effective than the "old math" in developing arithmetic skills?

11 A compendium of actuarial tables published in 1958 shows that of all 45-year-old men in a certain occupational grouping 7% die before reaching age 55. A more recent survey of 2000 men in the age and occupational group reveals that percentage to be 5%. At level $\alpha = .01$, does the recent survey indicate that the probability of death in that group has been significantly reduced since 1958?

12 One urban affairs sociologist claims that 55% of the adult population of a particular major city have been victimized by a criminal at one time or another.

    a Does a random sample of 18 residents of whom 8 have been victimized tend to refute his claim at level of significance $\alpha = .05$?

    b Does a random sample of 180 residents of whom 80 have been victimized tend to refute the claim at level $\alpha = .05$?

13 As part of an analysis of overdue accounts which eventually require legal action to force payment, a corporation accountant discovers that 45 out of 130 corporate accounts required legal action in the past, while 214 of 1070 individual accounts required such action. Can the accountant conclude that a corporate account is significantly more likely to require legal action than an individual account? Use level $\alpha = .01$.

14 A political poll concerning the election in which Able is running against Baker shows that 520 of 1000 urban voters favor Able, while 240 of 500 rural voters favor Able. Can we conclude from the results of the poll that Able is significantly more popular at level $\alpha = .05$ in urban areas than in rural areas?

15 The following data record the number of cases of infectious diseases reported in Scotland in 1972, as reported in both the 1973 and 1974 editions of the "Annual Abstract of Statistics."[1] (The 1974 edition presumably contained revised and, hopefully, more accurate data.)

Disease	Number of Cases Reported in 1972	
	1973 ed.	*1974 ed.*
Typhoid	30	30
Erysipelas	120	123
Scarlet fever	570	565
Whooping cough	230	226
Dysentery	3472	3188
Ophthalmia neonatorum	46	44
Pneumonia	3364	2959
Puerperal fever	52	53
Tuberculosis	1719	1655

At level $\alpha = .05$, can we say that significant differences exist between the numbers of cases reported in the two editions of the compendium?

[1]Central Statistical Office (United Kingdom), Her Majesty's Stationery Office, London, 1973 (p. 73) and 1974 (p. 74).

LINEAR REGRESSION
AND CORRELATION

In a study of the efficiency of automobile engines with respect to consumption of gasoline, one particular 3990-pound car was operated at various speeds and the gasoline consumption (in miles per gallon) at each speed level was carefully measured. The data of Table 6.1 express the results of the experiment. Looking at the data, we can immediately draw one conclusion: At higher speeds, there is less efficient use of gasoline; namely, the car gets fewer miles per gallon of gas. As we have noted several times already, however, a set of data contains within it information of precision much greater than the sort given by the above general statement. In particular, from a detailed analysis of the data, we might be able to answer, with some degree of accuracy, the following four questions:

1 Is there a simple mathematical relationship (i.e., a relationship expressible by a simple formula) between speed, in miles per hour, and efficiency, in miles per gallon?
2 If so, what is this relationship?
3 How much of the variation in efficiency level can be attributed to variations in

TABLE 6.1
**Efficiency of an
Automobile Engine**

Speed, miles/hour	Efficiency, miles/gallon
30	20
40	18
50	17
60	14
70	11

speed; i.e., among all those factors that tend to affect gasoline efficiency, how important a role is played by the speed of the car?

4 Can we use the data to predict the efficiency at other speeds, for example 25, 55, or 80, and how accurate will our predictions be?

The techniques of regression and correlation were developed for the purpose of answering the four questions above. We now proceed to the development of these techniques.

SECTION 6.A THE GEOMETRY OF THE STRAIGHT LINE

Suppose we put the data of Table 6.1 on a scattergram of the sort introduced in Chap. 1. The scattergram of Fig. 6.1 communicates to the viewer information beyond that contained in the earlier statement, that "at higher speeds, there is less efficient use of gasoline." From the scattergram, we can see that there seems to be a definite linear trend which, while not perfect, does describe the general characteristics of the relationship between speed and efficiency.

If the scattergram of a set of data indicates a linear trend, the next logical step is to specify the unique straight line which most closely approximates all the data points simultaneously. Figure 6.2 shows three possible candidates for the straight line which best fits the data. To select from among all the possible straight lines the one best-fitting line, it is necessary to analyze more carefully the numerical relationship between speed and efficiency presented by Table 6.1.

As it turns out, every straight line has its own equation, which is a mathematical description of the line's slope and location. For example, let's consider the line that joins the two points in Fig. 6.3. One of the points is labeled (1, 2) to indicate that it lies at a perpendicular distance 1 to the right of the vertical axis and a perpendicular distance 2 above the horizontal axis. The other is labeled (5, 4), because it lies 5 units to the right of the vertical axis and 4 units above the horizontal axis. [*Note*: A point labeled $(-4, -3)$ would lie 4 units to the *left* of the vertical axis and 3 units *below* the horizontal axis.]

The line joining (1, 2) with (5, 4) also contains several other points—infinitely

FIGURE 6.1
Scattergram of speed versus efficiency.

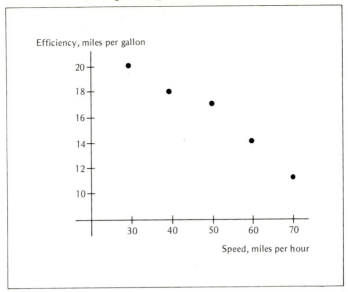

many, to be sure. In fact, a line can be viewed as a collection of points all "glued" together in a special configuration. If a typical point is labeled (x, y) when it lies x units to the right of the vertical axis and y units above the horizontal axis, some of these points

FIGURE 6.2
Possible best-fitting lines.

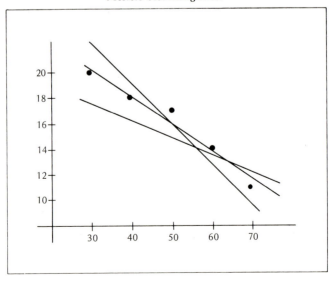

FIGURE 6.3
A typical straight line.

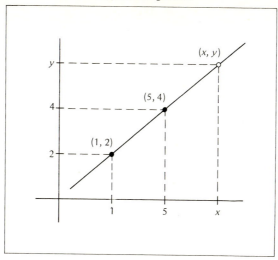

(x, y) happen to fall exactly on the line, and some do not. Actually, relatively few of the points (x, y) do fall on the line in question—most do not. When (x, y) does fall on the line, two triangles are formed as in Fig. 6.4, a smaller triangle inside a larger triangle. When (x, y) falls off the line, no such "nested" triangles are formed, but a situation like that in Fig. 6.5 occurs instead.

In Figure 6.4, where (x, y) falls on the line joining (1, 2) with (5, 4), the triangles formed are similar. "Similar to what?," you may ask. Well, you might recall from plane geometry that two triangles are said to be similar if they have the same shape, although they need not be the same size. In such triangles, corresponding sides are proportional; for example, in Fig. 6.4, we know that A is to $B + D$ as C is to D. Algebraically, this means that (x, y) lies on the line if and only if[1]

$$\frac{A}{B + D} = \frac{C}{D}$$

Now the geographic layout of Fig. 6.4 indicates that $A = y - 2$, being the distance from 2 to y on the vertical axis. Similarly, $B + D = x - 1$, $C = 4 - 2 = 2$, and $D = 5 - 1 = 4$. Therefore, making these substitutions, we get that

$$\frac{y - 2}{x - 1} = \frac{2}{4}$$

Using cross-multiplication and other algebraic techniques, we find that (x, y) lies on the

[1]For example suppose $x = 9$ and $y = 6$. Then $A = 4$, $B = 4$, $C = 2$, and $D = 4$. Substituting, we get $4/(4 + 4) = 2/4$, which is true, so that (9, 6) lies on the line. On the other hand, suppose $x = 10$ and $y = 7$. Then $A = 5$, $B = 5$, $C = 2$, and $D = 4$. Substituting, we get $5/(5 + 4) = 2/4$, which is *not* true, so (10, 7) does not lie on the line.

FIGURE 6.4
(x, y) on the line.

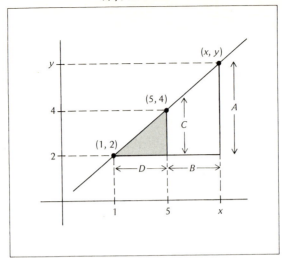

line if and only if, in sequence,

$$4(y - 2) = 2(x - 1)$$
$$4y - 8 = 2x - 2$$
$$4y = 2x + 6$$

and, finally

$$y = \tfrac{1}{4}(2x + 6)$$
$$= 0.5x + 1.5$$
$$= 1.5 + 0.5x$$

FIGURE 6.5
(x, y) off the line.

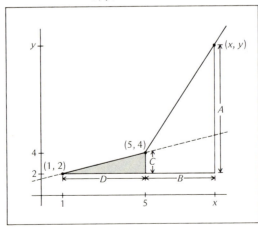

This means that (x, y) lies on the line joining $(1, 2)$ with $(5, 4)$ if and only if

$$y = 1.5 + 0.5x$$

That equation is called "the equation of the line joining $(1, 2)$ with $(5, 4)$." The equation specifies which points (x, y) are on the line and which are not, for it says that (x, y) is on the line if and only if y is 1.5 more than half of x. For example, $(5, 4)$ is on the line because $4 = 1.5 + (0.5)(5)$; $(9, 6)$ is on the line because $6 = 1.5 + (0.5)(9)$; and $(11, 8)$ is not on the line because $8 \neq 1.5 + (0.5)(11)$.

In the same way, we could start with any two points and work out the equation of the line joining them. Because the geometric and algebraic techniques used would be the same, we would come up with an equation of the form

$$y = a + bx$$

where a and b are numbers. Therefore every straight line has an equation of the form $y = a + bx$.

Conversely, from any equation of the form $y = a + bx$, we can draw the corresponding straight line. For example, consider the equation

$$y = -2 + 3x$$

If we can find any two points on the line represented by this equation, we can connect them in order to obtain the graph of the line. We can find points by picking x values at random and then working out the corresponding y values. If we choose $x = 0$, then $y = -2 + (3)(0) = -2 + 0 = -2$, and so the point $(0, -2)$ is on the line. If, next, we select $x = 2$, then $y = -2 + (3)(2) = -2 + 6 = 4$, so that $(2, 4)$ lies on the line. Connecting the points $(0, -2)$ and $(2, 4)$, as in Fig. 6.6, we get a picture of the line. Notice that an individual choosing $x = 1$ and $x = 3$, instead of $x = 0$ and $x = 2$, would obtain the same line by use of the points $(1, 1)$ and $(3, 7)$. All these points lie on the line whose equation is $y = -2 + 3x$.

FIGURE 6.6
The line $y = -2 + 3x$.

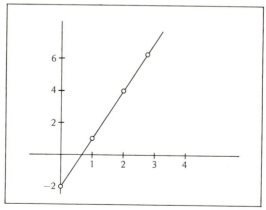

1 For each of the following pairs of points, determine the equation of the straight line passing through them:
 a (1, 2) and (5, 3).
 b (0, −2) and (2, 1).
 c (1, 9) and (5, 5).

2 For each of the following pairs of points, determine the equation of the straight line passing through them:
 a (1, 2) and (5, 6).
 b (0, −2) and (2, 5).
 c (1, 9) and (5, 8).

3 For each of the following equations, find two points on the corresponding straight line and draw the line:
 a $y = 6 + 2x$.
 b $y = 8 − 3x$.
 c $y = −3 + 4x$.

4 For each of the following equations, find two points on the corresponding straight line and draw the line:
 a $y = 8 + 3x$.
 b $y = 6 − 2x$.
 c $y = −4 + 3x$.

5 Determine the equation of the straight line passing through the points (8.1, 2.3) and (3.2, 1.8), and find a third point lying on that line.

6 Find two points lying on the straight line having equation $y = 7.8 − 2.2x$, and draw the line.

SECTION 6.B THE REGRESSION LINE

Let's return now to our original objective of finding the line that best fits the points of Fig. 6.1. Such a line is called the "regression line" of the points. We know from the discussion above that the regression line has an equation $y = a + bx$, where a and b are particular numbers. Therefore, in order to find the equation of the regression line, we have only to determine the numerical values of a and b.

Because we are going to have to get very precise in our search for the numerical values of a and b, we must first specify exactly what we mean by the phrase "best-fitting line." In particular, we must decide which line would best fit the purposes for which we want to use the line eventually.

The data of Table 6.1 relate the gasoline efficiency of the engine to the speed of the car. The regression line would be a mathematical relationship between speed and efficiency, and we would most likely use such a relationship to predict efficiency at various speeds. In that situation it is traditional, algebraically speaking, to label speed by x, the "independent" variable, and efficiency by y, the "dependent" variable. Then the

FIGURE 6.7
Vertical deviations of data points from the line $y = a + bx$.

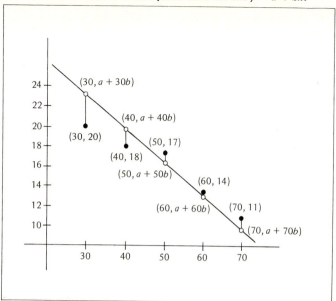

equation $y = a + bx$ expresses the manner in which efficiency "depends on" speed. It is important to note that a statement such as this does not necessarily imply any causal dependence, like a cause-and-effect relationship, but merely an algebraic connectión between the numbers representing efficiency and the numbers representing speed. We agree then to label efficiency as y and speed as x in accordance with the procedures just discussed, and, in general, we shall always use y to denote the variable to be predicted.

A prediction of efficiency, based on the regression line $y = a + bx$, will be as accurate as possible only if we choose a and b in such a way that the vertical distances between the data points and the line (known as "vertical deviations") are collectively as small as possible. This situation is illustrated in Fig. 6.7, where the original data points are compared with the location predicted for them by a possible regression line $y = a + bx$. By analogy with the development of the mean and standard deviation in Chap. 2, we consider the regression line as a sort of "mean" or "balancing line" of the data points. By analogy with the balancing property of the mean, it would be reasonable to have the positive and negative vertical deviations of the data points from the line cancel each other. As in the analysis leading to the standard deviation, then, the real deviation of the points from the line will have to be calculated using absolute values or squares in order to eliminate the cancellation effects of points falling above and below the line. As in the case of the mean absolute deviation discussed briefly in Chap. 2, there are some situations when mean-absolute-deviation regression lines would be appropriate,[1] but,

[1]See, for example, the articles listed in the bibliography.

by and large, it is more often useful to deal with squared deviations. This is especially true where we have reason to believe that one or both columns of data come from normally distributed populations, but even in nonnormal situations there are Chebyshev-type estimates based on the squared deviations.

A typical set of n data pairs (x_1, y_1), (x_2, y_2), ..., (x_n, y_n) to be fitted to a line $y = a + bx$ will have its sum of squared deviations from the line calculated as in Table 6.2. The "predicted locations" in Table 6.2 are those open circles lying on the line in Fig. 6.7. From Table 6.2, we see that the sum of squared deviations, SSD, is

$$SSD = \sum_{k=1}^{n} (y_k - a - bx_k)^2$$

Now, for a particular set of data points, the set given in Table 6.1, for example, we want to find the particular numerical values of a and b for which the equation $y = a + bx$ will be the best-fitting line. More precisely, we want to find the values of a and b that result in the smallest possible value of SSD, the sum of squared deviations. Then the line $y = a + bx$ will be, on the average, the best possible predictor. The regression line resulting from this method of reasoning is often called "the least-squares line." For different sets of data points, we would expect to come up with different numerical values of a and b, because the regression lines would most likely be different in each case. So what we really need are formulas for a and b that we can use to compute their numerical values from the data.

As it turns out, the expression for the sum of squared deviations SSD contains the solution to our problem. As a and b change, the numerical value of SSD increases and decreases. We are interested in those values of a and b for which SSD is as small as possible (and can only increase, not decrease, if we vary a and b). Happily, the determination of those values of a and b from the data is not too difficult, although the algebra is somewhat complicated. The primary algebraic technique is "completing the square," and for those of you who are familiar with the algebra and who would like to follow the derivation of the expressions for a and b, the details appear in Supplement 6.1. Here we shall only quote the resulting formulas and show how to apply them.

TABLE 6.2
Calculation of the Sum of Squared Deviations

Actual Data Points		Predicted Locations $y = a + bx$	Deviations $y - (a + bx)$	Squared Deviations $(y - a - bx)^2$
x	y			
x_1	y_1	$a + bx_1$	$y_1 - (a + bx_1)$	$(y_1 - a - bx_1)^2$
x_2	y_2	$a + bx_2$	$y_2 - (a + bx_2)$	$(y_2 - a - bx_2)^2$
......				
x_n	y_n	$a + bx_n$	$y_n - (a + bx_n)$	$(y_n - a - bx_n)^2$

$$\text{Sum of squared deviations } (SSD) = \sum_{k=1}^{n} (y_k - a - bx_k)^2$$

The upshot of the algebraic analysis in the small print is that a and b can be determined from the data using the formulas

$$b = \frac{n\Sigma xy - (\Sigma x)(\Sigma y)}{n\Sigma x^2 - (\Sigma x)^2} \qquad (6.1)$$

$$a = \frac{\Sigma y - b\Sigma x}{n} \qquad (6.2)$$

It should be noted that the numerical value of b must be calculated first, because it is itself used in the calculation of a. Another important observation is that Σx^2 is not the same as $(\Sigma x)^2$. For example, if $x_1 = 2$, $x_2 = 5$, and $x_3 = 1$, we would have

$$\Sigma x^2 = 2^2 + 5^2 + 1^2 = 4 + 25 + 1$$

but
$$(\Sigma x)^2 = (2 + 5 + 1)^2 = 8^2 = 64$$

Let us now proceed to the use of these formulas to solve the speed/efficiency problem for which the data of Table 6.1 were gathered.

The formulas for a and b involve sums of several quantities, namely the following:

Σx = sum of the x values of the data points
Σy = sum of the y values of the data points
Σx^2 = sum of the squares of the x values
Σxy = sum of the products of each x multiplied by
its corresponding y

The simplest way to keep track of all the calculations necessary to obtain the regression line is by means of a table of the sort illustrated in Table 6.3. In our table, we will also include the calculation of Σy^2, the sum of squares of the y values, which will turn out to be needed in estimating the possible error inherent in using the regression line to predict future occurrences. Using the sums obtained in Table 6.3, we can complete the calculation of a and b by inserting the sums into their proper places in the formulas. This is done as follows:

TABLE 6.3
Preliminary Calculations in Linear Regression

	Speed x, miles/hour	Efficiency y, miles/gallon	x^2	y^2	xy
	30	20	900	400	600
	40	18	1600	324	720
	50	17	2500	289	850
	60	14	3600	196	840
	70	11	4900	121	770
Sums $n = 5$	250	80	13,500	1330	3780

$$b = \frac{n\Sigma xy - (\Sigma x)(\Sigma y)}{n\Sigma x^2 - (\Sigma x)^2} = \frac{5(3780) - (250)(80)}{5(13,500) - (250)^2}$$

$$= \frac{18,900 - 20,000}{67,500 - 62,500} = \frac{-1100}{5000} = -.22$$

$$a = \frac{\Sigma y - b\Sigma x}{n} = \frac{80 - (-.22)(250)}{5}$$

$$= \frac{80 + 55}{5} = 27$$

Having calculated $a = 27$ and $b = -.22$, we therefore know that the equation of the regression line

$$y = a + bx$$

is, in this case,

$$y = 27 - .22x$$

It is often useful, for purposes of visual communication and analysis, to superimpose the graph of the regression line upon the scattergram of the data. The scattergram for the speed/efficiency problem appears in Fig. 6.1. We superimpose the graph of the line

$$y = 27 - .22x$$

on the scattergram by choosing two points on the line and connecting them as in the discussion relating to Fig. 6.6. If we choose $x = 30$, then $y = 27 - .22x = 27 - (.22)(30) = 20.4$, so that $(30, 20.4)$ is on the regression line. Choosing $x = 60$, the corresponding $y = 27 - (.22)(60) = 13.8$, so that $(60, 13.8)$ is a second point on the regression line. The two points, $(30, 20.4)$ and $(60, 13.8)$, are graphed together with the regression line in Fig. 6.8.

SUPPLEMENT 6.1 Algebraic Derivation of the Linear Regression Formulas

As explained in the regular text, we must find the values of a and b that make

$$SSD = \sum_{k=1}^{n} (y_k - a - bx_k)^2$$

as small as possible. Using the binomial formula $(A - B)^2 = A^2 - 2AB + B^2$ and other algebraic operations such as factoring, we have that

$$SSD = \sum_{k=1}^{n} (y_k - a - bx_k)^2 = \sum_{k=1}^{n} [(y_k - bx_k) - a]^2$$

$$= \sum_{k=1}^{n} [(y_k - bx_k)^2 - 2a(y_k - bx_k) + a^2]$$

$$= \sum_{k=1}^{n} (y_k - bx_k)^2 - 2a \sum_{k=1}^{n} (y_k - bx_k) + \sum_{k=1}^{n} a^2$$

$$= \sum_{k=1}^{n} (y_k - bx_k)^2 + \left[na^2 - 2a \left(\sum_{k=1}^{n} y_k - b \sum_{k=1}^{n} x_k \right) \right]$$

FIGURE 6.8
Superposition of regression line upon scattergram.

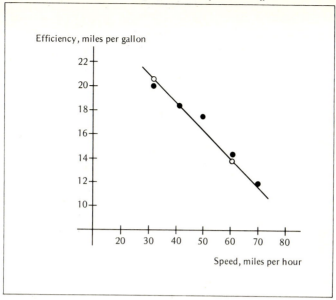

Now take a look at the last quantity in the parentheses. If we abbreviate the item

$$\frac{\sum_{k=1}^{n} y_k - b \sum_{k=1}^{n} x_k}{n}$$

by B, the quantity appears as $a^2 - 2aB$, which looks almost like $a^2 - 2aB + B^2 = (a - B)^2$. In fact, $a^2 - 2aB = (a - B)^2 - B^2$; i.e.,

$$a^2 - 2a\left(\frac{\sum_{k=1}^{n} y_k - b \sum_{k=1}^{n} x_k}{n}\right) = \left(a - \frac{\sum_{k=1}^{n} y_k - b \sum_{k=1}^{n} x_k}{n}\right)^2 \\ - \left(\frac{\sum_{k=1}^{n} y_k - b \sum_{k=1}^{n} x_k}{n}\right)^2$$

We can therefore write SSD as

$$SSD = \sum_{k=1}^{n} (y_k - bx_k)^2 + n\left(a - \frac{\sum_{k=1}^{n} y_k - b \sum_{k=1}^{n} x_k}{n}\right)^2 - n\left(\frac{\sum_{k=1}^{n} y_k - b \sum_{k=1}^{n} x_k}{n}\right)^2$$

Now, note that the middle term in the expression for SSD is the only term where an a appears. Note also that the middle term is never negative, because it is the product of n times a square; therefore, the smallest possible value of the middle term is 0. In view of the fact that we are trying to find the values of a and b which make SSD as small as possible, it would be important to choose a in such a way that the middle term becomes 0, its smallest possible value. We can accomplish this by choosing

$$a = \frac{\sum_{k=1}^{n} y_k - b \sum_{k=1}^{n} x_k}{n}$$

Therefore, the above expression is what we use as a formula for computing a. It remains now to find b, and again we use the method of completing the square.

With the middle term now equal to 0, we have that

$$SSD = \sum_{k=1}^{n} (y_k - bx_k)^2 - n \left(\frac{\sum_{k=1}^{n} y_k - b \sum_{k=1}^{n} x_k}{n} \right)^2$$

$$= \sum_{k=1}^{n} (y_k^2 - 2bx_k y_k + b^2 x_k^2)$$

$$- \frac{1}{n}\left[\left(\sum_{k=1}^{n} y_k\right)^2 - 2b \sum_{k=1}^{n} x_k \sum_{k=1}^{n} y_k + b^2 \left(\sum_{k=1}^{n} x_k\right)^2 \right]$$

$$= \frac{1}{n}\left[n\sum_{k=1}^{n} y_k^2 - 2bn \sum_{k=1}^{n} x_k y_k + b^2 n \sum_{k=1}^{n} x_k^2 \right.$$

$$\left. - \left(\sum_{k=1}^{n} y_k\right)^2 + 2b \sum_{k=1}^{n} x_k \sum_{k=1}^{n} y_k - b^2 \left(\sum_{k=1}^{n} x_k\right)^2 \right]$$

$$= \frac{1}{n}\left\{ b^2\left[n\sum_{k=1}^{n} x_k^2 - \left(\sum_{k=1}^{n} x_k\right)^2 \right] \right.$$

$$\left. - 2b\left(n\sum_{k=1}^{n} x_k y_k - \sum_{k=1}^{n} x_k \sum_{k=1}^{n} y_k \right) + \left[n\sum_{k=1}^{n} y_k^2 - \left(\sum_{k=1}^{n} y_k\right)^2 \right] \right\}$$

$$= \frac{\left[n\sum_{k=1}^{n} x_k^2 - \left(\sum_{k=1}^{n} x_k\right)^2 \right]}{n}\left[b^2 - 2b \frac{n\sum_{k=1}^{n} x_k y_k - \sum_{k=1}^{n} x_k \sum_{k=1}^{n} y_k}{n\sum_{k=1}^{n} x_k^2 - \left(\sum_{k=1}^{n} x_k\right)^2} \right.$$

$$\left. + \frac{1}{n}\left[n\sum_{k=1}^{n} y_k^2 - \left(\sum_{k=1}^{n} y_k\right)^2 \right] \right]$$

Now setting

$$A = \frac{n\sum_{k=1}^{n} x_k y_k - \sum_{k=1}^{n} x_k \sum_{k=1}^{n} y_k}{n\sum_{k=1}^{n} x_k^2 - \left(\sum_{k=1}^{n} x_k\right)^2}$$

we note that the first term of SSD contains the item $b^2 - 2bA$. But by the binomial formula, we know that $b^2 - 2bA = (b - A)^2 - A^2$. Making this substitution into the expression for SSD, we have

$$SSD = \frac{n \sum\limits_{k=1}^{n} x_k^2 - \left(\sum\limits_{k=1}^{n} x_k\right)^2}{n} \left\{ \left[b - \frac{n \sum\limits_{k=1}^{n} x_k y_k - \left(\sum\limits_{k=1}^{n} x_k\right)\left(\sum\limits_{k=1}^{n} y_k\right)}{n \sum\limits_{k=1}^{n} x_k^2 - \left(\sum\limits_{k=1}^{n} x_k\right)^2} \right]^2 \right.$$

$$\left. - \left[\frac{n \sum\limits_{k=1}^{n} x_k y_k - \sum\limits_{k=1}^{n} x_k \sum\limits_{k=1}^{n} y_k}{n \sum\limits_{k=1}^{n} x_k^2 - \left(\sum\limits_{k=1}^{n} x_k\right)^2} \right]^2 \right\} + \frac{1}{n}\left[n \sum\limits_{k=1}^{n} y_k^2 - \left(\sum\limits_{k=1}^{n} y_k\right)^2 \right]$$

$$= \frac{n \sum\limits_{k=1}^{n} x_k^2 - \left(\sum\limits_{k=1}^{n} x_k\right)^2}{n} \left[b - \frac{n \sum\limits_{k=1}^{n} x_k y_k - \sum\limits_{k=1}^{n} x_k \sum\limits_{k=1}^{n} y_k}{n \sum\limits_{k=1}^{n} x_k^2 - \left(\sum\limits_{k=1}^{n} x_k\right)^2} \right]^2$$

$$- \frac{n \sum\limits_{k=1}^{n} x_k^2 - \left(\sum\limits_{k=1}^{n} x_k\right)^2}{n} \left[\frac{n \sum\limits_{k=1}^{n} x_k y_k - \sum\limits_{k=1}^{n} x_k \sum\limits_{k=1}^{n} y_k}{n \sum\limits_{k=1}^{n} x_k^2 - \left(\sum\limits_{k=1}^{n} x_k\right)^2} \right]^2$$

$$+ \left[\sum\limits_{k=1}^{n} y_k^2 - \frac{1}{n}\left(\sum\limits_{k=1}^{n} y_k\right)^2 \right]$$

Now the first term of *SSD* above is never negative, in view of the fact that it is basically the product of a squared quantity and

$$n \sum\limits_{k=1}^{n} x_k^2 - \left(\sum\limits_{k=1}^{n} x_k\right)^2$$

which is never negative (because of the Schwarz inequality of Chapter 2). Therefore the smallest possible value of the first term is 0, which occurs when

$$b = \frac{n \sum\limits_{k=1}^{n} x_k y_k - \sum\limits_{k=1}^{n} x_k \sum\limits_{k=1}^{n} y_k}{n \sum\limits_{k=1}^{n} x_k^2 - \left(\sum\limits_{k=1}^{n} x_k\right)^2}$$

and so this choice of *b* also gives the smallest possible value of *SSD*. We then use the above expression as the formula for *b* leading to the smallest possible value of *SSD*, the attainment of which is our goal. Choosing *a* and *b*, then, on the basis of the above formulas, we obtain that the smallest possible value of *SSD* is given by

$$SSD = - \frac{\left(n \sum\limits_{k=1}^{n} x_k y_k - \sum\limits_{k=1}^{n} x_k \sum\limits_{k=1}^{n} y_k\right) b}{n} + \sum\limits_{k=1}^{n} y_k^2 - \frac{1}{n}\left(\sum\limits_{k=1}^{n} y_k\right)^2$$

$$= -b \sum\limits_{k=1}^{n} x_k y_k + \frac{b \sum\limits_{k=1}^{n} x_k \sum\limits_{k=1}^{n} y_k}{n} + \sum\limits_{k=1}^{n} y_k^2 - \frac{1}{n}\left(\sum\limits_{k=1}^{n} y_k\right)^2$$

$$= \sum_{k=1}^{n} y_k^2 - \sum_{k=1}^{n} y_k \frac{\sum_{k=1}^{n} y_k - b \sum_{k=1}^{n} x_k}{n} - b \sum_{k=1}^{n} x_k y_k$$

$$= \sum_{k=1}^{n} y_k^2 - a \sum_{k=1}^{n} y_k - b \sum_{k=1}^{n} x_k y_k$$

This is the smallest possible value for the sum of the squared deviations for the data points $(x_1, y_1), (x_2, y_2), \ldots, (x_n, y_n)$ from a straight line.

EXERCISES 6.B

1 The manager of an independent supermarket would like to know the relationship, if there is one, between the amount of display space occupied by a local brand of tuna and the dollar value of weekly sales of that item. The amount of space occupied was varied over a period of 6 weeks, and the following data were obtained:

Week Beginning	Display Space, square feet	Total Sales, hundreds of dollars
Jun 1	5	7
Jun 8	2	5
Jun 15	5	7
Jun 22	8	8
Jun 29	2	4
Jul 6	8	9

a Draw a scattergram of the data.
b Find the equation of the straight line which best fits the data.
c Superimpose the graph of the regression line upon the scattergram.

2 An agricultural research organization tested a particular chemical fertilizer to try to find whether an increase in the amount of fertilizer used would lead to a corresponding increase in the food supply. They obtained the following data based on seven plots of arable land:

Pounds of Fertilizer	Bushels of Beans
2	4
1	3
3	4
2	3
4	6
5	5
3	5

a Draw a scattergram of the data.
b Find the equation of the regression line that would be used to predict the number of bushels of beans obtained from a number of pounds of fertilizer.
c Superimpose the graph of the regression line upon the scattergram.

3 One psychologist suspects that there is a connection between the rate of inflation in the economy and the rate of divorce in the general population. In an attempt to find a way to predict the divorce rate from the inflation rate, she collects the following data from records of the past several years:

Inflation rate,%	2	4	5	7	10	12
Divorce rate per 1000 of population	3	7	10	15	25	30

a Draw a scattergram of the data.
b Find the equation of the regression line.
c Superimpose the graph of the regression line upon the scattergram.

4 A demographer presented the following data to support his theory that high protein diets tend to reduce fertility levels:

Country	Taiwan	Japan	Italy	West Germany	United States	Sweden
Protein in diet, grams/day	5	10	15	40	60	70
Birth rate per 100 population	4	3	3	2	2	1

a Draw a scattergram of the data.
b Find the equation of the regression line.
c Superimpose the graph of the regression line upon the scattergram.

SECTION 6.C STRENGTH OF THE LINEAR RELATIONSHIP

By use of the formulas for a and b, a regression line can be found for any set of paired data. Unfortunately, therefore, we can find a regression line even for a set of data that is fundamentally nonlinear, that is to say, a set of data that really does not follow a straight line. Some examples of nonlinear data are presented in Table 6.4, and their scattergrams are illustrated in Fig. 6.9. It is often difficult, especially with large numbers of data, to decide from either the table of data or the scattergram whether the data are truly linear or not. Furthermore, even if the underlying relationship really is a linear one, it would be unreasonable to expect all the data points to fall exactly on the straight line. And who's to say how far they are allowed to fall from the line without abandoning the belief that the underlying relationship is really linear?

To deal with the problem presented in the preceding paragraph, what is needed is a way of measuring the extent to which a set of data can be considered linear. In particular, we need a way of measuring the strength of the relationship between the data and the regression line calculated from the data. We now proceed to the development of such a measure, which will permit us to give an informed answer to the question: Does the regression line present a valid pictorial representation of the behavior of the data?

Let's look again at the typical set of n data pairs $(x_1, y_1), (x_2, y_2), \ldots, (x_n, y_n)$, whose

TABLE 6.4
Some Nonlinear Sets of Data

(a) Parabolic		(b) Logarithmic		(c) Exponential	
x	y	x	y	x	y
8	9	1	1	5	2
1	16	3	4	3	3
6	1	8	8	1	10
3	4	14	9.5	2	5
5	0	20	10	4	2
2	10	10	9	6	1
		5	6		
		2	3		

sum of squared deviations (*SSD*) we analyzed in Table 6.2. In general, y varies; that is to say, the y values, $y_1, y_2, \ldots, y_n$ are not all the same. If all the numbers $y_1, y_2, \ldots, y_n$ were the same, they would each be equal to the mean of the group, $\bar{y}$. Therefore we can consider the sum of squared deviations from the mean,

$$TV = \sum_{k=1}^{n} (y_k - \bar{y})^2$$

as a measure of the total variation of y. For example, in Table 6.5, we analyze the variation inherent in the engine efficiency data of Table 6.1. From the fourth column of Table 6.5, we can see that the total variation of engine efficiency is

$$TV = \Sigma(y - \bar{y})^2 = 50$$

The sum of squared deviations, *SSD*, mentioned earlier in this chapter, is actually the total variation of the y values away from the regression line, because it is the sum of the

FIGURE 6.9
Scattergrams of nonlinear data.

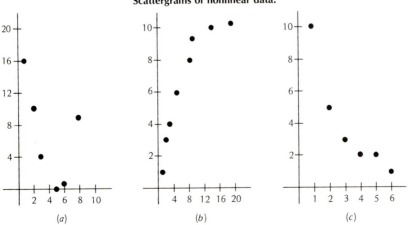

(a)   (b)   (c)

TABLE 6.5
Analysis of the Variation of Engine Efficiency
Regression Line: $y = 27 - 0.22x$

Actual Data Points		Total Variation		Variation from Line		Variation with Line
x	y	$\bar{y}$	$(y - \bar{y})^2$	$a + bx$	$(y - a - bx)^2$	$(a + bx - \bar{y})^2$
30	20	16	16	20.4	0.16	19.36
40	18	16	4	18.2	0.04	4.84
50	17	16	1	16.0	1.00	0.00
60	14	16	4	13.8	0.04	4.84
70	11	16	25	11.6	0.36	19.36
Sums	80		50		1.60	48.40

squared deviations of the actual y values from their values $a + bx$ predicted by the regression line. In the speed/efficiency example,

$$SSD = \Sigma(y - a - bx)^2 = 1.60$$

as can be seen from the sixth column of Table 6.5.

Well, the total variation in engine efficiency as recorded in the data is 50, while the variation away from the regression line is 1.60. The difference $50 - 1.60 = 48.40$ indicates that there is still substantial variation remaining as part of the total—variation that is not part of the variation away from the regression line. To what can we attribute this remaining variation?

Let's look at the problem from another point of view. Suppose there were no variation at all in y. Then every y value would equal $\bar{y}$, and we would have the total variation $TV = \Sigma(y - \bar{y})^2 = 0$. In fact, however, there is variation in y. Suppose the only variation in y were due to the influence of the regression line $y = a + bx$. Then every y_k would equal its corresponding $a + bx_k$. The resulting total variation would then be

$$TV = \Sigma(y - \bar{y})^2 = \Sigma(a + bx - \bar{y})^2$$

since y and $a + bx$ would always be the same. The quantity $\Sigma(a + bx - \bar{y})^2$ is therefore the variation in y that can be attributed to the effect of the regression line. We denote this term as

$$VR = \Sigma(a + bx - \bar{y})^2$$

where VR means "variation due to regression." In the far right column of Table 6.5, we have the calculation that

$$VR = \Sigma(a + bx - \bar{y})^2 = 48.40$$

in our engine efficiency example. In the example, therefore, we have that

$$TV = SSD + VR$$

because $50 = 1.60 + 48.40$.

In words, we can express these assertions as follows:

Total variation of y = variation away from regression line

+ variation due to the influence of regression

If you think about it for a while, you might agree that it's quite logical for the total variation to divide up in this manner.

In our example, then, SSD and VR together account for the entire variation of y, TV. Because $1.60 + 48.40 = 50$, there is no variation left to be attributed to anything else. We might ask the question: Does this happen in every possible example, or do we have a special case here? The answer is that TV *always* equals SSD plus VR. It is an algebraic fact that

$$\sum_{k=1}^{n} (y_k - \bar{y})^2 = \sum_{k=1}^{n} (y_k - a - bx_k)^2 + \sum_{k=1}^{n} (a + bx_k - \bar{y})^2$$

for all sets of paired data. Algebraic details of the reasons behind this general fact are presented to the interested reader in Supplement 6.2.

Because TV is the total variation in y and VR is the variation in y that can be attributed to the influence of the regression line, the ratio

$\dfrac{VR}{TV}$ = proportion of the variation in y that can be attributed to the influence of the regression line

VR/TV is often referred to as "the proportion of the variation in y explained by regression," or, simply, the "explained variation." In the engine efficiency example,

$$\frac{VR}{TV} = \frac{48.40}{50} = 0.968 = 96.8\%$$

so that 96.8% of the variation in engine efficiency can be explained on the basis of its linear relationship $y = 27 - 0.22x$ with speed.

The quantity VR/TV is called the "coefficient of linear determination" because it measures the extent to which variation in y is determined by its linear relationship with x. When we say "determined" here, we mean determined algebraically and numerically only; we have not proved, nor can we prove conclusively using mathematics alone, the existence of any type of cause-and-effect relationship between x and y.

The coefficient of determination is the measure that we have been seeking in this section—a measure of the strength of the linear relationship $y = a + bx$.

As it turns out, the coefficient of determination is rarely calculated by using the expression VR/TV. To calculate VR/TV requires (1) a table like Table 6.3 of preliminary calculations; (2) the calculation of the regression line using the formulas for a and b; and finally (3) a table similar to Table 6.5. This is more work than is really necessary to get the coefficient of determination. A particular disadvantage is that we must compute the regression line first, although one purpose of the coefficient of determination is to find out whether the data are really linear. It seems unreasonable to have to compute the regression line in order to find out whether we should bother to compute the regression line.

Fortunately, it is possible to calculate the coefficient of determination directly from a table like Table 6.3 by means of a relatively straightforward formula. First we calculate

●
$$r = \frac{n\Sigma xy - (\Sigma x)(\Sigma y)}{\sqrt{n\Sigma x^2 - (\Sigma x)^2}\sqrt{n\Sigma y^2 - (\Sigma y)^2}} \qquad (6.3)$$

which is called the "coefficient of linear correlation." (In the next section, we will discuss additional uses of the correlation coefficient.) Then the coefficient of linear determination is given by

$$\frac{VR}{TV} = r^2$$

Algebraic details of the discussion explaining why VR/TV and r^2 are one and the same are presented to the interested reader in Supplement 6.3. Here we will show how to use r^2 as the coefficient of determination.

In the continuing engine efficiency example, Table 6.3 contains all the basic information needed to compute the correlation coefficient r. The calculation is made as follows:

$$
\begin{aligned}
r &= \frac{n\Sigma xy - (\Sigma x)(\Sigma y)}{\sqrt{n\Sigma x^2 - (\Sigma x)^2}\sqrt{n\Sigma y^2 - (\Sigma y)^2}} \\
&= \frac{5(3780) - (250)(80)}{\sqrt{5(13,500) - (250)^2}\sqrt{5(1330) - (80)^2}} \\
&= \frac{18,900 - 20,000}{\sqrt{67,500 - 62,500}\sqrt{6650 - 6400}} = \frac{-1100}{\sqrt{5000}\sqrt{250}} \\
&= \frac{-1100}{(70.71)(15.81)} = -.984
\end{aligned}
$$

Because $r = -.984$, the coefficient of determination is then given by

$$\frac{VR}{TV} = r^2 = (-.984)^2 = .968 = 96.8\%$$

The reader should observe that we, of course, get exactly the same value for VR/TV using the correlation coefficient that we obtained earlier by the direct method of Table 6.5. In practice, the correlation coefficient is almost always used as a bridge to the coefficient of determination. There are several reasons for this, two of which are: (1) further valuable information obtainable from the correlation coefficient, and (2) the simplicity of the computation of r from Table 6.3 and the close similarity between the components of r and the components of b used in calculating the regression line.

Before proceeding to the next topic, let's calculate the coefficients of determination for the three sets of nonlinear data appearing in Table 6.4 and Fig. 6.9. As we shall see, they contrast markedly with the 96.8% of the engine efficiency example.

For the parabolic data of Table 6.4(a), we first make the preliminary calculations appearing in Table 6.6. To compute the coefficient of determination, i.e., the proportion of variation in y that can be attributed to the linear relationship with x based on the regression line, we then calculate r. We have

TABLE 6.6
Preliminary Calculations for Parabolic Data

x	y	x^2	y^2	xy
8	9	64	81	72
1	16	1	256	16
6	1	36	1	6
3	4	9	16	12
5	0	25	0	0
2	10	4	100	20
25	40	139	454	126
$n = 6$				

$$r = \frac{n\Sigma xy - (\Sigma x)(\Sigma y)}{\sqrt{n\Sigma x^2 - (\Sigma x)^2}\sqrt{n\Sigma y^2 - (\Sigma y)^2}}$$

$$= \frac{6(126) - (25)(40)}{\sqrt{6(139) - (25)^2}\sqrt{6(454) - (40)^2}}$$

$$= \frac{-244}{\sqrt{209}\sqrt{1124}} = \frac{-244}{(14.46)(33.53)} = -.503$$

Therefore, $r^2 = (-.503)^2 = .253 = 25.3\%$ of the variation in y can be attributed to the relationship with x given by the regression line. It follows that the regression line fails to account for almost 75% of the variation in y. These numbers provide a strong hint that x and y are really not linearly related. Therefore no attempt should be made to find the best-fitting straight line, for even that one will not fit very well. The next step in the analysis of such data would be to inspect the scattergram for a recognizable nonlinear trend in the data. Because the scattergram of Fig. 6.9 indicates that the data are parabolic in shape, we would apply the techniques of parabolic regression in order to find the best-fitting parabola. Various techniques of "curvilinear regression" will be presented in Chap. 7.

For the logarithmic data of Table 6.4(b) the preliminary calculations appear in

TABLE 6.7
Preliminary Calculations for Logarithmic Data

x	y	x^2	y^2	xy
1	1	1	1	1
3	4	9	16	12
8	8	64	64	64
14	9.5	196	90.25	133
20	10	400	100	200
10	9	100	81	90
5	6	25	36	30
2	3	4	9	6
63	50.5	799	397.25	536
$n = 8$				

Table 6.7. We have

$$r = \frac{n\Sigma xy - (\Sigma x)(\Sigma y)}{\sqrt{n\Sigma x^2 - (\Sigma x)^2}\sqrt{n\Sigma y^2 - (\Sigma y)^2}}$$

$$= \frac{8(536) - (63)(50.5)}{\sqrt{8(799) - (63)^2}\sqrt{8(397.25) - (50.5)^2}}$$

$$= \frac{1106.5}{\sqrt{2423}\sqrt{627.75}} = \frac{1106.5}{(49.22)(25.05)} = .897$$

Therefore the coefficient of determination is

$$r^2 = (.897)^2 = .805 = 80.5\%$$

indicating that the regression line accounts for all but about 20% of the variation in y. It should be observed that the logarithmic data are considerably more linear than the parabolic data having $r^2 = 25.3\%$, but appreciably less linear than the linear engine efficiency data where $r^2 = 96.8\%$.

Finally, considering the exponential data of Table 6.4(c), we make the preliminary calculations in Table 6.8. As usual, we first compute the correlation coefficient r:

$$r = \frac{n\Sigma xy - (\Sigma x)(\Sigma y)}{\sqrt{n\Sigma x^2 - (\Sigma x)^2}\sqrt{n\Sigma y^2 - (\Sigma y)^2}} = \frac{6(53) - (21)(23)}{\sqrt{6(91) - (21)^2}\sqrt{6(143) - (23)^2}}$$

$$= \frac{-165}{\sqrt{105}\sqrt{329}} = \frac{-165}{(10.25)(18.14)} = -.888$$

Therefore the coefficient of determination is given by $r^2 = (-.888)^2 = .788 = 78.8\%$, again more linear than the parabolic data but less linear than the linear data.

One more remark: In addition to its use in simplifying the calculation of the coefficient of determination, the correlation coefficient r immediately gives information about the direction of the trend of the data. In particular, if r turns out to be negative, this means that the regression line slopes downward to the right, in the manner of the scattergrams in Figs. 6.8 and 6.9c. If, on the other hand, r is positive, the line slopes upward to the right, as in Figs. 6.3, 6.6, and 6.9b. The reason for this association of r with slope is the close mathematical relationship between the formulas for r and b. In fact,

TABLE 6.8
Preliminary Calculations for Exponential Data

x	y	x^2	y^2	xy
5	2	25	4	10
3	3	9	9	9
1	10	1	100	10
2	5	4	25	10
4	2	16	4	8
6	1	36	1	6
21	23	91	143	53
$n = 6$				

$$r = \frac{n\Sigma xy - (\Sigma x)(\Sigma y)}{\sqrt{n\Sigma x^2 - (\Sigma x)^2}\sqrt{n\Sigma y^2 - (\Sigma y)^2}} = \frac{b\sqrt{n\Sigma x^2 - (\Sigma x)^2}}{\sqrt{n\Sigma y^2 - (\Sigma y)^2}}$$

so that the algebraic sign of r (+ or −) is always the same as the sign of b. But, for $y = a + bx$, when b (and so r) is negative, y grows smaller as x grows larger, and so the line slopes downward. Analogously, if b (and so r) is positive, y grows larger as x grows larger, and so the line slopes upward. Therefore, taking note of the sign of r and its square, the coefficient of determination, we can form a good mental picture of the regression line and how well the data fit it.

SUPPLEMENT 6.2   Algebraic Analysis of the Total Variation

$$TV = \text{total variation of } y = \sum_{k=1}^{n} (y_k - \bar{y})^2$$

$$= \sum_{k=1}^{n} [(y_k - a - bx_k) + (a + bx_k - \bar{y})]^2$$

$$= \sum_{k=1}^{n} [(y_k - a - bx_k)^2 + 2(y_k - a - bx_k)(a + bx_k - \bar{y}) + (a + bx_k - \bar{y})^2]$$

$$= \sum_{k=1}^{n} (y_k - a - bx_k)^2 + \sum_{k=1}^{n} (a + bx_k - \bar{y})^2$$

$$+ 2\sum_{k=1}^{n} [(y_k - a - bx_k)(a + bx_k - \bar{y})]$$

$$= SSD + VR + 2\sum_{k=1}^{n} [(y_k - a - bx_k)(a + bx_k - \bar{y})]$$

We now show that the third term in the above expression is *always* 0, no matter what the data, so that $TV = SSD + VR$ for every set of data points. From the formula for a in Sec. 6.B of this chapter, we know that

$$a = \frac{\sum_{k=1}^{n} y_k - b\sum_{k=1}^{n} x_k}{n} = \frac{\sum_{k=1}^{n} y_k}{n} - b\frac{\sum_{k=1}^{n} x_k}{n} = \bar{y} - b\bar{x}$$

from which it follows that $\bar{y} = a + b\bar{x}$, and so

$$a + bx_k - \bar{y} = a + bx_k - (a + b\bar{x}) = b(x_k - \bar{x})$$

Therefore,

$$2\sum_{k=1}^{n} [(y_k - a - bx_k)(a + bx_k - \bar{y})] = 2\sum_{k=1}^{n} [(y_k - a - bx_k) b(x_k - \bar{x})]$$

$$= 2b\sum_{k=1}^{n} (x_k y_k - ax_k - bx_k^2 - \bar{x}y_k + a\bar{x} + b\bar{x}x_k)$$

$$= 2b\left(\sum_{k=1}^{n} x_k y_k - a\sum_{k=1}^{n} x_k - b\sum_{k=1}^{n} x_k^2 - \bar{x}\sum_{k=1}^{n} y_k + na\bar{x} + b\bar{x}\sum_{k=1}^{n} x_k\right)$$

In view of the fact that $n\bar{x} = \sum_{k=1}^{n} x_k$, we know that $a\sum_{k=1}^{n} x_k = na\bar{x}$, and resulting cancellation and substitution yield that

$$2 \sum_{k=1}^{n} [(y_k - a - bx_k)(a + bx_k - \bar{y})]$$

$$= \frac{2b}{n} \left\{ n \sum_{k=1}^{n} x_k y_k - \sum_{k=1}^{n} x_k \sum_{k=1}^{n} y_k - b \left[n \sum_{k=1}^{n} x_k^2 - \left(\sum_{k=1}^{n} x_k \right)^2 \right] \right\}$$

$$= 0$$

because $b = \dfrac{n \sum_{k=1}^{n} x_k y_k - \sum_{k=1}^{n} x_k \sum_{k=1}^{n} y_k}{n \sum_{k=1}^{n} x_k^2 - \left(\sum_{k=1}^{n} x_k \right)^2}$

It follows from these algebraic calculations that for all sets of data points

$$\sum_{k=1}^{n} (y_k - \bar{y})^2 = \sum_{k=1}^{n} (y_k - a - bx_k)^2 + \sum_{k=1}^{n} (a + bx_k - \bar{y})^2$$

i.e., $TV = SSD + VR$.

EXERCISES 6.C

1 Based on the data presented in Exercise 6.B.1, what proportion of the variation in total sales can be explained by a linear relationship between total sales and display space?

2 From the data of Exercise 6.B.2, how much of the variation in yield of beans can be attributed to a linear relationship with the amount of fertilizer?

3 On the basis of the data given in Exercise 6.B.3 and the linear relationship described there, how much of the variation in the divorce rate can be ascribed to corresponding variations in the inflation rate?

4 Using the demography data of Exercise 6.B.4, what percentage of the variation in birth rates can be explained in terms of variations in the protein content of national diets and their linear relationship with the national birth rate?

5 Coronado Nautotronics of San Diego, bidding for the contract to produce the radar displays for the Navy's new fleet of patrol hydrofoils, collects the following information in an attempt to determine the cost curve for the radar displays:

Quantity produced	10	50	100	160	200	320	630	800
Total cost of production run	7.0	8.5	9.0	9.4	9.5	10.0	10.5	10.8

a Draw a scattergram of the data.
b Calculate the coefficient of linear determination.

6 An economist wants to determine the daily demand equation for rolled steel in a small industrial town. She collects the following data relating the price with the quantity of rolled steel that can be sold at that price:

Tons of Rolled Steel That Can Be Sold	Price per Ton, hundreds of dollars
1.0	50.10
2.0	12.60
2.5	8.00
4.0	3.20
5.0	2.00
6.3	1.25

a Draw a scattergram of the data.
b Calculate the coefficient of linear determination.

7 In an attempt to develop an aptitude test measuring an individual's aptitude for pursuing a career in computer programming, a psychologist first compares mathematical aptitude scores of already employed computer programmers with their job performance ratings. The data follow:

Person	A	B	C	D	E	F	G
Math aptitude score	2	5	0	4	3	1	6
Job performance rating	8	5	1	8	9	5	1

a Draw a scattergram of the data.
b Calculate the coefficient of linear determination.
c What proportion of variation in job performance ratings can be explained on the basis of a linear relationship with mathematical aptitude scores?

8 Consider the data on days lost due to labor disputes of Exercise 2.B.14. What proportion of the variation in worker days lost per year can be explained on the basis of a linear relationship between the worker days lost and the year?

9 The following data[1] compare the number of cases of various infectious diseases reported in Scotland in 1963 and 1973:

Disease	1963	1973
Typhoid	236	22
Erysipelas	197	133
Scarlet fever	1646	621
Whooping cough	5683	131
Dysentery	6152	1557
Ophthalmia neonatorum	75	49
Pneumonia	6675	2925
Puerperal fever	396	26
Tuberculosis	3024	1522

[1]Central Statistical Office (United Kingdom), *Annual Abstract of Statistics 1974*, p. 74, Her Majesty's Stationery Office, London, 1974.

What proportion of the variation in 1973 cases can be attributed to corresponding variations in 1963 cases by way of a linear relationship?

SECTION 6.D CORRELATION ANALYSIS

In the previous section, we have shown how the coefficient of linear correlation r can be used to determine both the strength of a possible linear relationship between the corresponding x and y values and the general direction of the slope of the regression line. The correlation coefficient (for historical reasons, sometimes referred to as "the Pearson product-moment correlation coefficient") can also be used in statistical inference to test whether or not the population from which a sample of data points was drawn can reasonably be considered at a given significance level to be linear in nature. That is, we will use r as part of a test statistic to test a hypothesis dealing with linear relationships existing within populations.

As we have noted before in many situations of statistical inference, it is necessary for the validity of the correlation test of linearity that both the x and y values have been drawn from populations that can reasonably be assumed to have normal distributions. If this assumption cannot validly be made in any given problem, we must, as we have pointed out earlier in analogous discussions, use a nonparametric alternative to the correlation test.

Under the required assumption of normality, then, we have an underlying population coefficient of linear correlation ρ (pronounced "row") of which r is merely a sample estimate based on n data points. We want to test the hypothesis

$$H : \rho = 0 \text{ (no correlation—the data is nonlinear)}$$

against the alternative

$$A : \rho \neq 0 \text{ (significant correlation—the data exhibit a significant linear trend)}$$

Note that the verbalizations of hypothesis and alternative are consistent with the role played by r in the computation of the coefficient of linear determination. If ρ were near 0, r would most likely be small, and so $VR/TV = r^2$ would represent a very small percentage of the total variation of y. This would indicate that the relationship between x and y, if one exists, is probably nonlinear. On the other hand, a significantly large value of ρ, in either the positive or negative direction, would be consistent with a high percentage of the variation of y's being accounted for by the regression line, and so rejection of $H : \rho = 0$ in favor of $A : \rho \neq 0$ would tend to indicate the existence of a significant linear relationship in the underlying population. For simplicity, we can abbreviate our testing problem as

$$H : \rho = 0 \text{ (nonlinear)}$$

versus

$$A : \rho \neq 0 \text{ (linear)}$$

To test H against A, we use the test statistic

$$t = \frac{r\sqrt{n-2}}{\sqrt{1-r^2}} \qquad (6.4)$$

which has been shown by advanced mathematical techniques to have the t distribution with $n - 2$ degrees of freedom. We would therefore reject H (nonlinearity) in favor of A (linearity) at significance level α if

$$|t| > t_{\alpha/2}(n-2)$$

To illustrate the use of the correlation test for linearity, let's apply the technique to each of the four sets of data discussed in the previous section. Suppose we consider various values of α simultaneously so that we may compare the sets of data with regard to their linearity. To be able to apply the correlation test, we must implicitly make appropriate assumptions of normality in the underlying populations, assumptions that may not always be valid in applied work. From the previous section, let's recall the information in Table 6.9. It is important to observe that r will always fall between -1 and $+1$ for any set of data. This is due to the fact that r^2 is a proportion and so must always lie between 0 and 1. A value of r outside the range -1 to $+1$ signals the presence of a computational error.

Once we have obtained the values of r in Table 6.9, the next step in carrying out the test of linearity is to look up the comparison values $t_{\alpha/2}(n - 2)$. We obtain these from Table A.4 in the back of the book for three levels of α, and compile them in Table 6.10.

It remains only to calculate, from Table 6.9, the numerical value of the test statistic (6.4) and to compare it against the corresponding t value listed in Table 6.10. For the engine efficiency example, we have that

$$t = \frac{r\sqrt{n-2}}{\sqrt{1-r^2}} = \frac{(-.984)\sqrt{5-2}}{\sqrt{1-(-.984)^2}} = \frac{(-.984)\sqrt{3}}{\sqrt{1-0.968}} = \frac{(-.984)(1.732)}{\sqrt{.0317}}$$

$$= \frac{(-.984)(1.732)}{(.178)} = -9.566$$

We are to reject

$$H : \rho = 0 \text{ (nonlinear)}$$

in favor of

$$A : \rho \neq 0 \text{ (linear)}$$

TABLE 6.9
Correlation Coefficients of Sets of Data

Data Set	n	r
Engine efficiency	5	$-.984$
Parabolic	6	$-.503$
Logarithmic	8	$.897$
Exponential	6	$-.888$

TABLE 6.10

Comparison t Values for Correlation Test

$$t_{\alpha/2}(n-2)$$

n	$n-2$	$\alpha = .10$	$\alpha = .05$	$\alpha = .01$
5	3	2.353	3.182	5.841
6	4	2.132	2.776	4.604
8	6	1.943	2.447	3.707

if

$$|t| > t_{\alpha/2}(n-2)$$

In the engine efficiency case, $|t| = 9.566$ exceeds $t_{\alpha/2}(3)$ for all reasonable levels of α, as can be seen from the top row of numbers in Table 6.10. This means that whatever your selected level of α, within a reasonable range, you would be led to reject H in favor of A, concluding that in the engine efficiency problem efficiency is related to speed by a linear relationship. (That relationship is, of course, the one given by the regression line $y = 27 - 0.22x$ of Sec. B.)

For the parabolic data, Table 6.9 indicates that

$$t = \frac{r\sqrt{n-2}}{\sqrt{1-r^2}} = \frac{(-.503)\sqrt{6-2}}{\sqrt{1-(-.503)^2}} = \frac{(-.503)\sqrt{4}}{\sqrt{1-.253}} = \frac{(-.503)(2)}{\sqrt{.747}}$$

$$= \frac{-1.006}{.864} = -1.164$$

Here $|t| = 1.164$, which fails to exceed $t_{\alpha/2}(4)$ recorded in Table 6.10 for any reasonable level of α. Therefore, we would fail to reject H at every reasonable α, concluding that we have no evidence to indicate the presence of a linear relationship in the underlying population.

For the logarithmic data, we see from Table 6.9 that

$$\frac{r\sqrt{n-2}}{\sqrt{1-r^2}} = \frac{(.897)\sqrt{8-2}}{\sqrt{1-(.897)^2}} = 4.971$$

Here $|t| = 4.971$, which exceeds the values of $t_{\alpha/2}(6)$ recorded in Table 6.10 for every reasonable α. Therefore we would reject H and conclude, at every reasonable level α, that there is significant evidence of a linear relationship between x and y in the underlying population. Recalling that when $\alpha = .01$, there is a 1% chance of rejecting H when it's really true, the fact that $|t| = 4.971$ for the logarithmic data of Fig. 6.9b means that if the underlying population were strongly nonlinear, there would be less than a 1% chance of having $|t|$ as high as 4.971. Because $|t|$ does turn out as high as 4.971, we know that the logarithmic data cannot be considered as strongly nonlinear. In fact, looking at its scattergram in Fig. 6.9b, we see that it is only slightly nonlinear. Although a logarithmic curve would describe the data best, our correlation test of linearity shows that the regression line would not be very bad.

A somewhat similar situation prevails for the exponential data whose scattergram appears in Fig. 6.9c. Here

$$\frac{r\sqrt{n-2}}{\sqrt{1-r^2}} = \frac{(-.888)\sqrt{6-2}}{\sqrt{1-(-.888)^2}} = -3.862$$

so that $|t| = 3.862$. The comparison t values in Table 6.10 for $n = 6$ show that we should reject H if $\alpha = .10$ or $\alpha = .05$, but we should not reject H if $\alpha = .01$. We therefore have one more example of how the subjective choice of α influences the outcome of a statistical decision process.

If it is known that the x and y data points come from underlying populations that are normally distributed, the correlation test of linearity should be performed before any calculation of the regression line is made. Such a procedure will protect against the waste of time and effort involved in calculating a regression line that will be essentially useless in case the data are later judged to be nonlinear. When it is not reasonable to assume normality in the underlying populations, the best we can do is to use the coefficient of linear determination. Of course, in the latter situation, we cannot obtain the degree of precision in significance levels that is available in the case of normality.

SUPPLEMENT 6.3   Formula for the Coefficient of Correlation

Our purpose in this supplement is to show that

$$\frac{VR}{TV} = r^2$$

where

$$VR = \sum_{k=1}^{n} (a + bx_k - \bar{y})^2$$

$$TV = \sum_{k=1}^{n} (y_k - \bar{y})^2$$

and

$$r = \frac{n\sum_{k=1}^{n} x_k y_k - \sum_{k=1}^{n} x_k \sum_{k=1}^{n} y_k}{\sqrt{n\sum_{k=1}^{n} x_k^2 - \left(\sum_{k=1}^{n} x_k\right)^2}\sqrt{n\sum_{k=1}^{n} y_k^2 - \left(\sum_{k=1}^{n} y_k\right)^2}}.$$

We first reduce TV and VR to their most basic components:

$$TV = \sum_{k=1}^{n} (y_k - \bar{y})^2 = \sum_{k=1}^{n} (y_k^2 - 2\bar{y}y_k + \bar{y}^2) = \sum_{k=1}^{n} y_k^2 - 2\bar{y} \sum_{k=1}^{n} y_k + n\bar{y}^2$$

$$= \sum_{k=1}^{n} y_k^2 - \frac{2}{n}\left(\sum_{k=1}^{n} y_k\right)^2 + \frac{1}{n}\left(\sum_{k=1}^{n} y_k\right)^2$$

$$= \frac{1}{n}\left[n \sum_{k=1}^{n} y_k^2 - \left(\sum_{k=1}^{n} y_k\right)^2\right]$$

$$VR = \sum_{k=1}^{n} (a + bx_k - \bar{y})^2 = \sum_{k=1}^{n} (a + bx_k - a - b\bar{x})^2 = \sum_{k=1}^{n} [b(x_k - \bar{x})]^2$$

$$= b^2 \sum_{k=1}^{n} (x_k - \bar{x})^2 = b^2 \sum_{k=1}^{n} (x_k^2 - 2\bar{x}x_k + \bar{x}^2)$$

$$= b^2 \left(\sum_{k=1}^{n} x_k{}^2 - 2\bar{x} \sum_{k=1}^{n} x_k + n\bar{x}^2 \right)$$

$$= \frac{b^2}{n} \left[n \sum_{k=1}^{n} x_k{}^2 - 2 \left(\sum_{k=1}^{n} x_k \right)^2 + \left(\sum_{k=1}^{n} x_k \right)^2 \right]$$

$$= \frac{b^2}{n} \left[n \sum_{k=1}^{n} x_k{}^2 - \left(\sum_{k=1}^{n} x_k \right)^2 \right]$$

$$= \frac{\left(n \sum_{k=1}^{n} x_k y_k - \sum_{k=1}^{n} x_k \sum_{k=1}^{n} y_k \right)^2}{n \left[n \sum_{k=1}^{n} x_k{}^2 - \left(\sum_{k=1}^{n} x_k \right)^2 \right]}$$

inserting the expression for b obtained in Supplement 6.1. It follows that

$$\frac{VR}{TV} = \frac{\left(n \sum_{k=1}^{n} x_k y_k - \sum_{k=1}^{n} x_k \sum_{k=1}^{n} y_k \right)^2}{\left[n \sum_{k=1}^{n} x_k{}^2 - \left(\sum_{k=1}^{n} x_k \right)^2 \right] \left[n \sum_{k=1}^{n} y_k{}^2 - \left(\sum_{k=1}^{n} y_k \right)^2 \right]} = r^2$$

EXERCISES 6.D

1   At level of significance $\alpha = .05$, can we consider the relationship between display space and total sales as discussed in Exercises 6.B.1 and 6.C.1 to be a linear one?

2   At significance level $\alpha = .05$, is it reasonable to assume that the relationship between amount of fertilizer and bushels of beans studied earlier in Exercises 6.B.2 and 6.C.2 is linear?

3   At level $\alpha = .01$, does it make sense to use a linear equation to describe the relationship between inflation rate and divorce rate based on the data analyzed in Exercises 6.B.3 and 6.C.3?

4   Using a level of significance of $\alpha = .01$, does the relationship between protein content of diet and national birth rate based on the data of Exercises 6.B.4 and 6.C.4 appear to be linear?

5   At level $\alpha = .05$, does the cost curve based on the data of Exercise 6.C.5 seem to be linear?

6   Can the demand curve of Exercise 6.C.6 be considered linear at level $\alpha = .05$?

7   At level of significance $\alpha = .10$, can we validly use a linear equation to describe the relationship between mathematical aptitude and job performance in computer programming based on the data of Exercise 6.C.7?

SECTION 6.E PREDICTION USING THE REGRESSION LINE

To complete our discussion of linear regression and correlation, we now tackle the fourth question raised at the beginning of the chapter: Can we use the speed/efficiency

TABLE 6.11
Predicted Efficiency Levels for Various Speeds

Speed x, miles/hour	Computation 27 − .22x	Predicted Efficiency y, miles/gallon
25	27 − (.22)(25)	21.5
55	27 − (.22)(55)	14.9
80	27 − (.22)(80)	9.4

data to predict the efficiency at other speeds, for example, 25, 55, or 80; and how accurate will our predictions be? We answer the first part of the question in the affirmative, and we illustrate a method of confidence intervals in response to the second part.

We first recall from Sec. 6.B that the regression line relating speed x and engine efficiency y has equation $y = 27 - .22x$. In Fig. 6.8, we have represented this relationship graphically in a manner that illustrates the linear trend of variation in engine efficiency relative to variations in speed. What this means is that, insofar as we are able to determine from the five data points, when the speed is x miles/hour, our best estimate of the corresponding engine efficiency level will be $27 - .22x$ miles/gallon. In Table 6.11, we use the equation of the regression line to predict the efficiency level for the speeds 25, 55, and 80. We see that at 25 miles/hour we predict an efficiency of 21.5 miles/gallon, at 55 miles/hour we predict 14.9 miles/gallon, and at 80 miles/hour we predict 9.4 miles/gallon. For any desired value x of speed, a predicted value y of engine efficiency can be worked out in the same way, namely, by using the regression equation $y = 27 - .22x$.

Now that we have predicted, for example, efficiency to be 14.9 miles/gallon when the speed is 55 miles/hour, it is important to know how accurate our prediction can claim to be. This problem is comparable to the one discussed in Chap. 4 where we were trying to estimate the true mean μ of a normally distributed population from a sample of n data points $x_1, x_2, \ldots, x_n$. There we concluded, as you may recall, that the sample mean

$$\bar{x} = \frac{1}{n} \sum_{k=1}^{n} x_k$$

is used as an estimate of μ, with the accuracy of the estimate expressed by means of a confidence interval; we can be $(1 - \alpha)100\%$ sure that

$$\bar{x} - t_{\alpha/2}(n-1)\frac{s}{\sqrt{n}} \leq \mu \leq \bar{x} + t_{\alpha/2}(n-1)\frac{s}{\sqrt{n}}$$

where

$$s = \sqrt{\frac{\sum_{k=1}^{n} (x_k - \bar{x})^2}{n-1}}$$

is the sample standard deviation of the n data points. By analogy with the earlier procedure, we develop confidence intervals for the accuracy of predictions based on the regression line.

If, for each value of x, we can reasonably assume that the possible y values are normally distributed about the predicted value of y lying on the regression line, we can set up the desired confidence intervals. From Fig. 6.8, we can see that the assumption of normality is not unreasonable; two pieces of evidence for normality are the facts that (1) some data points fall above the regression line, while others fall below it, and (2) most points fall close to the line, while relatively few fall far away. The normality assumption indicates, for $x = 55$ and its predicted value $y = 14.9$, that the true engine efficiency for a speed of 55 miles/hour is as likely to be above 14.9 miles/gallon as below it, but more likely to be near 14.9 miles/gallon than far away from it.

The number that plays the role of the sample standard deviation, informing us how variable the possible values of y are likely to be from the predicted value, is called the standard error of the estimate and is given by the formula

$$s_e = \sqrt{\frac{\sum_{k=1}^{n} (y_k - a - bx_k)^2}{n - 2}}$$

The numerator $\sum_{k=1}^{n} (y_k - a - bx_k)^2$, what we have called SSD, is analogous to $\sum_{k=1}^{n} (x_k - \bar{x})^2$ for a single set of data points. From some algebraic transformations of SSD worked out at the end of Supplement 6.1, we obtain a more easily usable formula for s_e, which is as follows:

● $$s_e = \sqrt{\frac{\Sigma y^2 - a\Sigma y - b\Sigma xy}{n - 2}} \tag{6.5}$$

Using the preliminary calculations in Table 6.3 and the facts that $a = 27$ and $b = -0.22$, we can compute the standard error of the estimate for the speed/efficiency problem. We have

$$s_e = \sqrt{\frac{\Sigma y^2 - a\Sigma y - b\Sigma xy}{n - 2}} = \sqrt{\frac{1330 - (27)(80) - (-.22)(3780)}{5 - 2}}$$

$$= \sqrt{\frac{1330 - 2160 - (-831.6)}{3}} = \sqrt{\frac{1330 - 2160 + 831.6}{3}}$$

$$= \sqrt{\frac{1.6}{3}} = \sqrt{.533} = .730$$

Because the numerator of the term under the square-root sign is SSD, namely, a sum of squares, it can never be negative. A negative number under the square-root sign in s_e is a signal of an error in arithmetic.

Assuming, then, that y varies in a normally distributed manner about its predicted value, we can write down the formula for a $(1 - \alpha)100\%$ confidence interval for the true value of y at a given value of x. We denote by x_0 the value of x at which we are interested in predicting y. In the speed/efficiency example, we could take $x_0 = 55$ because we want to predict the engine efficiency at a speed of 55 miles/hour. The predicted engine efficiency is then $a + bx_0 = 27 - (.22)(55) = 14.9$. If we denote the true value of y as y_0, when $x = x_0$, we can be $(1 - \alpha)100\%$ sure that

$$a + bx_0 - E_0 \le y_0 \le a + bx_0 + E_0 \qquad (6.6)$$

where the possible error E_0 is given by

$$E_0 = t_{\alpha/2}(n - 2)s_e\sqrt{1 + \frac{1}{n} + \frac{n(x_0 - \bar{x})^2}{n\Sigma x^2 - (\Sigma x)^2}} \qquad (6.7)$$

In predicting efficiency when the speed is 55 miles/hour, we need the following components for working out a 90% confidence interval for the efficiency:

$$x_0 = 55$$
$$a + bx_0 = 27 - (.22)(55) = 14.9$$

$$\left.\begin{array}{l} \alpha = .10 \\[2mm] \dfrac{\alpha}{2} = \dfrac{.10}{2} = .05 \\[2mm] n = 5 \end{array}\right\} \qquad t_{\alpha/2}(n - 2) = t_{.05}(3) = 2.353$$

$$s_e = .730$$

$$\left.\begin{array}{l} \bar{x} = \dfrac{\Sigma x}{n} = \dfrac{250}{5} = 50 \end{array}\right\} \qquad n(x_0 - \bar{x})^2 = 5(55 - 50)^2 = 5(5)^2 = 125$$

$$n\Sigma x^2 - (\Sigma x)^2 = 5(13{,}500) - (250)^2 = 5000$$

using the calculations of Table 6.3 wherever necessary. We then have

$$\begin{aligned} E_0 &= t_{\alpha/2}(n - 2)\, s_e\sqrt{1 + \frac{1}{n} + \frac{n(x_0 - \bar{x})^2}{n\Sigma x^2 - (\Sigma x)^2}} \\ &= (2.353)(.730)\sqrt{1 + \frac{1}{5} + \frac{125}{5000}} \\ &= (1.718)\sqrt{1 + .2 + .025} = (1.718)\sqrt{1.225} \\ &= (1.718)(1.108) = 1.90 \end{aligned}$$

Therefore we can be 90% sure that

$$a + bx_0 - E_0 \le y_0 \le a + bx_0 + E_0$$

namely, that

$$14.9 - 1.9 \le y_0 \le 14.9 + 1.9$$

or, finally,

$$13.0 \le y_0 \le 16.8$$

In words, we can state our conclusion as follows: If the car under study were to be driven at a speed of 55 miles/hour, we could be 90% sure that it would be getting between 13.0 and 16.8 miles/gallon of gasoline.

The key component of the formula for the error E_0 is the term

$$n(x_0 - \bar{x})^2$$

This term is the only part of E_0 that involves the number x_0, the value of x in which we are interested. The quantity $(x_0 - \bar{x})^2$ is the squared distance between x_0 and $\bar{x}$, and so the term $n(x_0 - \bar{x})^2$ is small for values of x_0 that are close to $\bar{x}$, and larger for values of x_0 that are farther away from $\bar{x}$. This means that the farther x_0 is from $\bar{x}$, the larger E_0 itself is. In fact, as we consider values of x_0 farther and farther away from $\bar{x}$, the possible error E_0 grows in size and so our estimate of y_0 correspondingly becomes increasingly unreliable. This situation, which is illustrated graphically in Fig. 6.10, points up the fact that a prediction based on the regression line is most accurate only when it involves numbers well within the range of the original data points, and it loses accuracy as it proceeds beyond that range.

For example, let's consider the accuracy of the three predictions recorded in Table 6.11. For 90% confidence intervals, all components of the formula for E_0 are the same as the above calculation for $x_0 = 55$, except the various values of x_0 itself. In particular, we have

$$E_0 = (2.353)(0.730)\sqrt{1 + \frac{1}{5} + \frac{5(x_0 - 50)^2}{5000}}$$

for all possible 90% confidence intervals involved in the speed/efficiency problem. When $x_0 = 55$, we have already shown above that $E_0 = 1.9$. Other values of x_0 appearing in Table 6.11 are $x_0 = 25$ and $x_0 = 80$. We calculate E_0 for both these values. For $x_0 = 25$,

$$E_0 = (1.718)\sqrt{1 + .2 + \frac{5(25 - 50)^2}{5000}} = (1.718)\sqrt{1 + .2 + .625}$$

$$= (1.718)\sqrt{1.825} = (1.718)(1.35) = 2.32$$

FIGURE 6.10
90% confidence intervals for
predicting efficiency at various speeds.

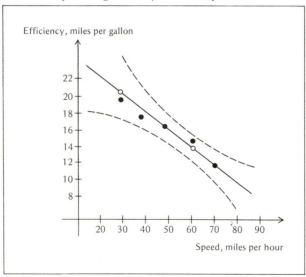

TABLE 6.12
Errors of Prediction for Various Speeds

Speed x_0	Predicted Efficiency	Error E_0
25	21.5	2.32
55	14.9	1.90
80	9.4	2.49

and for $x_0 = 80$,

$$E_0 = (1.718)\sqrt{1 + .2 + \frac{5(80 - 50)^2}{5000}} = (1.718)\sqrt{1 + .2 + .9}$$
$$= (1.718)\sqrt{2.1} = (1.718)(1.45) = 2.49$$

We summarize these results in Table 6.12.

The increasing size of the 90% confidence interval as x_0 moves away from the central location of the original data is illustrated in Fig. 6.10. Note that the confidence "band" is at its narrowest for $x_0 = \Sigma x/n = 50$, where it extends a length 1.88 in both directions, and as x_0 lies farther and farther away from 50, the "band" widens more and more.

EXERCISES 6.E

1 Based on the data of Exercise 6.B.1, find a 90% confidence interval for predicting the weekly sales of the brand of coffee involved if the allotted display space is eventually set at 6 square feet.

2 Using the data appearing in Exercise 6.B.2, find
   a A 90% confidence interval for predicting the number of bushels of beans harvested from a plot of land on which 6 pounds of fertilizer have been used.
   b An 80% confidence interval for the harvest of beans if no fertilizer is used.

3 If next year's inflation rate looks as though it will be 8%, use the data of Exercise 6.B.3 to calculate a 95% confidence interval for the divorce rate next year.

4 The typical protein content of the daily diet in Greece is 18 grams. Find a 90% confidence interval for the birth rate in Greece basing your analysis on the data of Exercise 6.B.4.

5 The regression line $y = a + bx$, calculated as it is from a set of data points, n x's and n y's, should really be viewed as a sample estimate of a true linear relationship existing between random variables represented by the x and y data points. If we denote this true relationship by $y = \alpha + \beta x$, then α is called the true intercept and β is called the true slope of the regression line. The numbers a and b are then estimates of these true values α and β. The question then arises: How good an estimate of α is a and how good an estimate of β is b? The answers are given by the following confidence intervals:
   We can be $(1 - \alpha)100\%$ sure that

$$a - E_a \leq \alpha \leq a + E_a$$

where
$$E_a = t_{\alpha/2}(n-2)s_e\sqrt{\frac{1}{n} + \frac{n \cdot \bar{x}^2}{n(\Sigma x^2) - (\Sigma x)^2}}$$

and we can be $(1 - \alpha)100\%$ sure that
$$b - E_b \leq \beta \leq b + E_b$$

where
$$E_b = t_{\alpha/2}(n-2)s_e\sqrt{\frac{n}{n(\Sigma x^2) - (\Sigma x)^2}}$$

[Unfortunately, since a knowledge of b is necessary for the calculation of a in formula (6.2), it would not be proper to calculate both the above confidence intervals using the same set of data. We would need two independently selected sets of data in order to get independent confidence intervals for both α and β.] Based on the data of Exercise 6.B.1, find a 90% confidence interval for the true intercept α used for predicting weekly coffee sales.

6   From the data of Exercise 6.B.2, find an 80% confidence interval for the true slope β of the regression line used in predicting the size of the bean harvest.

7   Use the data of Exercise 6.B.3 to find a 90% confidence interval for the true slope β for predicting next year's divorce rate.

8   Find a 90% confidence interval for the true intercept α for predicting birth rates using the data of Exercise 6.B.4.

SECTION 6.F A COMPLETELY WORKED-OUT EXAMPLE

The many new topics introduced in this chapter often appear in practice as merely the components of a single problem and its comprehensive solution. This essential unity among the various techniques has perhaps been hidden from you, as they have been well scattered over several pages of text. In the final section of this chapter, we present an example of a regression-correlation problem, completely solved in a logical sequence of steps from beginning to end.

Example 6.1   Bacteria Growth   A biologist conducts an experiment aimed at determining how the growth rate of a certain type of bacteria increases with the passage of time. She sets up six cultures of the bacteria, studies each one's growth for a different time period, and records their growth rates at the conclusion of the respective time periods. The data follow:

Culture	A	B	C	D	E	F
Time period, hours	3	5	1	7	5	9
Growth rate, units per hour	10	13	8	13	11	15

The following questions are of interest to the biologist:

a   What proportion of the variation in growth rates among the bacteria can be attributed to a linear relationship with corresponding variations in the time periods?

b What does the scattergram of the data look like?

c At significance level $\alpha = .05$, can we consider the growth rate to be linearly related to the length of the time period?

d If so, what is the equation of the regression line that will predict growth rates for various given time periods?

e Find a 90% confidence interval for the growth rate of a culture of the bacteria after a time period of 12 hours.

SOLUTION

a To find the proportion of variation in growth rates that can be explained on the basis of variations in the time period, we must compute the coefficient of linear determination. We first set up the table of preliminary calculations (Table 6.13).

Here the growth rate, the variable to be predicted, is denoted by y, while time period is denoted by x.

The coefficient of determination is r^2, where

$$r = \frac{n\Sigma xy - (\Sigma x)(\Sigma y)}{\sqrt{n\Sigma x^2 - (\Sigma x)^2}\sqrt{n\Sigma y - (\Sigma y)^2}}$$

$$= \frac{6(384) - (30)(70)}{\sqrt{6(190) - (30)^2}\sqrt{6(848) - (70)^2}}$$

$$= \frac{2304 - 2100}{\sqrt{1140 - 900}\sqrt{5088 - 4900}} = \frac{204}{\sqrt{240}\sqrt{188}}$$

$$= \frac{204}{(15.5)(13.7)} = .960$$

Therefore $r^2 = (.960)^2 = .922 = 92.2\%$ of the variation in growth rate can be attributed to whatever linear relationship may exist between growth rate and time period.

b The scattergram of the growth rate data appears in Fig. 6.11.

c We test the hypothesis

$$H : \rho = 0 \quad \text{(nonlinear)}$$

versus
$$A : \rho \neq 0 \quad \text{(linear)}$$

at level $\alpha = .05$ by calculating

$$t = \frac{r\sqrt{n - 2}}{\sqrt{1 - r^2}}$$

and rejecting H if $|t| > t_{\alpha/2}(n - 2) = t_{.025}(4) = 2.776$. We have

$$t = \frac{r\sqrt{n - 2}}{\sqrt{1 - r^2}} = \frac{(.960)\sqrt{6 - 2}}{\sqrt{1 - (.960)^2}} = \frac{(.960)\sqrt{4}}{\sqrt{1 - .9216}} = \frac{1.92}{\sqrt{.0784}}$$

$$= \frac{1.92}{.28} = 6.857$$

Therefore $|t| = 6.857 > 2.776$, and so we reject H and conclude at level $\alpha = .05$ that growth rate and length of time period seem to be linearly related.

FIGURE 6.11
Scattergram of growth rate data.

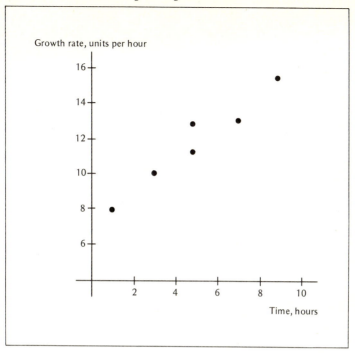

d The regression line has equation $y = a + bx$, where y represents growth rate (the variable to be predicted), and x represents length of time period. Here

$$b = \frac{n\Sigma xy - (\Sigma x)(\Sigma y)}{n\Sigma x^2 - (\Sigma x)^2} = \frac{204}{240} = .85$$

where we have obtained the quantities $n\Sigma xy - (\Sigma x)(\Sigma y) = 204$ and $n\Sigma x^2 - (\Sigma x)^2 = 240$, not by computation, but by merely looking their values up among the

TABLE 6.13
Preliminary Calculations for Growth Rate Example

Time Period x, hours	Growth Rate y, units/hour	x^2	y^2	xy
3	10	9	100	30
5	13	25	169	65
1	8	1	64	8
7	13	49	169	91
5	11	25	121	55
9	15	81	225	135
Sums 30 n = 6	70	190	848	384

calculations we made to find the correlation coefficient r. Notice that the numerator in the formula for b is exactly the same as the numerator in the formula for r, while the denominator of b is one of the factors appearing under a square-root sign in the expression for r. In view of the relationship between b and r, therefore, we are able to simplify the calculation of b somewhat. Then

$$a = \frac{\Sigma y - b\Sigma x}{n} = \frac{70 - (.85)(30)}{6} = \frac{70 - 25.5}{6}$$

$$= \frac{44.5}{6} = 7.42$$

The regression line therefore has equation $y = 7.42 + .85x$. For illustrative purposes, as well as to provide a check on our calculations we often superimpose the graph of the regression line upon the scattergram. To do this, we choose two values of x within the range of the original data, say $x = 4$ and $x = 10$. If $x = 4$, then $y = 7.42 + (.85)(4) = 10.82$, so that $(4, 10.82)$ lies on the regression line, and if $x = 10$, then $y = 7.42 + (.85)(10) = 15.92$, so that $(10, 15.92)$ lies on the regression line. The regression line is shown superimposed on the scattergram in Fig. 6.12.

e To find a 90% confidence interval for the growth rate y_0 of a culture of the bacteria after a time period of 12 hours, we set $\alpha = .10$ and $x_0 = 12$, and then proceed to accumulate all the components of the confidence interval formulas

FIGURE 6.12
Regression line superimposed on scattergram.

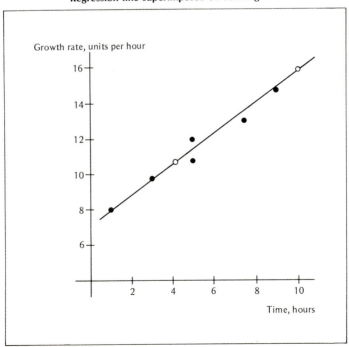

$$a + bx_0$$

and
$$E_0 = t_{\alpha/2}(n-2)s_e\sqrt{1 + \frac{1}{n}} = \frac{n(x_0 - \bar{x})^2}{n\Sigma x^2 - (\Sigma x)^2}$$

We list the components as follows:

$$a + bx_0 = 7.42 + (.85)(12) = 17.62$$
$$t_{\alpha/2}(n-2) = t_{.05}(4) = 2.132$$
$$s_e = \sqrt{\frac{\Sigma y^2 - a\Sigma y - b\Sigma xy}{n-2}} = \sqrt{\frac{848 - (7.42)(70) - (.85)(384)}{6-2}}$$
$$= \sqrt{\frac{848 - 519.4 - 326.4}{4}} = \sqrt{\frac{2.2}{4}} = \sqrt{.55} = .742$$

$$n(x_0 - \bar{x})^2 = 6(12 - \tfrac{30}{6})^2 = 6(12-5)^2 = 6(7)^2 = 6(49) = 294$$
$$n\Sigma x^2 - (\Sigma x)^2 = 240 \qquad \text{(from the expression for } r\text{)}$$
$$E_0 = (2.132)(.742)\sqrt{1 + \tfrac{1}{6} + \tfrac{294}{240}} = (1.582)\sqrt{1 + .167 + 1.225}$$
$$= (1.582)\sqrt{2.392} = (1.582)(1.547) = 2.45$$

The 90% confidence interval is then

$$a + bx_0 - E_0 \le y_0 \le a + bx_0 + E_0$$
$$17.62 - 2.45 \le y_0 \le 17.62 + 2.45$$
$$15.17 \le y_0 \le 20.07$$

We can therefore be 90% sure that after 12 hours a culture of the bacteria will be growing at a rate somewhere between 15.17 and 20.07 units per hour.

EXERCISES 6.F

1 To check on the possibility that the slope of a delta region can be accurately predicted from a knowledge of the typical size of the stones found there, a geographer gathered stones from various locations in a delta region of West Baffin Island. The typical size of the stones found at each location was then compared with the cotangent of the angle of slope of the delta there. (A larger cotangent corresponds to a smaller angle of slope.) The resulting data follow:

Typical Size of Stones (diameter in meters)	Slope of Delta (1% of cotangent of angle)
0.5	3.0
3.0	0.5
1.0	2.0
2.5	1.0
2.0	1.5

a What proportion of the variation in delta slope can be explained on the basis of a linear relationship with size of stones?

b Draw a scattergram of the data.

c At a level of significance $\alpha = .02$, can the relationship between size of stones and slope of delta be considered linear?

d Determine the equation of the regression line which predicts delta slope in terms of stone size, and superimpose the graph of the regression line upon the scattergram.

e Find a 90% confidence interval for 1% of the cotangent of the angle of slope at a location where the typical stone has diameter 1.5 meters.

2 The manager of a salmon cannery suspects that the demand for his product is closely related to the disposable income of his target region. To test out his suspicion, he collects the following data for 1976:

Region	Disposable Income, millions of dollars	Sales Volume, thousands of cases
A	10	1
B	20	3
C	40	4
D	50	5
E	30	2

a What percentage of the variation in sales volume among regions can be attributed to a linear relationship with disposable income of the regions?

b Draw a scattergram of the data.

c At level $\alpha = .05$, do the data indicate the existence of a linear relationship between disposable income and sales volume?

d Develop the regression equation that can be used to predict sales volume in a region from a knowledge of the region's disposable income, and superimpose the graph of the regression line upon the scattergram.

e Find an 80% confidence interval for the sales volume in a region whose disposable income is $25,000,000.

3 A sociologist employed by HEW is assigned the task of finding the relationship, if there is any, between the level of an individual's formal education and whether or not that person is currently unemployed. One of her methods of investigation centers on a comparison between the various levels of formal education and the unemployment rate among individuals at that educational level. In the data that follow, the years 0 to 8 are grade school, 9 to 12 are high school, 13 to 16 are college, 17 and 18 are master's degree study, and 19 to 24 are doctoral degree study.

Educational Level, years	Unemployment Rate, %
4	30
8	20
12	6
14	10
18	3
24	6

The researcher would like to use the data to develop a method of predicting a person's chances of being unemployed from that person's educational level.

a What proportion of the variation in unemployment rate among the various educational levels can be accounted for by a linear relationship with educational level.

b Draw a scattergram of the data.

c At level $\alpha = .05$, can the sociologist conclude that the relationship between educational level and unemployment rate is a linear one?

d Find the equation of the regression line for predicting unemployment rate from educational level, and superimpose the graph of the regression line upon the scattergram.

e Find a 90% confidence interval for the unemployment rate among persons having 10 years of formal education.

4 A pharmaceutical researcher must know, as precisely as possible, the effect that a new drug will have on the human pulse rate. To investigate this effect, he administers different dosages of the drug to each of seven randomly selected patients, and he notes the increase in their pulse rates one hour later. The data follow:

Dosage, cm³	1.5	2.0	2.5	1.5	3.0	2.0	2.5
Change in pulse rate	8	10	14	10	15	11	12

a On the basis of a linear relationship, what percentage of the variation in pulse rate change can be attributed to variations in dosage?

b Construct a scattergram of the data.

c At level $\alpha = .05$, is there sufficient evidence to conclude that the relationship is really linear?

d Determine the equation of the regression line capable of predicting pulse rate change from a knowledge of the dosage, and superimpose the graph of the regression line upon the scattergram.

e Find a 90% confidence interval for the change in pulse rate corresponding to a dosage of 4.0 cm³.

5 In a study of the advertising budgets of small businesses, a consultant to the Small Business Administration collects the following data relating the size of a business's advertising budget with that business's total sales volume:

Advertising budget, hundreds of dollars	6	4	10	1	7	5	7
Total sales volume, thousands of dollars	50	60	100	30	60	60	40

a On the basis of the given data, what proportion of variation among total sales volumes of businesses can be explained by a linear relationship between the advertising budget and the sales volume?

b Draw a scattergram of the data.

c At level $\alpha = .10$, can the relationship described by the data be considered linear?

d Can the relationship be considered linear at a level of significance of $\alpha = .05$?

e Find the equation of the straight line which best predicts sales volume from a knowledge of the advertising budget, and superimpose the graph of the line upon the scattergram.

f Calculate a 90% confidence interval for the total sales volume of a small business that spends 800 dollars on advertising.

g Find a 90% confidence interval for the total sales volume of a small business that spends 300 dollars on advertising.

6 The following data[1] compare the Gross National Products (GNP's) of Japan and the United States, measured in billions of dollars at 1973 exchange rates:

Year	Japan GNP	U.S. GNP
1955	23.9	403.7
1960	43.1	511.4
1965	88.4	696.3
1970	197.2	993.3
1971	225.0	1068.8
1972	299.4	1155.2
1973	415.7	1289.1

a What proportion of the variation in the U.S. GNP over the years indicated can be explained by reference to a possible linear relationship with Japan's GNP?

b Draw a scattergram of the data.

c Test for the existence of linear relationship at a level of significance of $\alpha = .05$.

d Calculate the equation which best predicts the U.S. GNP in terms of the Japan GNP.

e If Japan's GNP were to grow to 600 billion dollars, at 1973 exchange rates, sometime in the future, find a 90% confidence interval for the U.S. GNP at that time.

SUMMARY AND DISCUSSION

In developing the techniques introduced in Chap. 6, our goal has been to discover the relationship between two sets of data for the purpose of eventually predicting one of the quantities from a knowledge of the other. The simplest and most fundamental measure of that relationship is, as we have seen, the coefficient of linear determination, which indicates the extent to which a straight line having equation $y = a + bx$ describes the data. A statistic closely tied to the coefficient of linear determination, Pearson's correlation coefficient r, can be used to test hypotheses involving the strength of a linear relationship in cases when the data points come from a normally distributed population. If our indicators seem to show that a linear relationship adequately describes the data, the next step would be to specify the regression line, the line that most closely approximates all the original data points simultaneously. After we have calculated the best-fitting (least-squares) line, we have shown how to use the line for the purpose of

[1] I. Frank and R. Hirono (eds), "How the United States and Japan See Each Other's Economy," p. 11, Committee for Economic Development, New York, 1974.

predicting one quantity from the other. Analogous techniques for analyzing sets of data that exhibit nonlinear relationships will be developed in Chap. 7.

BIBLIOGRAPHY

More Detailed Treatments of Linear Regression
Chapman, D. G., and R. A. Schaufele: "Elementary Probability Models and Statistical Inference," pp. 208–266, Xerox, Waltham, Mass., 1970.

Dixon, W. J., and F. J. Massey, Jr.: "Introduction to Statistical Analysis," 3d ed., pp. 193–236, McGraw-Hill, New York, 1969.

Regression Based on Mean Absolute Deviation
Rao, M. R., and V. Srinivasan: A Note on Sharpe's Algorithm for Minimizing the Sum of Absolute Deviations in a Simple Regression Problem, *Management Science*, October 1972, pp. 222–225 ("Erratum": July 1973, pp. 1334–1335).

Schlossmacher, E. J.: An Iterative Technique for Absolute Deviations Curve Fitting, *Journal of the American Statistical Association*, December 1973, pp. 857–859.

Sharpe, W. F.: Mean-Absolute-Deviation Characteristic Lines for Securities and Portfolios, *Management Science*, October 1971, pp. 1–13.

Special Topic
Searle, S. R.: Correlation Between Means of Parts and Wholes, *The American Statistician*, April 1969, pp. 23–24.

SUPPLEMENTARY EXERCISES

1 As part of an analysis of the relationship between smoking and absenteeism, the following data were obtained, relating an individual's number of packs smoked per day with the individual's number of days absent from his or her job.

Individual	A	B	C	D	E	F	G	H
Number of packs smoked per day	0.5	1.5	2.0	0.5	0.0	1.0	3.5	0.0
Number of days absent per year	4	8	15	0	3	10	20	0

a What proportion of the variation in absenteeism among individuals can be explained on the basis of a linear relationship between absenteeism and amount of smoking?

b Draw a scattergram of the data.

c At significance level $\alpha = .05$, is it reasonable to assume, from the data collected, that the relationship between smoking and absenteeism is linear?

d Calculate the equation of the regression line for predicting an individual's level of absenteeism from that individual's level of smoking, and superimpose the graph of the regression line upon the scattergram.

  e Find an 80% confidence interval for the number of days that an employee who smokes five packs a day will be absent.

2 A sociological study undertaken to investigate the relationship between the population density of a metropolitan area and that area's crime rate yields the following data:

Metropolitan Area	#1	#2	#3	#4	#5	#6
Population density, thousands per square mile	20	15	30	5	10	10
Crimes reported per 10,000 population	10	8	12	1	4	5

  a According to the results of the study, what proportion of variation in the crime rate among various metropolitan areas can be ascribed to variations in population density by way of a linear relationship?
  b Draw a scattergram of the data.
  c Decide using a level of significance $\alpha = .05$ whether or not the relationship between population density and crime rate, as described by the data, can reasonably be considered to be linear.
  d Determine the equation of the regression line which best accounts for the relationship between population density and crime rate, and superimpose its graph upon the scattergram.
  e Find a 90% confidence interval for the crime rate in a metropolitan area having population density of 25,000 per square mile.

3 Some investors think that prices on the New York Stock Exchange are related to prices on the London Gold Exchange. To test out this theory, one financial analyst compared the Dow-Jones industrial average with the price of gold in London on five randomly selected days. He came up with the following data in an attempt to find out whether he could predict changes in the Dow-Jones average from the behavior of the London gold prices which close several hours earlier because of time zone differences.

London Gold Prices	Dow-Jones Industrial Average
2.00	560
1.90	600
1.60	840
1.80	700
1.70	840

  a According to the data, how much of the variation in the Dow-Jones average is attributable to a linear relationship with the price of gold in London?

b Draw a scattergram of the data.

c At level $\alpha = .01$, do the data substantiate the existence of a linear relationship between the Dow-Jones average and the price of gold in London?

d Calculate the regression line for predicting the Dow-Jones average from the price of London gold, and superimpose the graph of the regression line upon the scattergram.

e Find an 80% confidence interval for the Dow-Jones average if the price of gold in London were to rise to 2.20.

4 A meteorologist studying the relationship between altitude and temperature accumulates the following data at six military airfields around the state:

Airfield	A	B	C	D	E	F
Altitude, hundreds of feet above sea level	9	70	7	4	50	20
High temperature, °F	90	50	100	110	70	80

a What proportion of the variation in temperature can be explained by a linear relationship with altitude?

b Draw a scattergram of the data.

c At level $\alpha = .05$, can we assert that the relationship between altitude and high temperature is a linear one?

d Determine the equation of the regression line expressing high temperature in terms of altitude, and superimpose its graph upon the scattergram.

e Find an 80% confidence interval for the high temperature at an altitude of 3500 feet above sea level.

5 A manufacturer of chemicals for use in the pharmaceuticals industry conducts a study of the time it takes after receipt of an order for the chemicals to reach the customer's plant. Because some of the chemicals deteriorate over time, an accurate estimate of the transit time is needed. The following data compare the rail distance from the manufacturer to the customer with the transit time of the most recent shipment to that customer:

Customer	#1	#2	#3	#4	#5	#6
Rail distance, hundreds of miles	5	4	1	3	2	5
Transit time, days	6.0	5.0	3.0	4.5	4.0	5.5

a Find the proportion of variation in transit times that can be attributed to a linear relationship with the rail distances

b Draw a scattergram of the data.

c At level $\alpha = .05$, would we be justified in using a linear equation to predict the transit time from a knowledge of the rail distance?

d Find the equation of the regression line, and superimpose its graph upon the scattergram.

e Customer 7 is located 700 miles away from the manufacturer. Find a 90% confidence interval for the time it will take a shipment of chemicals to arrive at his warehouse.

6　A botanist has collected the following data on the height (in centimeters) of a certain plant at several ages (in weeks after germination):

Age	1	2	4	6	7
Height	10	30	30	40	50

a What proportion of the variation in height seems to be attributable to the age of the plant by way of a linear relationship between height and age?
b Draw a scattergram of the data.
c Decide at level $\alpha = .05$ whether or not the relationship between age and height can be considered linear.
d Determine the equation of the regression line which expresses height of the plant in terms of its age, and superimpose the graph of the regression line upon the scattergram.
e Find an 80% confidence interval for the height of the plant 10 weeks after germination.

7　Some economists believe that as the prime interest rate increases, the value of stocks declines. The following data compare the prime rate with the value of General Nucleonics common stock at five randomly selected times during the past several years:

Prime rate, %	10	6	5	7	12
GN stock prices, dollars per share	60	80	90	70	50

a What proportion of the variation in GN stock prices can be accounted for by a linear relationship with the prime interest rate?
b Draw a scattergram of the data.
c At level $\alpha = .05$, can the relationship between the prime rate and the price of GN stock be considered linear?
d Determine the equation of the regression line which best predicts GN stock prices from the prime rate, and superimpose the graph of the regression line upon the scattergram.
e Find an 80% confidence interval for the value of GN stock if the prime rate were to rise to 15%.

8　One sociologist believes that an increase in the divorce rate foreshadows an increase in the crime rate. He obtains the following data on the divorce and crime rates per 10,000 population for five metropolitan areas:

Area	Divorce Rate	Crime Rate
A	2	1
B	6	4
C	8	6
D	4	3
E	10	6

a What proportion of the variation in the crime rate from area to area can be explained on the basis of a linear relationship between the crime rate and the divorce rate?

b Draw a scattergram of the data.

c At a level of significance of $\alpha = .05$, is it reasonable to consider crime rate and divorce rate to be linearly related?

d Determine the equation of the regression line which best predicts crime rate from divorce rate, and superimpose the graph of the regression line upon the scattergram.

e Find an 80% confidence interval for the crime rate in a metropolitan area where the divorce rate is 5 per 10,000 population.

9 A record of maintenance cost is assembled on six identical metal-stamping machines of different ages in an attempt to find out the relationship, if there is any, between the age of a machine (in years) and the monthly maintenance cost (in dollars) required to keep it in peak operating condition. Such information would be useful in deciding when it is most economical to replace the old machines with new ones. Data on six randomly selected machines follow:

Machine Serial No.	901	887	923	927	891	906
Age, years	2	1	3	4	4	6
Maintenance cost, dollars per month	10	10	30	30	40	50

a What proportion of the variation in maintenance cost can be attributed, by the way of a linear relationship, to variations in age?

b Draw a scattergram of the data.

c At level $\alpha = .05$, can the relationship between maintenance cost and machine age be considered linear?

d Find the regression equation which predicts monthly maintenance cost in terms of machine age, and superimpose the graph of the regression line upon the scattergram.

e Find a 90% confidence interval for predicting the monthly maintenance cost of a 7-year-old machine.

10 The following table gives the wholesale price index and the index of industrial production in the United States on July 1 of each year of a 5-year period. Knowledge of the relationship between the two indices would be helpful to an economist attempting to forecast business conditions.

Year	Wholesale Prices	Industrial Production
#1	80	110
#2	80	90
#3	90	80
#4	80	100
#5	70	120

a What proportion of variation in the index of industrial production seems to be explained by a possible linear relationship with the wholesale price index?

b Draw a scattergram of the data.

c At level $\alpha = .10$, decide whether or not it is reasonable to consider the two indices to be linearly related.

d Find the equation of the regression line that can be used to predict the index of industrial production from a knowledge of the wholesale price index, and superimpose the graph of the regression line upon the scattergram.

e Determine an 80% confidence interval for the index of industrial production when the wholesale price index stands at 100.

11 The following data[1] compare over a 14-year period in the recent past expenditures for new homes in the United States with non–real-estate mortgage debt:

Annual Data Billions of Dollars, Seasonally Adjusted

Year	Mortgage Debt for Non-Real-Estate Purposes	Aggregate Outlays for New Homes
1960	.1	19.9
1961	3.6	17.4
1962	4.7	19.1
1963	8.4	18.7
1964	10.3	19.1
1965	10.8	19.0
1966	10.5	18.4
1967	11.6	15.8
1968	12.1	20.8
1969	14.4	21.2
1970	14.0	18.1
1971	18.3	26.6
1972	25.4	35.6
1973	28.0	39.9

a What proportion of the variation in aggregate outlays for new homes can be explained by a linear relationship with non–real-estate mortgage debt?

b Draw a scattergram of the data.

c At level $\alpha = .05$, can we consider the two economic measurements to be linearly related?

d Determine the regression equation which predicts aggregate outlays for new homes from a knowledge of non–real-estate mortgage debt.

e Find a 90% confidence interval for aggregate outlays for new homes in a year when the non–real-estate mortgage debt amounts to 20.0 billion dollars.

[1]The Conference Board, *Statistical Bulletin*, December 1974, p. 13.

CURVILINEAR AND MULTIPLE REGRESSION

During our study of linear regression in Chap. 6, we encountered, from time to time, sets of data which could not reasonably be described by a straight line. In particular, we presented parabolic, logarithmic, and exponential data in Table 6.4 and graphed their obviously nonlinear scattergrams in Fig. 6.9. Our objective in this chapter is to develop techniques for analyzing these and other patterns of data which are not adequately explained in terms of linear regression.

SECTION 7.A THE LOGARITHMIC TRANSFORMATION METHOD

Logarithmic and exponential data, as illustrated in scattergrams *b* and *c* of Fig. 6.9, can be studied by using linear regression techniques if the raw data are first subjected to a "logarithmic transformation."

In order to understand this transformation method, it is necessary to recall some facts about logarithms that you may have come in contact with in your study of algebra. (Here we will work with "base 10" or "common" logarithms only.) By the statement $y = \log x$ we mean that $10^y = x$, namely, that $y = \log x$ is the power to which we raise 10 in order to get x. Logarithms have the following three "basic properties":

1 $\log (ab) = \log a + \log b$
2 $\log (a/b) = \log a - \log b$
3 $\log (a^b) = b \log a$

We will make use of these "basic properties" extensively in this section.

The logarithmic data of Table 6.4*b* provides an illustration of "logarithmic regression" (as contrasted with the "linear regression" of Chap. 6).

Example 7.1   Musical Aptitude   Two researchers have devised what they claim is a psychological test of musical aptitude. Because the test is one of aptitude rather than skill, they feel that practice and study should not strongly affect the test scores. On the basis of data collected by repeated offerings of the test, the researchers discovered that there is a dramatic rise in an individual's score as the period of study increases from 1 to 10 months. After that, however, the scores level off and do not seem to improve significantly with practice. The following pattern of scores versus months of practice resulted from an administration of the test to eight persons:

Months of practice x	1	3	8	14	20	10	5	2
Test score y	1	4	8	9.5	10	9	6	3

The researchers need the answers to the following questions:

a What does the scattergram of the data look like, and to what general class of curves does the regression curve belong?
b How much of the variation in test scores can be attributed to the curvilinear relationship with months of practice?
c What is the regression equation for predicting the test score from a knowledge of the number of months of practice?
d What would be the best prediction for the test score of someone who has practiced for 30 months?

SOLUTION

a The scattergram appears in Fig. 7.1, and it indicates that a logarithmic regression curve is most appropriate for an analysis of the data. A logarithmic regression curve has the equation

$$y = a + b \log x \qquad (7.1)$$

where y, the variable to be predicted, is the test score, and x is the months of practice.

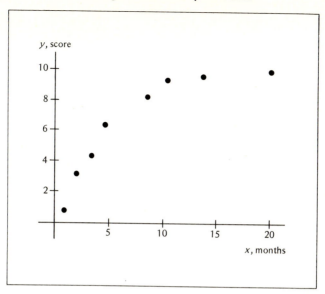

FIGURE 7.1
Scattergram of musical aptitude data.

b If we denote $X = \log x$, we can express

$$y = a + b \log x$$

as

$$y = a + bX$$

In this form, we can recognize the logarithmic regression equation as a *linear regression equation* between y and $X = \log x$. We can therefore use some of the techniques developed in Chap. 6 in our study of logarithmic regression. In particular, we can compute the coefficient of linear determination between X and y in order to find how much of the variation in test scores can be attributed to their logarithmic relationship with months of practice. We use the formula for r in Eq. (6.3) with x replaced by $X = \log x$. The logarithms of the x values are obtained from Table A.7 of the Appendix. The preliminary calculations appear in Table 7.1. We obtain, using the results of Table 7.1, that

$$r = \frac{n\Sigma Xy - (\Sigma X)(\Sigma y)}{\sqrt{n\Sigma X^2 - (\Sigma X)^2}\sqrt{n\Sigma y^2 - (\Sigma y)^2}}$$

$$= \frac{8(47.1282) - (5.8273)(50.5)}{\sqrt{8(5.6285) - (5.8273)^2}\sqrt{8(397.25) - (50.5)^2}}$$

$$= \frac{82.7470}{\sqrt{11.0706}\sqrt{627.75}} = \frac{82.747}{(3.327)(25.055)}$$

$$= .993$$

It follows that the proportion of variation in test scores that can be explained on the

TABLE 7.1
Preliminary Calculations for Musical Aptitude Data
(Logarithmic Regression)

x	$X = \log x$	y	X^2	y^2	Xy
1	.0000	1.0	.0000	1.00	.0000
3	.4771	4.0	.2276	16.00	1.9084
8	.9031	8.0	.8156	64.00	7.2248
14	1.1461	9.5	1.3135	90.25	10.8880
20	1.3010	10.0	1.6926	100.00	13.0100
10	1.0000	9.0	1.0000	81.00	9.0000
5	.6990	6.0	.4886	36.00	4.1940
2	.3010	3.0	.0906	9.00	.9030
$n = 8$	5.8273	50.5	5.6285	397.25	47.1282

basis of a *logarithmic* relationship with months of practice is

$$r^2 = (.993)^2 = .986 = 98.6\%$$

This result contrasts very favorably with the 80.5% of the variation that can be explained on the basis of linear relationship, as we calculated in Sec. 6.C using the same data. An r^2 of 98.6% indicates that we can expect reasonably accurate predictions if we use a logarithmic regression curve.

c To compute a and b in the logarithmic regression equation,

$$y = a + b \log x$$

or

$$y = a + bX$$

we simply apply formulas (6.1) and (6.2) to the calculated results of Table 7.1. We have

$$b = \frac{n\Sigma Xy - (\Sigma X)(\Sigma y)}{n\Sigma X^2 - (\Sigma X)^2} = \frac{82.7470}{11.0706} = 7.47$$

and

$$a = \frac{\Sigma y - b\Sigma X}{n} = \frac{50.5 - (7.47)(5.8273)}{8} = \frac{6.97}{8} = .87$$

Therefore the logarithmic regression equation is

$$y = .87 + 7.47 \log x$$

To graph the regression curve, we choose a few values of x within the range of the original data and compute their corresponding y's. The points on the curve are computed in Table 7.2, and the graph of the regression curve is superimposed on the scattergram in Fig. 7.2.

d To predict the test score of an individual who has practiced for 30 months, we merely set $x = 30$ in the regression equation and then compute y. We then get

$$y = .87 + 7.47 \log 30$$
$$= .87 + (7.47)(1.4771)$$
$$= .87 + 11.03 = 11.9$$

FIGURE 7.2
The logarithmic regression curve.

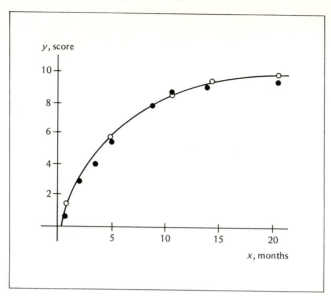

from which we conclude that for an individual with 30 months of practice, our best prediction would be a test score of 11.9. (What we are really saying by the prediction is that the average test score of all those having 30 months of practice would most likely be 11.9.)

You will notice that we have not discussed the test for linearity (using the correlation coefficient) or the confidence intervals for predictions, as done in Sec. 6.F. The reason is that these methods require the basic data sets (here the X's and the y's) to come from normally distributed populations. We cannot be sure that the X's are normally distributed unless we have previously tested specifically for normality, using the techniques of Sec. 9.C. Even if we are sure that the original x's are normally distributed, their logarithms (namely, the X's) will not be.

In the next example, we analyze the exponential data of Table 6.4c and Fig. 6.9c.

TABLE 7.2
Points on the Logarithmic Regression Curve

x	$X = \log x$	$y = .87 + 7.47 \log x$
1	0.0000	.87
5	0.6990	6.09
10	1.0000	8.34
15	1.1761	9.66
20	1.3010	10.59

Example 7.2 Demand for Color Televisions Annual demand for one brand of color TV in the San Francisco–Oakland metropolitan area drops off rapidly as prices rise, but there always seems to be some market for them no matter what the price. Sales data collected recently related prices and sales figures as follows:

Price, hundreds of dollars	5	3	1	2	4	6
Sales, thousands of TV's	2	3	10	5	2	1

We need answers to the following questions:

a What does the scattergram of the data look like, and to what general class of curves does the regression curve of demand belong?

b How much of the variation in sales can be attributed to the appropriate curvilinear relationship with the prices?

c What is the best-fitting regression equation for predicting sales from a knowledge of the price?

d What would be the best estimate of sales of color TV's carrying a price of 350 dollars?

SOLUTION

a The scattergram appears in Fig. 7.3 and indicates that an exponential curve is most appropriate for an analysis of the data. There are two basic types of exponential curves: the "simple exponential regression" equation is of the form

$$y = a \cdot b^x \tag{7.2}$$

and the "power regression" equation has the form

$$y = ax^b \tag{7.3}$$

Which of these equations better describes a particular set of data can be discovered by comparing the coefficients of determination of each.[1] We carry out that process in answer to question b.

b We compute the coefficient of determination corresponding to each type of exponential regression in an attempt to find which kind better fits the data. Looking first at the simple exponential regression equation

$$y = a \cdot b^x$$

---

[1]Those students having some acquaintance with technical graph paper may recognize that simple exponential regression is appropriate for data that follow a straight line when graphed on semilog paper, while power regression should be used for data following a straight line on log-log paper. The coefficient of determination, however, provides a mathematically more precise method of distinguishing between the two.

FIGURE 7.3
Scattergram of color TV demand data.

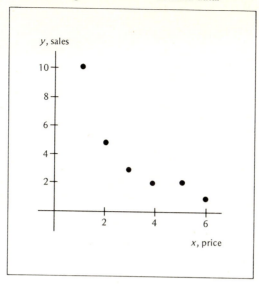

we take logarithms of both sides and apply the basic properties of logarithms listed at the beginning of this section:

$$\log y = \log (a\ b^x)$$
$$= \log a + \log (b^x)$$
If we set
$$= \log a + x \log b$$

$$Y = \log y$$
$$A = \log a$$
and
$$B = \log b$$

our simple exponential regression equation can be written in the form

$$Y = A + Bx$$

which we can recognize as *linear* between x and $Y = \log y$. As in the previous example, we can compute the coefficient of linear determination between x and $Y = \log y$ in the usual way. We present the preliminary calculations in Table 7.3. Using the results of Table 7.3, we obtain that

$$r = \frac{n\Sigma xY - (\Sigma x)(\Sigma Y)}{\sqrt{n\Sigma x^2 - (\Sigma x)^2}\sqrt{n\Sigma Y^2 - (\Sigma Y)^2}}$$

$$= \frac{6(6.5383) - (21)(2.7781)}{\sqrt{6(91) - (21)^2}\sqrt{6(1.8974) - (2.7781)^2}}$$

$$= \frac{-19.1103}{\sqrt{105}\sqrt{3.67}} = \frac{-19.11}{(10.25)(1.91)} = -.974$$

TABLE 7.3
Preliminary Calculations for Color TV Demand Data
Simple Exponential Regression

x	y	$Y = \log y$	x^2	y^2	xY
5	2	.3010	25	.0906	1.5050
3	3	.4771	9	.2276	1.4313
1	10	1.0000	1	1.0000	1.0000
2	5	.6990	4	.4886	1.3980
4	2	.3010	16	.0906	1.2040
6	1	.0000	36	.0000	.0000
21		2.7781	91	1.8974	6.5383
$n = 6$					

It follows that the proportion of variation in sales that can be explained on the basis of a *simple exponential* relationship with price

$$r^2 = (-.974)^2 = .949 = 94.9\%$$

Again this value of r^2 represents an improvement over the coefficient of determination for the *linear* relationship of 78.8% that we found in Sec. 6.C.

Before we adopt the simple exponential regression approach, however, we ought to check out the coefficient of determination corresponding to the power regression equation $y = ax^b$. Logarithmic transformation of this equation yields

$$\log y = \log (ax^b)$$
$$= \log a + \log x^b$$
$$= \log a + b \log x$$
or
$$Y = A + bX$$

Here we have a linear equation between $X = \log x$ and $Y = \log y$. The preliminary calculations for the coefficient of determination appear in Table 7.4. The results of Table 7.4 yield that

TABLE 7.4
Preliminary Calculations for Color TV Demand Data
Power Regression

x	y	$X = \log x$	$Y = \log y$	X^2	Y^2	XY
5	2	.6990	.3010	.4886	.0906	.2104
3	3	.4771	.4771	.2276	.2276	.2276
1	10	.0000	1.0000	.0000	1.0000	.0000
2	5	.3010	.6990	.0906	.4886	.2104
4	2	.6021	.3010	.3625	.0906	.1812
6	1	.7782	.0000	.6056	.0000	.0000
$n = 6$		2.8574	2.7781	1.7749	1.8974	.8296

$$r = \frac{n\Sigma XY - (\Sigma X)(\Sigma Y)}{\sqrt{n\Sigma X^2 - (\Sigma X)^2}\sqrt{n\Sigma Y^2 - (\Sigma Y)^2}}$$

$$= \frac{6(.8296) - (2.8574)(2.7781)}{\sqrt{6(1.7749) - (2.8574)^2}\sqrt{6(1.8974) - (2.7781)^2}}$$

$$= \frac{-2.9605}{\sqrt{2.4847}\sqrt{3.667}} = \frac{-2.96}{(1.58)(1.91)} = -.981$$

so that the proportion of variation in sales that can be explained on the basis of a *power* relationship with price is

$$r^2 = (-.981)^2 = .962 = 96.2\%$$

This value of r^2 exceeds not only the value of 78.8% that we obtained earlier for linear regression but also the value of 94.9% corresponding to simple exponential regression. We therefore conclude the best-fitting curvilinear relationship is the power regression one, $y = ax^b$, and it accounts for 96.2% of the variations in sales based on variations in price.

c   To find a and b in the power regression equation

$$y = ax^b$$
or
$$Y = A + bX$$

we simply apply formulas (6.1) and (6.2) to the results calculated in Table 7.4. We get b and $A = \log a$ as follows:

$$b = \frac{n\Sigma XY - (\Sigma X)(\Sigma Y)}{n\Sigma X^2 - (\Sigma X)^2} = \frac{-2.9605}{2.4847} = -1.19$$

and $A = \dfrac{\Sigma Y - b\Sigma X}{n} = \dfrac{2.7781 - (-1.19)(2.8574)}{6} = \dfrac{6.18}{6} = 1.03$

Using Table A.7 of the Appendix, we can find the value of a from $A = \log a = 1.03$. Elementary facts about logarithms show that a is between 10 and 100, for $A = \log a$ is between 1 and 2. Looking up .03 in the body of Table A.7, we then see that

$$a = 10.7$$

(If you have a sufficiently advanced pocket calculator, you can find a as follows: $a = 10^A = 10^{1.03} = 10.7$.) Therefore the power regression equation is

$$y = 10.7x^{-1.19}$$

Some points on the power regression curve are listed in Table 7.5, and its graph is superimposed on the scattergram in Fig. 7.4.

d   To estimate the sales level corresponding to a price of 350 dollars, we note first that the price data is expressed in hundreds of dollars. Therefore a price of 350 dollars translates into an x value of 3.5. The estimate of sales (in thousands of TV's) can then

TABLE 7.5
Points on the Power Regression Curve

x	$x^{-1.19}$	$y = 10.7x^{-1.19}$
1	1.0000	10.70
3	.2705	2.89
5	.1473	1.58
7	.0987	1.06

FIGURE 7.4
The power regression curve.

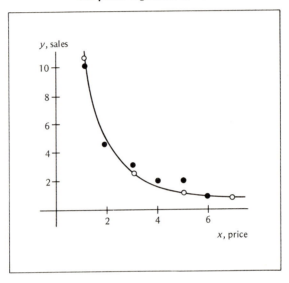

be obtained by setting $x = 3.5$ in the regression equation $y = 10.7 x^{-1.19}$. We get

$$y = 10.7(3.5)^{-1.19} = (10.7)(.225) = 2.41$$

using which we estimate sales at 2410 television sets.

EXERCISES 7.A

1 Based on the Coronado Nautotronics data of Exercise 6.C.5, answer the following questions:
   a What does the scattergram of the data look like, and to what general class of curves does the regression curve belong?
   b How much of the variation in total cost of production run can be attributed to its curvilinear relationship with the quantity produced?
   c What is the regression equation for the cost curve predicting total cost in terms of the quantity produced?

d What would be the best estimate for the total cost of a production run of 500 radar displays?

2 Using the data of Exercise 6.C.6:
  a Draw the scattergram of the data, and determine the general class of curves to which the demand equation for rolled steel belongs.
  b Calculate relevant coefficients of determination, and find the proportion of variation in price that can be explained on the basis of the appropriate curvilinear relationship with the quantity that can be sold (the demand).
  c Find the best-fitting regression equation for predicting the price of rolled steel from the demand, and superimpose the regression curve upon the scattergram.
  d Predict the price required to dispose of 3 tons of rolled steel.

3 A behavioral biologist believes that performance of a laboratory rat on an intelligence test depends, to a large extent, on the amount of protein in the rat's daily diet. To check out the theory, he accumulated the following data after working with 10 rats:

Rat	A	B	C	D	E	F	G	H	I	J
No. of units of protein daily	12	12	10	10	10	3	2	20	20	30
Score on standard test	20	40	30	80	50	50	90	30	40	40

  a Draw the scattergram of the data, and determine the general class of curves to which the regression curve belongs.
  b Establish the most appropriate curvilinear relationship on the basis of a comparison between pertinent coefficients of determination.
  c Calculate the equation of the best-fitting regression curve, and superimpose its graph on the scattergram.
  d Predict the test score of rat K who gets 30 units of protein per day.

4 A sociologist, undertaking a preliminary study of the effects on society of a proposed rescheduling of the 36-hour work week into three 12-hour days, measures the productivity of employees in situations where an experimental 10-hour day is in operation. She collects the following data, which accords with the law of diminishing returns, on one randomly selected employee each hour:

Employee	M	N	P	Q	R	S	T	U	V	W
Hour	1	2	3	4	5	6	7	8	9	10
Accumulated productivity	3	6	8	9	10	11	11	12	12	13

  a Draw the scattergram, and use it to find the general class of curves to which the regression curve belongs.
  b How much of the variation in accumulated productivity can be attributed to the appropriate curvilinear relationship with the hour of the day?

c Find the regression equation that predicts accumulated productivity at each hour of the day, and superimpose its graph on the scattergram.

d Predict the accumulated productivity for the 12-hour day.

SECTION 7.B PARABOLIC REGRESSION

Parabolic data, such as those whose scattergram appears in Fig. 6.9a, can be easily recognized by the fact that the data reverse their trend in the middle of the course. If the curve starts in a downward direction, then it later turns upward, but if it starts in an upward direction, it later turns downward. A general parabolic regression equation has the form

$$y = a + bx + cx^2 \qquad (7.4)$$

In the next example, we illustrate the computation and use of the parabolic regression equation, with the data of Table 6.4a.

Example 7.3 Repair of Trucks   A trucking company wants to determine the typical repair needs of its trucks during each year of their useful life. It chooses six of its trucks at random and records their ages (in years) and the number of days they were inoperable because of repairs that year. The following data indicate that a truck seems to require a large number of repairs early in its lifetime, perhaps to get the "bugs" out, and again toward the end of its life as its parts begin to wear out.

Age of truck, years	8	1	6	3	5	2
Days lost due to repairs	9	16	1	4	0	10

The company wants to know the answers to the following questions:

a What does the scattergram of the data look like, and to what general class of curves does the regression curve belong?
b What is the best-fitting regression equation for predicting the number of days lost due to repairs from a knowledge of the age of the truck?
c What would be the best estimate of days lost by a 10-year-old truck?

SOLUTION

a As the scattergram in Fig. 7.5 shows, a parabolic regression equation would be the appropriate way of analyzing the truck repair data. (We will bypass the coefficient of parabolic determination which is somewhat more difficult to compute and also not as necessary since the data are obviously parabolic in shape.)
b To find the best-fitting parabolic regression equation of the form given in formula (7.4), we must compute the values of a, b, and c. It can be shown, using calculus,

FIGURE 7.5
Scattergram of truck repair data.

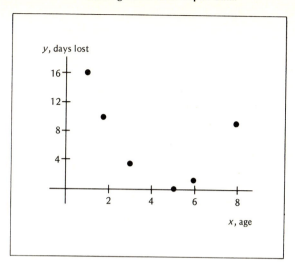

that a, b, and c are the three unknowns whose values can be determined by solving the following three simultaneous equations:

$$an + b\Sigma x + c\Sigma x^2 = \Sigma y$$
$$a\Sigma x + b\Sigma x^2 + c\Sigma x^3 = \Sigma xy \qquad (7.5)$$
$$a\Sigma x^2 + b\Sigma x^3 + c\Sigma x^4 = \Sigma x^2 y$$

We work out the components of these simultaneous equations in Table 7.6. Inserting the results of the calculations into Eq. (7.5), we have the three simultaneous equations in three unknowns:

$$6a + 25b + 139c = 40$$
$$25a + 139b + 889c = 126$$
$$139a + 889b + 6115c = 704$$

TABLE 7.6
Preliminary Calculations for Truck Repair Data
Parabolic Regression

x	y	x^2	x^3	x^4	xy	$x^2 y$
8	9	64	512	4096	72	576
1	16	1	1	1	16	16
6	1	36	216	1296	6	36
3	4	9	27	81	12	36
5	0	25	125	625	0	0
2	10	4	8	16	20	40
25	40	139	889	6115	126	704
$n = 6$						

There are many ways of solving three simultaneous equations (as you may have learned in your algebra courses), but none of them are particularly easy. We will present one method here, and your instructor undoubtedly knows of several others. We first calculate the "determinant,"

$$D = \begin{vmatrix} 6 & 25 & 139 \\ 25 & 139 & 889 \\ 139 & 889 & 6115 \end{vmatrix} = 29{,}040$$

by multiplying diagonally as follows:

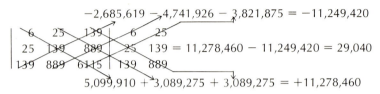

$$-2{,}685{,}619 - 4{,}741{,}926 - 3{,}821{,}875 = -11{,}249{,}420$$

$$139 = 11{,}278{,}460 - 11{,}249{,}420 = 29{,}040$$

$$5{,}099{,}910 + 3{,}089{,}275 + 3{,}089{,}275 = +11{,}278{,}460$$

(For example $6 \times 139 \times 6115 = 5{,}099{,}910$.) Now to calculate a, we replace the first column of the determinant D by the column to the right of the equal signs in the simultaneous equations, and we calculate the determinant

$$N_a = \begin{vmatrix} 40 & 25 & 139 \\ 126 & 139 & 889 \\ 704 & 889 & 6115 \end{vmatrix}$$

$$-13{,}601{,}984 - 31{,}612{,}840 - 19{,}262{,}250 = -64{,}477{,}074$$

$$= \begin{vmatrix} 40 & 25 & 139 \\ 126 & 139 & 889 \\ 704 & 889 & 6115 \end{vmatrix} \quad 139 = +738{,}672$$

$$+33{,}999{,}400 + 15{,}646{,}400 + 15{,}569{,}946 = +65{,}215{,}746$$

Then
$$a = \frac{N_a}{D} = \frac{738{,}672}{29{,}040} = 25.44$$

To calculate b, we replace the second column of D by the column to the right of the equal signs in the simultaneous equations, and we get

$$N_b = \begin{vmatrix} 6 & 40 & 139 \\ 25 & 126 & 889 \\ 139 & 704 & 6115 \end{vmatrix} = -292{,}402$$

so that
$$b = \frac{N_b}{D} = \frac{-292{,}402}{29{,}040} = -10.07$$

Similarly

$$N_c = \begin{vmatrix} 6 & 25 & 40 \\ 25 & 139 & 126 \\ 139 & 889 & 704 \end{vmatrix} = 29{,}062$$

so that
$$c = \frac{29{,}062}{29{,}040} = 1.00$$

TABLE 7.7
Points on the Parabolic Regression Curve

x	x^2	$y = 25.44 - 10.07x + x^2$
2	4	9.30
4	16	1.16
6	36	1.02
8	64	8.88

FIGURE 7.6
The parabolic regression curve.

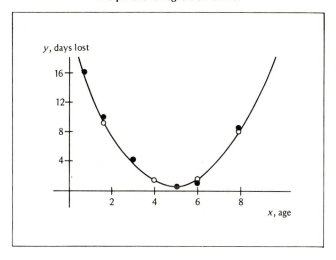

Using the values obtained for a, b, and c, we can write the parabolic regression equation $y = a + bx + cx^2$ as

$$y = 25.44 - 10.07x + 1.00x^2$$

Table 7.7 lists some points on the regression curve, and the graph of the curve is superimposed on the scattergram in Fig. 7.6.

c To estimate the days lost ("downtime") due to repairs by a 10-year-old truck, we merely set $x = 10$ in the regression equation. We get

$$y = 25.44 - 10.07x + x^2$$
$$= 25.44 - (10.07)(10) + (10)^2$$
$$= 24.74$$

So we would predict approximately 25 days lost by a 10-year-old truck.

EXERCISES 7.B

1 The data of Exercise 6.C.7 are to be used to measure an individual's aptitude for computer programming.

a Draw a scattergram of the data, and determine the general class of curves to which the regression curve belongs.

b Find the best-fitting regression equation for predicting job performance rating from mathematical aptitude, and superimpose its graph on the scattergram.

c Predict the job performance rating of an individual whose mathematical aptitude score is 4.5.

2 Demand for Napoleon rubies increases as the price goes down, because more people are able to buy them. When the price is high, demand is also high since they are a much sought-after status symbol. At moderate prices, however, not too many are sold, because the price is too high for many people but yet not high enough to result in demand as status symbols. In particular, some recent data relating price and demand of these items went as follows:

Price per carat, hundreds of dollars	0.5	1	3.5	5	7	8
Demand, thousands	160	130	40	50	130	200

a Draw the scattergram, and determine the general class of curves to which the regression curve belongs.

b Calculate the equation of the best-fitting regression curve, and superimpose its graph on the scattergram.

c Predict the demand for Napoleon rubies if the price is set at 600 dollars per carat.

3 To determine the properties of an antibiotic used in treating 10-year-old children for stomach infections, a medical researcher conducts a study of various dosage levels and corresponding times until recovery. The data that follow may reflect the fact that too small a dosage is insufficient to destroy the infection whereas too large a dosage impedes recovery by introducing side effects.

Hourly dosage, grains	1	2	3	4	5	6
Time to recovery, hours	13	10	8	9	12	16

a Construct the scattergram to determine the general class of curves to which the regression curve belongs.

b Calculate the equation of the appropriate regression curve, and superimpose its graph on the scattergram.

c Predict the time to recovery corresponding to a dosage of 3.2 grains.

4 In a study of the efficiency of automobile engines with respect to consumption of gasoline, one subcompact car was operated at various speeds, and the fuel consumption (in miles per gallon) was carefully measured. The data are as follows:

Speed, miles/hour	10	20	30	40	50	60
Efficiency, miles/gallon	15	20	30	35	25	20

a Draw the scattergram to find the general class of curves to which the regression curve belongs.

b Find the equation of the appropriate regression curve, and superimpose its graph on the scattergram.

c Estimate the efficiency at 70 miles/hour.

SECTION 7.C MULTIPLE REGRESSION

In many situations of applied interest, there are two or even more factors which affect the numerical value of a variable that we may be working with. All the regression equations that we have studied up to now, however, express y in terms of only one factor, namely, x. Using the techniques of multiple regression, we can find equations for y in terms of two or more factors. (Such factors are called "regressors.") In this section, for example, we will concentrate on "multiple linear regression" in the presence of two regressors x and w, so that our general multiple linear regression equation will be

$$y = a + bx + cw \qquad (7.6)$$

Consider the following example:

Example 7.4 Department Store Volume In Portland, Oregon, it rains all day, every day, from October to April. The vice-president in charge of advertising at one small department store claims that the store's daily sales volume can be predicted from a knowledge of the day's advertising costs and the morning's total precipitation. A search of accounting and weather data reveals the following information for six randomly selected days last year:

	Jan. 12	Mar. 4	Apr. 2	Aug. 14	Oct. 27	Dec. 6
Advertising costs x, hundreds of dollars	2	3	5	4	6	10
Morning precipitation w, tenths of an inch	5	3	1	0	1	5
Sales volume y, thousands of dollars	1	9	19	21	25	33

We would like to use the data to answer the following questions. (Unfortunately, with three variables x, w, and y, it is not so practical to construct scattergrams or to draw the graph of the regression equation. The graph of the multiple linear regression equation will be a plane instead of a line, and so the geometric interpretation would be somewhat complicated.)

a How much of the variation in sales volume y can be attributed to a linear relationship with advertising costs x alone?

b How much of the variation in sales volume y can be attributed to a linear relationship with precipitation w alone?

c How much of the variation in sales volume y can be explained by a multiple linear relationship with advertising costs x and precipitation w together?

d What is the best-fitting multiple linear regression equation?

e If the store spent 700 dollars for today's advertising, and .6 inch of precipitation fell this morning, what would be the best prediction for today's sales volume?

SOLUTION   To answer questions $a, b, c,$ and d, we need the calculations appearing in Table 7.8.

a The answer to this question is the coefficient of linear determination between x and y. We calculate the correlation coefficient

$$r_{xy} = \frac{n\Sigma xy - (\Sigma x)(\Sigma y)}{\sqrt{n\Sigma x^2 - (\Sigma x)^2}\sqrt{n\Sigma y^2 - (\Sigma y)^2}} = \frac{6(688) - (30)(108)}{\sqrt{6(190) - (30)^2}\sqrt{6(2598) - (108)^2}}$$

$$= \frac{888}{\sqrt{240}\sqrt{3924}} = \frac{888}{(15.492)(62.642)} = .9150$$

Therefore

$$r_{xy}^2 = (.9150)^2 = .8372 = 83.7\%$$

of the variation in y can be accounted for by a linear relationship with x alone.

b To calculate the correlation coefficient between w and y

$$r_{wy} = \frac{n\Sigma wy - (\Sigma w)(\Sigma y)}{\sqrt{n\Sigma w^2 - (\Sigma w)^2}\sqrt{n\Sigma y^2 - (\Sigma y)^2}} = \frac{6(241) - (15)(108)}{\sqrt{6(61) - (15)^2}\sqrt{6(2598) - (108)^2}}$$

$$= \frac{-174}{\sqrt{141}\sqrt{3924}} = \frac{-174}{(11.874)(62.642)} = -.2339$$

TABLE 7.8
Preliminary Calculations for Department Store Data
Multiple Linear Regression

x	w	y	x^2	w^2	y^2	xw	xy	wy
2	5	1	4	25	1	10	2	5
3	3	9	9	9	81	9	27	27
5	1	19	25	1	361	5	95	19
4	0	21	16	0	441	0	84	0
6	1	25	36	1	625	6	150	25
10	5	33	100	25	1089	50	330	165
30	15	108	190	61	2598	80	688	241
$n = 6$								

Therefore the proportion of the variation in y that can be accounted for by a linear relationship with w alone is

$$r_{wy}^2 = (-.2339)^2 = .0547 = 5.5\%$$

c In order to compute the proportion of variation in y that can be attributed to a multiple linear relationship with x and w together, we calculate the "coefficient of multiple determination"

● $$R^2 = r_{xy}^2 + (1 - r_{xy}^2)r_{wy:x}^2 \qquad (7.7)$$

or, equivalently,

● $$R^2 = r_{wy}^2 + (1 - r_{wy}^2)r_{xy:w}^2 \qquad (7.8)$$

where the "coefficients of partial correlation" are

● $$r_{wy:x} = \frac{r_{wy} - r_{xw}r_{xy}}{\sqrt{(1 - r_{xw}^2)(1 - r_{xy}^2)}} \qquad (7.9)$$

and

● $$r_{xy:w} = \frac{r_{xy} - r_{xw}r_{wy}}{\sqrt{(1 - r_{xw}^2)(1 - r_{wy}^2)}} \qquad (7.10)$$

Using formula (7.7) to compute the coefficient of multiple determination R^2 requires a computation of $r_{wy:x}$ by formula (7.9). We carry this out by first finding r_{xw}.

$$r_{xw} = \frac{n\Sigma xw - (\Sigma x)(\Sigma w)}{\sqrt{n\Sigma x^2 - (\Sigma x)^2}\sqrt{n\Sigma w^2 - (\Sigma w)^2}} = \frac{6(80) - (30)(15)}{\sqrt{240}\sqrt{141}}$$

$$= \frac{30}{(15.492)(11.874)} = .1631$$

Therefore, formula (7.9) yields that

$$r_{wy:x} = \frac{r_{wy} - r_{xw}r_{xy}}{\sqrt{(1 - r_{xw}^2)(1 - r_{xy}^2)}} = \frac{(-.2339) - (.1631)(.9150)}{\sqrt{[1 - (.1631)^2][1 - (.9150)^2]}}$$

$$= \frac{-.3831}{\sqrt{(1 - .0266)(1 - .8372)}} = \frac{-.3831}{\sqrt{(.9734)(.1628)}}$$

$$= \frac{-.3831}{\sqrt{.1585}} = \frac{-.3831}{.3981} = -.9623$$

It then follows from formula (7.7) that

$$R^2 = r_{xy}^2 + (1 - r_{xy}^2)r_{wy:x}^2$$

$$= .8372 + (.1628)(-.9623)^2$$

$$= .8372 + (.1628)(.9260) = .9880$$

$$= 98.8\%$$

We conclude, then, that x (advertising costs) and w (precipitation) together account for 98.8% of the variation in y (sales volume).[1]

d To find the best-fitting multiple regression equation of the form

$$y = a + bx + cw$$

of Eq. (7.6) above, we solve the simultaneous equations,

$$na + b\Sigma x + c\Sigma w = \Sigma y$$
● $$a\Sigma x + b\Sigma x^2 + c\Sigma xw = \Sigma xy \qquad\qquad (7.11)$$
$$a\Sigma w + b\Sigma xw + c\Sigma w^2 = \Sigma wy$$

Inserting the results of the calculations in Table 7.8 into Eq. (7.11), we obtain

$$6a + 30b + 15c = 108$$
$$30a + 190b + 80c = 688$$
$$15a + 80b + 61c = 241$$

We solve these equations as in Sec. 7.B:

$$D = \begin{vmatrix} 6 & 30 & 15 \\ 30 & 190 & 80 \\ 15 & 80 & 61 \end{vmatrix} = 5490$$

$$N_a = \begin{vmatrix} 108 & 30 & 15 \\ 688 & 190 & 80 \\ 241 & 80 & 61 \end{vmatrix} = 18{,}630$$

$$N_b = \begin{vmatrix} 6 & 108 & 15 \\ 30 & 688 & 80 \\ 15 & 241 & 61 \end{vmatrix} = 21{,}738$$

$$N_c = \begin{vmatrix} 6 & 30 & 108 \\ 30 & 190 & 688 \\ 15 & 80 & 241 \end{vmatrix} = -11{,}400$$

Therefore

$$a = \frac{N_a}{D} = \frac{18{,}630}{5490} = 3.39$$

$$b = \frac{N_b}{D} = \frac{21{,}738}{5490} = 3.96$$

$$c = \frac{N_c}{D} = \frac{-11{,}400}{5490} = -2.08$$

[1]Note that although x by itself accounts for only 83.7% *and* w by itself accounts for only 5.5%, together they account for more than 83.7% + 5.5%. This shows that the *interaction* between x and w also contributes something to the explanation of variations in y. Correspondingly, there may also be situations where x and w together account for *less* than the sum of their individual coefficients of linear determination.

Having obtained the above numerical values for a, b, c, we can write the best-fitting multiple regression equation as

$$y = 3.39 + 2.96x - 2.08w$$

e   The best prediction for today's sales volume given that 700 dollars was spent on advertising and 0.6 inches of precipitation fell this morning can be obtained by setting $x = 7$ (since advertising costs are listed in hundreds of dollars) and $w = 6$ (since precipitation is listed in tenths of an inch) into the multiple regression equation. We get

$$
\begin{aligned}
y &= 3.39 + 2.96x - 2.08w \\
&= 3.39 + (3.96)(7) - (2.08)(6) \\
&= 3.39 + 27.72 - 12.48 \\
&= 18.63
\end{aligned}
$$

Because y is expressed in thousands of dollars, we would predict the day's sales volume to be 18,630 dollars.

EXERCISES 7.C

1   Some behavioral biologists seem to think that performance of a laboratory rat on an intelligence test depends, to a large extent, on two factors: previous experience with such tests and amount of protein in the rat's daily diet. For 10 rats we have the following data:

Rat	Hours Spent in Intelligence Testing x	No. of Units of Protein Daily w	Score on Standard Test y
A	1	12	20
B	3	12	40
C	2	10	30
D	10	10	80
E	5	10	50
F	10	3	50
G	12	2	90
H	1	20	30
I	2	20	40
J	1	30	40

a   How much of the variation among test scores can be attributed to a linear relationship with the number of hours spent in intelligence testing?
b   How much of the variation among test scores can be explained by a linear relationship with varying amounts of protein in the diet?
c   How much of that variation can be accounted for by a multiple linear relationship with both variables?
d   Find the best-fitting multiple regression equation.

e Predict the test score of rat K who gets 30 units of protein per day and has spent 5 hours in intelligence testing.

2 In a study of factors that motivate inventors, two psychologists studied the backgrounds of six randomly selected researchers of age 40 attending an engineering convention. They collected the following data:

Researcher	1	2	3	4	5	6
Average hours of television watched per day: ages 8–12 (x)	1	2	3	1	1	2
Number of part-time jobs held: ages 13–19 (w)	20	10	5	15	10	20
Number of patents awarded (y)	25	15	5	20	15	20

a How much of the variation among the researchers in number of patents awarded can be explained by a linear relationship with their early television viewing patterns?
b How much of the variation in number of patents awarded can be attributed linearly to variations in number of part-time jobs held as a teenager?
c How much of the variation is accounted for by a multiple linear relationship with both factors?
d Find the best-fitting multiple regression equation.
e Predict the number of patents awarded by age 40 to a researcher who watched no television as a child but held 30 part-time jobs as a teenager.

3 Two factors that affect the price of sardines in your local grocery store are the availability of sardines in the fishing areas and the costs of transportation. The following data were collected over a recent 6-year period:

Year	1971	1972	1973	1974	1975	1976
Estimated size of sardine population of Norway in trillions x	30	30	40	20	20	30
Transportation cost per gross of cans in dollars w	6	6	8	10	15	15
Price in cents of 3-oz. can to consumer y	28	32	30	40	45	40

a Determine the proportions of variation in price that can be attributed to separate linear relationships with each of the factors, x and w, respectively.

b What proportion of the variation in price can be explained by a multiple linear relationship with both factors?

c Calculate the multiple regression equation, and use it to predict the price of a 3-ounce can of sardines in 1977 when the sardine population is estimated to be 25 trillion and transportation costs are 18 dollars per gross of cans.

4 An agricultural research organization tested various mixtures of chemical and organic fertilizer to try to find how variations in the amounts of fertilizer induce variations in food production. The following data were collected from seven virtually identical plots of farmland:

Pounds of Chemical Fertilizer Used	Pounds of Organic Fertilizer Used	Bushels of Beans Harvested
1	7	120
2	6	140
3	5	150
4	4	180
5	3	200
6	2	190
7	1	220

a Determine the proportions of variation in harvest that can be attributed to separate linear relationships with each of the two fertilizers.

b Calculate the proportion of variation in harvest that can be explained on the basis of a multiple linear relationship with both fertilizers.

c What is the best-fitting multiple linear regression equation for bushels of beans in terms of both kinds of fertilizer?

d Predict the harvest when 8 pounds of chemical and no pounds of organic fertilizer are used.

e Predict the harvest when 8 pounds of organic and no pounds of chemical fertilizer are used.

f Predict the harvest when no fertilizer of either kind is used.

5 By analogy with the correlation test for linearity of Sec. 6.D, there is a test for "multiple linearity" based on the coefficient of multiple linear determination R^2. (As with the correlation test, this test for multiple linearity is valid only for normally distributed data.) Instead of the t distribution used in Sec. 6.D, we use here the F distribution of Table A.5. To test the hypothesis

H : There is no significant multiple linearity

against the alternative

A : There is significant multiple linearity

we compute

$$F = \frac{R^2(n-3)}{2(1-R^2)}$$

and we reject H at level α if $F > F_\alpha(3, n-3)$. Using this procedure, test the behavioral biology data of Exercise 1 of this section for multiple linearity at level $\alpha = .05$.

6 Test the psychological data of Exercise 2 for multiple linearity at level $\alpha = .01$.

7 Test the sardine price data of Exercise 3 for multiple linearity at level $\alpha = .01$.

8 Test the agricultural research data of Exercise 4 for multiple linearity at level $\alpha = .05$.

SUMMARY AND DISCUSSION

Our study of regression, begun in Chap. 6, has been completed here in Chap. 7. Using a logarithmic transformation, we have adapted the linear regression methods of Chap.6 to analyses of curvilinear data which follow logarithmic, simple exponential, and power curves. The coefficient of linear determination was used to measure the extent to which a set of data follows each of those types of curves and even to choose the type of curve that best fits the data. For data that are parabolic in shape, the algebraic technique of setting up and solving three simultaneous equations in three unknowns yields the parabolic regression equation, and that same technique also turned out to be appropriate for solving problems involving multiple linear regression, where a quantity under study depends on more than one factor. The coefficient of multiple determination, which measures the degree to which a multiple linear regression equation describes a set of data, is based on the intertwining relationships among various coefficients of linear determination. Much of what we have done in Chap. 7 grew out of the concepts introduced in Chap. 6 and serves to illustrate that the versatility of those concepts transcends the straight-line data they were originally developed to analyze.

BIBLIOGRAPHY

Curvilinear Regression
Siegel, D. G.: Several Approaches for Measuring Average Rates of Change for a Second Degree Polynomial, *The American Statistician*, February 1975, pp. 36–37.

Smith, J. H.: Families of Transformations for Use in Regression Analysis, *The American Statistician*, June 1972, pp. 59–61.

Multiple Regression and Correlation
Crocker, D. C.: Some Interpretations of the Multiple Correlation Coefficient, *The American Statistician*, April 1972, pp. 31–33.

Fleiss, J. L., and J. M. Tanur: A Note on the Partial Correlation Coefficient, *The American Statistician*, February 1971, pp. 43–45.

Heward, J. H., and P. M. Steele: "Business Control through Multiple Regression Analysis," Gower Press, Plymouth, England, 1972.

Mullet, G. M.: Graphical Illustration of the Simple (Total) and Partial Regression, *The American Statistician*, December 1972, pp. 25–27.

Weiss, N. S.: A Graphical Representation of the Relationships between Multiple Regression and Multiple Correlation, *The American Statistician*, April 1970, pp. 25–29.

SUPPLEMENTARY EXERCISES

1 As part of an experimental study of soil erosion, a geographer constructed a laboratory model of a river valley and used it to illustrate the changing width of a river as time passes. He obtained the following data:

Day	1	3	5	10	20	30
Width, feet	.20	.35	.40	.50	.60	.65

  a Draw the scattergram of the data, and use it to determine the mathematical form of the regression curve of the width of the river expressed in terms of the passage of time.
  b Compare the coefficient of linear determination of the data with the coefficient of determination for the appropriate curvilinear relationship.
  c Calculate the equation of the regression curve, and superimpose its graph upon the scattergram.
  d Predict the width of the model river on the 40th day.

2 A sociologist gathered the following data on ten randomly selected employees of a large corporation, comparing their respective salary levels with the number of dependents they have:

Employee	A	B	C	D	E	F	G	H	I	J
No. of dependents (not including self)	3	7	7	0	8	2	4	5	1	1
Salary level, thousands of dollars	60	15	10	10	5	40	50	40	25	15

  a Construct the scattergram of the data, and use it to determine the mathematical form of the regression curve for predicting salary level based on number of dependents.
  b Calculate the coefficient of linear determination to find out whether or not it would be reasonable to use a linear regression equation to analyze these data.
  c Find the appropriate regression equation, and superimpose its graph on the scattergram.
  d Estimate the salary level of an employee who has six dependents.

3 The sociologist of Supplementary Exercise 2 also obtained data on those ten employees regarding their years of experience in the corporation. These data follow:

Employee	A	B	C	D	E	F	G	H	I	J
Years of experience	10	6	3	2	5	10	10	9	8	5
Salary level, thousands of dollars	60	15	10	10	5	40	50	40	25	15

    a Draw the scattergram, and find the general class of curves to which the regression curve belongs.

    b Contrast the coefficient of linear determination of the data with the coefficient of determination for the appropriate curvilinear relationship.

    c Compute the equation of the appropriate regression curve, and superimpose its graph on the scattergram.

    d Estimate the salary level of an employee with 4 years' experience.

4 If the sociologist of Exercises 2 and 3 above were to combine all her data into one set, it would appear as follows:

Employee	A	B	C	D	E	F	G	H	I	J
No. of dependents	3	7	7	0	8	2	4	5	1	1
Years of experience	10	6	3	2	5	10	10	9	8	5
Salary level, thousands of dollars	60	15	10	10	5	40	50	40	25	15

    a Calculate the proportion of variation in salary level that can be attributed to a multiple linear relationship with the two factors, number of dependents and years of experience.

    b Test for multiple linearity at level $\alpha = .05$.

    c Compute the multiple regression equation for predicting salary level from a knowledge of the other two factors.

    d Use the regression equation to predict the salary level of an employee with six dependents and 4 years of experience.

5 As part of a study of voting patterns, a political scientist collects data on six randomly selected 40-year-old individuals concerning their educational level, total family income, and frequency of voting. The data are as follows:

Years of formal education x	8	12	18	10	6	14
Family income in thousands of dollars w	12	20	16	16	8	18
No. of national and state elections voted in y	7	12	16	12	4	16

a Calculate the coefficient of multiple determination of y in terms of x and w, using formula (7.7).

b Calculate the coefficient of multiple determination of y in terms of x and w, using formula (7.8), and observe that the answer is the same as in part a.

c Test the data for multiple linearity at level $\alpha = .05$.

d Find the multiple regression equation for predicting y in terms of x and w.

e Estimate the number of national and state elections in which a person having 16 years of formal education and a family income of 20,000 dollars voted.

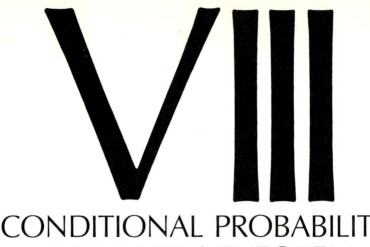

CONDITIONAL PROBABILITY
AND BAYES' THEOREM

In earlier sections of the text, we have often referred to the "probability" of various occurrences. Until now, however, we have used probability notions and concepts primarily as a vehicle for undertaking statistical analyses such as confidence intervals and hypothesis testing. For example, we have defined the significance level of a statistical test as the probability of a Type I error, i.e., the probability of rejecting the hypothesis H when it's really true. Our point of view in this chapter will be different. Here we will be studying directly the probabilities of various events, and we will be calculating the probabilities of related events based on some collected information or data. By "probability of an event," we usually mean the proportion of times that the event can be expected to occur over a long period of time; sometimes, however, events cannot be repeated (for example, in attempting to find the probability that an earthquake will destroy Los Angeles), and then the probability will have to mean our view of the chances of the occurrence of the event, based on prior information in our possession.

SECTION 8.A EVENTS

The collection of all possible outcomes of an experiment, survey, or other method of data collection is called the "sample space" of the experiment. We shall use the letter S as an abbreviation for the sample space. To simplify the explanation of the concepts involved, we will illustrate all the basic concepts of probability by use of a "die," sometimes called a probability cube, a picture of which can be found in Fig. 8.1. (The plural of die is dice.) We will make a detailed analysis of the following experiment for the purpose of introducing the concepts of probability: We roll the die once only, and we note the number of dots on the side facing upward when the die comes to a stop. If the die comes to a stop as in Fig. 8.1, with the one-dot side facing upward, we say we have rolled a "1." The sample space of this experiment, the collection of all possible outcomes, consists of the numbers 1, 2, 3, 4, 5, and 6. We therefore write

$$S = \{1, 2, 3, 4, 5, 6\}$$

By an "event," we mean a collection of some but not necessarily all of the possible outcomes of an experiment. For example, in the probability cube experiment, we can consider the event that we roll an even number. This event can be symbolized by

$$E = \{2, 4, 6\}$$

Other events that may be of interest are:

$$L = \{1, 2\} = \text{we roll a very low number}$$
$$H = \{5, 6\} = \text{we roll a very high number}$$
$$M = \{3, 4\} = \text{we roll a middle number}$$
$$A = \{1, 2, 3, 4\} = \text{we roll a not-too-high number}$$

There is an "arithmetic of events," a way of combining events, that is somewhat analogous to addition and multiplication of numbers. Those readers having a background in "set theory" or the "new mathematics" might recognize some of the concepts of the arithmetic of events. We will use the following arithmetic operations of combining events:

1 The "intersection of two events" is the set consisting of all outcomes that are simul-

FIGURE 8.1
Die (probability cube).

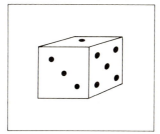

taneously in both the events. For example, the intersection of events E and L above, abbreviated $E \cap L$ where the symbol $\cap$ is pronounced "intersect," is

$$E \cap L = \{2\}$$

because 2 is the only outcome found in both E and L.

2   The "union of two events" is the set consisting of all outcomes that are in either one or both of the events. For example, the union of events E and L, abbreviated $E \cup L$ where the symbol $\cup$ is pronounced "union," is

$$E \cup L = \{1, 2, 4, 6\}$$

because 1 is in L, 4 and 6 are in E, and 2 is in both L and E.

3   The "complement of an event" is the set consisting of all those outcomes in the sample space S that are not in the event itself. For example, the complement of the event A above, abbreviated A^c which is pronounced "A-complement," is

$$A^c = \{5, 6\}$$

because $A = \{1, 2, 3, 4\}$, and $S = \{1, 2, 3, 4, 5, 6\}$, so that 5 and 6 are the outcomes in S which are not in A. In view of the fact that we have $H = \{5, 6\}$ above, the reader should notice that $A^c = H$, because A^c and H consist of exactly the same outcomes.

The above three operations together comprise the basic structure of the arithmetic of events. One more item is also used extensively in this arithmetic, namely, the empty event. The "empty event," denoted by the symbol ϕ (the Greek letter, lowercase "phi," pronounced "fee" or "fie"), is the event that consists of no outcomes at all. It plays a role in the arithmetic of events analogous to the role played by the number zero in ordinary arithmetic of numbers. For example, we have the following equations involving the empty event ϕ:

$L \cap H = \phi$ because there are no outcomes that are in both L and H, as $L = \{1, 2\}$ and $H = \{5, 6\}$.

$A \cap A^c = \phi$ because there are no outcomes that are in both A and A^c, as A^c consists of exactly those outcomes that are not in A.

$S^c = \phi$ because S^c consists of exactly those outcomes in S that are not in S. Such outcomes cannot exist, since they would have to be in S and not in S simultaneously.

The two events L and H are said to be "mutually exclusive" or "disjoint" if $L \cap H = \phi$. In fact, if C and D are any two events such that $C \cap D = \phi$, then they are said to be disjoint events.

By means of the Venn diagrams introduced in Chap. 1, we can illustrate the concepts of intersection, union, complement, and disjoint events. We do this in Figs. 8.2, 8.3, 8.4, and 8.5, respectively.

There is an important relationship among events which we will need to use when we get ready to solve problems in applied fields of work. This relationship is an equation expressed in language of the arithmetic of events and involves the three concepts of intersection, union, and complement. This important relationship can be explained as

FIGURE 8.2
Intersection of events.

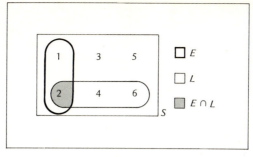

FIGURE 8.3
Union of events.

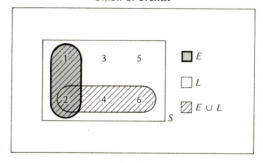

FIGURE 8.4
Complement of an event.

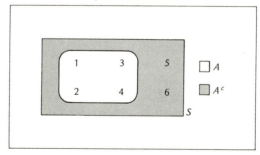

FIGURE 8.5
Disjoint events.

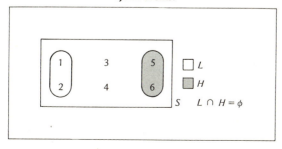

FIGURE 8.6
$$E = (E \cap F) \cup (E \cap F^c).$$

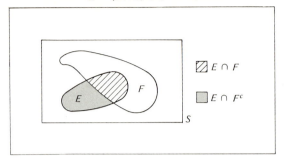

follows: Suppose we have two events, labeled E and F, as in the Venn diagram of Fig. 8.6. Because E is not contained completely inside F, part of E lies within F while the rest of E lies outside F. The part of E within F is really the intersection of E and F, namely $E \cap F$. Therefore E is a union of two sets: One set is $E \cap F$, and the other is the piece of E lying outside F. How can we express in symbols the piece of E lying outside F? Well, every outcome in that piece is in E, but is outside F. Recalling that F^c is the set of outcomes lying outside F, we see that an outcome outside F can be considered to be inside F^c. Therefore outcomes in E, but outside F, can be viewed as being simultaneously in E and in F^c. The set of such outcomes can be denoted as $E \cap F^c$. We summarize this discussion in the following way:

$$E \cap F = \text{the set of outcomes in } E \text{ that are also in } F$$
$$E \cap F^c = \text{the set of outcomes in } E \text{ that are not in } F$$

As we have already pointed out, those two sets together comprise E, as an outcome in E must be either in F or not in F. (In fact, every outcome in the sample space is either in F or not in F.) Therefore E is the union of the two sets $E \cap F$ and $E \cap F^c$. We can express this in probability symbolism as follows:

● $$E = (E \cap F) \cup (E \cap F^c) \qquad (8.1)$$

When we begin to use probability methods to solve problems in the various applied fields of interest, we shall need to use the above equation and some consequences of it to be developed in a little while.

There is actually a somewhat more useful form of Eq. (8.1). Suppose that the sample space S is divided into n sets, called $F_1, F_2, \ldots, F_n$, which are disjoint (no outcome lies in more than one of these events) and together comprise the entirety of S (each outcome lies in one of the F_k's). If E is another event, it must necessarily cut across some or all of the F_k's since there is nowhere else for it to go. The event E can then be viewed as the union of n pieces, illustrated in Fig. 8.7:

$$E \cap F_1 = \text{the piece of } E \text{ lying in } F_1$$
$$E \cap F_2 = \text{the piece of } E \text{ lying in } F_2$$
$$\cdots\cdots\cdots\cdots\cdots\cdots\cdots\cdots\cdots\cdots$$
$$E \cap F_n = \text{the piece of } E \text{ lying in } F_n$$

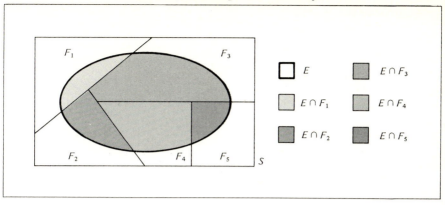

Therefore the expanded version of the important relationship discussed above can be written as

● $\qquad E = (E \cap F_1) \cup (E \cap F_2) \cup \cdots \cup (E \cap F_n)$ $\qquad$ (8.2)

We shall also have occasion to use this formula in developing techniques of solving problems of applied interest.

Now that we have discussed what events are and have developed some ways of mathematically analyzing them, the next step is to calculate their probabilities. We devote the next section to the study of probabilities of events.

EXERCISES 8.A

1 Consider the experiment of tossing a pair of dice, one of them red and the other green. The outcome of the experiment is considered to be the total number of dots facing upward on both dice together.

a Explain why the sample space of this experiment is the set $S = \{2, 3, 4, 5, 6, 7, 8, 9, 10, 11, 12\}$.

b Draw a Venn diagram of the sample space and the following events:

$$W = \{7, 11\}$$
$$L = \{2, 3, 12\}$$
$$E = \{2, 4, 6, 8, 10, 12\}$$
$$D = \{3, 5, 7, 9, 11\}$$

c List the outcomes of the following events:

$W \cup L$	$W^c \cap L^c$
$W \cap L$	E^c
W^c	$E^c \cap D$
L^c	$E^c \cup D$
$W^c \cup L^c$	

2　a In the situation of Exercise 8.A.1, list the outcomes of each of the events $(L \cap E)^c$ and $L^c \cup E^c$.

  b Construct a Venn diagram which illustrates the fact that for any two events A and B it is always true that $(A \cap B)^c = A^c \cup B^c$.

3 Consider the experiment of rolling three dice simultaneously, one red, the second green, and the third white. The outcome of the experiment is considered to be the total number of dots facing upward on all three dice together.

  a List the sample space of this experiment.

  b Draw a Venn diagram of the sample space and each of the following events:

$$W = \{7, 11, 17\}$$
$$L = \{3, 4, 12, 13, 18\}$$
$$E = \text{the even numbers of the sample space}$$
$$D = \text{the odd numbers of the sample space}$$

  c List the outcomes of the following events:

$$L \cup E \qquad\qquad E^c$$
$$(L \cup E)^c \qquad L^c \cap E^c$$
$$L^c$$

  d Construct a Venn diagram which illustrates the fact that for any two events A and B it is always true that $(A \cup B)^c = A^c \cap B^c$.

SECTION 8.B PROBABILITIES OF EVENTS

By the probability of an event which can be repeated over and over again, we mean the proportion of times that the event can be expected to occur relative to the number of times the experiment is repeated. Probabilities of events that cannot be repeated must be handled differently, and these will be postponed until the next section. In the case of the probability cubes (dice) discussed in Sec. 8.A, we are talking about a repeatable experiment, rolling the die.

A die is said to be "fair" if each of the six sides has the same probability of being in the upward position after a roll. Because there are six possible outcomes in the sample space $S = \{1, 2, 3, 4, 5, 6\}$, we say that each outcome has probability one-sixth (1/6). In particular, we have recorded in Table 8.1 the facts that $P(\{1\}) = 1/6$, $P(\{2\}) = 1/6$, $P(\{3\}) = 1/6$, $P(\{4\}) = 1/6$, $P(\{5\}) = 1/6$, and $P(\{6\}) = 1/6$, where the symbols $P(\ )$ are pronounced "the probability of." These facts can be interpreted to mean that in a large number of rolls of the die we would expect each outcome to occur about one-sixth of the time.

Let's take a look at the events we worked with in Sec. 8.A. We recall the following events:

$$E = \{2, 4, 6\} = \text{we roll an even number}$$
$$L = \{1, 2\} = \text{we roll a low number}$$
$$H = \{5, 6\} = \text{we roll a high number}$$

TABLE 8.1
Probabilities for a Fair Die

Outcome, number of dots facing upward	Proportion of Rolls in Which Outcome Occurs	Probability of Outcome
1	1/6	1/6
2	1/6	1/6
3	1/6	1/6
4	1/6	1/6
5	1/6	1/6
6	1/6	1/6

$M = \{3, 4\}$ = we roll a middle number

$A = \{1, 2, 3, 4\}$ = we roll a not-too-high number

What do we mean by $P(E)$, the probability of E? Well, E is the event that we roll an even number, and since there are three even numbers among the six numbers available, we can expect to roll an even number about three-sixths of the time. For this reason, we can agree that

$$P(E) = \frac{3}{6} = \frac{1}{2}$$

Let's look at $P(E)$ another way. The event E is composed of three distinct outcomes, each of which occurs one-sixth of the time on the average. Therefore the event E will occur that one-sixth of the time that $\{2\}$ occurs, that one-sixth of the time that $\{4\}$ occurs, and that one-sixth of the time that $\{6\}$ occurs. From this analysis, we can see that

$$P(E) = P(\{2\}) + P(\{4\}) + P(\{6\}) = \frac{1}{6} + \frac{1}{6} + \frac{1}{6} = \frac{3}{6} = \frac{1}{2}$$

Both the analyses used above are correct methods of calculating the probability of the event E. Similar calculations give us the facts that

$$P(L) = \frac{2}{6} = \frac{1}{3}$$

$$P(H) = \frac{2}{6} = \frac{1}{3}$$

$$P(M) = \frac{2}{6} = \frac{1}{3}$$

$$P(A) = \frac{4}{6} = \frac{2}{3}$$

Now that we have settled on a method for calculating the probabilities of events such as E, L, H, M, and A, we turn our attention to finding the probabilities of various combinations of these events. For example we denote by $E \cap L$ the event that the roll of the die results in a number which is both even and low. Because $E \cap L = \{2\}$, it is reasonable to agree that

$$P(E \cap L) = P(\{2\}) = \frac{1}{6}$$

In a similar manner, we find that

$$P(E \cap A) = P(\{2, 4\}) = \frac{2}{6} = \frac{1}{3}$$

$$P(L \cap A) = P(\{1, 2\}) = P(L) = \frac{1}{3}$$

$$P(L \cap H) = P(\phi) = 0$$

The last assertion that $P(L \cap H) = 0$ can be considered as a statement that there is no chance of having the roll result in a number that is both low and high.

Now that we know that $P(L \cap H) = 0$, among other things, what about $P(L \cup H)$? The event $L \cup H$ is the event that a roll of the die results in either a low or a high (or both, if that were possible) number. As we know, a low number will occur about one-third of the time, and a high number will occur about one-third of the time. It therefore seems reasonable to expect that the goal of obtaining an extreme number (low or high) would be attained about two-thirds of the time.

Translating that statement into mathematical symbols gives the assertion that

$$P(L \cup H) = P(L) + P(H) = \frac{1}{3} + \frac{1}{3} = \frac{2}{3}$$

This estimate is corroborated by the following calculations, using the direct method of counting outcomes:

$$P(L \cup H) = P(\{1, 2, 5, 6\}) = \frac{4}{6} = \frac{2}{3}$$

Suppose we apply the same method of analysis to the problem of calculating $P(E \cup A)$, the probability of rolling a number that is either even or not-too-high or both. We know that an even number occurs about one-half of the time, while a not-too-high number occurs about two-thirds of the time. It definitely cannot be true, however, that

$$P(E \cup A) = P(E) + P(A) = \frac{1}{2} + \frac{2}{3} = \frac{3}{6} + \frac{4}{6} = \frac{7}{6}$$

because we would be saying that $E \cup A$ could be expected to occur about seven out of every six times on the average, which is more than 100% of the time. Probabilities can never exceed one, for, if they did, the events involved would have to be occurring more than 100% of the time. What, then, went wrong with the statement that $P(E \cup A) = P(E) + P(A)$? If we look at the Venn diagram in Fig. 8.8, we observe that the union $E \cup A$ consists of five outcomes. It therefore seems reasonable to believe that $P(E \cup A) = 5/6$.

FIGURE 8.8
Probability of the union of events.

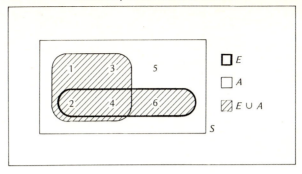

This belief is substantiated by the direct method of counting outcomes, namely

$$P(E \cup A) = P(\{1, 2, 3, 4, 6\}) = \frac{5}{6}$$

How, then, did we manage to come up with the allegation that $P(E \cup A) = 7/6$? If we look closely at the statement

$$P(E \cup A) = P(E) + P(A) = \frac{3}{6} + \frac{4}{6} = \frac{7}{6}$$

we will see, hidden inside it, the following statement:

$$P(E \cup A) = P(E) + P(A)$$
$$= P(\{2, 4, 6\}) + P(\{1, 2, 3, 4\})$$
$$= \frac{3}{6} + \frac{4}{6} = \frac{7}{6}$$

Take a look at what we have done. We have blatantly counted the outcomes 2 and 4 twice: once as part of E and once as part of A. In short, while computing the probability of $E \cup A = \{1, 2, 3, 4, 6\}$, we have counted the outcomes 1, 3, and 6 once each and the outcomes 2 and 4 twice each. Therefore we have implicitly been assuming that $E \cup A$ *contains seven outcomes,* e.g. $\{1, 2, 2, 3, 4, 4, 6\}$, where in fact it contains only five.

Now that we have discovered the mistake, how do we rectify it? How do we correct the erroneous statement that $P(E \cup A) = P(E) + P(A)$? Well, when we were using $P(E) + P(A)$ to calculate $P(E \cup A)$, we were counting twice all those outcomes that are in both E and A, namely, all those outcomes in $E \cap A = \{2, 4\}$. But we want to count these outcomes only once. We have therefore counted the outcomes in $E \cap A$ once too often. We can correct the mistake by subtracting the overcount of 2 and 4, an amount equal to $P(E \cap A) = P(\{(2, 4\}) = 2/6 = 1/3$. This subtraction will correct the mistake of counting

$E \cap A$ twice instead of once. We can therefore write that

$$P(E \cup A) = P(E) + P(A) - P(A \cap E)$$
$$= \frac{3}{6} + \frac{4}{6} - \frac{2}{6} = \frac{5}{6}$$

which is the correct probability. We can summarize this development by the following.

General Rule: If C and D are any two events, then

$$\bullet \qquad\qquad P(C \cup D) = P(C) + P(D) - P(C \cap D) \qquad\qquad (8.3)$$

Now that we have established the above general rule, let's go back and try to figure out how we managed to get away with the statement that

$$P(L \cup H) = P(L) + P(H) = \frac{1}{3} + \frac{1}{3} = \frac{2}{3}$$

If we apply the general rule to $L \cup H$, we should really be writing

$$P(L \cup H) = P(L) + P(H) - P(L \cap H)$$

What about $P(L \cap H)$? Well, as it turns out, $L \cap H = \phi$ because none of the possible outcomes can be found in both L and H. Therefore $P(L \cap H) = P(\phi) = 0$. It follows that

$$P(L \cup H) = P(L) + P(H) - P(L \cap H)$$
$$= \frac{1}{3} + \frac{1}{3} - 0 = \frac{2}{3}$$

and so the missing term $P(L \cap H)$ had no effect on the calculation of $P(L \cup H)$, and therefore we did not even realize that we were leaving it out.

We can now modify the general rule to include cases such as that of L and H.

Special Rule   If C and D are disjoint events, namely events for which $C \cap D = \phi$, then

$$P(C \cup D) = P(C) + P(D)$$

because the remaining term $P(C \cap D)$ is zero.

Having developed the basic rules of finding probabilities of events, we should now return to a further discussion of the important relationship (8.1): If E and F are any two events, we discovered that

$$E = (E \cap F) \cup (E \cap F^c)$$

What, then, can we say about $P(E) = P([(E \cap F) \cup (E \cap F^c)])$? Well, notice that $E \cap F$ is contained entirely inside F, because $E \cap F$ is composed only of those outcomes which are both in E and in F. On the other hand $E \cap F^c$ is contained entirely inside of F^c, because $E \cap F^c$ consists of those outcomes that are both in E and in F^c. If there were some outcomes in the intersection $(E \cap F) \cap (E \cap F^c)$, they would have to be in $E \cap F$ (therefore in F) and also in $E \cap F^c$ (therefore in F^c). Those outcomes would then have to be both in F and F^c. But there cannot be any outcomes which are simultaneously in both F and F^c because F^c consists only of those outcomes that are not in F. Therefore there cannot be any outcomes that are in $(E \cap F) \cap (E \cap F^c)$. It follows that $(E \cap F) \cap (E \cap F^c) = \phi$, and the special rule above then implies that

$$P([(E \cap F) \cup (E \cap F^c)]) = P(E \cap F) + P(E \cap F^c)$$

We can therefore write that

● $$P(E) = P(E \cap F) + P(E \cap F^c) \tag{8.4}$$

We will find the above equation to be of great value in discussing Bayes' theorem in Sec. 8.F.

By analogous reasoning, we can study the expanded version (8.2) of relationship (8.1). We conclude that

$$P(E) = P(E \cap F_1) + P(E \cap F_2) + \cdots + P(E \cap F_n)$$

More compactly, we express this in the form

● $$P(E) = \sum_{k=1}^{n} P(E \cap F_k) \tag{8.5}$$

Now that we have introduced the fundamental concepts of calculating probabilities of events, we proceed to the study of "conditional probability," the method of updating our estimate of the probability of an event as new information involving the event becomes known to us.

EXERCISES 8.B

1 Consider the experiment of tossing a pair of fair dice, one of them red and the other green. Because there are six possible ways for each of the dice to turn up (namely, 1, 2, 3, 4, 5, or 6), there are 36 possible ways for the pair to turn up. For example, we can get 1 on the red and 3 on the green, or 6 on the red and 4 on the green, or 3 on the red and 1 on the green, or 5 on the red and 5 on the green, etc. Since the dice are fair, each of these 36 possible ways is equally likely and so has probability 1/36 of turning up. To calculate the probability of rolling a total of 4 dots on the two dice together, we notice that 3 of the 36 possibilities have a total of 4 dots, namely, (1) 1 on the red and 3 on the green, (2) 2 on the red and 2 on the green, and (3) 3 on the red and 1 on the green. Therefore the probability of rolling a 4 with a pair of fair dice is 3/36.

a Find the probability of rolling a 7.
b Find the probability of rolling an 11.
c Find the probability of the event $W = \{7, 11\}$.
d Find the probability of the event $L = \{2, 3, 12\}$.
e Find the probability of the event $E = \{2, 4, 6, 8, 10, 12\}$.
f Find the probability of the event $D = \{3, 5, 7, 9, 11\}$.
g Find $P(L \cap D)$.

h Find $P(L \cup D)$.
i Find $P(W \cap L)$.
j Find $P(W \cup L)$.
k Find $P(E \cap D)$.
l Find $P(E \cup D)$.

2 If A and B are two events such that $P(A) = 1/2$, $P(B) = 2/3$, and $P(A \cap B) = 1/3$, then determine the following probabilities:
a $P(A^c)$
b $P(B^c)$
c $P(A^c \cap B)$
d $P(A^c \cap B^c)$

3 If Q and R are two events such that $P(Q) = 1/2$, $P(R) = 3/8$, and $P(Q \cap R) = 1/4$, then determine the following probabilities:
a $P(Q^c)$
b $P(R^c)$
c $P(Q^c \cap R)$
d $P(Q^c \cap R^c)$

4 Explain why it is impossible to have two events U and V such that $P(U) = 2/3$, $P(V) = 4/5$, and $P(U \cap V) = 1/4$.

5 Explain why it is impossible to have two events T and W such that $P(T) = 1/8$, $P(W) = 1/5$, and $P(T \cup W) = 1/2$.

6 No matter what the events A and B are, show that $P(A \cup B)$ can never be larger than $P(A) + P(B)$.

SECTION 8.C UPDATING PROBABILITIES OF EVENTS

In most questions in applied fields which involve uncertainty, the researcher or manager is called upon to provide an estimate of the probability of some uncertain event. Often the event of interest is not of the sort that can be repeated over and over again; therefore, it is not possible to estimate its probability by finding the relative frequency of its occurrence. The usual procedure for finding the probability of an event would then be to gather as many as possible of the relevant facts about the situation, try to determine what effect each of the facts has upon the probability, and then come up with an estimate of the desired probability.

As an example, suppose we want to estimate the probability of the event I that a particular individual will sustain a very serious injury within the week.

Actuarial statistics indicate that every week about one out of every million individuals sustains a very serious injury. In probabilistic terminology, this means that

$$P(I) = 0.000001$$

Now, after making this estimate of $P(I)$, suppose that new information comes in about the particular individual under discussion. Say, for example, that we have the information P that the individual is a licensed pilot of small airplanes. Now, the injury rates for pilots of small planes are somewhat higher than those of the general public. Perhaps, then,

$$P(I \mid P) = 0.0001$$

namely, one out of ten thousand. The symbol $P(I \mid P)$ is pronounced "the probability of I given P," and it means the probability of I updated so as to reflect the new information contained in statement P. In particular, $P(I \mid P)$ is the probability that an individual will sustain a very serious injury within the week if that individual is a licensed pilot of small airplanes. Updated probabilities such as $P(I \mid P)$ are technically referred to as "conditional probabilities" because they measure the probabilities of events updated to take changing conditions into account.

In most cases of applied interest, new information about a changing situation continues to flow in and the probabilities must be continually updated. Suppose the following fact about our current problem becomes known:

C = the individual regularly pilots a crop-dusting plane

Now, it turns out that piloting a crop-dusting plane is somewhat more hazardous than piloting other small planes because crop-dusting planes commonly fly about 10 to 20 feet above the ground with trees, wires, etc., in their paths. It would therefore be reasonable to update the probability of injury to

$$P(I \mid C) = 0.001$$

or one chance out of a thousand. More information flows in, with each piece of information having an effect on the probability of injury. The bits of information and their effects on the probability of injury are listed in Table 8.2. It is to be noted that some pieces of information tend to increase the probability of I, while other pieces tend to decrease it. As can be seen from the material contained in Table 8.2, the conditional probability fluctuates up and down as conditions change. The events P, C, F, B, and MB tend to exert upward pressure on the probability of I, while the events H and MH tend to exert downward pressure.

To illustrate the calculation of more concrete conditional probabilities, let's return to the probability cube example of the last two sections. As the reader may recall, we have been analyzing the following five events:

$$
\begin{array}{ll}
E = \{2, 4, 6\} & P(E) = 1/2 \\
L = \{1, 2\} & P(L) = 1/3 \\
H = \{5, 6\} & P(H) = 1/3 \\
M = \{3, 4\} & P(M) = 1/3 \\
A = \{1, 2, 3, 4\} & P(A) = 2/3
\end{array}
$$

We know that the probability of L is $1/3$. Suppose, now, that we have received new information to the effect that the roll of the die resulted in the occurrence of the event A. What is the updated probability of L in view of the new information that A has occurred?

TABLE 8.2

Effect of New Information on the Probability of Injury

I = Event That an Individual Will Sustain a Very Serious Injury within the Week

$P(I)$ = .000001 (at 12:00 noon)

Time of Receipt of Information, P.M.	Gist of New Information	Symbol	Updated Probability
1:00	He is licensed to pilot small planes.	P	$P(I \mid P)$ = .0001
2:00	He regularly pilots crop dusters.	C	$P(I \mid C)$ = .001
3:00	He fell out of his plane while crop-dusting.	F	$P(I \mid F)$ = .99
3:30	There was an extremely large haystack below the plane.	H	$P(I \mid H)$ = .05
3:45	There was a bull sleeping in the haystack.	B	$P(I \mid B)$ = .75
3:47	He missed the bull.	MB	$P(I \mid MB)$ = .05
3:48	He missed the haystack.	MH	$P(I \mid MH)$ = .99

We come up with the numerical value of $P(L \mid A)$ by reasoning as follows: The fact that A has occurred means that the number of dots on the upward facing side of the die was either 1, 2, 3, or 4. The possibility that 5 or 6 might have been rolled was excluded by the occurrence of event A. We have no information asserting that any of the numbers 1, 2, 3, and 4 was more likely to occur than any of the others, so the nature of the probability cube requires that the chances of each of these numbers be updated equally to 1/4. This is due to the fact that there are now only four possible outcomes, taking the new information into account that the outcomes 5 and 6 definitely did not occur. Since the only possible outcomes were 1, 2, 3, and 4, the updated probability of $L = \{1, 2\}$ is then

$$P(L \mid A) = \frac{2}{4} = \frac{1}{2}$$

because L includes two of the four possible outcomes. While the original probability of L was 1/3, we see that the conditional probability of L given A is 1/2. Therefore the occurrence of A has *increased* the probability of L because $P(L \mid A) > P(L)$. We say that the event A is *favorable* to the event L.

If, instead of the occurrence of A, we were given the information that the event H has occurred, we would be interested in calculating $P(L \mid H)$, the updated probability of L. The information that H has occurred automatically means that the number rolled on the die was definitely either a 5 or a 6. There is no longer any possibility that a 1, a 2, a 3, or a 4 was rolled. But $L = \{1, 2\}$, so that there is no chance that event L could've occurred. It follows that

$$P(L \mid H) = 0$$

Because the original probability of L was 1/3 and the conditional probability of L, given H, is 0, we can see that $P(L \mid H) < P(L)$. We say that the event H is *unfavorable* to the event L because the occurrence of H has decreased, quite substantially in this case, the probability of L.

Suppose, finally, that the information we were given was that the event E has occurred. The occurrence of E would mean that the actual number rolled was either 2, 4, or 6. The only possible outcome which would therefore result in the occurrence of L would be the 2. There is no chance that L's outcome 1 would occur because the occurrence of E specifically excludes the occurrence of 1, 3, and 5. Therefore the numbers 2, 4, and 6 would each be assigned an updated probability of 1/3, since those are now the only three possible outcomes. The occurrence of 2 would imply the occurrence of L, while the occurrence of 4 or 6 would not. Therefore the updated probability of L, in view of the information that E has occurred, is

$$P(L \mid E) = 1/3$$

because L includes one of the three possible outcomes 2, 4, and 6. We note something unusual here: the original probability of L was 1/3, while the conditional probability of L given E also turned out to be 1/3. Therefore, it seems that the occurrence of E *did not affect* in any way the chances of L, because $P(L \mid E) = P(L)$. We describe this situation by saying that the event L is *independent* of the event E.

We summarize the types of relationships between the original and the updated probabilities in Table 8.3.

In many cases of applied interest, the conditional probabilities will not be as easy to visualize as they were in our analysis of the roll of a probability.cube. In such instances it will be very useful to have a formula for computing conditional probabilities in terms of original probabilities. In order to develop such a formula, let's try to figure out what the conditional probability of C given D really is. Consider the Venn diagram on the left side of Fig. 8.9. The original probability of C can be viewed as the proportion of the sample space S that is occupied by C. If C is a large part of S, the probability of C will be high; but if C is a small part of S, the probability of C will be small. What happens to C after D occurs? This situation is illustrated on the right side of Fig. 8.9.

The information that D has occurred has the effect of reducing the sample space to D, because the outcomes not in D can no longer be considered as possible outcomes. The probability of C in this situation is then the proportion of the "new sample space" D that is occupied by C. The part of D that is occupied by C is the set $C \cap D$. Therefore, if $C \cap D$ is a large part of D, the conditional probability of C given D will be large; but if $C \cap D$ is a small part of D, the conditional probability of C given D will be small. To

TABLE 8.3
Conditional Probability Relationships

Verbal Expression	Mathematical Expression
C is favorable to D.	$P(D \mid C) > P(D)$
C is unfavorable to D.	$P(D \mid C) < P(D)$
D is independent of C.	$P(D \mid C) = P(D)$

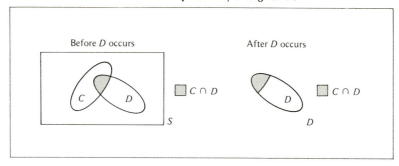

express the conditional probability as the proportion of D occupied by $C \cap D$, we have the following formula for conditional probability:

●
$$P(C \mid D) = \frac{P(C \cap D)}{P(D)}$$
(8.6)

Before illustrating the use of this formula in applied work, let's use it to recalculate the conditional probabilities that we found directly for the roll of a die. Recall that

$$L = \{1, 2\}$$
$$A = \{1, 2, 3, 4\}$$
$$H = \{5, 6\}$$
$$E = \{2, 4, 6\}$$

Then, using the formula for conditional probability, we can make the following calculations:

$$P(L \mid A) = \frac{P(L \cap A)}{P(A)} = \frac{P(\{1, 2\})}{P(\{1, 2, 3, 4\})} = \frac{1/3}{2/3} = \frac{1}{2}$$

$$P(L \mid H) = \frac{P(L \cap H)}{P(H)} = \frac{P(\phi)}{P(\{5, 6\})} = \frac{0}{1/3} = 0$$

$$P(L \mid E) = \frac{P(L \cap E)}{P(E)} = \frac{P(\{2\})}{P(\{2, 4, 6\})} = \frac{1/6}{1/2} = \frac{2}{6} = \frac{1}{3}$$

Naturally, these results are exactly the same as we obtained earlier.
    If in the formula for conditional probability

$$P(C \mid D) = \frac{P(C \cap D)}{P(D)}$$

we multiply both sides of the equation by $P(D)$, we see that

$$P(C \mid D)P(D) = \frac{P(C \cap D)}{P(D)} \cdot P(D) = P(C \cap D)$$

as the two $P(D)$'s cancel out. The resulting expression

$$\bullet \qquad P(C \cap D) = P(C \mid D)P(D) \qquad\qquad (8.7)$$

is very useful in applications, perhaps just as useful as the formula for conditional probability itself.

As an application of the above expression, consider the following example.

Example 8.1 A Bond Referendum An elementary school bond referendum which requires a majority vote to pass is put before the voters in a school district where 30% of the voters have children in elementary school and 70% do not. Polls indicate that 90% of those with children in elementary school will vote for the bond issue, while only 20% of those without children in elementary school will do so. If the polls are right, will the bond issue pass?

SOLUTION   We need to compactly label some events. The important ones are

F = the event that a voter voted for the bond issue
E = the event that a voter had children in elementary school

In the language of these events we know that $P(E) = .30$ and $P(E^c) = .70$ for the school district under study. The polls indicate that $P(F \mid E) = .90$ and $P(F \mid E^c) = .20$. What we want to know is $P(F)$, the proportion of voters who voted for the bond issue. By formula (8.4), we know that

$$P(F) = P(F \cap E) + P(F \cap E^c)$$

Now using formula (8.7), we can calculate each of the terms $P(F \cap E)$ and $P(F \cap E^c)$. We have that

$$P(F \cap E) = P(F \mid E)P(E) = (.90)(.30) = .27$$

and $\qquad\qquad P(F \cap E^c) = P(F \mid E^c)P(E^c) = (.20)(.70) = .14$

It follows that

$$P(F) = P(F \cap E) + P(F \cap E^c) = (.27) + (.14) = .41$$

and so the indications are that the bond issue will get 41% of the votes, which is short of the majority required for passage.

We can use a tree diagram of the sort introduced in Sec. 1.C to illustrate the necessary calculations. The diagram appears in Fig. 8.10.

Example 8.2 Automobile Insurance An insurance company issues three types of automobile policies: Type G for good risks, Type M for moderate risks, and Type B for bad risks. The company's clients are 20% Type G, 40% Type M, and 40% Type B. Accident statistics reveal that a Type G driver has probability .01 of causing an accident in a 12-month period, a Type M driver has probability .02, and a Type B driver has probability .08. What proportion of the company's clients will cause an accident in the next 12 months?

FIGURE 8.10
Tree diagram of the bond referendum example.

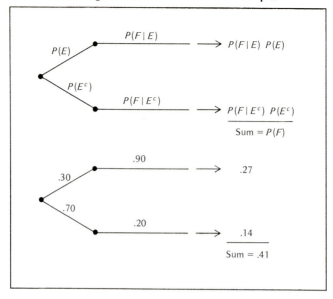

SOLUTION   The pertinent events are as follows:

A = the event that a client will cause an accident
   in the next 12 months
G = the event that a client is a Type G driver
M = the event that a client is a Type M driver
B = the event that a client is a Type B driver

The answer to the question is $P(A)$, the proportion of clients who will cause an accident in the next 12 months. From formula (8.5), we know that

$$P(A) = P(A \cap G) + P(A \cap M) + P(A \cap B)$$

Applying the conditional probability expression (8.7), we get that

$$P(A) = P(A \mid G)P(G) + P(A \mid M) P(M) + P(A \mid B)P(B)$$

The information given in the example translates into the following facts:

$$
\begin{array}{ll}
P(G) = .20 & P(A \mid G) = .01 \\
P(M) = .40 & P(A \mid M) = .02 \\
P(B) = .40 & P(A \mid B) = .08
\end{array}
$$

Therefore

$$
\begin{aligned}
P(A) &= (.01)(.20) + (.02)(.40) + (.08)(.40) \\
&= .002 + .008 + .032 \\
&= .042
\end{aligned}
$$

FIGURE 8.11
Tree diagram of the accident insurance example.

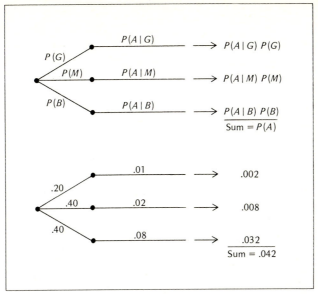

so that 4.2% of the company's clients can be expected to cause an accident in the next 12 months. The calculation is illustrated by a tree diagram in Fig. 8.11.

EXERCISES 8.C

1   Consider again the experiment of tossing a pair of fair dice, one of them red and the other green. (This situation was previously discussed in Exercises 8.A.1 and 8.B.1.)
   a Defining the events $W = \{7, 11\}$ and $D = \{3, 5, 7, 9, 11\}$, calculate the probability $P(W \mid D)$.
   b Defining the events $L = \{2, 3, 12\}$ and $E = \{2, 4, 6, 8, 10, 12\}$, calculate the probability $P(L \mid E)$.
   c Find $P(D \mid W)$.
   d Find $P(E \mid L)$.
   e Find $P(W \mid E)$.
   f Find $P(L \mid D)$.

2   A jar contains eight green balls and two purple balls. Balls are drawn from the jar according to the following scheme: Whenever a green ball is drawn, it is replaced by a new green ball before the next draw, but whenever a purple ball is drawn, it is replaced by three new purple balls before the next draw. For example, if a purple ball is drawn on the first draw, then there will be eight green balls and four purple balls in the jar awaiting the second draw.
   a Find the probability that the first ball drawn is green.
   b If the first ball drawn is green, find the probability that the second ball drawn will also be green.

c Find the probability that the first ball drawn is purple.

d If the first ball drawn is purple, find the probability that the second ball drawn is green.

e Find the probability that the second ball drawn will be green.

3 In a learning experiment on a T maze, a rat is equally likely to turn left (receiving an electric shock) or right (receiving food) on his first time through. If he received food the first time, he has probability 3/5 of turning right again the second time. If, on the other hand, he received a shock the first time, he has probability 4/5 of turning right the second time.

a What is the probability that the rat will turn right both times?

b What is the probability that the rat will turn left on the first try but right on the second?

c What is the probability that the rat will turn left both times?

d What is the probability that the rat will receive food the second try?

4 One psychologist has studied the incidence of grimacing and rheumatic disease in schizophrenic patients. In a study of 2000 such patients, he found that 100 had rheumatic disease, and of those 100, 40 also had grimacing. Another 360 had grimacing but no rheumatic disease.

a What was the probability that a patient had grimacing?

b What was the conditional probability that a patient had grimacing given that the patient had rheumatic disease?

c Is grimacing independent of rheumatic disease in schizophrenic patients?

5 The probability that a certain type of rocket launching will be successful is 0.8.

a If launchings are attempted until one is successful, what is the probability that no more than two attempts will be necessary?

b If launchings are attempted until two are successful, what is the probability that no more than four attempts are necessary?

c Which probability is larger? Can you give a verbal explanation of the reasons why this should be so?

6 If $P(A) = P(B)$, show that $P(A \mid B) = P(B \mid A)$ must also be true.

SECTION 8.D INDEPENDENT EVENTS

In the preceding section, we have noticed that $P(C \mid D)$, the conditional probability of C given D, could be either less than, equal to, or greater than the original probability of C. In those cases where $P(C \mid D)$ is different from $P(C)$, the knowledge of the occurrence of D has had an influence on our view of the likelihood of the occurrence of C. We can describe this situation by saying that the event C "depends on" the event D, in the sense that the occurrence of D affects the probability of C. Suppose, however, that $P(C \mid D)$ is exactly the same, numerically speaking, as $P(C)$. This means that the occurrence of D has had no influence at all on the probability of C. In this situation, we can say that the event C is "independent" of the event D, in the sense that the occurrence of D does not affect the probability of C. We have the following formal definition: The event C is said to be independent of the event D if $P(C \mid D) = P(C)$.

The concept of independent events is often mistaken for the concept of disjoint events. In fact, however, independence and disjointness are extreme opposites of each other. If the events C and D are disjoint, then $C \cap D = \phi$, so that, by the formula for conditional probability,

$$P(C \mid D) = \frac{P(C \cap D)}{P(D)} = \frac{P(\phi)}{P(D)} = \frac{0}{P(D)} = 0$$

no matter what the original probability of C. What this means is that, even if $P(C)$ is very high, the occurrence of D virtually eliminates all chance of C's occurrence because C and D are mutually exclusive. Therefore C depends very heavily on D, as the conditional probability of C, given D, is very small compared with the original probability of C. In summary, if C and D are disjoint, i.e. if $C \cap D = \phi$, then $P(C \mid D) \neq P(C)$ because $P(C \mid D) = 0$, so that C is *not* independent of D. Therefore, instead of having similar meanings, disjointness and independence are really at opposite ends in meaning.

A very useful fact about independent events is that if C is independent of D, then D is independent of C. Specifically, if the occurrence of C does not affect the probability of D, then the occurrence of D would not affect the probability of C. Mathematically, this fact can be expressed as follows: If $P(C \mid D) = P(C)$, then $P(D \mid C) = P(D)$. The justification of this fact involves the use of the formulas for conditional probability (8.6) and (8.7). If $P(C \mid D) = P(C)$, then

$$P(D \mid C) = \frac{P(D \cap C)}{P(C)} = \frac{P(C \cap D)}{P(C)} = \frac{P(C \mid D)P(D)}{P(C)} = \frac{P(C)P(D)}{P(C)} = P(D)$$

where we have replaced $P(C \mid D)$ by its equal $P(C)$. Because of this symmetry of meaning of the word "independence," we can refer to C and D as a pair of independent events. It is not necessary to specify which one is independent of the other because the algebra above shows that if the former is independent of the latter, the latter will also be independent of the former.

There is an algebraic expression of the meaning of independence which reflects this symmetry. The independence of C and D means that $P(C) = P(C \mid D)$, while the general formula for conditional probability means that

$$P(C \mid D) = \frac{P(C \cap D)}{P(D)}$$

Combining these algebraic expressions, we see that

$$P(C) = \frac{P(C \cap D)}{P(D)}$$

which, by cross-multiplication, becomes

$$P(C)P(D) = P(C \cap D)$$

It then follows that the events C and D are independent if and only if

$$P(C \cap D) = P(C)P(D) \tag{8.8}$$

The above expression of independence is the one most useful in the solution of applied problems. It forms the basic theoretical foundation for the chi-square test of independence, to be discussed in the next chapter. In addition, it underlies the calculation of the table of binomial distribution (Table A.2), for a sequence of independent repeated events often generates binomially distributed data. The development of Table A.2 will be discussed in Sec. 8.E.

In dealing with more than two independent events, we need an expanded version of the above equation. We will say that n events $C_1, C_2, \ldots, C_n$ are independent if every event is independent not only of every other event but also of every possible combination of all other events. It turns out that

● $$P(C_1 \cap C_2 \cap \cdots \cap C_n) = P(C_1)P(C_2) \cdots P(C_n) \qquad (8.9)$$

if the events $C_1, C_2, \ldots, C_n$ are independent. This means that the probability that all the events will occur is given by the product of all their original probabilities.

As illustrations of the use of these computing formulas for the probabilities of independent events, we look at the following two examples.

Example 8.3   Coin Tossing   What is the probability that a fair coin will fall heads on ten consecutive tosses?

SOLUTION   Denote by H_k the event that the coin falls heads on the kth toss, i.e.,

$H_1 =$ the event that the coin falls heads on the
first toss
$H_2 =$ the event that the coin falls heads on the
second toss
. . . .
$H_{10} =$ the event that the coin falls heads on the
tenth toss

Then, because the coin does not remember which side it landed on each time, the events $H_1, H_2, H_3, H_4, H_5, H_6, H_7, H_8, H_9,$ and H_{10} are independent. Each of these events has probability 1/2 because the coin is fair, so that heads and tails are equally likely. Therefore the probability of heads on all ten tosses is

$$P(H_1 \cap H_2 \cap H_3 \cap H_4 \cap H_5 \cap H_6 \cap H_7 \cap H_8 \cap H_9 \cap H_{10})$$
$$= P(H_1)P(H_2)P(H_3)P(H_4)P(H_5)P(H_6)P(H_7)P(H_8)P(H_9)P(H_{10})$$
$$= \left(\frac{1}{2}\right)\left(\frac{1}{2}\right)\left(\frac{1}{2}\right)\left(\frac{1}{2}\right)\left(\frac{1}{2}\right)\left(\frac{1}{2}\right)\left(\frac{1}{2}\right)\left(\frac{1}{2}\right)\left(\frac{1}{2}\right)\left(\frac{1}{2}\right)$$
$$= \frac{1}{1024} = 0.000977$$

using the expanded version (8.9) of the independence formula. Therefore the chances of tossing ten heads in a row with a fair coin are slightly less than one in a thousand. (Because coin tossing is one of the prime examples of a process which generates binomially distributed data, we could have read the desired probability directly from Table A.2 which has itself been constructed by means of the procedure we have just worked out.)

Example 8.4   A Political Poll   Candidates Able and Baker, running for the same office, are facing an electorate of several million persons. A national poll claims that two out of every three voters favor Able over Baker. Suppose we want to test the poll's validity in our district of several thousand persons. We choose a random sample of five voters, and we ask ourselves the following question: If it were really true that two out of three favored Able, what is the probability that in our sample of five voters only one of the five would favor Able?

SOLUTION   The relevant events are the following:

A_1 = the event that the 1st voter in our sample favors Able
A_2 = the event that the 2nd voter in our sample favors Able
. . . .
B_1 = the event that the 1st voter in our sample favors Baker
. . . .
B_5 = the event that the 5th voter in our sample favors Baker

If it were really true that two out of three voters favor Able, the probabilities would be as follows:

$$P(A_1) = P(A_2) = P(A_3) = P(A_4) = P(A_5) = \frac{2}{3}$$

and $$P(B_1) = P(B_2) = P(B_3) = P(B_4) = P(B_5) = \frac{1}{3}$$

We are interested in $P(E)$, the probability that exactly one of the five voters selected favors Able. Because of the large number of voters from whom the five are to be selected, it is reasonable to consider the five voters selected as independent of one another. The following events together comprise E, the event that exactly one of the five voters favors Able:

$A_1^* = A_1 \cap B_2 \cap B_3 \cap B_4 \cap B_5$ = the event that the 1st voter favors Able while the other four favor Baker

$A_2^* = B_1 \cap A_2 \cap B_3 \cap B_4 \cap B_5$ = the event that the 2nd voter favors Able while the other four favor Baker

. . . .

$A_5^* = B_1 \cap B_2 \cap B_3 \cap B_4 \cap A_5$ = the event that the fifth voter favors Able while the other four favor Baker

Because $E = A_1^* \cup A_2^* \cup A_3^* \cup A_4^* \cup A_5^*$, where the latter five events are disjoint, an expanded version of the special rule following formula (8.3) implies that $P(E) = P(A_1^*) + P(A_2^*) + P(A_3^*) + P(A_4^*) + P(A_5^*)$. It remains now only to compute the probabilities of the events A_k^*. Because the voters are assumed to be independent of one another, we have, by the expanded formula (8.9) for independent events, that

$$P(A_1^*) = P(A_1)P(B_2)P(B_3)P(B_4)P(B_5)$$

$$= \left(\frac{2}{3}\right)\left(\frac{1}{3}\right)\left(\frac{1}{3}\right)\left(\frac{1}{3}\right)\left(\frac{1}{3}\right) = \frac{2}{243} = .00823$$

$$P(A_2^*) = \left(\frac{1}{3}\right)\left(\frac{2}{3}\right)\left(\frac{1}{3}\right)\left(\frac{1}{3}\right)\left(\frac{1}{3}\right) = \frac{2}{243} = .00823$$

· ·

$$P(A_5^*) = \left(\frac{1}{3}\right)\left(\frac{1}{3}\right)\left(\frac{1}{3}\right)\left(\frac{1}{3}\right)\left(\frac{2}{3}\right) = \frac{2}{243} = .00823$$

The five events A_1^*, A_2^*, A_3^*, A_4^*, and A_5^* each have probability .00823. It follows that the probability that exactly one of the five voters favors Able is

$$P(E) = .00823 + .00823 + .00823 + .00823 + .00823$$
$$= .04115 \approx 4\%$$

Therefore, if two out of every three voters really favored Able, the chances are only about 4% that exactly one of five randomly selected voters would favor Able. If we were to choose a sample resulting in only one of the five favoring Able, it would then be reasonable to assume that in our district it is probably not true that two out of three favor Able.

EXERCISES 8.D

1 Suppose we toss a fair coin six times. Determine the following probabilities and compare your answers with the appropriate numbers listed in Table A.2 for the binomial distribution with parameters $n = 6$ and $p = \frac{1}{2}$:
a The probability that none of the six tosses results in a head.
b The probability that exactly one of the six tosses results in a head.
c The probability that exactly two of the six tosses result in heads.
d The probability that exactly three of the six tosses result in heads.
e The probability that exactly four of the six tosses result in heads.
f The probability that exactly five of the six tosses result in heads.
g The probability that all six tosses result in heads.
h Show that your answers to a through g add up to one.

2 Suppose we roll a fair die six times. Determine the following probabilities and compare your answers with the appropriate numbers listed in Table A.2 for the binomial distribution with parameters $n = 6$ and $p = \frac{1}{6}$:
a The probability that none of the six rolls results in a 5.
b The probability that exactly one of the six rolls results in a 5.
c The probability that exactly two of the six rolls result in 5s.
d The probability that exactly three of the six rolls result in 5s.
e The probability that exactly four of the six rolls result in 5s.
f The probability that exactly five of the six rolls result in 5s.
g The probability that all six rolls result in 5s.
h Show that your answers to a through g add up to one.

3 A community within a small urban area has two ambulances available, one at the

north end of town and the other at the south end. The two ambulances operate independently of each other, but due to differences in mechanical condition and traffic patterns, they have probabilities 0.9 and 0.5, respectively, of arriving at an accident scene within 10 minutes.

a If an accident requires both ambulances, what is the probability that they both arrive within 10 minutes?

b What is the probability that the one from the north end arrives within 10 minutes but the one from the south end does not?

c What is the probability that both take longer than 10 minutes to arrive on the scene?

4 A small machine shop is considering purchase of an oscilloscope with four vital parts, all of which work properly or fail independently of the others. The machine fails to work properly if two or more of these four parts fail. The shop's tests have yielded the following data on operating probabilities:

Part	Probability of Working Properly
A	0.9
B	0.9
C	0.8
D	0.5

a What is the probability that all four parts work properly simultaneously?

b What is the probability that A fails while B, C, and D all work properly?

c What is the probability that D fails while A, B, and C all work properly?

d What is the probability that the oscilloscope works properly?

5 Suppose it is really true that 60% of the voters are for repeal of a certain local tax law. If we were to select a random sample of seven voters, what would be the probability that a majority of the sample would be against repeal, thus making it appear that the antirepeal forces are in the lead? Compare your solution, using probability theory, with that obtained in Example 3.2.

6 Show that the events C and D are independent if and only if $P(C \mid D) = P(C \mid D^c)$.

SECTION 8.E THE TABLE OF THE BINOMIAL DISTRIBUTION

When we introduced the binomial probability distribution in Sec. 3.B, we mentioned that a method of calculating binomial probabilities directly would have to await the development of the techniques of probability theory. Now, with the discussion of independent events in Sec. 8.D, we have completed this development. Accordingly, we are now ready to present a full explanation of how the binomial probabilities used in Chap. 3 were computed and, more importantly, how the table of the binomial distribution, Table A.2, was generated. We begin with some facts about "permutations" and "combinations," two concepts whose meaning will soon be made clear.

Suppose we have a jar containing n billiard balls, each having a number from 1 to

n, inclusive, painted on its surface. (For an ordinary game of pool, n would be 15.) Out of this set of n balls in the jar, suppose we have to draw k of them in sequence, where k is a number less than or equal to n. (For example, if $n = 15$, we could take $k = 8$ in order to select 8 out of the 15 pool balls in the jar.) The question we want to ask first is: How many possible different "sequences" of k balls can be selected from the n balls in the jar? As a clarification of what a sequence is, we should point out that, if you draw ball 3 first, ball 8 second, and ball 6 third and I draw ball 6 first, ball 3 second, and ball 8 third, our sequences are different although we both have drawn the same three balls. A sequence depends not only on the balls that are drawn but also on the order in which they are drawn. We turn to the tree diagrams of Sec. 1.C to help us answer the question of how many different sequences of size k can be drawn from a set of n elements. In Fig. 8.12, we illustrate the various sequences of size 3 that can be drawn from a jar containing five balls.

As Fig. 8.12 shows, there are 60 possible sequences of size 3 that can be drawn from a set of five elements. As you can imagine, the number of sequences rapidly becomes astronomical as the number of elements involved grows. Clearly we need a formula for the number of sequences so that we don't have to draw a tree diagram every time.

Fortunately the tree diagram of Fig. 8.12 contains within it the seeds of the formula we need. Notice that we have five options on the first draw, four on the second (for we are already holding the one element we had drawn first), and three on the third (for we are holding two elements drawn previously). Therefore, for each of the five options on the first draw, we have four options on the second, yielding $5 \times 4 = 20$ different sequences resulting from the first two draws. Having in hand these 20 two-element sequences, we then have three options for the third draw. So each of the 20 two-element sequences can be augmented by one of three remaining elements, resulting in 60 possible three-element sequences. In summary, we can say that the total number of possible different sequences of size 3 from a set of five elements is

$$5 \times 4 \times 3 = 60$$

Now, what about drawing k elements in sequence from a set of n elements? Well, on the first draw we have n options, on the second we have $n - 1$, on the third we have $n - 2, \ldots$, and on the kth draw, we have $n - (k - 1) = n - k + 1$ options. Therefore the total number of possible different sequences of size k from a set of n elements is given by

● $$P(n, k) = n \times (n - 1) \times \cdots \times (n - k + 1) \qquad (8.10)$$

Here $P(n, k)$ is read: "the number of permulations of n elements taken k at a time." The word permutation is the technical term for sequences in this context. If we insert $n = 5$ and $k = 3$ into the formula for $P(n, k)$, we get

$$n - k + 1 = 5 - 3 + 1 = 3$$

so that
$$\begin{aligned} P(5, 3) &= n \times (n - 1) \times \cdots \times (n - k + 1) \\ &= 5 \times 4 \times 3 \\ &= 60 \end{aligned}$$

FIGURE 8.12
Drawing sequences of size 3 from a set of 5 elements.

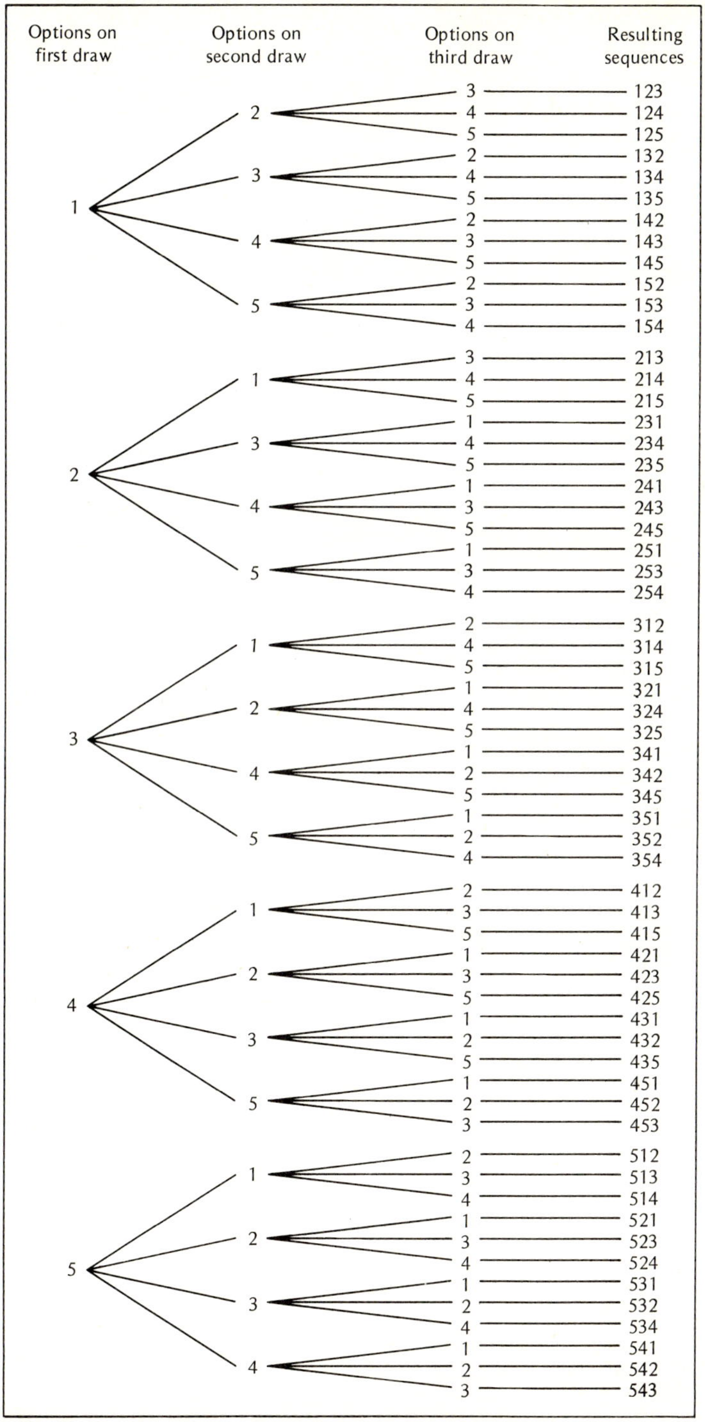

Options on first draw	Options on second draw	Options on third draw	Resulting sequences
		3	123
	2	4	124
		5	125
		2	132
	3	4	134
		5	135
1		2	142
	4	3	143
		5	145
		2	152
	5	3	153
		4	154
		3	213
	1	4	214
		5	215
		1	231
	3	4	234
		5	235
2		1	241
	4	3	243
		5	245
		1	251
	5	3	253
		4	254
		2	312
	1	4	314
		5	315
		1	321
	2	4	324
		5	325
3		1	341
	4	2	342
		5	345
		1	351
	5	2	352
		4	354
		2	412
	1	3	413
		5	415
		1	421
	2	3	423
		5	425
4		1	431
	3	2	432
		5	435
		1	451
	5	2	452
		3	453
		2	512
	1	3	513
		4	514
		1	521
	2	3	523
		4	524
5		1	531
	3	2	532
		4	534
		1	541
	4	2	542
		3	543

which is, of course, the same result obtained from the tree diagram of Fig. 8.12.

At this point, we can introduce some new mathematical symbolism so that we can simplify the formulas to come. We define $n!$ (pronounced "n-factorial") to be

$$n! = n \times (n - 1) \times (n - 2) \times \cdots \times 1$$

For example

$$3! = 3 \times 2 \times 1 = 6$$
$$5! = 5 \times 4 \times 3 \times 2 \times 1 = 120$$
$$6! = 6 \times 5 \times 4 \times 3 \times 2 \times 1 = 720$$

Furthermore, $6! = 6 \times (5!) = 6 \times 5 \times 4 \times (3!)$. Using this notation, we can see that

$$
\begin{aligned}
P(n, k) &= n \times (n - 1) \times \cdots \times (n - k + 1) \\
&= \frac{n \times (n - 1) \times \cdots \times (n - k + 1) \times (n - k) \times (n - k - 1) \times \cdots \times 1}{(n - k) \times (n - k - 1) \times \cdots \times 1} \\
&= \frac{n!}{(n - k)!}
\end{aligned}
$$

because the number $(n - k) \times (n - k - 1) \times \cdots \times 1$ cancels out of both the numerator and the denominator. If we try out the permutation formula

● $$P(n, k) = \frac{n!}{(n - k)!} \qquad (8.11)$$

on $n = 5$ and $k = 3$, we get

$$P(5, 3) = \frac{5!}{(5 - 3)!} = \frac{5!}{2!} = \frac{120}{2} = 60$$

exactly the same result we obtained earlier, of course. If we are drawing $k = 8$ balls at random out of a set of $n = 15$, there would be

$$P(15, 8) = \frac{15!}{(15 - 8)!} = \frac{15!}{7!} = \frac{1,307,674,368,000}{5040} = 259,459,200$$

different sequences obtainable, justifying our earlier description of the number of possible sequences as astronomical.

Now we are ready to ask a second question: How many possible different "subsets" of k balls can be selected from the n balls in the jar? By a subset here, we mean simply the collection of the k balls drawn without regard to the order in which they are drawn. Now, are there more sequences of size k or more subsets of size k? Clearly, there are more sequences than subsets because one subset can be drawn in several different orders, each order counting as a separate sequence. In particular, each subset of size k can be drawn as one of

$$P(k, k) = k \times (k - 1) \times \cdots \times 1 = k!$$

different sequences, using formula (8.10) with $n = k$. Therefore, because

$$
\begin{array}{ccc}
\text{Number of subsets} & \text{number of sequences} & \text{number of sequences} \\
\text{of size } k & \times \quad \text{per subset} & = \quad \text{of size } k
\end{array}
$$

we see that

$$\frac{\text{Number of subsets}}{\text{of size } k} = \frac{\text{number of sequences of size } k}{\text{number of sequences per subset}}$$

$$= \frac{P(n, k)}{P(k, k)} = \frac{\dfrac{n!}{(n - k)!}}{\dfrac{k!}{}}$$

$$= \frac{n!}{(n - k)!} \cdot \frac{1}{k!} = \frac{n!}{k!(n - k)!}$$

The technical term for the word subset in this context is "combination," so that we can express "the number of combinations of n elements taken k at a time" by the combination formula

● $$C(n, k) = \frac{n!}{k!(n - k)!} \qquad (8.12)$$

Using the combination formula, let's calculate the number of subsets of size 3 of a set of five elements. Here $n = 5$ and $k = 3$, so that there are

$$C(5, 3) = \frac{5!}{3!(5 - 3)!} = \frac{5!}{(3!)(2!)} = \frac{120}{(6)(2)} = 10$$

possible subsets. An examination of the tree diagram in Fig. 8.12 confirms this, the 10 different subsets being $\{1, 2, 3\}, \{1, 2, 4\}, \{1, 2, 5\}, \{1, 3, 4\}, \{1, 3, 5\}, \{1, 4, 5\}, \{2, 3, 4\}, \{2, 3, 5\}, \{2, 4, 5\}, \{3, 4, 5\}$. Each of the 60 sequences is merely a permutation of one of these 10 combinations.

Now, what does all this have to do with the binomial distribution? To find out, let's take a look at the following situation, which we discussed earlier in Example 3.3 of Sec. 3.B.

Example 8.5   A Political Poll   Suppose it is really true that 60% of the voters are for repeal of a certain local tax law. If we were to select a random sample of seven voters, what would be the probability that a majority of the sample would be against repeal, thus making it appear that the antirepeal forces are in the lead?

SOLUTION   We denote by F the number of voters in the sample that are for repeal, as we did in Example 3.3. Then the probability

$$P(\text{a majority of the sample is against repeal}) = P(F \leq 3)$$

We now proceed to calculate $P(F \leq 3)$.[1] First of all, we see that

$$P(F \leq 3) = P(F = 0) + P(F = 1) + P(F = 2) + P(F = 3)$$

Let's look at $P(F = 3)$. For F to equal 3, the random sample of seven voters must contain three who favor repeal and four who oppose repeal. One way to have $F = 3$ would be to have $F_1 \cap F_2 \cap F_3 \cap A_4 \cap A_5 \cap A_6 \cap A_7$ (using the notation of Example 8.4), where F

[1]In Example 3.3, we noted that F was a binomial random variable with parameters $n = 7$ and $p = 60$. Table A.2 then showed that $P(F \leq 3) = .2898$.

indicates "for repeal" and A indicates "against repeal." Because the seven randomly selected voters act independently, these seven events are independent so that

$$P(F_1 \cap F_2 \cap F_3 \cap A_4 \cap A_5 \cap A_6 \cap A_7) = P(F_1)P(F_2)P(F_3)P(A_4)P(A_5)P(A_6)P(A_7)$$
$$= (.60)(.60)(.60)(.40)(.40)(.40)(.40)$$
$$= (.60)^3(.40)^4$$
$$= (.216)(.0256) = .0055296$$

because the probability of a voter's being for repeal is .60 and the probability of being against repeal is therefore .40.

Now this is only one way to fill the seven positions in the random sample with three F's and four A's and so to have $F = 3$. There are several other ways to have $F = 3$; in fact, there is one way corresponding to each subset of three positions in which to put the F's. The number of ways of having $F = 3$ is therefore equal to the number of subsets of three of a set of seven elements. By formula (8.12), this number is

$$C(7, 3) = \frac{7!}{3!4!} = \frac{5040}{(6)(24)} = 35$$

So there are 35 ways to get $F = 3$, and each of these has probability .0055296. It follows that

$$P(F = 3) = (35)(.0055296) = .19354$$

From the reasoning we used we can derive a general rule of the binomial distribution: If X is a random variable having the binomial distribution with parameters n and p, then

● $$P(X = j) = C(n, j)(p)^j(1 - p)^{n-j} \qquad (8.13)$$

For $P(F = 3)$, we have $j = 3$, $n = 7$, and $p = .60$, so that

$$P(F = 3) = C(7, 3)(.60)^3(.40)^{7-3}$$
$$= (35)(.60)^3(.40)^4 = .19354$$

Using formula (8.13), we can complete the calculation:

$$P(F = 2) = C(7, 2)(.60)^2(.40)^{7-2}$$
$$= \left(\frac{7!}{2!5!}\right)(.60)^2(.40)^5$$
$$= (21)(.36)(.01024) = .07741$$
$$P(F = 1) = C(7, 1)(.60)^1(.40)^6$$
$$= (7)(.60)(.004096) = .01720$$
$$P(F = 0) = C(7, 0)(.60)^0(.40)^7$$
$$= (1)(1)(.0016384) = .00164$$

From the results of these calculations, we find that

$$P(F \leq 3) = P(F = 0) + P(F = 1) + P(F = 2) + P(F = 3)$$
$$= .00164 + .01720 + .07741 + .19354$$
$$= .2898$$

which is the same answer obtained earlier from Table A.2.

To summarize the above computational procedure, we observe that the entry in Table A.2 corresponding to parameters n and p is the number

$$P(X \leq k) = \sum_{j=0}^{k} P(X = j) = \sum_{j=0}^{k} C(n, j)p^j(1 - p)^{n-j} \qquad (8.14)$$

As a further illustration of the techniques by which Table A.2, the table of the binomial distribution, was developed, let's calculate the entry in Table A.2 corresponding to $n = 9$, $k = 4$, and $p = .55$. According to formula (8.14), this entry should be

$$P(X \leq 4) = \sum_{j=0}^{4} C(9, j)(.55)^j(1 - .55)^{9-j}$$

$$= C(9, 0)(.55)^0(.45)^9 + C(9, 1)(.55)^1(.45)^8 + C(9, 2)(.55)^2(.45)^7$$
$$+ C(9, 3)(.55)^3(.45)^6 + C(9, 4)(.55)^4(.45)^5$$
$$= (1)(1)(.000757) + (9)(.55)(.001682) + (36)(.3025)(.003737)$$
$$+ (84)(.166375)(.008304) + (126)(.091506)(.018453)$$
$$= .00076 + .00833 + .04070 + .11605 + .21276$$
$$= .3786$$

A glance at Table A.2 reveals that the entry for $n = 9$, $k = 4$, and $p = .55$ is indeed .3786. In the above computation, we have used the facts that

$$C(9, 0) = \frac{9!}{0!9!} = 1 \qquad \text{(NOTE: } 0! = P(0, 0) = 1 \text{ because}$$
there is only one sequence of
0 elements in a set of 0 elements.)

$$C(9, 1) = \frac{9!}{1!8!} = 9$$

$$C(9, 2) = \frac{9!}{2!7!} = 36$$

$$C(9, 3) = \frac{9!}{3!6!} = 84$$

$$C(9, 4) = \frac{9!}{4!5!} = 126$$

EXERCISES 8.E

1   A corporation has offices in 10 major cities of South America, and the chief operating officer would like to visit five of these on her tour of the area next month.
  a How many different groups of five cities can she arrange to visit? Is this a problem of permutations or combinations?
  b If she is willing to visit any five of the ten cities, how many different routes can her travel agent consider in his planning of the trip? Is this a problem of permutations or combinations?

2   The United States Senate has 100 members. A pollster attempting to predict the outcome of the voting on a spending bill wants to choose a sample of 20 senators on which to base his prediction.

a How many different samples of 20 senators can the pollster select? Is this a problem of permutations or combinations?

b If the pollster chooses the 20 members of his sample by walking from one senator's office to another until he has covered 20 offices, how many different routes can he follow in knocking on 20 of the 100 office doors? Is this a problem of permutations or combinations?

3 Four-letter "words" in which no letter appears more than once can be formed by choosing in order four out of the 26 letters of the English alphabet.

a How many four-letter "words"[1] with no letters appearing more than once are there in the English language?

b Is this a problem of permutations or combinations?

c If we allow letters to appear more than once, how many four-letter "words" are there in the English language?

4 To conduct an experiment in an attempt to find out whether any of several self-proclaimed clairvoyants actually have ESP ("extrasensory perception"), a psychologist takes 20 index cards and marks 5 of them with a red "X." The cards are then laid on a table face down and the candidate for clairvoyancy is asked to select the five marked cards without peeking.

a How many different groups of five cards is it possible for the candidate to select?

b Of these, how many contain the five marked cards?

c If the candidate does not really have ESP, what is the probability that he will pick exactly the five marked cards?

5 In a study of the life-styles of 1000 families, a sociologist needs a sample of 10 families. How many different samples can she choose?

6 An accountant is assigned the task of verifying a group of 400 transactions. How many samples of 15 transactions each are available for his detailed analysis?

7 Calculate the entry in Table A.2 corresponding to $n = 5$, $k = 3$, and $p = .85$.

8 Calculate the entry in Table A.2 corresponding to $n = 11$, $k = 4$, and $p = .35$.

9 Calculate the entry in Table A.2 corresponding to $n = 16$, $k = 13$, and $p = .70$.

10 For any two positive whole numbers n and k, where k is not larger than n, show that it is always true that $C(n, k) = C(n, n - k)$.

SECTION 8.F BAYES' THEOREM

From the applied point of view, Bayes' theorem provides an extremely precise way of developing and specifying relationships between events. From the technical point of view, it is a culmination point of the theory of conditional probability. To use it effectively requires an understanding of all the basic aspects of conditional probability covered so far in this chapter. But, when properly applied, Bayes' theorem often brings order out of a chaos of intertwined relationships and sometimes even gives results so

---

[1] Of course, not all of these are real words. Consider "lorf," for example.

surprising that they would not even have been considered possible by a researcher unacquainted with the procedure. (One illustration of this sort of situation can be found in Example 8.6 below.)

Well, after such a spectacular buildup, what exactly is Bayes' theorem? It looks innocent enough on its surface, perhaps even a little cold and aloof. It goes as follows:

If A and B are any two events, then

$$P(B \mid A) = \frac{P(A \mid B)P(B)}{P(A \mid B)P(B) + P(A \mid B^c)P(B^c)} \tag{8.15}$$

To justify the validity of Bayes' theorem, we need to know only the following three equations involving conditional and ordinary probability, all of which have already been discussed earlier in sections of this chapter:

$$P(C \mid D) = \frac{P(C \cap D)}{P(D)} \tag{8.6}$$

$$P(E) = P(E \cap F) + P(E \cap F^c) \tag{8.4}$$

$$P(C \cap D) = P(C \mid D)\, P(D) \tag{8.7}$$

Using these three equations as appropriate, we have

$$P(B \mid A) = \frac{P(B \cap A)}{P(A)} = \frac{P(A \cap B)}{P(A \cap B) + P(A \cap B^c)}$$

$$= \frac{P(A \mid B)P(B)}{P(A \mid B)P(B) + P(A \mid B^c)P(B^c)}$$

where we have also used the fact that $B \cap A$ and $A \cap B$ are one and the same, the event consisting of those outcomes which are in both A and B. So there it is—Bayes' theorem. Now, what good is it? Consider, in turn, the following examples.

Example 8.6   A Medical Test   A certain disease can be detected by a blood test in 98% of those who have it. Unfortunately, the test also has probability .01 of asserting that a person has the disease when he really doesn't. On the average, about .5% (one-half of one percent) of those routinely tested actually have the disease. If the test asserts that you have the disease, what is the probability that you really have it?

SOLUTION   Before proceeding with the solution, it would be constructive for you to guess the answer to the question. Check the box below whose percentage you feel best represents your chances of really having the disease if the results of the blood test indicate that you do.

99% 98%   95% 90%   80% 50%   33% 20%   15% 10%
☐   ☐   ☐   ☐   ☐   ☐   ☐   ☐   ☐   ☐

Now that you've checked off your guess, let's proceed to the formal analysis of the problem. First we must specify the events involved. These seem to be:

D = the event that you have the disease
T = the event that the test says you have the disease

First of all, we should note that D and T are two distinctly different events, although there are some relationships between them. We would worry more about D than about T; we are not concerned with T in itself but only because of its relationship with D. In fact, what we want to determine is the strength of this relationship $P(D\,|\,T)$, the conditional probability of your having the disease given that the blood test says you have it. (Just because you might test positive does not necessarily guarantee that you have the disease. The skin test for tuberculosis is notoriously inaccurate in this regard; many, many more people have positive results on the skin test than the number who actually have tuberculosis.)

Now that we have established that $P(D\,|\,T)$ is the answer to the question, let's see what information we are given. The statement that .5% of those routinely tested actually have the disease means that $P(D) = .005$. This is the original probability of D before the test results are known. The fact that the disease is detected by the test in 98% of those who have it can be translated mathematically as $P(T\,|\,D) = .98$. Finally, $P(T\,|\,D^c)$, the probability of testing positive given you do not have the disease, is .01.

In Table 8.4, we summarize the known information and desired information. The primary characteristic apparent from Table 8.4 is that we know the conditional probabilities of T given some facts about D, but we want to know the conditional probability of D given T. For purposes of comparison, let's take a look at what Bayes' theorem says:

$$P(B\,|\,A) = \frac{P(A\,|\,B)P(B)}{P(A\,|\,B)P(B) + P(A\,|\,B^c)P(B^c)}$$

As is apparent from the above formula, if we know the conditional probabilities of A given some facts about B, we can insert them into the formula to obtain the conditional probability of B given A. Note that Bayes' theorem has the effect of interchanging the events involved in conditional probabilities.

Note also that what we need to answer our question is a formula for interchanging the events involved in conditional probabilities. As it turns out, Bayes' theorem gives exactly the formula we need. If we replace B by D and A by T in Bayes' theorem, we see from the information in Table 8.4 that

$$P(D\,|\,T) = \frac{P(T\,|\,D)P(D)}{P(T\,|\,D)P(D) + P(T\,|\,D^c)P(D^c)}$$
$$= \frac{(.98)(.005)}{(.98)(.005) + (.01)(.995)} = \frac{.0049}{.0049 + .00995}$$
$$= \frac{.0049}{.01485} = .33 = 33\%$$

Here we have used the fact that $P(D^c) = 1 - P(D)$: If .5% have the disease, then 99.5% do not.

The answer to our question is therefore that $P(D\,|\,T) = .33$. This means that if the blood test asserts you have the disease the chances are 33% that you really have it. In view of the facts that $P(T\,|\,D) = .98$ while $P(T\,|\,D^c) = .01$, the value of .33 for $P(D\,|\,T)$ might be considered low. (Virtually no one guesses anything near 33% as the answer.) However, a low answer becomes more reasonable when we consider that the .98 refers

TABLE 8.4

TABLE 8.4
Information Relating the Disease
and the Blood Test

Desired Information	Known Information
$P(D\,|\,T)$	$P(T\,|\,D) = .98$
	$P(T\,|\,D^c) = .01$
	$P(D) = .005$

to 98% of a group which comprises .5% of the population, while the .01 stands for 1% of the other 99.5%. A modified tree diagram illustrating the fact that $P(D\,|\,T) = 33\%$ is presented in Fig. 8.13. The diagram shows that of 100,000 people tested for the disease 500 (.5%) have the disease while the other 99,500 do not. Of the 500, 98% or 490 test positive, while of the 99,500, 1% or 995 test positive. Therefore $490 + 995 = 1485$ persons test positive. Of these 1485, only 490 or 33% actually have the disease.

Tree diagrams of the sort introduced in Chap. 1 can also be applied to this problem. The relevant tree diagram is presented in Fig. 8.14. The interpretation of the tree diagram is as follows: Of all persons tested, a proportion .01485 test positive. Part of this proportion, an amount .00490, comes from among those having the disease, while the rest, .00995, do not have the disease. Of those testing positive, then, a fraction

FIGURE 8.13
Status of 100,000 persons given blood tests.

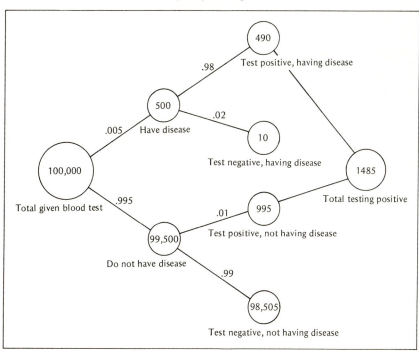

FIGURE 8.14
Tree diagram of disease–blood test example.

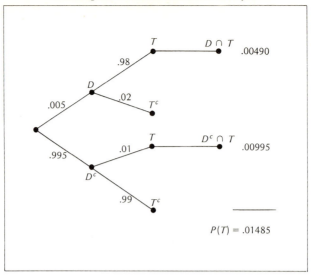

$$\frac{.00490}{.01485} = 0.33 = 33\%$$

actually have the disease. This, of course, is the same result as obtained earlier.

Finally, we can display the data on a Venn diagram, which can then be used to calculate the probability of your having the disease given that your blood test results are positive. The Venn diagram appears in Fig. 8.15 and makes use of the following facts from Table 8.4:

$$P(T \cap D) = P(T \mid D)P(D)$$
$$= (.98)(.005)$$
$$= .0049$$
$$P(T \cap D^c) = P(T \mid D^c)P(D^c)$$
$$= (.01)(.995)$$
$$= .00995$$
$$P(T) = P(T \cap D) + P(T \cap D^c) = .0049 + .00995 = .01485$$

We can now consider the discussion of the disease–blood test example to be complete.

Example 8.7 Political Analysis   The voters in a particular state senatorial district are registered as 45% Republicans, 40% Democrats, and 15% independents. As it turns out, there are three candidates in the race, a Republican, a Democrat, and an independent. Post-mortem analysis of the election results indicates that the Republican candidate obtained the votes of 80% of the registered Republicans, 10% of the registered Demo-

FIGURE 8.15
Venn diagram of disease–blood test example.

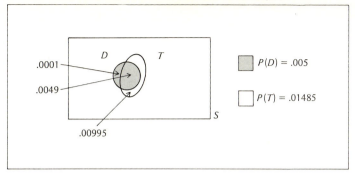

crats, and 15% of the registered independents. What percentage of those who voted Republican were registered Democrats?

SOLUTION   The first step is to specify the events involved in this problem. We have four basic events:

RR = the event that a voter is a registered Republican
RD = the event that a voter is a registered Democrat
RI = the event that a voter is a registered independent
VR = the event that a voter voted Republican

We want to know $P(RD \mid VR)$, the conditional probability of being a registered Democrat given that the voter voted Republican. The probabilities available from the data are:

$$P(RR) = .45 \qquad P(VR \mid RR) = .80$$
$$P(RD) = .40 \qquad P(VR \mid RD) = .10$$
$$P(RI) = .15 \qquad P(VR \mid RI) = .15$$

We apply an expanded version of Bayes' theorem, which is based on formulas (8.2) and (8.5). The expanded version asserts that

$$P(RD \mid VR) = \frac{P(VR \mid RD)P(RD)}{P(VR \mid RR)P(RR) + P(VR \mid RD)P(RD) + P(VR \mid RI)P(RI)}$$
$$= \frac{(.10)(.40)}{(.80)(.45) + (.10)(.40) + (.15)(.15)} = \frac{.04}{.4225} = .095$$

Therefore 9.5% of those who voted Republican were registered Democrats.

It would be interesting also to know how many of those who voted Republican were registered independents and, for that matter, how many were registered Republicans. To compute these, we write

$$P(RI \mid VR) = \frac{P(VR \mid RI)P(RI)}{P(VR \mid RR)P(RR) + P(VR \mid RD)P(RD) + P(VR \mid RI)P(RI)}$$

$$= \frac{(.15)(.15)}{(.80)(.45) + (.10)(.40) + (.15)(.15)} = \frac{.0225}{.4225}$$

$$= .053$$

and
$$P(RR \,|\, VR) = \frac{P(VR \,|\, RR)\, P(RR)}{P(VR \,|\, RR)P(RR) + P(VR \,|\, RD)P(RD) + P(VR \,|\, RI)P(RI)}$$

$$= \frac{(.80)(.45)}{(.80)(.45) + (.10)(.40) + (.15)(.15)} = \frac{.36}{.4225} = .852$$

Therefore, we conclude that of all those persons voting for the Republican candidate in this election 85.2% were registered Republicans, 9.5% were registered Democrats, and the remaining 5.3% were registered independents.

Before we present a final example illustrating the practical application of Bayes' theorem, let's briefly discuss the details of the expanded version of the theorem. We have a collection of n events, say $B_1, B_2, \ldots, B_n$ which are pairwise disjoint (no outcome is in more than one of the events) and together comprise the sample space S (each outcome is in at least one of these events); i.e., every outcome is in one and only one of the events $B_1, B_2, \ldots, B_n$. Now suppose some other event A occurs. What is the conditional probability of each B_j given A? Well, using formulas (8.2) and (8.5), we can write the expanded version as follows:

●
$$P(B_j \,|\, A) = \frac{P(B_j \cap A)}{P(A)} = \frac{P(A \cap B_j)}{\sum\limits_{k=1}^{n} P(A \cap B_k)} = \frac{P(A \,|\, B_j)P(B_j)}{\sum\limits_{k=1}^{n} P(A \,|\, B_k)P(B_k)} \tag{8.16}$$

Note that the denominator is the same for all the events B_j. Only the numerator changes along with B_j. We have taken

$$B_1 = RR$$
$$B_2 = RD$$
$$B_3 = RI$$
$$A = VR$$

Example 8.8 Weather Forecasting Long meteorological experience indicates that a particular geographic region under study has rain 20% of the time, snow 10% of the time, cloudiness 30% of the time, and fair weather 40% of the time during November. Looking back at the weather bureau's predictions and comparing the predictions with what had actually happened, we see that for rainy days the weather bureau had predicted rain 90% of the time and clouds 10% of the time. For those days on which it snowed, the prediction had been snow 80% of the time, clouds 10% of the time, and fair weather 10% of the time. For those days which turned out to be cloudy, the weather service had forecast clouds 60% of the time, rain 20%, snow 10%, and fair weather 10%. For those days which had fair weather, the prediction was fair weather 70% of the time, rain 10%, clouds 10%, and snow 10%. For tomorrow, a typical November day, the bureau forecasts fair weather. What is the probability that it will snow?

SOLUTION   The first problem is to define the events involved, and the second is to organize the data. We will be working with the following events:

R = the event that it rains tomorrow
S = the event that it snows tomorrow
C = the event that it is cloudy tomorrow
F = the event that it is fair tomorrow
PF = the event that the weather bureau predicts fair weather for tomorrow

We obtain the following probabilities from the data:

$$P(R) = .20 \qquad P(PF \mid R) = 0$$
$$P(S) = .10 \qquad P(PF \mid S) = .10$$
$$P(C) = .30 \qquad P(PF \mid C) = .10$$
$$P(F) = .40 \qquad P(PF \mid F) = .70$$

For example, $P(PF \mid C)$ is the conditional probability that the weather bureau predicted fair weather given that the day itself was cloudy; this probability is .10 because for cloudy days the bureau had forecast fair weather 10% of the time.

We want to compute $P(S \mid PF)$, the conditional probability that it will snow tomorrow given that the weather bureau predicts fair weather. By the expanded version (8.16) of Bayes' theorem, we get that

$$P(S \mid PF) = \frac{P(PF \mid S)P(S)}{P(PF \mid R)P(R) \ + P(PF \mid S)P(S) + P(PF \mid C)P(C) + P(PF \mid F)P(F)}$$

$$= \frac{(.10)(.10)}{(0)(.20) + (.10)(.10) + (.10)(.30) + (.70)(.40)}$$

$$= \frac{.01}{.32} = .03125 = 3.125\%$$

Therefore, if the bureau predicts fair weather, the chances are 3.125% that it will snow.

EXERCISES 8.F

1   An elementary school bond referendum is defeated at the polls in a school district where 30% of the voters have children in elementary school and 70% do not. It is estimated on the basis of residential areas that 90% of those with children in school voted for the bond issue, while only 20% of those without children in school did so. To find out whether or not it would be worthwhile for the school board to send explanatory literature home with the children, it is necessary to determine the composition of the group that voted "no." What percentage of those who voted against the bond issue did not have children in elementary school?

2   An insurance company issues three types of automobile insurance policies: Type G for good risks, Type M for moderate risks, and Type B for bad risks. Statistical studies show that according to the company's criteria licensed drivers are 20% Type G, 40% Type M, and 40% Type B. Accident statistics reveal that a Type G driver has

probability .01 of causing an accident in a 12-month period, a Type M driver has probability .02, and a Type B driver has probability .08. If a driver applying for insurance has caused an accident within the past 12 months,

a What is the probability he is a Type B risk?
b What is the probability he is a Type M risk?
c What is the probability he is a Type G risk?

3 On the average about 5% of the patients assigned to a state hospital which specializes in the treatment of respiratory diseases have tuberculosis. All persons entering this hospital are given chest x-rays, and past data indicate that 90% of those with tuberculosis have positive x-rays (substantiating presence of the disease), while only 2% of those without tuberculosis have positive x-rays. If a patient has a positive x-ray, what are the chances that he really has tuberculosis?

4 A national sociological survey indicates that 15% of all adults are college graduates, while 85% are not. Furthermore, 80% of all college graduates own their own homes, while only 30% of nongraduates are homeowners. What proportion of homeowners are college graduates?

5 Patients manifesting symptoms of certain psychological disorders are often tested for the Archimedes spiral aftereffect. Of the patients tested, accumulated data indicate that 40% have a functional illness, 30% are brain-damaged, and the other 30% have neither of these disorders. Of known functionally ill patients, 80% have the aftereffect when tested, while only 30% of brain-damaged patients and 10% of the others have the aftereffect. If a patient has the aftereffect, what is the probability that he is functionally ill?

6 If interest rates are headed downward, the probability is 70% that the stock market will go up. If interest rates are likely to go up, the chances are 80% that the market will go down. If interest rates are remaining stable, the chances are 60% that the market will go up. Past history indicates that interest rates go down 30% of the time, go up 20% of the time, and remain stable 50% of the time. If the stock market rises today, what are the prospects for a lower interest rate?

7 Of those who voted in a recent election, 10% were college graduates, 50% were high-school graduates, and the remaining 40% were grade-school graduates. Alphonse, the victorious candidate, received 80% of the college-graduate vote, 70% of the high-school-graduate vote, and 40% of the grade-school-graduate vote. What percentage of those who voted for Alphonse were college graduates?

8 Of used car and major appliance buyers, 80% are good credit risks. Of good credit risks, 70% have charge accounts at at least one department store, while 40% of bad credit risks have charge accounts. If a prospective buyer has a charge account at a major department store, what is the probability that she is a bad credit risk anyway?

SUMMARY AND DISCUSSION

In Chap. 8, we returned to probability theory. Our direct concern has been the updating of probabilities of events, taking account of newly available information. We saw that the presence of new information represents a change in circumstances which can increase or decrease or have no effect on probabilities of particular events. If the new information does not affect the probability of an event in which we are interested, we say that our event is independent of the new information. Independent events have special properties, and we devoted a full section to a discussion of them. We then used the properties of independent events to generate the entries of Table A.2, the table of the binomial distribution. Finally, we presented a complete development of Bayes' theorem, an extremely useful method of describing the extent of interrelationships between events. Bayes' theorem, which illuminates the effect of one event upon the probability of another, combines all aspects of the theory of conditional probability into a method of wide applicability.

BIBLIOGRAPHY

General Discussions of Probability
Fine, T. L.: "Theories of Probability: An Examination of Foundations," Academic Press, New York, 1973.

Maistrov, L. E. (S. Kotz, tr.): "Probability Theory: An Historical Sketch," Academic Press, New York, 1974.

Weaver, W.: "Lady Luck," Doubleday, New York, 1963.

Independent Events
Huff, B. W.: Another Definition of Independence, *Mathematics Magazine*, September 1971, pp. 196–197.

Bayes' Theorem
Aitchison, J.: A Geometrical Version of Bayes' Theorem, *The American Statistician*, December 1971, pp. 45–46.

Morgan, B. W.: "An Introduction to Bayesian Statistical Inference," Prentice-Hall, Englewood Cliffs, N.J., 1968.

SUPPLEMENTARY EXERCISES

1 Decide whether each of the following statements is always true or sometimes false:
   a $P(A) + P(A^c) = 1$
   b $P(A \mid B) + P(A^c \mid B) = 1$
   c $P(A \mid B) + P(A \mid B^c) = 1$
   d $P(A \mid B) + P(A^c \mid B^c) = 1$

2 Anode Electronics, Betatron Atomics, and Cobalt Combinatorics are working feverishly but independently of each other to develop a foolproof nuclear test detector. Their respective probabilities of success within 5 years are estimated as 0.4, 0.4, and 0.3. If those estimates are correct, what is the probability that at least one of them successfully develops the detector within 5 years?

3 You are offered the choice of one of the following two games to play:
    a You roll six fair dice, and you win if you roll at least one 6.
    b You roll twelve fair dice, and you win if you roll at least two 6s.
    Which game gives you the greater probability of winning?

4 You are offered the choice of one of the following two games to play:
    a You roll a fair die 4 times, and you win if you roll at least one 6.
    b You roll a pair of fair dice 24 times, and you win if you roll at least one 12.
    Which game gives you the greater probability of winning?

5 A national sociological survey indicates that 15% of all adults are college graduates, while 85% are not. Furthermore, 80% of all college graduates own their own homes, while only 30% of nongraduates are homeowners. What percentage of adults own their own homes?

6 If we roll a fair tetrahedron (a four-sided, rather than a six-sided, die) once, the sample space for the experiment is $S = \{1, 2, 3, 4\}$ and all four outcomes are equally likely. Consider the events

$$A = \{1, 2\}$$
$$B = \{1, 3\}$$
$$C = \{1, 4\}$$

    a Show that A, B, and C are pairwise independent, namely, that A is independent of B, B is independent of C, and C is independent of A.
    b Show, however, that A, B, and C do not constitute a set of independent events. Accomplish this by showing that $P(A \cap B \cap C)$ is not equal to $P(A)P(B)P(C)$.
    c Show that A is not independent of the event $B \cap C$.

7 If A and B are events such that $P(A) = 1/2$, $P(B) = 2/3$, and $P(A|B) = 1/4$, then calculate
    a $P(A^c|B)$
    b $P(A|B^c)$
    c $P(B|A)$
    d $P(A^c|B^c)$

8 A major corporation has several hundred projects in the initial stages of development. These projects are classified into three groups: Group I, for those having a 90% chance of eventually turning a profit, Group II, for those having a 70% chance of turning a profit, and Group III, for those having a 50% chance of turning a profit. Of the corporation's projects, 30% are in Group I, 50% are in Group II, and 20% are in Group III. What percentage of the corporation's projects will eventually turn a profit?

9 Three years out of every 10, one southern state suffers both coastal hurricanes and inland flooding. Four years out of every 10, it suffers coastal hurricanes but no inland flooding.
    a What percentage of years does the area suffer coastal hurricanes?
    b For this year, the weather service has forecast coastal hurricanes. What is the probability that inland flooding will also occur?

10 The U.S. House of Representatives has 435 members. How many different samples of 25 members each are available to a political scientist interested in sampling members' opinions?

11 How many five-letter "words,"[1] with no letter appearing more than once, can be constructed from the English alphabet?

12 As a result of a sociological study of family educational level, it is discovered that 35% of today's college students or graduates aged 18 to 25 have at least one parent with a college degree. Of those in the age group who have never attended college, only 10% have at least one parent with a degree. Of those aged 18 to 25, 30% are college students or graduates. What are the chances that a person aged 18 to 25 will be a college student or graduate if neither parent has a college degree?

13 About 6% of hospital patients exhibiting a particular combination of symptoms actually have cancer. Three research scientists propose a test for cancer which they claim has the following detection rates: If a patient has cancer, the test will say so with probability .95, while if the patient does not have cancer, the test will say *that* with probability .95. Assuming that those claimed detection rates are the correct ones, what is the probability that a patient tested has cancer if the test indicates that he does?

14 In a recent election, the local Democratic candidate received 50% of the vote, the Republican candidate got 40%, while an independent candidate obtained 10%. At the same time, ballot proposition 23 received the votes of 20% of those who voted Democratic, 90% of those who voted Republican, and 90% of those who voted independent.
a What proportion of those favoring proposition 23 also voted Republican?
b What proportion of the voters favored proposition 23?

15 A stop-smoking clinic advertises that in a test region last August 80% of those who tried and were able to stop smoking had participated in its program, while only 30% of those who tried and failed to stop smoking had participated. Research by the Consumer Fraud Division supports these assertions but also reveals that of all smokers in the region who tried to "kick the habit" last August, 90% had failed, while only 10% succeeded. What proportion of the clinic's customers actually were able to stop smoking?

16 A psychological test is designed to separate entering college freshmen into good prospects and not-so-good prospects. Among those who later performed satisfactorily during the year, 80% had passed the test. Among the students who did unsatisfactory work in their first year, only 40% had passed the test. On the whole, 70% of the students tested did satisfactory work that year.
a What percentage of entering freshmen passed the psychological test?
b If a student passed the test, what is the probability that she went on to do satisfactory work?

17 The noxious oxides of nitrogen comprise 20% of all pollutants in the air by weight in a certain metropolitan area. Automobile exhaust accounts for 70% of those noxious oxides but only 10% of all other pollutants in the air. Of the pollution contributed by automobile exhaust, what percentage are noxious oxides of nitrogen?

[1]Of course, not all of these are real words. Consider "vlopt" for example.

18 By his own admission, an unusually candid student knew only 50% of the material covered in his geography class. The final exam was a multiple-choice test having four possibilities for the answer to each question, and it covered in great detail all aspects of the course. On the half of the test covering material the student knew, he naturally got all the answers right; on the other half, he guessed one of the four options.

a What proportion of the questions did the student answer correctly?

b If he answered question 473 correctly, what is the probability that he really knew the answer to it?

19 In a recent small war, one side used both heat-seeking missiles and laser-guided rockets to shoot down enemy aircraft. In fact, the objects launched at enemy planes were 80% of the heat-seeking type and 20% of the laser-guided variety. The heat-seeking missiles struck their targets 40% of the time, while the laser-guided rockets scored hits 90% of the time. What percentage of enemy aircraft shot down were struck by heat-seeking missiles and what percentage by laser-guided rockets?

20 An independent supermarket receives 60% of its fruit from Foster's Fruit Fields and the other 40% from Gregory's Giant Groves. On the average, about 25% of Foster's shipments arrive overripe, while only 16% of Gregory's do. Because the farmers issue credits when their fruit arrives overripe, the market keeps strict records on the subject. What percentage of the market's overripe fruit comes from Foster, and what percentage from Gregory?

21 A test for a certain rare disease is capable of detecting the disease in 97% of all afflicted individuals. However, when entirely healthy persons are tested, 5% of them are incorrectly diagnosed as having the disease; and when persons who have other, milder diseases are tested, 10% of them are incorrectly diagnosed as having the rare disease. In reality, only about 1% of the population has the rare disease; 3% have other, milder diseases; and the remaining 96% are entirely healthy.

a What proportion of the population is diagnosed by the test as having the rare disease?

b If a person is diagnosed as having the rare disease, what is the probability that he really has it?

22 A dealer in used and rebuilt appliances has the opportunity to purchase three refrigerators at an exceptionally good price. There is a chance, however, that one or more of the refrigerators might require extensive repairs which would wipe out anything gained by the low purchase price. In situations like this, experience has shown the dealer that the chances are about 1/24 that none of the three will really be in acceptable condition, about 1/8 that only one will be acceptable, about 1/3 that two will be acceptable, and about 1/2 that all three will be acceptable. Because of the limitations of time and cost, the dealer decides to pick one at random (with each being equally likely to be picked) out of the group of three and to subject it to a thorough inspection. If that one turns out to be acceptable, he agrees to buy all three. If he does decide to buy all three (after testing one), what are the chances that they will all be in acceptable condition?

IX

TESTING FOR INDEPENDENCE AND MEASURING DEPENDENCE

One of the major objectives of the statistical analysis of a set of data is to discover and verify relationships existing between components of real-life situations. The search for and measurement of relationships were our primary goals in our earlier study of z scores, two-sample t and z tests, regression and correlation, and conditional probability. The present chapter continues this development by focusing on yet other aspects of the problem. In this chapter, we first look at a set of cross-classified data with a view toward determining whether two characteristics are independent of each other, and, if not, how dependent on each other they are. We then study a set of data with the objective of discovering the probability distribution of the underlying population. The major statistical method for analyzing both problems is a technique called the chi-square test.

As an example of the situations we will deal with in this chapter, consider the following problem: A survey was conducted to evaluate the relative effectiveness of two allergy remedies which had been administered in a small community. The treatments were provided in the spring, free of charge to those wishing to avail themselves of them. Some people received the first treatment, an allergy shot, while others preferred the second treatment, an allergy capsule, and the remaining persons received no treatment.

A random sample of 1000 local inhabitants the following fall yielded the results presented in Table 9.1. The figures in Table 9.1 are interpreted as follows: Of the 1000 people surveyed, 350 had received no treatment. Of those 350, 44 had manifested severe allergy symptoms while the remaining 306 had not. Of those in the sample, 150 had taken the allergy capsule treatment, and of these, 19 had severe symptoms while 131 escaped them. Finally, 500 persons out of the 1000 had been treated with an allergy shot, with 37 reporting severe symptoms and 463 managing to avoid them.

The main questions we will try to answer in this chapter are the following two:

1 Is the successful avoidance of severe allergy symptoms independent of the receipt of treatment?
2 If not, to what extent does a successful avoidance of severe symptoms depend on receipt of treatment?

In effect, what we are trying to do by answering these questions is to measure the effectiveness of the treatment in controlling the symptoms.

For the theoretical structure upon which the practical techniques of this chapter are built we turn to Sec. 8.D. Two events A and B are called independent events if $P(A \mid B) = P(A)$, or, equivalently, $P(B \mid A) = P(B)$. The definition means that the occurrence of one of the events does not affect the probability of the other. According to equation (8.8), events A and B are independent if and only if $P(A \cap B) = P(A)P(B)$. There is therefore only one way for the events to be independent, and that way is to have $P(A \cap B)$ equal to $P(A)P(B)$. Otherwise, the events are said to be dependent. Unfortunately, dependent is a less precise word than independent. If the events are independent, then $P(A \mid B)$ *must* be the same as $P(A)$, but if the events are dependent, $P(A \mid B)$ must be different from $P(A)$.

But there are many ways of being different, and only one way of being the same. Our first goal in this chapter will be to develop a statistical test that will tell us (on the basis of data displayed in a "contingency table" of the sort illustrated in Tabie 9.1) when two characteristics are independent and when they are dependent. In cases that turn out to be independent, we are done, for we have established the lack of any relationship between the two. In those cases that turn out to exhibit a dependence relationship, however, the problem remains of trying to find what sort of dependence relationship

TABLE 9.1
Contingency Table of Allergy Treatment Data

Treatment Record \ Allergy Record	Severe Symptoms	Mild or No Symptoms	Row Totals
No treatment	44	306	350
Allergy capsule	19	131	150
Allergy shot	37	463	500
Column totals	100	900	1000

exists. Our second goal, then, will be to introduce a method of measuring the direction and extent of dependence relationships.

The third section of this chapter develops a method of analyzing a set of data points with a view toward determining its probability distribution. Our interest will be aimed primarily at testing a set of data to see whether it has the binomial or normal distribution.

SECTION 9.A THE CHI-SQUARE TEST OF INDEPENDENCE

We shall now apply the chi-square test of independence to answer question 1 above, whether successful avoidance of severe allergy symptoms is independent of the receipt of treatment. The basic principle underlying the chi-square test goes as follows: We calculate a hypothetical contingency table, usually called the "expected contingency table," showing how the data would have turned out if avoidance of severe symptoms was independent of receipt of treatment. Then we compare this expected contingency table with the actual contingency table showing how the data really turned out. If the actual table looks very similar to the expected table, we would conclude that the two characteristics (namely, "avoidance of severe symptoms" and "receipt of treatment") are probably independent. If, on the other hand, the actual table looks quite different from the expected table, we would have to say that the two characteristics are probably not independent.

Two questions immediately present themselves. The first is: How do we calculate the expected contingency table? The second is: How different must the expected and actual contingency tables be in order to convince us that the characteristics represented are really not independent? We will soon answer the first by showing how to calculate the expected contingency table. We then will answer the second by introducing the chi-square statistic as a measure of the difference between the two contingency tables, and we will reject the hypothesis of independence if the chi-square value exceeds a corresponding number in the chi-square table (Table A.6). The chi-square statistic measures the difference between two contingency tables in a manner analogous to the way the two-sample t-statistic measures the difference between two sample means.

We now proceed to the calculation of the expected contingency table of the allergy treatment data of Table 9.1. We are dealing with the following five events:

$$NT = \text{the event that a person received no treatment}$$
$$AC = \text{the event that a person received an allergy capsule}$$
$$AS = \text{the event that a person received an allergy shot}$$
$$SS = \text{the event that a person reported severe symptoms}$$
$$MS = \text{the event that a person reported mild or no symptoms}$$

From Table 9.1, we can find the probabilities of these events:

$P(NT) = \dfrac{350}{1000}$, because 350 out of the 1000 received no treatment

$P(AC) = \dfrac{150}{1000}$, because 150 out of the 1000 received the capsule

$P(AS) = \dfrac{500}{1000}$, because 500 out of the 1000 received the shot

$P(SS) = \dfrac{100}{1000}$, because 100 out of the 1000 reported severe symptoms

$P(MS) = \dfrac{900}{1000}$, because 900 out of the 1000 reported mild or no symptoms

From the actual contingency table on the left side of Table 9.2, we see that the event $NT \cap SS$ consists of 44 persons. This means that there were exactly 44 persons who both received no treatment and also exhibited severe symptoms. Let's now compute the expected number of persons in the event $NT \cap SS$ in the case of independence. If NT and SS are independent events, then from Eq. (8.8) we know that the proportion of persons in $NT \cap SS$ is given by

$$P(NT \cap SS) = P(NT)P(SS) = \left(\frac{350}{1000}\right)\left(\frac{100}{1000}\right)$$

$$= \frac{35,000}{1,000,000} = \frac{35}{1000}$$

Therefore, if 35/1000 of the persons involved in the experiment are expected to be in the event $NT \cap SS$, this means that 35/1000 of the 1000 persons, namely,

$$\frac{35}{1000} \times 1000 = 35$$

persons are expected to be in $NT \cap SS$. This argument accounts for the number 35 in the upper left corner of the expected contingency table of Table 9.2.

We can streamline the above calculation by noting in advance that the number of persons in $NT \cap SS$ will be $P(NT \cap SS) \times 1000$. Therefore, we can run the entire calculation directly as follows:

$$P(NT \cap SS) \times 1000 = P(NT)P(SS)(1000)$$

$$= \left(\frac{350}{1000}\right)\left(\frac{100}{1000}\right)(1000)$$

$$= \frac{350 \times 100}{1000} = 35$$

because of the cancellation of the last 1000 with one of those in the denominator.

We now proceed to calculate the expected number of persons in $AC \cap SS$ under the assumption of independence. We have

$$P(AC \cap SS) \times 1000 = P(AC)P(SS)(1000)$$

$$= \left(\frac{150}{1000}\right)\left(\frac{100}{1000}\right)(1000)$$

$$= \frac{150 \times 100}{1000} = 15$$

	Actual Contingency Table			Expected Contingency Table		
	SS	MS	Row Σ		SS	MS	Row Σ
NT	44	306	350	NT	35	315	350
AC	19	131	150	AC	15	135	150
AS	37	463	500	AS	50	450	500
Column Σ	100	900	1000	Column Σ	100	900	1000

This calculation explains the appearance of the 15 in the expected contingency table of Table 9.2 at the intersection of the row headed AC and the column headed SS.

The remaining cells of the expected contingency table are filled as follows:

$$P(AS \cap SS) \times 1000 = P(AS)P(SS)(1000)$$
$$= \left(\frac{500}{1000}\right)\left(\frac{100}{1000}\right)(1000)$$
$$= \frac{500 \times 100}{1000} = 50$$

$$P(NT \cap MS) \times 1000 = P(NT)P(MS)(1000)$$
$$= \left(\frac{350}{1000}\right)\left(\frac{900}{1000}\right)(1000)$$
$$= \frac{350 \times 900}{1000} = 315$$

$$P(AC \cap MS) \times 1000 = P(AC)P(MS)(1000)$$
$$= \left(\frac{150}{1000}\right)\left(\frac{900}{1000}\right)(1000)$$
$$= \frac{150 \times 900}{1000} = 135$$

$$P(AS \cap MS) \times 1000 = P(AS)P(MS)(1000)$$
$$= \left(\frac{500}{1000}\right)\left(\frac{900}{1000}\right)(1000)$$
$$= \frac{500 \times 900}{1000} = 450$$

For several reasons, including the ability to have a check on the calculation, it is useful to notice that the row and column totals in the expected contingency table are exactly the same as those in the actual contingency table. For example, in Table 9.2, the top rows in each of the contingency tables sum to 350, as

$$44 + 306 + 350$$

and

$$35 + 315 = 350$$

This indicates that the process leading to the expected contingency table merely rearranges the survey results to make them conform with the requirements of independent events.

Because the row and column sums must be the same in both contingency tables, it is possible to get away with computing only two of the six "expected frequencies." This is, in fact, the way the "pro's" (professional statisticians, of course) do it. In Table 9.3, for example, only the 35 and 15 have been calculated, using the formula for independent events. The remaining four cells have been filled by subtracting those two numbers from the row and column totals. For example,

$$350 - 35 = 315$$
$$150 - 15 = 135$$
$$100 - 35 - 15 = 50$$
$$500 - 50 = 450$$

The hand calculation of the components of the expected contingency table can be considerably simplified by means of this procedure. The number of cells for which the entire computation must be carried out is called the "degrees of freedom" of the contingency table. To find the degrees of freedom, we use the formula

●

$$df = (r - 1)(c - 1) \qquad (9.1)$$

where
r = number of rows in the table
c = number of columns in the table

The formula arises from the fact that it would not be necessary to fill directly those cells belonging to the last row or the last column, for these could be obtained indirectly by subtraction. Therefore, the block of cells to be filled by computation consists of one less row and one less column than the original contingency table, and so contains $(r - 1)(c - 1)$ boxes. The contingency table for the allergy treatment data has three rows

TABLE 9.3
Expected Frequency Computations

	SS	MS	Row Σ
NT	35	**315**	350
AC	15	**135**	150
AS	**50**	**450**	500
Column Σ	100	900	1000

$$\frac{350 \times 100}{1000} = 35$$

$$\frac{150 \times 100}{1000} = 15$$

and two columns, so that $r = 3$ and $c = 2$. It follows that $df = (3 - 1)(2 - 1) = (2)(1) = 2$, so that it is necessary to fill directly only two[1] of the six boxes.

To decide the question of the dependence of the severity of the symptoms on the nature of the treatment, it remains only to measure the difference between the actual and expected contingency tables of Table 9.2. We do this by means of the chi-square statistic. If we denote by f_k the number (actual frequency) appearing in the kth cell of the actual contingency table and by e_k the number (expected frequency) appearing in the corresponding cell of the expected contingency table, then the chi-square statistic is given by the formula

●
$$\chi^2 = \sum_{k=1}^{n} \frac{(f_k - e_k)^2}{e_k} \tag{9.2}$$

where $n = rc$ is the number of boxes in each table. If the classifications of treatment and severity of symptoms are truly independent, this fact would be indicated by a situation where each e_k was relatively close to its corresponding f_k. Therefore, in case of independence, the numerical value of χ^2 would be small. It follows that a large value of χ^2 would tend to deny independence and so would indicate a significant dependence relationship between treatment and severity of symptoms. What value of χ^2 would be the cutoff point between large and small, between an indication of dependence and a confirmation of independence? The cutoff point for a significance level of α in testing

$$H : \text{Independence}$$

versus

$$A : \text{Dependence}$$

would be $\chi_\alpha^2(df)$, a number to be found in Appendix Table A.6. That is, we would reject H (independence) in favor of A (dependence) at level α if $\chi^2 > \chi_\alpha^2(df)$.

We now carry out the chi-square test of independence for the allergy treatment situation. Using a significance level of $\alpha = .05$ and recalling that $df = (r - 1)(c - 1) = 2$ in this case, we see from Table A.6 that $\chi_{.05}^2(2) = 5.991$. Therefore we will reject H at level $\alpha = .05$ and conclude that the severity of symptoms depends on the treatment only if $\chi^2 > 5.991$. We now proceed to the computation of χ^2. The calculations in Table 9.4 implement the formula for χ^2 above. The columns headed f_k and e_k are filled by the numbers appearing in the proper cells of the contingency tables of Table 9.2. From Table 9.4, it follows that

$$\chi^2 = \sum_{k=1}^{n} \frac{(f_k - e_k)^2}{e_k}$$
$$= \frac{(44 - 35)^2}{35} + \frac{(19 - 15)^2}{15} + \frac{(37 - 50)^2}{50} + \frac{(305 - 315)^2}{315}$$
$$+ \frac{(131 - 135)^2}{135} + \frac{(463 - 450)^2}{450}$$

[1]Not any two at all, but two chosen strategically!

TABLE 9.4
Calculation of χ^2 for the Allergy Treatment Data

Cell	f_k	e_k	$f_k - e_k$	$(f_k - e_k)^2$	$\dfrac{(f_k - e_k)^2}{e_k}$
$NT \cap SS$	44	35	9	81	2.3143
$AC \cap SS$	19	15	4	16	1.0667
$AS \cap SS$	37	50	-13	169	3.3800
$NT \cap MS$	306	315	-9	81	.2571
$AC \cap MS$	131	135	-4	16	.1185
$AS \cap MS$	463	450	13	169	.3756
Sums	1000	1000	0		$\chi^2 = 7.5122$

$$= 2.3143 + 1.0667 + 3.3800 + .2571 + .1185 + .3756$$
$$= 7.5122$$

Therefore $\chi^2 = 7.512 > 5.991$, and so we reject H (independence) at level $\alpha = .05$. We conclude that, according to the data, there seems to be a dependence relationship between the type of treatment and the severity of the symptoms.

As a further remark on this particular example, we compare the actual and expected contingency tables in Table 9.2 seeking the reason behind the rejection of the hypothesis of independence. One of the points we notice is that under independence we would have expected 50 persons who had received allergy shots to have exhibited severe allergy symptoms. As shown by the actual data, however, only 37 individuals of that category reported such symptoms. On the other hand, more than the expected number of persons in the remaining two treatment categories showed severe symptoms. This analysis tends to indicate that the allergy shot was a more effective treatment than the allergy capsule and than the lack of any treatment. Therefore, the dependence relationship whose existence is asserted by the chi-square test seems to imply that the allergy shot treatment reduces the severity of symptoms, while the allergy capsule and the lack of treatment do not.

As a second application of the chi-square test of independence, consider Example 9.1.

Example 9.1   Effectiveness of a New Drug   Suppose that the U-Feel Better (UFB) Pharmaceutical Company claims that it has developed a drug which can ease mental depression. To check on the claim, the Food and Drug Administration (FDA) sets up a test involving 500 persons with mental depression. A random sample of 300 of them (the experimental group) are treated with the new drug, while the remaining 200 (the control group) are given only sugar pills. All 500 are given the impression that they are receiving the new treatment. All 500 persons are later asked whether the treatment they received made them feel better, made them feel worse, or had no effect. The resulting data are presented in Table 9.5.

TABLE 9.5
Contingency Table of Mental Depression Drug Data

Reaction / Treatment	Felt Better	No Effect	Felt Worse	Row Totals
New drug	50	240	10	300
Sugar pill	25	160	15	200
Column totals	75	400	25	500

SOLUTION   The statistician for the FDA considers $\alpha = .05$ to be a reasonable level of significance in testing

$$H : \text{Independence (the drug is not effective)}$$

versus

$$A : \text{Dependence (the drug may be effective)}$$

Here the hypothesis of independence means that a person's reaction does not depend on whether he was treated with the drug or with a sugar pill. If the treatment made no difference, we would be led to believe that the new drug was no more effective than a sugar pill and therefore not effective. Using the contingency table of Table 9.5, the statistician further observes that the degrees of freedom are

$$df = (r - 1)(c - 1) = (2 - 1)(3 - 1) = (1)(2) = 2$$

because there are two rows and three columns in the table. The chi-square test of independence, then, will have us reject H in favor of A at level $\alpha = .05$ if $\chi^2 > \chi_{.05}^2(2) = 5.991$. That number can be found in Appendix Table A.6. We now proceed to the calculation of χ^2.

The following events will be involved in the computational aspects of this problem:

$$ND = \text{the event that a person was treated with the new drug}$$
$$SP = \text{the event that a person was treated with a sugar pill}$$
$$FB = \text{the event that a person felt better after treatment}$$
$$NE = \text{the event that a person had no effect from the treatment}$$
$$FW = \text{the event that a person felt worse after treatment}$$

We calculate the expected contingency table in the quickest way possible and present the results in Table 9.6. We have under the assumption of independence that

$$P(ND \cap FB) \times 500 = \frac{300 \times 75}{500} = 45$$

$$P(ND \cap NE) \times 500 = \frac{300 \times 400}{500} = 240$$

TABLE 9.6
Actual and Expected Contingency Tables of
Mental Depression Drug Data

Actual Contingency Table					Expected Contingency Table				
	FB	NE	FW	Row Σ		FB	NE	FW	Row Σ
ND	50	240	10	300	ND	45	240	15	300
SP	25	160	15	200	SP	30	160	10	200
Column Σ	75	400	25	500	Column Σ	75	400	25	500

and we fill the remaining cells by subtracting these numbers from the row and column totals where called for. On the basis of the information presented in Table 9.6, we calculate the numerical value of χ^2, using the procedure outlined in Table 9.7.

Table 9.7 gives us the result that $\chi^2 = 5.556$, and it follows that $\chi^2 = 5.556 < 5.991$, so we cannot reject the hypothesis of independence at level $\alpha = .05$. The results of the FDA, then, tend to refute the company's claim that their new drug relieves mental depression.

At this point in the discussion, the statistician for the pharmaceutical company challenges the conclusions of the FDA analysis. She suggests instead that all persons feeling no effect be eliminated from further consideration, and only those persons indicating a change after the treatment be included in the computations. Such a procedure is common, especially in advertising, where results of a survey are presented as being based on all those responding to the question, or all those expressing an opinion, or all those indicating a preference, or some similar criterion for throwing away an often substantial portion of the data.

The company's statistician anchors her analysis on the contingency table illustrated in Table 9.8. This contingency table has $r = 2$ rows and $c = 2$ columns, so that the number of degrees of freedom is

$$df = (r - 1)(c - 1) = (2 - 1)(2 - 1) = 1$$

TABLE 9.7
Calculation of χ^2 for the Mental Depression Drug Data

Cell	f_k	e_k	$f_k - e_k$	$(f_k - e_k)^2$	$\dfrac{(f_k - e_k)^2}{e_k}$
ND ∩ FB	50	45	5	25	.5556
SP ∩ FB	25	30	−5	25	.8333
ND ∩ NE	240	240	0	0	.0000
SP ∩ NE	160	160	0	0	.0000
ND ∩ FW	10	15	−5	25	1.6667
SP ∩ FW	15	10	5	25	2.5000
Sums	500	500	0		$\chi^2 = 5.5556$

TABLE 9.8
The Company's View of Mental Depression Drug Data

Reaction / Treatment	Felt Better	Felt Worse	Row Totals
New drug	50	10	60
Sugar pill	25	15	40
Column totals	75	25	100

In the company view then, the hypothesis H (independence) should be rejected at level $\alpha = .05$ in favor of A (dependence) if χ^2 turns out to be larger than $\chi^2_{.05}(1) = 3.841$, as available in Appendix Table A.6. In Table 9.9, we compare the actual and expected contingency tables based on the pharmaceutical company's view of the mental depression drug data.

The expected contingency table at the right side of Table 9.9 gives an especially clear illustration of the way in which such tables represent the independence of the two classifications, treatment and reaction to treatment. We observe first that 75 of the 100 persons under study (3 out of every 4) have reported that they felt better, instead of worse, after "treatment." If reaction to treatment is truly independent of which treatment was received, this same proportion of persons feeling better (namely, 3 out of every 4) should be observed in each of the treatment groups. Therefore, of the 60 given the new drug, 45 should have reported feeling better, while of the 40 given sugar pills, 30 should have reported feeling better. The expected contingency table reflects these proportions exactly. The extent to which figures in the actual contingency table differ from them specifies the extent to which treatment and reaction are not independent. In Table 9.10, we calculate the numerical value of χ^2.

According to Table 9.10, $\chi^2 = 5.556$. Because $\chi^2 = 5.556 > 3.841$, we are led to reject the hypothesis of independence at significance level $\alpha = .05$. Based on the company's view of the data, then, we would conclude that reaction does depend on treatment. The fact that more persons (50) actually felt better than the number of persons (45) that would be expected to feel better in case of independence tends to indicate that the new drug is somewhat effective in relieving mental depression.

TABLE 9.9
Actual and Expected Contingency Tables of Mental Depression Drug Data Based on the Company's View of the Statistics

Actual Contingency Table				Expected Contingency Table			
	FB	FW	Row Σ		FB	FW	Row Σ
ND	50	10	60	ND	45	15	60
SP	25	15	40	SP	30	10	40
Column Σ	75	25	100	Column Σ	75	25	100

TABLE 9.10

The Company's Calculation of χ^2 for the Mental Depression Drug Data

Cell	f_k	e_k	$f_k - e_k$	$(f_k - e_k)^2$	$\dfrac{(f_k - e_k)^2}{e_k}$
$ND \cap FB$	50	45	5	25	.5556
$SP \cap FB$	25	30	−5	25	.8333
$ND \cap FW$	10	15	−5	25	1.6667
$SP \cap FW$	15	10	5	25	2.5000
Sums	100	100	0		$\chi^2 = 5.5556$

What lessons can be drawn from comparing these two competing analyses of the mental depression drug data? When all 500 persons participating in the survey were included in the statistical analysis, the χ^2 value of 5.556 did not permit us to reject the hypothesis of independence at level $\alpha = .05$, and so we had to conclude that there was no evidence to substantiate the effectiveness of the new drug.

On a strictly mathematical level, removing the 400 persons and the two cells $ND \cap NE$ and $SP \cap NE$ from the contingency tables resulted in a lowering of the degrees of freedom from 2 to 1. This caused the rejection value χ_α^2 to drop from 5.991 to 3.841. Therefore, as 5.556 does not exceed 5.991 but does exceed 3.841, the decisions are not to reject independence in the first case, but to reject it in the second case. On this basis of purely practical considerations, the results of the mathematical analysis make good sense. If 400 of the 500 persons, including 240 of the 300 given the drug, a substantial majority of those participating in the survey, reported that the treatment they were given had no effect, then it cannot be considered likely that the new drug is really effective. Removing these 400 persons from consideration is equivalent to ignoring the fact that so many of those given the new drug were unaffected by it. After ignoring such a substantial percentage of those participating in the study, we should not be surprised at any change of conclusion because whatever result happened to emerge would lack credibility. The procedure suggested by the company's statistician, then, is invalid because it excludes 80% of the data from consideration and bases its decision on a selected 20% of the data.

One final technical comment: It is a requirement of the proper usage of the chi-square test that there be a sufficiently large number of data to guarantee that each expected frequency (e_k) be 5 or more. The specification that $e_k \geq 5$ is to be considered as an empirical rule of thumb, although it is supported by theoretical facts. Naturally, the more data there are available, the more reliable our statistical analysis will be. The requirement that $e_k \geq 5$ in every case merely puts some control on the number of data points considered acceptable.

EXERCISES 9.A

1 The following data record the positions of 400 members of the U.S. House of Representatives on repeal of a certain farm law:

Position / Party	For Repeal	Against Repeal
Republican	140	10
Democratic	100	150

At significance level $\alpha = .01$, do the data indicate that there is a dependence relationship between party affiliation and position on repeal of the law?

2 After a proposal was introduced in the city council to set up a car-free area in the downtown business district, a local newspaper surveyed 500 persons who regularly made use of that section of town, either for business or pleasure purposes. The objective of the survey was to find out whether persons who used their cars for transportation downtown felt differently about the proposal than those who ordinarily used public transportation. Each respondent was classified according to his or her usual method of transportation downtown and his or her opinion on the proposal for a car-free area. The results of the survey are as follows:

	Method of Transportation	
	Public Bus	Private Car
Favor proposal	200	100
Oppose proposal	100	100

At level $\alpha = .05$, can we conclude that a person's opinion of the proposal depends, to some extent, on his or her method of transportation downtown?

3 An economist conducted a study of the effect of the presence of local water resources on local vegetable prices in 1000 localities west of the Mississippi River. The resulting data presented below relate a locality's availability of water with average retail vegetable prices in that locality.

Water Resources	Retail Vegetable Prices		
	Low	Moderate	High
Scarce	10	70	120
Satisfactory	40	400	60
Abundant	150	130	20

Are we justified in asserting at level $\alpha = .05$ that local vegetable prices depend to a significant extent on local water resources?

4 One psychologist has studied the incidence of grimacing and rheumatic disease in schizophrenic patients. In a study of 2000 such patients, he collected the following data:

Grimacing / Rheumatic Disease	Present	Not Present
Present	40	60
Not Present	360	1540

At level $\alpha = .05$, does there seem to be a dependence between grimacing and rheumatic disease in schizophrenic patients?

5 A psychologist wants to know whether the fact of being male or female affects the ability to break the smoking habit. To answer the question, she selects a random sample of 70 men and 30 women from lists of participants in stop-smoking clinics and obtains the following data:

Status	Sex	
	Male	Female
Still not smoking	45	15
Returned to smoking	25	15

At significance level $\alpha = .01$, can the psychologist assert that ability to stop smoking is independent of sex?

6 After an epidemic of a certain strain of flu, 1000 schoolchildren were examined for aftereffects. It turned out that 600 of the children had received medical treatment for the flu, while the other 400 had their cases unattended. The following data relates the occurrence of aftereffects with the receipt of medical treatment:

Treatment	Aftereffects	
	Noticeable	Not Noticeable
Was treated	225	375
Was not treated	175	225

At level $\alpha = .01$, can it be asserted that the treatment was of significant value in reducing aftereffects of the disease?

7 In a behavioral study of rats, each of 200 rats was classified by two characteristics: tail length and social behavior. The results appear in the following table:

Social Behavior	Tail Length		
	Short	Medium	Long
Aggressive	14	32	14
Normal	27	45	28
Submissive	9	23	8

Do the results of the study indicate at significance level $\alpha = .05$ that social behavior of rats depends on tail length?

8 A study was recently conducted within the Los Angeles school system aimed at determining whether or not an elementary pupil's ability to do mathematics was dependent on the teacher's ability to do algebra. One thousand students were selected at random from the system, and each was classified by his or her own ability to do elementary school mathematics and by the ability of his or her teacher to do algebra. The resulting data follow:

Teacher's Ability in Algebra	Pupil's Ability in Mathematics		
	Low	Average	High
Low	70	10	20
Average	260	330	110
High	70	60	70

Do the results of the study confirm at level $\alpha = .05$ that a pupil's ability to do mathematics depends on the teacher's ability to do algebra?

9 The following data[1] give the number of cases of various infectious diseases reported in 1973 in the United Kingdom:

Region	Disease				
	Typhoid	Scarlet Fever	Whooping Cough	Dysentery	Tuber-culosis
England and Wales	256	12,034	2437	8032	11,156
Scotland	22	621	131	1557	1,522
Northern Ireland	0	355	191	138	296

At level $\alpha = .01$, do the data indicate that type of disease depends significantly on the region in which it is reported?

SECTION 9.B THE INDEX OF PREDICTIVE ASSOCIATION

Our objective in the present section will be to develop a technique of measuring the strength and direction of dependence relationships revealed in a contingency table. When the hypothesis of independence is rejected by the chi-square test, the question arises of how dependent the classifications are and what manner of dependence exists. To answer this deeper question, we might be tempted to assume that the larger the chi-square value, the more heavily dependent the classifications are.

Unfortunately, this argument is not strictly correct. The chi-square test is really a measure of *how certain we are that the classifications are not independent*, not a measure of *how dependent the classifications themselves are*. Furthermore, the chi-square statistic does not address itself to the problem of establishing the nature and direction of any dependence relationship that may exist. Many methods of adequately measuring dependence using data displayed in a contingency table have been developed. Here we will study one of the most useful, the Goodman-Kruskal index of predictive association. The index of predictive association measures how valuable it is to know a person's category in one classification when we want to predict his category in the other.

[1]Central Statistical Office (United Kingdom), "Annual Abstract of Statistics, 1974," p. 74, Her Majesty's Stationery Office, London, 1974.

Example 9.2   Voting in Congress   Let's take a look at the farm bill vote data of Table 9.11. The set of data records the positions of 400 members of the U.S. House of Representatives on repeal of a certain farm law. The results show that of 150 Republicans in the survey, 140 favor repeal while the remaining 10 oppose it. Of 250 Democrats in the chamber, only 100 favor repeal while 150 oppose it. If we apply the chi-square test of independence to the contingency table of Table 9.11, we obtain the conclusion, at all reasonable levels of significance, that there is a dependence relationship between party affiliation and position on repeal of the law. Let's try to find out how strong the relationship is. In particular, we ask the question: If we know to which party a legislator belongs, how well can we predict his position on the bill? To answer this question we use the index of predictive association.

SOLUTION   Suppose we choose one legislator at random out of the 400 and try to predict his position. Because 240 out of the 400 legislators are for repeal while only 160 are against (according to the data of Table 9.11), our best decision would be to say that the selected legislator is for repeal. Using such a decision process, we would have a probability of $160/400 = 40\%$ of making an error in our decision. In arriving at this decision that the selected legislator is for repeal, we have not used any information regarding the party affiliation of that legislator.

Now, because the chi-square test indicates the presence of a dependence relationship, we might be able to reduce our error probability by making use of information regarding the party affiliation of the selected legislator. Suppose, for example, we know that the legislator is a Republican. Table 9.11 shows that 140 of the 150 Republicans favor repeal, while the remaining 10 oppose it.

Our best strategy then would be to assert that the Republican legislator favors repeal; we would be correct in 140 of the 150 cases, and we would be in error in 10 of the 150 cases. If, on the other hand, we know that the legislator is a Democrat, our best strategy would be to assert that the legislator opposes repeal, because Table 9.11 shows that 150 of the 250 Democrats oppose repeal and only the remaining 100 favor it. In making such a decision, we would therefore be correct in 150 cases and incorrect in 100. Knowing the party affiliation of the legislator, then, our eventual decision would be correct in $140 + 150 = 290$ of the 400 cases, and erroneous in $10 + 100 = 110$ of them. This gives us a probability of error of $110/400 = 27.5\%$. Information about party affiliation has therefore permitted us to reduce our probability of error in predicting a

TABLE 9.11
Farm Bill Vote Data

Party \ Position	For Repeal	Against Repeal	Row Totals
Republican	140	10	150
Democratic	100	150	250
Column totals	240	160	400

legislator's position on the bill from 40 to 27.5%, a decline of 12.5%. This 12.5% reduction represents a proportion $12.5/40 = .3125$ of the original error probability of 40%. Therefore, knowledge of party affiliation has the effect of reducing our chances of an erroneous prediction by a factor of 31.25%.

The index of predictive association, which we will denote by the Greek letter λ (pronounced "lambda"), is the number $.3125 = 31.25\%$. It measures the extent of dependence between party affiliation and position on repeal by signifying that in predicting a legislator's position on repeal we can reduce our probability of making an erroneous prediction by a factor of almost one-third, from 40% down to 27.5%. We present in Table 9.12 the formal structure of the discussion leading to the numerical value of λ for the farm bill vote data for Table 9.11.

Before attempting to handle more complex contingency tables, it would be useful to have a computing formula for λ. Such a formula would increase our efficiency, for it would allow an almost immediate calculation of λ without the necessity of again going through the whole argument just presented from beginning to end. With a view toward developing a formula for λ, we will set up a typical contingency table and attach symbolic labels to some of its important components. The precise mathematical justification for the formula for λ will be presented in Supplement 9.1.

A model of a typical contingency table having n rows and m columns appears in Table 9.13. For specific examples of actual contingency tables based on this model, the reader should look at Tables 9.1 (a 3×2 table), 9.5 (a 2×3 table), 9.8 (a 2×2 table), and 9.11 (a 2×2 table). We further define the following items:

$$LR_1 = \text{the largest number in row } R_1$$
$$LR_2 = \text{the largest number in row } R_2$$
$$\cdots\cdots\cdots\cdots\cdots\cdots\cdots\cdots\cdots\cdots$$
$$LR_n = \text{the largest number in row } R_n$$
$$L(\Sigma C) = \text{the largest of the column totals}$$

For an example of how to determine these items, let's take a look at Table 9.11. Here $LR_1 = 140$ because row R_1 consists of the numbers 140 and 10, $LR_2 = 150$ because row R_2 consists of the numbers 100 and 150, and $L(\Sigma C) = 240$ because $\Sigma C_1 = 240$ and $\Sigma C_2 = 160$. We denote by $\Sigma(LR)$ the sum of the largest numbers in each row, i.e.,

$$\Sigma(LR) = LR_1 + LR_2 + \cdots + LR_n$$

For the data of Table 9.11, we have that

$$\Sigma(LR) = LR_1 + LR_2 = 140 + 150 = 290$$

We are now ready to present the formula for λ, the Goodman-Kruskal index of predictive association:

● $$\lambda = \frac{\Sigma(LR) - L(\Sigma C)}{T - L(\Sigma C)} \tag{9.3}$$

In words, λ measures the relative decrease in the error when one is predicting the

TABLE 9.12

Calculation of Index of Predictive Association for Farm Bill Vote Data

Without Knowledge of Party Affiliation

Best Decision	Number of Erroneous Cases	Total Number of Cases	Proportion of Erroneous Cases
For repeal	160	400	$\frac{160}{400} = 40\%$

With Knowledge of Party Affiliation

Party Affiliation	Best Decision	Number of Erroneous Cases	Total Number of Cases	Proportion of Erroneous Cases
Republican	For repeal	10	150	
Democratic	Against repeal	100	250	
Sums		110	400	$\frac{110}{400} = 27.5\%$

$$\lambda = \frac{40 - 27.5}{40} = \frac{12.5}{40} = 0.3125 = 31.25\%$$

TABLE 9.13
A Typical n × m Contingency Table

Rows \ Columns	C_1	C_2	...	C_m	Row Totals
R_1	...	...	...	...	ΣR_1
R_2	...	...	...	...	ΣR_2
....	...	...	...	...	...
R_n	...	...	...	...	ΣR_n
Column totals	ΣC_1	ΣC_2	...	ΣC_m	T

column of a member of the population under study knowing the member's row as opposed to not knowing the member's row. For the data of Table 9.11, we have that

$$\Sigma(LR) = LR_1 + LR_2 = 140 + 150 = 290$$
$$L(\Sigma C) = \text{larger of } \{240, 160\} = 240$$
$$T = 400$$

Therefore

$$\lambda = \frac{290 - 240}{400 - 240} = \frac{50}{160} = 0.3125 = 31.25\%$$

which is, of course, the same answer as obtained earlier. Table 9.14 presents an organized procedure for working out λ from data displayed in a contingency table. It is vital to remember that λ deals only with predicting the column when the row is known. This means that the contingency table must be constructed in such a way that the classification to be predicted is one of the columns, not one of the rows. If necessary, the rows and columns will have to be interchanged. Our next example will give an illustration of this situation.

Example 9.3   Water Resources and Vegetable Prices   Consider the data of Table 9.15 relating a locality's availability of water resources with average retail vegetable prices in

TABLE 9.14
Computations Leading to λ for the Farm Bill Vote Data

Party \ Position	For Repeal	Against Repeal	Row Totals	Largest of Row
Republican	140	10	150	$140 = LR_1$
Democratic	100	150	250	$150 = LR_2$
Column totals	240	160	$400 = T$	$290 = \Sigma(LR)$

Largest column total $L(\Sigma C) = 240$

$$\lambda = \frac{\Sigma(LR) - L(\Sigma C)}{T - L(\Sigma C)} = \frac{290 - 240}{400 - 240} = \frac{50}{160} = 31.25\%$$

TABLE 9.15
Water Resources–Vegetable Prices Data

Water Resources	Retail Vegetable Prices			Row Totals
	Low	Moderate	High	
Scarce	10	70	120	200
Satisfactory	40	400	60	500
Abundant	150	130	20	300
Column totals	200	600	200	1000

that locality. The data resulted from an economist's study of the effect of the presence of local water resources on local vegetable prices in 1000 localities west of the Mississippi River. The chi-square test of Sec. 9.A indicates, at all reasonable levels of significance, that there is a dependence relationship existing between the level of retail vegetable prices and the availability of water resources. The next step therefore is to measure the extent of the existing dependence by calculating the index of predictive association.

SOLUTION   By analogy with our procedures in connection with the farm bill vote data of Table 9.14, we develop the needed calculations in Table 9.16. These calculations are aimed at producing λ, the relative decrease in our error of predicting a locality's level of vegetable prices based on a knowledge of the locality's water resources as compared with the error we make with no knowledge of water resources. Table 9.16 shows that $\Sigma(LR) = 670$ and $L(\Sigma C) = 600$, so that

TABLE 9.16
Computations Leading to λ for the Water Resources–Vegetable Prices Data

Water Resources	Retail Vegetable Prices			Row Totals	Largest of Row
	Low	Moderate	High		
Scarce	10	70	120	200	$120 = LR_1$
Satisfactory	40	400	60	500	$400 = LR_2$
Abundant	150	130	20	300	$150 = LR_3$
Column totals	200	600	200	$1000 = T$	$670 = \Sigma(LR)$

Largest column total $L(\Sigma C) = 600$

$$\lambda = \frac{\Sigma(LR) - L(\Sigma C)}{T - L(\Sigma C)} = \frac{670 - 600}{1000 - 600} = \frac{70}{400} = 0.175 = 17.5\%$$

TABLE 9.17
**Computations Leading to λ for the Water Resources–Vegetable Prices Data
(for Predicting Water Resources from a Knowledge of Vegetable Prices)**

Retail Vegetable Prices \ Water Resources	Scarce	Satisfactory	Abundant	Row Totals	Largest of Row
Low	10	40	150	200	150
Moderate	70	400	130	600	400
High	120	60	20	200	120
Column totals	200	500	300	$1000 = T$	$670 = \Sigma(LR)$

Largest column total $L(\Sigma C) = 500$

$$\lambda = \frac{\Sigma(LR) - L(\Sigma C)}{T - L(\Sigma C)} = \frac{670 - 500}{1000 - 500} = \frac{170}{500} = 0.34 = 34\%$$

$$\lambda = \frac{\Sigma(LR) - L(\Sigma C)}{T - L(\Sigma C)} = \frac{670 - 600}{1000 - 600} = \frac{70}{400} = 0.175 = 17.5\%$$

This means that a knowledge of the local water resources will reduce our error of predicting vegetable prices by a factor of 17.5%.

Suppose, on the other hand, that we want to use knowledge of a locality's vegetable prices in order to predict the local availability of water resources. By what factor will such knowledge decrease our error of prediction? In order to answer this question by applying our formula (9.3) for λ, it is first necessary to interchange the rows and columns of the contingency table of Table 9.15. This interchange is required because as we have constructed λ, it describes the error in predicting the column based on a knowledge of the row. Table 9.17 contains the computations for the λ value of the interchanged contingency table.

The result of our computations in Table 9.17 to the effect that λ = 34% means that a knowledge of local vegetable prices will reduce our error by a factor of 34% (more than one-third) when we try to predict the level of local water resources.

Now that we have shown how to calculate and interpret the significance of the index of predictive association, it would be useful to list a few important facts about λ:

1 λ is always between 0 and 1; a negative value for λ or a value greater than 1 is a sure indication of a computational error.
2 λ = 0 only when knowledge of the row of a member of the population gives no help at all in predicting the member's column.
3 λ = 1 only when knowledge of a member's row allows us to predict the column with certainty.

TABLE 9.18
Computations Leading to λ for the Allergy Treatment Data

Treatment Record \ Allergy Record	Severe Symptoms	Mild or No Symptoms	Row Totals	Largest of Row
No treatment	44	306	350	$306 = LR_1$
Allergy capsule	19	131	150	$131 = LR_2$
Allergy shot	37	463	500	$463 = LR_3$
Column totals	100	900	$1000 = T$	$900 = \Sigma(LR)$

Largest column total $L(\Sigma C) = 900$

$$\lambda = \frac{\Sigma(LR) - L(\Sigma C)}{T - L(\Sigma C)} = \frac{900 - 900}{1000 - 900} = \frac{0}{100} = 0 = 0\%$$

Example 9.4 Allergy Treatment Referring to the allergy treatment data of Table 9.1, we recall from Sec. 9.A that the chi-square test indicated the existence of a dependence relationship between the type of treatment and the severity of symptoms. It therefore would be useful to compute λ, the proportional reduction in our error of predicting severity of symptoms having knowledge of the treatment as opposed to having no knowledge of the treatment.

SOLUTION From Table 9.1 we construct Table 9.18, which sets up the computation of λ, and we reach the surprising result that

$$\lambda = \frac{\Sigma(LR) - L(\Sigma C)}{T - L(\Sigma C)} = \frac{900 - 900}{1000 - 900} = 0$$

This seems to say that despite the chi-square test's confirmation of an existing dependence relationship knowledge of a person's treatment will not at all improve our ability to predict that person's symptoms accurately.

Why is this so? Well, for every possible treatment (namely, none, the capsule, and the shot) our best decision would be to predict mild or no symptoms because the great majority of persons receiving each treatment did show mild or no symptoms. On the other hand, even if we had no knowledge of an individual's treatment, our best guess would also be to predict mild or no symptoms because 900 of the 1000 persons under study reported mild or no symptoms. Therefore, either with or without knowledge of the treatment, we would predict mild or no symptoms. It follows that knowledge of the treatment will not result in any change in our prediction and will therefore result in no reduction in our error of prediction. The factor $\lambda = 0$ is, consequently, an accurate measure of the value of knowledge of the treatment in predicting the severity of the symptoms.

What, then, about a dependence relationship whose existence is indicated by the chi-square test? Well, if it really exists (recall that there is a small probability that a Type I error was made by the chi-square test and so a chance that the asserted dependence relationship does not really exist), it would have to be a relationship that did not involve the ability to improve our prediction of an individual's symptoms by using knowledge about the treatment (in Example 9.4). It is important to remember that there are many ways of having a dependence relationship—the only thing dependence means is that the two classifications are not independent.

EXERCISES 9.B

1 Using the data of Exercise 9.A.6, by what factor can the error of predicting aftereffects be reduced if we take account of information concerning whether or not the child received medical treatment?

2 Based on the data of Exercise 9.A.8, to what extent can we reduce our error of predicting a pupil's ability in mathematics by finding out what the teacher's ability in algebra is?

3 To what degree can we reduce our error in predicting the social behavior of a rat if we base our decision on the length of the rat's tail? Use the data of Exercise 9.A.7. (It is first necessary to interchange the rows with the columns because of the way the data are presented.)

4 The following data record the positions of 400 mayors with regard to whether or not they would favor oil drilling within their city limits or close by offshore if the oil royalties were to go into the city's treasury. The mayors are classified according to regions of the country in which their cities are located:

Region of Country	Position on Oil Drilling	
	Favor	Oppose
East	50	150
Midwest	75	25
South	40	10
West	35	15

By what factor can our error of predicting a mayor's position on the issue be lessened if we take into account the region of the country in which the mayor's city is located?

5 A survey of 1000 persons concerning their views on a proposed tax cut shows that the persons surveyed can be cross-classified according to income level and view on the issue as follows:

Income Level	Oppose Tax Cut	Favor Tax Cut	Don't Know
Lower	100	25	75
Middle	75	400	225
Upper	25	75	0

To what extent can we reduce our error of predicting a person's views on the tax cut issue if we make use of information concerning the person's income level?

6 From the data provided in Exercise 9.A.9, to what extent can we reduce our error of predicting the exact type of infectious disease under discussion if we know the region in which it was reported?

7 As part of a study of whether or not pension plans are a major factor in collective bargaining discussions, the following data[1] were collected on the number of workers covered by collective bargaining agreements in which pension plans are mentioned and in which they are not.

Industry	Number of Workers, in thousands	
	Plans Mentioned	Not Mentioned
Mining	242	86
Construction	908	164
Manufacturing	6281	2857
Utilities and communications	1042	228
Transportation	898	388
Wholesale trade	340	139
Retail trade	158	282
Finance, insurance, and real estate	78	656
Services	191	118

By how much can we reduce our error of predicting whether or not a worker's collective bargaining agreement mentions a pension plan if we know the worker's industry as opposed to not knowing the industry?

SUPPLEMENT 9.1 The Goodman-Kruskal Index of Predictive Association

In this supplement, we shall develop the algebraic justification of the formula for λ, namely, the formula

$$\lambda = \frac{\Sigma(LR) - L(\Sigma C)}{T - L(\Sigma C)}$$

The problem is to pick a member of the population at random and to guess the column to which the member belongs. If we have no additional information about the member in question, our best guess would be to say that the member belongs to the modal column, i.e., the column having the most members, namely the column for which the column total ΣC is largest. We are referring to the typical $n \times m$ contingency table of Table 9.13. If we guess the column for which ΣC is largest, we will be correct in $L(\Sigma C)$ cases and incorrect in the remaining $T - L(\Sigma C)$ cases.

[1] J. Namias, "Handbook of Selected Sample Surveys in the Federal Government," p. 210, St. Johns University Press, New York, 1969.

Therefore our probability of error will be

$$\frac{T - L(\Sigma C)}{T}$$

Suppose now that we know the row to which the member belongs. Now, what do we predict the column to be? Well, we construct a table like Table 9.12. If we know that the member in question belongs to row R_1, for example, our best strategy would be to guess the column of that member to be the one corresponding to the largest number of row R_1, because that column contains more members of row R_1 than any other column. Therefore, for members belonging to row R_1, our guess would be correct in LR_1 cases and incorrect in $\Sigma R_1 - LR_1$ cases. Overall, we have the following results of our decision process:

Row to Which Member Belongs	Number of Correct Guesses	Number of Incorrect Guesses
R_1	LR_1	$\Sigma R_1 - LR_1$
R_2	LR_2	$\Sigma R_2 - LR_2$
. . . .		
R_n	LR_n	$\Sigma R_n - LR_n$
Totals	$\Sigma(LR)$	$T - \Sigma(LR)$

Therefore, knowing the member's row, we would make $\Sigma(LR)$ correct guesses and $T - \Sigma(LR)$ incorrect guesses. Therefore our probability of error would be

$$\frac{T - \Sigma(LR)}{T}$$

It follows that knowledge of the member's row will reduce our probability of error in guessing that member's column by an amount

$$\frac{T - L(\Sigma C)}{T} - \frac{T - \Sigma(LR)}{T} = \frac{\Sigma(LR) - L(\Sigma C)}{T}$$

Compared with the original probability of error, this signifies a reduction in error by a factor of

$$\frac{\dfrac{\Sigma(LR) - L(\Sigma C)}{T}}{\dfrac{T - L(\Sigma C)}{T}} = \frac{\Sigma(LR) - L(\Sigma C)}{T - L(\Sigma C)}$$

The latter quantity is what we have denoted by λ, the proportional decrease in probability of error if we know the row as opposed to not knowing the row.

SECTION 9.C TESTING THE PROBABILITY DISTRIBUTION OF A POPULATION

In addition to its role in deciding whether two classifications are independent or dependent, the chi-square test can also be used to study the probability distribution of the population from which a random sample has been selected. When used in this type of problem, the chi-square test is referred to as a "goodness-of-fit" test. As we have seen in earlier chapters, it is often important to know whether or not we are dealing with, for example, a normally distributed population. If the population is normal, the more accurate t tests for differences and linearity can be used, but if the population is not normal, we have no choice but to apply a nonparametric test. In questions involving proportions we can proceed with full confidence only if we know that we are working with data from a binomially distributed population. In this section, our goal will be to determine from a set of data whether or not the underlying population can reasonably be considered to have the binomial or normal distribution.

Example 9.5 Problem Readers To decide how to distribute reading specialists throughout the four elementary schools under its jurisdiction, a local school board collects data on the number of problem readers in the fourth-grade classrooms at each school. The data appear in Table 9.19. The board would like to test at level $\alpha = .05$ the hypothesis that problem readers are uniformly distributed throughout the four schools.

SOLUTION   The board wants to find out whether or not it is reasonable to make reading specialist assignments under the assumption that all four schools have the same number of problem readers. (The question can even be viewed as one involving independence—whether or not the number of problem readers is independent of the school.) We compare the actual table of data, Table 9.19, with the expected table in the case of uniform distribution. Because the total number of problem readers in the survey is $28 + 15 + 31 + 26 = 100$, the uniform distribution would give each school an allotment of 25 problem readers. From this calculation, we can set up the comparison of actual and expected tables appearing in Table 9.20.

We use the chi-square statistic to measure the difference between these tables, a procedure similar to that introduced in Sec. 9.A for the purpose of testing independence. In particular, we are testing the hypothesis

H : Uniform distribution of problem readers
A : Nonuniform distribution

Our rejection rule for this type of chi-square test is to reject H at level α if $\chi^2 > \chi_\alpha^2(df)$, where df $= m - 1$, for m, the number of cells appearing in the table of data of the sort illustrated in Table 9.19.

Noting that $m = 4$, we would reject the hypothesis H of uniformity at level $\alpha = .05$

TABLE 9.19
Distribution of Problem Readers in Schools

School	Washington	Jefferson	Lincoln	Wilson
Number of problem readers in the fourth grade	28	15	31	26

TABLE 9.20
Actual and Expected Data on Problem Readers

	Actual Table of Data					Expected Table of Data			
Wa.	Je.	Li.	Wi.	Sum	Wa.	Je.	Li.	Wi.	Sum
28	15	31	26	100	25	25	25	25	100

if $\chi^2 > \chi_{.05}^2(3) = 7.815$. For computing the actual numerical value of χ^2, we set up a table of calculations like that in Tables 9.4, 9.7, and 9.10. The relevant calculations appear in Table 9.21, from which we see that

$$\chi^2 = \sum_{k=1}^{m} \frac{(f_k - e_k)^2}{e_k} = 0.3600 + 4.0000 + 1.4400 + 0.0400 = 5.8400$$

Because we agreed to reject H if χ^2 turned out to be larger than $\chi_{.05}^2(3) = 7.815$, our value of $\chi^2 = 5.8400$ does not permit us to reject H at level $\alpha = .05$. Therefore, the data seem to indicate at level $\alpha = .05$ that the hypothesis of uniform distribution of problem readers is a reasonable one. More specifically, the significance level of 5% means that if problem readers were truly uniformly distributed among the four schools, the chances exceed 5% of getting a χ^2 value even larger than the 5.8400 that we got, and therefore this assumption of uniformity can be considered somewhat probable. The chances of getting a χ^2 value larger than 7.815, however, are not larger than 5%; so if we had gotten a $\chi^2 > 7.815$, we would have been led to reject H and to assume that problem readers were not uniformly distributed.

Our next example illustrates a problem that can be solved by applying the chi-square test for the binomial distribution.

Example 9.6 A Political Ballot Study Some political scientists believe that among several candidates for the same position the one listed first has a built-in advantage. In an attempt to measure the extent of this advantage, one political scientist set up the

TABLE 9.21
Calculation of χ^2 for the Problem Reader Data

Cell	f_k	e_k	$f_k - e_k$	$(f_k - e_k)^2$	$\dfrac{(f_k - e_k)^2}{e_k}$
Wa.	28	25	3	9	.3600
Je.	15	25	−10	100	4.0000
Li.	31	25	6	36	1.4400
Wi.	26	25	1	1	.0400
Sums	100	100	0		$\chi^2 = 5.8400$

TABLE 9.22
Ballot Results on Listing of Candidates

Number of contests in which first-listed candidate was preferred	0	1	2	3	4
Number of ballots	35	170	350	330	115

following experiment: She invented eight names which were not politically or otherwise prominent (like yours and mine) and listed them in pairs, randomly alternating them as "candidates" for four county council seats. She then conducted a mock political poll by asking 1000 randomly selected voters to "vote" for their choices for the seats and hypothesized that 60% of the votes cast would go to the candidate listed first for each seat. The ballots were classified according to how many of each voter's four choices were the candidates listed first in their contests, and the resulting data appear in Table 9.22. At level $\alpha = .01$, do the results of the mock poll support the assertion that 60% of votes cast go to candidates listed first?

SOLUTION   The hypothesis we want to test can be rephrased as follows: a voter has probability $0.60 = 60\%$ of voting for the first-listed candidate in any given contest, independently of how that voter voted in any other contest. We are really testing the data of Table 9.22 for the binomial distribution with parameters $n = 4$ and $p = 0.60$. To set up the expected table of data, we therefore base our calculations on the portion of Table A.2, the table of the binomial distribution, which appears in Table 9.23.

We can now use the information in Table 9.23 in order to calculate the expected table of data which corresponds to the actual data of Table 9.22. What we need are the theoretical binomial proportions $P(F = 0)$, $P(F = 1)$, $P(F = 2)$, $P(F = 3)$, and $P(F = 4)$, referring to the number of ballots involved in each of the five cells of Table 9.22.

From the proportions listed in Table 9.23, we can calculate that

$$P(F = 0) = P(F \leq 0) = .0256$$
$$P(F = 1) = P(F \leq 1) - P(F \leq 0) = .1792 - .0256 = .1536$$
$$P(F = 2) = P(F \leq 2) - P(F \leq 1) = .5248 - .1792 = .3456$$

TABLE 9.23
Binomial Proportions for $n = 4$, $p = .60$
F = number of votes for first-listed candidates

k	$P(F \leq k)$
0	.0256
1	.1792
2	.5248
3	.8704
4	1.0000

TABLE 9.24
Actual and Expected Data on Listing of Candidates

Actual Table of Data					Expected Table of Data				
0	1	2	3	4	0	1	2	3	4
35	170	350	330	115	25.6	153.6	345.6	345.6	129.6

$$P(F = 3) = P(F \leq 3) - P(F \leq 2) = .8704 - .5248 = .3456$$
$$P(F = 4) = P(F \leq 4) - P(F \leq 3) = 1.0000 - .8706 = .1296$$

Because the survey involved the preferences of a total of $35 + 170 + 350 + 330 + 115 = 1000$ voters, we would multiply each of the above proportions by 1000 to get the expected frequencies of each cell. This procedure would yield the table of expected frequencies appearing in Table 9.24.

Now that we are ready to calculate the value of χ^2 for the data on listing of candidates preferred by voters, it would be appropriate to formalize the hypothesis and alternative under consideration. We are testing the following hypothesis about a parameter of a binomial distribution:

$$H : p = 0.60 \text{ (60\% of votes cast go to first-listed candidates)}$$

versus $\quad A : p \neq 0.60$

We compute χ^2, and we reject H at significance level α if $\chi^2 > \chi_\alpha^2(df)$, where again df $= m - 1$ for m the number of cells in the original table of data, Table 9.22 in this case. For the listing of candidates data, $m = 5$, so that we would reject H at level $\alpha = .01$ if $\chi^2 > \chi_{.01}^2(4) = 13.277$. In Table 9.25, we perform the calculation of χ^2 in the usual way, and we obtain the result that $\chi^2 = 7.608$. Because $\chi^2 = 7.608 < 13.277$, we cannot reject H at level $\alpha = .01$, and so we conclude that the survey data tend to support the assertion that 60% of the votes cast will indeed go to first-listed candidates.

It is possible to streamline the calculations required for this type of problem by

TABLE 9.25
Calculation of χ^2 for the Listing of Candidates Data

Cell	f_k	e_k	$f_k - e_k$	$(f_k - e_k)^2$	$\dfrac{(f_k - e_k)^2}{e_k}$
0	35	25.6	9.4	88.36	3.452
1	170	153.6	16.4	268.96	1.751
2	350	345.6	4.4	19.36	.056
3	330	345.6	−15.6	243.36	.704
4	115	129.6	−14.6	213.16	1.645
Sums	1000	1000.0	0		$\chi^2 = 7.608$

TABLE 9.26
**Table for the Chi-Square Goodness-of-Fit Test for
the Binomial Distribution**
Listing of Candidates Data, 1000 data points, $p = .60$

Cell	f_k	k	$P(F = k)$ $n = 4, p = .60$	$\times 1000 = e_k$	$f_k - e_k$	$(f_k - e_k)^2$	$\dfrac{(f_k - e_k)^2}{e_k}$
0	35	0	.0256	25.6	9.4	88.36	3.452
1	170	1	.1536	153.6	16.4	268.96	1.751
2	350	2	.3456	345.6	4.4	19.36	.056
3	330	3	.3456	345.6	−15.6	243.36	.704
4	115	4	.1296	129.6	−14.6	213.16	1.645
Sums	1000		1.0000	1000.0	0		$\chi^2 = 7.608$

combining Tables 9.22 to 9.25 into one single all-inclusive table. Without further comment, we illustrate the procedure in Table 9.26.

As a final example of the use of the chi-square test in figuring the probability distribution of a population, we present the chi-square goodness-of-fit test for the normal distribution. The concepts involved here are the same as those in the test for the binomial distribution. The expected frequencies are, of course, calculated on the basis of the table of the normal distribution, Table A.3, instead of Table A.2. Because the normal distribution is continuous instead of discrete, we must use "intervals" instead of "cells," but this change presents no additional problems in setting up a computational table analogous to Table 9.26.

Example 9.7   Demand for Electricity   Let's take a look at the electricity demand data of Sec. 1.A (Table 1.1) and test it for normality at significance level $\alpha = .05$.

SOLUTION   Before conducting the chi-square test of normality, it is necessary first to compute the mean and standard deviation of the set of data under study. This must be done in order to be able to arrive at the z scores we need for making use of Table A.3. As routine calculations will reveal, this set of data has sample mean $\bar{x} = 29.61$ and sample standard deviation $s = 3.493$.

The next step is to organize the data into a reasonable frequency distribution. For the electricity demand data, we have already constructed a frequency distribution, which appears in Table 1.2. We then modify the intervals to close up all gaps in the manner of Table 1.3, and next we extend the first and last intervals to $-\infty$ and ∞, respectively. Finally, we transform the intervals into z scores. The process of setting up the intervals for the chi-square test of normality is presented in detail in Table 9.27.

Proceeding from the z-score intervals obtained by the computations for Table 9.27, we now construct a table analogous to Table 9.26. The major components of the new table, Table 9.28, are the calculation of the proportions of normally distributed data falling in each of the z-score intervals (based on Table A.3, the table of the normal

TABLE 9.27
Adjustment of Intervals for Chi-Square Test of Normality
Electricity Demand Data

Original Interval	Adjusted Interval	Upper End Point x	$z = \dfrac{x - \bar{x}}{s}$	z Score Interval
20 to 22	$-\infty$ to 22.5	22.5	-2.04	$-\infty$ to -2.04
23 to 25	22.5 to 25.5	25.5	-1.18	-2.04 to -1.18
26 to 28	25.5 to 28.5	28.5	-0.32	-1.18 to -0.32
29 to 31	28.5 to 31.5	31.5	0.54	-0.32 to 0.54
32 to 34	31.5 to 34.5	34.5	1.40	0.54 to 1.40
35 to 37	34.5 to 37.5	37.5	2.26	1.40 to 2.26
38 to 40	37.5 to ∞	∞	∞	2.26 to ∞

NOTE: $z = \dfrac{x - \bar{x}}{s}$, where $\bar{x} = 29.61$ and $s = 3.493$.

distribution) and the calculation of the chi-square value for this problem. We are testing the hypothesis

H : daily electricity demand is normally distributed

versus　　　A : It is not normally distributed

In the test of the hypothesis of normality, our rejection rule is as follows: We reject H if $\chi^2 > \chi_\alpha^2(df)$, where $df = m - 3$, for m, the number of z-score intervals at the final stage (which is still to come) of the adjustment process.[1] Let's take $\alpha = .05$, but we will have to wait a while before we know the values of m and df.

A very important aspect of the calculations performed in connection with Table 9.28 appears in the columns of that table headed "combined intervals." You may recall from the last paragraph of Sec. 9.A that proper usage of the chi-square test requires that each expected frequency (e_k) be 5 or larger in magnitude. In the column of Table 9.28 headed e_k, all numbers exceed 5, except for 2.07 and 1.19. These two numbers are unacceptable values of e_k and therefore must be taken care of in some way before the chi-square process can continue. The best way to do this is to absorb their intervals into adjacent ones, thereby building the two large intervals $-\infty$ to -1.18 (having $f_k = 4 + 8 = 12$ and $e_k = 2.07 + 9.83 = 11.90$) and 1.40 to ∞ (having $f_k = 4 + 1 = 5$ and $e_k = 6.89 + 1.19 = 8.08$). This is what is done in the column of Table 9.28 headed "combined intervals." Now all e_k's exceed 5, and we can proceed with the operation.

It remains to compute df for the quantity $\chi_{.05}^2(df)$, as we have taken $\alpha = .05$. As we have mentioned earlier, $df = m - 3$, where m is the number of z-score intervals at the final stage. Now, although Table 9.28 began with 7 intervals, it finished up with only 5,

---

[1] The degrees of freedom (df) drop to $m - 3$ in testing for normality due to the fact that we need first to calculate $\bar{x}$ and s. Generally, we subtract from m the number of quantities needed from the data in order to compute the expected frequencies. For normality, the three quantities are $\bar{x}$, s, and the number of data points.

TABLE 9.28
Table for the Chi-Square Goodness-of-Fit Test for the Normal Distribution
Electricity Demand Data, 100 data points

z-Score Interval	f_k	Normal Proportion in Interval	$\times 100 = e_k$	Combined Intervals		$f_k - e_k$	$(f_k - e_k)^2$	$\dfrac{(f_k - e_k)^2}{e_k}$
				f_k	e_k			
$-\infty$ to -2.04	4	.0207	2.07	12	11.90	.10	.01	.0008
-2.04 to -1.18	8	.0983	9.83					
-1.18 to $-.32$	23	.2555	25.55	23	25.55	-2.55	6.5025	.2545
$-.32$ to $.54$	34	.3309	33.09	34	33.09	.91	.8281	.0250
$.54$ to 1.40	26	.2138	21.38	26	21.38	4.62	21.3444	.9983
1.40 to 2.26	4	.0689	6.89	5	8.08	-3.08	9.4864	1.1741
2.26 to ∞	1	.0119	1.19					
Sums	100	1.0000	100.00	100	100.00	0		$\chi^2 = 2.4527$

NOTE: The f_k's are taken from the frequency distribution of Table 1.2.

because of the absorption of 2 intervals by adjacent ones. Therefore $m = 5$, and so $df = m - 3 = 5 - 3 = 2$. It follows that $\chi_{.05}^2(df) = \chi_{.05}^2(2) = 5.991$.

We should therefore reject the hypothesis H of normality if $\chi^2 > 5.991$. As the calculations for Table 9.28 show, $\chi^2 = 2.4527 < 5.991$. Therefore we cannot reject H. The chi-square test of normality, then, supports the assertion that daily electricity demand is normally distributed.

EXERCISES 9.C

1 The 50 diameters of stones found in the Niger River delta, which are presented in Exercise 1.A.1, have sample mean 8.5 centimeters and sample standard deviation 4.87 centimeters. At significance level $\alpha = .05$, can the geographer conclude that the diameters are normally distributed?

2 In view of the fact that the learning times of Exercise 1.A.5 have sample mean 9.35 and sample standard deviation 7.68, is the psychologist justified at level $\alpha = .01$ in assuming that the times are normally distributed?

3 A city housing agency conducted a survey of residential units in order to be better able to forecast future housing needs. An analysis of 300 single family homes classified according to number of bedrooms yielded the following data:

Number of Bedrooms	Homes
1	70
2	75
3	70
4	50
5	35

At level $\alpha = .05$, does the survey indicate that the single family homes are distributed uniformly throughout the five number-of-bedroom categories?

4 A distributor of office equipment has six outlets serving different regions of the country. A random sample of the sales records for filing cabinets shows that the number of units sold by each outlet is as follows:

Region	Northeast	Midwest	Southeast	North Central	South Central	West Coast
Number of units sold	85	75	60	65	90	75

Do the records show that at level $\alpha = .01$ sales of filing cabinets are uniformly distributed throughout the six outlets?

5 The Mendelian theory of genetics implies that of a certain breed of cow one-sixteenth (6.25%) will be all brown, one-fourth (25%) will be brown with black spots, three-eighths (37.5%) will be brown with white spots, one-fourth (25%) will

be white with brown spots, and one-sixteenth (6.25%) will be all white. Such a distribution is said to have the proportions 1:4:6:4:1. The following data have been taken from a study of 400 cows of the breed under discussion:

All Brown	Brown with Black Spots	Brown with White Spots	White with Brown Spots	All White
30	105	160	90	15

Do the results of the study accord with the predictions of Mendelian theory at significance level $\alpha = .01$?

6 A seed company advertises that each of its seeds has an 80% chance of germinating. A detailed analysis of 50 of the company's seed packets, containing four seeds each, gives the following information:

Number of seeds that germinated	0	1	2	3	4
Number of packets	6	4	15	20	5

The company's claim of an 80% germination probability will be substantiated if the number of seeds germinated per packet has a distribution which is binomial with parameters $n = 4$ and $p = .80$. Is the company's claim substantiated at level $\alpha = .05$?

7 The current values of 500 common stocks which comprise the frequency distribution of Exercise 1.A.2 have sample mean $\bar{x} = 29.9$ dollars and sample standard deviation $s = 16.45$ dollars. Can we conclude at level $\alpha = .01$ that the stock prices are normally distributed?

8 Test at level $\alpha = .05$ whether or not the forecasts of the Consumer Price Index, given in Exercise 1.A.6, can reasonably be considered to have the normal distribution.

SUMMARY AND DISCUSSION

The bulk of Chap. 9 has dealt with various applications of the chi-square test. We first applied the test to contingency tables of cross-classified data to try to find out whether two descriptive classifications are independent of each other. Using the probabilistic definition of independent events discussed in Chap. 8, we set up a companion expected contingency table of how the data would've turned out had the classification been truly independent. Then we used the chi-square test to measure the difference between the actual and the expected contingency tables.

On the basis of the size of this difference, we made our decision as to whether or not the classifications could really be considered independent. Where the chi-square test indicated the existence of a dependence relationship, we presented one way of measuring the nature and extent of the dependence—the Goodman-Kruskal index of predictive association. Finally, we showed how the chi-square test can be used to check on the probability distribution of a set of data. We illustrated the chi-square goodness-of-fit test, as it is called, for the uniform, binomial, and normal distributions.

BIBLIOGRAPHY

Testing Independence
Grizzle, J. E.: Continuity Correction in the χ^2-Test for 2×2 Tables, *The American Statistician*, October 1967, pp. 28–32.

Mantel, N., and S. W. Greenhouse: What Is the Continuity Correction? *The American Statistician*, December 1968, pp. 27–30.

Measuring Dependence
Costner, H. L.: Criteria for Measures of Association, *American Sociological Review*, June 1965, pp. 341–353.

Goodman, L. A., and W. H. Kruskal: Measures of Association for Cross Classifications, *Journal of the American Statistical Association*, December 1954, pp. 732–764.

Goodman, L. A., and W. H. Kruskal: Measures of Association for Cross Classifications II: Further Discussion and References, *Journal of the American Statistical Association*, March 1959, pp. 123–163.

Testing for Normality
Carlson, J. A.: Are Price Expectations Normally Distributed? *Journal of the American Statistical Association*, December 1975, pp. 749–754.

SUPPLEMENTARY EXERCISES

1   One financial analyst believes that the stock market's Dow-Jones average (DJA) declines the week following an increase in the prime rate of interest at major banks. He feels, on the other hand, that the DJA rises the week after the prime rate goes down. To substantiate his opinion, he studies the data for the last 1000 weeks and obtains the following information relating the prime rate behavior one week with the DJA behavior the next:

Prime Rate	Dow-Jones Average a Week Later		
	Down	Same	Up
Down	100	50	250
Same	70	100	30
Up	330	50	20

a At level $\alpha = .01$, can the financial analyst conclude that changes in the DJA really depend on fluctuations in the prime rate of interest?

b By what factor can the error of predicting changes in the Dow-Jones average be reduced if we take into consideration the behavior of the prime rate a week earlier?

2　In a recent study of the use of corporate annual financial reports by prospective and current stockholders, it was hypothesized that college-educated investors were better able to understand a company's balance sheet than were those investors who had not had a college education. To check on the hypothesis, 400 randomly selected stockholders were interviewed and classified according to whether or not they had a college education and whether or not they had difficulty understanding the balance sheet. The results of the survey follow:

Stockholders	Had Difficulty Understanding the Company Balance Sheet	
	Yes	No
Had college education	145	55
No college education	155	45

a At level $\alpha = .05$, do the results of the survey indicate that college-educated investors are better able to understand the balance sheet?

b By what percentage can the error of predicting whether or not an investor had difficulty with the balance sheet be reduced by the knowledge of whether or not that investor was college-educated?

3　A dental study of 200 children, aimed at analyzing the effect of fluoride toothpaste on the number of cavities, yielded the following data on the change in the number of cavities each child had compared with the total at the last checkup:

	More Cavities	Same Number	Fewer Cavities
Fluoride toothpaste	5	5	40
Regular toothpaste	20	20	60
No toothpaste	20	20	10

a Decide at level $\alpha = .05$ whether or not the study shows that the change in number of cavities depends significantly on the type of toothpaste used.

b To what extent can our error of predicting the change in number of cavities be reduced by making use of information concerning the type of toothpaste involved?

4　Recent data on population trends show that the average age (in years) of a person moving to Alaska to live is 26.8 with a standard deviation of 19.08 based on a random sample of 50 immigrants. The age distribution of those 50 persons was as follows:

Age, in years	Number of Persons
0 up to but not including 10	5
10 up to but not including 20	15
20 up to but not including 30	20
30 up to but not including 60	6
60 or older	4

At level of significance $\alpha = .05$, do the data indicate that the ages are normally distributed?

5 When a city's water supply is fluoridated, there is some question whether the chemical is uniformly distributed throughout the city or whether certain neighborhoods wind up with greater (possibly dangerous) concentrations than others. A survey of five neighborhoods yielded the following data on units of fluoride in a gallon of water:

Neighborhood	#1	#2	#3	#4	#5
Units of fluoride	72	85	68	77	73

At level $\alpha = .05$ is it reasonable to assume that the chemical is uniformly distributed throughout the water supply?

6 If it were really true that a live birth was equally likely to be male or female, the number of girls born into a family of six children should logically be a binomial random variable with parameters $n = 6$ and $p = \frac{1}{2}$. A demographic survey of 300 families with 6 children each discovers the following information:

Number of girls	0	1	2	3	4	5	6
Number of families	5	30	70	90	65	30	10

Can we conclude from the data at level $\alpha = .01$ that the logical explanation above accords with the actually existing situation?

7 The following data[1] record the number of divorce petitions filed in England and Wales recently and classifies them by cause and year:

	Year				
Cause	1962	1964	1966	1968	1970
Adultery	15,366	18,778	21,541	26,011	36,474
Desertion	9,160	9,945	10,181	11,147	12,266
Behavior	5,791	7,840	9,846	12,753	17,534
Mental disorder	96	92	68	63	63
Physical disease	85	103	104	109	82
Cruelty	98	141	123	117	134

[1]Central Statistical Office (United Kingdom), "Annual Abstract of Statistics, 1973," p. 96, Her Majesty's Stationery Office, London, 1973.

a Test at level $\alpha = .01$ for independence of cause of divorce petition from year of petition.

b Knowing the cause of a petition, by what percentage can we reduce our error of predicting the year in which it was filed?

c Knowing the year in which a petition was filed, by what factor can we reduce our error of predicting its cause?

8 The production of canned fish (in millions of pounds) in the United States is given for several recent years by the following data:[1]

Canned Fish	1960	1965	1970	1971	1972
Salmon	136	174	183	168	96
Maine sardines	74	30	19	22	37
Tuna	301	358	438	439	620
Mackerel	42	32	8	17	14
Shrimp	14	16	25	22	24
Clams	45	68	65	71	75
Oysters	6	4	2	3	3
Others	464	440	606	587	732

a Conduct a test of independence at level $\alpha = .05$ between type of fish canned and the year it was canned.

b If we know the year in which a certain can of fish was produced, by what percentage can we reduce our error of predicting what type of fish it is?

[1]"Statistical Abstract of the United States, 1974," U.S. Department of Commerce, 1974. p. 653.

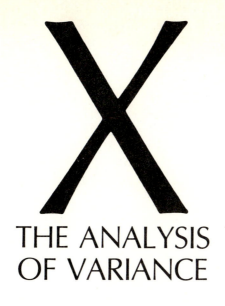

THE ANALYSIS
OF VARIANCE

In our discussion of testing statistical hypotheses, which appears in Chap. 5, we studied one-sample tests and two-sample tests about population means. We introduced some techniques valid for large samples and some for small samples, and we allowed for independent samples and for paired samples. However, as you may have noticed, we did not even once analyze a situation involving three or four or more samples. Yet, it is often necessary to run an experiment involving several samples. For example, an agricultural seed company interested in developing multipurpose seed might want to know whether a new variety of corn will produce similar or differing yields in widely scattered geographical regions, such as low plains, delta, mountains, and irrigated desert. The way to find out would be to plant several acres in each of the four regions and then compare the four mean yields. As a second example, an office executive might be considering five competing computerized storage systems for his company's office records, and he would want to know whether they are equally efficient or not. The problem could be analyzed by solving some sample information retrieval problems in each of the five systems and then comparing the five mean efficiency levels. As it turns out, none of the techniques studied in Chap. 5 are effective in studying either of these and similarly structured problems. We need a completely new method of analysis.

Before we proceed with the details of this new method, called the "analysis of variance," let's take a brief look at the reasons why our previous methods are inadequate for the study of these problems. Suppose we are working with the agricultural seed problem mentioned just above. We denote the true mean yields of the new variety of corn in the various geographical regions as follows:

$$\mu_L = \text{true mean yield in low plains}$$
$$\mu_D = \text{true mean yield in delta}$$
$$\mu_M = \text{true mean yield in mountains}$$
$$\mu_I = \text{true mean yield in irrigated desert}$$

We would then want to test at significance level α the hypothesis

$$H : \mu_L = \mu_D = \mu_M = \mu_I$$

against the alternative

$$A : \text{These true means are not all the same}$$

If we were to apply our previous techniques to this problem, the simplest thing we could do would be to run each of the following three tests using a two-sample test on each one:

1 $H : \mu_L = \mu_D$ versus $A : \mu_L \neq \mu_D$
2 $H : \mu_D = \mu_M$ versus $A : \mu_D \neq \mu_M$
3 $H : \mu_M = \mu_I$ versus $A : \mu_M \neq \mu_I$

If we were to reject H in any one of these three tests, we would be concluding that the true means are not all the same, and we would therefore have to reject the hypothesis

$$H : \mu_L = \mu_D = \mu_M = \mu_I$$

in favor of the alternative

$$A : \text{These true means are not all the same}$$

So far so good, but let's now talk about the significance levels of the tests. If we ran each of the three two-sample tests at significance level $\alpha = .05$, we would have a probability .05 of making a Type I error and a probability .95 of avoiding a Type I error on each one. This means that we would have a 5% chance of rejecting each of the three H's if they were really true.

In other words, we would have a 95% chance in each case of asserting that the two means in question were the same when they really were. Now suppose that the overall hypothesis

$$H : \mu_L = \mu_D = \mu_M = \mu_I$$

were really true. What would be the probability of avoiding a Type I error here; namely, what would be the probability of our deciding that H was really true? Well, to accept the overall H on the basis of the three two-sample tests we would have to accept those three

H's. But we have a 95% chance of accepting each such H, because all three would really be true if the overall H were. To be fair, all three two-sample tests would have to be conducted independently of one another, and so we would see that if the overall H were really true,

$$
\begin{aligned}
P(\text{accepting overall } H) &= P(\text{accepting all three two-sample } H\text{'s}) \\
&= P(\text{accepting first } H)P(\text{accepting} \\
&\quad \text{second } H)P(\text{accepting third } H) \\
&= (.95)(.95)(.95) \\
&= .857 \\
&= 85.7\%
\end{aligned}
$$

using properties of independent events discussed in Sec. 8.D. Therefore we have an 85.7% chance of accepting the overall H when it's really true if we use three two-sample tests. It automatically follows that we have a 14.3% chance of rejecting the overall H when it's really true. This means that the probability of a Type I error, the significance level of our test, will be .143.

We began with three two-sample tests each having significance level $\alpha = .05$, and we wound up with a test of our overall hypothesis H having significance level $\alpha = .143$. This is an unacceptably high level of significance, of course, for it indicates that we will be making a Type I error nearly one-seventh of the time when H is really true. This is what invalidates the method of compounding two-sample tests—it unfortunately compounds the probability of error. (To guarantee a significance level of .05 for the test of the overall hypothesis, it would be necessary to use a level $\alpha = .017$ on each of the three two-sample tests. Such a low level of α would push β, the probability of a Type II error, beyond any reasonable level.) Therefore, we need a method of testing the equality of several means by a single several-sample test. That method, the analysis of variance, is our subject for this chapter.

SECTION 10.A COMPONENTS OF THE VARIANCE

If we plant corn on five randomly selected acres in each of four types of geographical regions, we will have a total of 20 plots of corn involved in our experiment. Even if there is no real difference between the corn yields in differing geographical regions, it would be unreasonable to expect all 20 plots to give exactly the same yield. As we discussed earlier in Chap. 6, for example, this would be as unreasonable as expecting an entire set of data points to lie exactly on a straight line, even if the underlying relationship were truly linear. Therefore, we would expect to have some variation among the corn yields, and it is the extent of this variation that leads us to decide whether or not there is a significant difference in yield among different types of geographical regions.

Suppose that our experiment results in the data presented in Table 10.1. How are we to decide whether or not the true mean yields are the same for the four geographical regions? As can be seen from Table 10.1, we find that

TABLE 10.1
Corn Yields in Different Geographical Regions
Data Given in Hundreds of Bushels

Low Plains	Delta	Mountains	Irrigated Desert	Entire Set of Data
13	8	3	4	
6	2	5	8	
5	10	10	5	
6	9	6	6	
10	11	6	2	
Sums 40	40	30	25	135
Sample means 8	8	6	5	6.75

$$\bar{x}_L = \text{sample mean yield in low plains} = 8$$
$$\bar{x}_D = \text{sample mean yield in delta} = 8$$
$$\bar{x}_M = \text{sample mean yield in mountains} = 6$$
$$\bar{x}_I = \text{sample mean yield in irrigated desert} = 5$$

We notice immediately that these four sample means are not all the same. Before we draw any inferences regarding the hypothesis that the true means are identical, we must discuss two pertinent questions:

1 To what extent do the four sample means differ among themselves?
2 To what extent do the four sample means represent the true means of their respective groups?

As you will recall from the introduction to this chapter, we are looking for conditions under which we will be justified in rejecting the hypothesis

$$H : \mu_L = \mu_D = \mu_M = \mu_I$$

in favor of the alternative

$$A : \text{These true means are not all the same}$$

Our decision to accept or reject H will be made on the basis of the answers to the above two questions. We will be inclined to reject H if the sample means differ greatly among themselves and we think that they are accurate representations of the true means; otherwise, we will probably accept H. In Table 10.2, we illustrate the relationship between the answers to questions 1 and 2 above and our decision regarding the hypothesis H.

All that remains, therefore, is to analyze the data of Table 10.1 with a view toward answering the two questions above.

Let's deal with the first question involving the extent to which the four sample means differ among themselves. We have the sample means 8, 8, 6, and 5, each based

TABLE 10.2
When Should the Hypothesis $H : \mu_L = \mu_D = \mu_M = \mu_I$ Be Rejected in
Favor of A : These True Means Are Not All the Same?

Extent to Which Sample Means Represent True Means	Extent to Which Sample Means Differ among Themselves	
	Little	Much
Little	Cannot reject H	Cannot reject H
Much	Cannot reject H	Should reject H

on one of the four sets of data in Table 10.1. The extent to which these sample means differ among themselves can be measured by their variation from *their own* mean 6.75, which is also the "grand mean" of all 20 data points. We can denote this measure of variation by

$$VB = (8 - 6.75)^2 + (8 - 6.75)^2 + (6 - 6.75)^2 + (5 - 6.75)^2$$
$$= (1.25)^2 + (1.25)^2 + (-0.75)^2 + (-1.75)^2$$
$$= 1.5625 + 1.5625 + 0.5625 + 3.0625 = 6.75$$

where VB stands for "variation between the samples."

Before we investigate the significance of VB in detail, let's work with the second question above, which concerns the extent to which the four sample means represent the true means of their groups. To study this question, we can compute the variation within each sample from *its own* mean as follows (using the squared deviations from the mean as we have been doing since Sec. 2.C):

$$VW_L = (13 - 8)^2 + (6 - 8)^2 + (5 - 8)^2 + (6 - 8)^2 + (10 - 8)^2$$
$$= 25 + 4 + 9 + 4 + 4 = 46$$
$$VW_D = (8 - 8)^2 + (2 - 8)^2 + (10 - 8)^2 + (9 - 8)^2 + (11 - 8)^2$$
$$= 0 + 36 + 4 + 1 + 9 = 50$$
$$VW_M = (3 - 6)^2 + (5 - 6)^2 + (10 - 6)^2 + (6 - 6)^2 + (6 - 6)^2$$
$$= 9 + 1 + 16 + 0 + 0 = 26$$
$$VW_I = (4 - 5)^2 + (8 - 5)^2 + (5 - 5)^2 + (6 - 5)^2 + (2 - 5)^2$$
$$= 1 + 9 + 0 + 1 + 9 = 20$$

It follows that the "variation within the samples," denoted by VW, can be calculated as

$$VW = VW_L + VW_D + VW_M + VW_I$$
$$= 46 + 50 + 26 + 20 = 142$$

Before proceeding, let's take a look at what we have done so far. We have four samples of data, and we have defined two measures of variation:

$$VB = \text{variation between the samples}$$
and
$$VW = \text{variation within the samples}$$

We have developed these measures of variation for the purpose of testing the hypothesis

$$H : \mu_L = \mu_D = \mu_M = \mu_I$$

versus A : These true means are not all the same

Now, if H is true, there should be little variation *between* the samples, but if H is false, there should be a lot of variation *between* the samples. What does this have to do with variation *within* the samples? Well, as it turns out, there are only two types of variation—between samples and within samples. Therefore, when VB is small, VW is large, and when VB is large, VW is small. *When the hypothesis H is true, VB is small and VW is large, and when the alternative A is true, VB is large and VW is small.*

This is the basic principle upon which the method of analysis of variance is based. To see how it works in the example data of Table 10.1, it is necessary to compute the "total variation," denoted by TV, of the data and then to compare it with VB and VW. In Table 10.3, we compute the total variation of the data, and we obtain the result that $TV = 175.75$.

If we set

$$n = \text{number of data points in each sample}$$

TABLE 10.3
Calculation of the
Total Variation
Corn Yield Data

x	$x - \bar{x}$	$(x - \bar{x})^2$
13	6.25	39.0625
6	−.75	.5625
5	−1.75	3.0625
6	−.75	.5625
10	3.25	10.5625
8	1.25	1.5625
2	−4.75	22.5625
10	3.25	10.5625
9	2.25	5.0625
11	4.25	18.0625
3	−3.75	14.0625
5	−1.75	3.0625
10	3.25	10.5625
6	−.75	.5625
6	−.75	.5625
4	−2.75	7.5625
8	1.25	1.5625
5	−1.75	3.0625
6	−.75	.5625
2	−4.75	22.5625
135	0	175.7500

$$\bar{x} = \tfrac{135}{20} = 6.75$$
$$TV = 175.75$$

then the relationship among TV, VB, and VW is given by the equation

$$TV = nVB + VW$$

The mathematical details of the argument justifying this equation are presented in Supplement 10.1. Here we shall only show how it works in practice. For the data of Table 10.1, we have indicated that

$$TV = 175.75$$
$$n = 5$$
$$VB = 6.75$$
$$VW = 142$$

The equation $TV = nVB + VW$ then reduces to the assertion that

$$175.75 = (5)(6.75) + 142$$

which is a true statement since $(5)(6.75) = 33.75$. The formula

● $$TV = nVB + VW \qquad (10.1)$$

is called the "components of the variance formula." In the next two sections of this chapter, we shall see how some revised versions of it can be used to solve a wide variety of hypothesis testing problems. In the remainder of this section, however, we shall simply complete the discussion of the corn yield problem by formally testing the hypothesis

$$H : \mu_L = \mu_D = \mu_M = \mu_I$$

against the alternative

$$A : \text{These true means are not all the same}$$

To test H against A, we use the F test. This is the same F test that we used in Sec. 5.E to study the differences between standard deviations. It is reasonable to use the F test here because the quantities TV, VB, and VW are basically the squares of standard deviations. (In fact, the word "variance" of a set of data is simply the technical name for the square of the standard deviation.) All that is needed to make them into exact squares of standard deviations is to divide them by the proper numbers.

The rejection rule is as follows: We reject H in favor of A at level α if

$$F > F_\alpha(m - 1, m(n - 1))$$

where

● $$F = \frac{nVB/(m - 1)}{VW/m(n - 1)} \qquad (10.2)$$

In the expression for F, we have used the symbols

$$m = \text{number of samples involved in the problem}$$

and

$$n = \text{number of data points in each sample}$$

If we use a level of significance $\alpha = .05$, then we agree to reject H if

$$F > F_{.05}(3, 16) = 3.24$$

because $m = 4$ and $n = 5$ here. Now, recalling that

$$nVB = (5)(6.75) = 33.75$$
$$VW = 142$$
$$m - 1 = 4 - 1 = 3$$

and
$$m(n - 1) = 4(5 - 1) = 16$$

we have

$$F = \frac{nVB/(m - 1)}{VW/m(n - 1)} = \frac{(33.75)/3}{(142)/16} = \frac{(33.75)(16)}{(142)(3)} = 1.27$$

Since we agreed to reject H if $F > 3.24$, but actually it turned out that $F = 1.27 < 3.24$, we cannot reject H at level $\alpha = .05$. We therefore conclude that the data support the hypothesis that the mean corn yields do not differ significantly among the four geographical regions.

How were the components of the variance involved in the solution to the above problem? Well, the final decision was that F was not large enough to permit us to reject H. Looking at the formula for F, we can see that this means that VB was not large enough relative to VW to warrant the rejection of H in favor of A. In other words, the variation between the sample means was not much greater than the variation within the samples, and therefore we could not conclude that variations among the data points could be attributed largely to variations between the sample means. We were therefore led to believe that the sample means did not exhibit inordinately large differences among themselves.

Let's take a look at one more applied example in order to solidify our understanding of the concepts involved before proceeding to the "one-way analysis of variance" studied in the next section.

Example 10.1 Word Processing Systems In Table 10.4, we have listed the weekly costs of operating each of five competing word processing (WP) systems in offices of a major steel corporation. The data were taken on randomly selected weeks during a six-month trial period. Costs are only one factor influencing the corporation's choice of a WP system, and so the corporation would like to know whether or not costs run about the same for each of the five systems.

SOLUTION   We use the following symbols to denote the true means involved in this example:

$$\mu_U = \text{true mean cost of United Electronics system}$$
$$\mu_V = \text{true mean cost of Verbiage Control system}$$
$$\mu_W = \text{true mean cost of Western Words system}$$
$$\mu_X = \text{true mean cost of X-Pensive Palaver system}$$
$$\mu_Y = \text{true mean cost of Yorktown Machine system}$$

TABLE 10.4
Weekly Costs of Different *WP* Systems
Data Given in Hundreds of Dollars

United Electronics	Verbiage Control	Western Words	X-Pensive Palaver	Yorktown Machine	Entire Set of Data
12	12	3	14	16	
13	10	6	18	21	
7	6	10	18	15	
8	4	13	14	20	
Sums 40	32	32	64	72	240
Sample means 10	8	8	16	18	12

We want to test the hypothesis

$$H : \mu_U = \mu_V = \mu_W = \mu_X = \mu_Y$$

against the alternative

$$A : \text{These true means are not all the same}$$

Recalling that

$$m = \text{number of samples involved in the problem}$$

and

$$n = \text{number of data points in each sample}$$

we see that

$$m = 5 \quad \text{and} \quad n = 4$$

Since the general rule is to reject H in favor of A at level α if

$$F > F_\alpha(m - 1, m(n - 1))$$

we should reject H at level $\alpha = .05$ if

$$F > F_{.05}(4, 15) = 3.06$$

Therefore it remains only to compute

$$F = \frac{nVB/(m - 1)}{VW/m(n - 1)} = \frac{4VB/4}{VW/15} = \frac{15VB}{VW}$$

In the above formula

$$VB = \sum_{i=1}^{m} (\bar{x}_i - \bar{x})^2 = (10 - 12)^2 + (8 - 12)^2 + (8 - 12)^2 + (16 - 12)^2 + (18 - 12)^2$$
$$= 2^2 + 4^2 + 4^2 + 4^2 + 6^2 = 4 + 16 + 16 + 16 + 36 = 88$$

and

$$VW = VW_U + VW_V + VW_W + VW_X + VW_Y$$

where

$$VW_U = (12 - 10)^2 + (13 - 10)^2 + (7 - 10)^2 + (8 - 10)^2$$
$$= 2^2 + 3^2 + 3^2 + 2^2 = 4 + 9 + 9 + 4 = 26$$

$$VW_V = (12 - 8)^2 + (10 - 8)^2 + (6 - 8)^2 + (4 - 8)^2$$
$$= 4^2 + 2^2 + 2^2 + 4^2 = 16 + 4 + 4 + 16 = 40$$
$$VW_W = (3 - 8)^2 + (6 - 8)^2 + (10 - 8)^2 + (13 - 8)^2$$
$$= 5^2 + 2^2 + 2^2 + 5^2 = 25 + 4 + 4 + 25 = 58$$
$$VW_X = (14 - 16)^2 + (18 - 16)^2 + (18 - 16)^2 + (14 - 16)^2$$
$$= 2^2 + 2^2 + 2^2 + 2^2 = 4 + 4 + 4 + 4 = 16$$
$$VW_Y = (16 - 18)^2 + (21 - 18)^2 + (15 - 18)^2 + (20 - 18)^2$$
$$= 2^2 + 3^2 + 3^2 + 2^2 = 4 + 9 + 9 + 4 = 26$$

It therefore follows that

$$VW = VW_U + VW_V + VW_W + VW_X + VW_Y$$
$$= 26 + 40 + 58 + 16 + 26$$
$$= 166$$

and so
$$F = \frac{15VB}{VW} = \frac{(15)(88)}{166} = 7.95$$

Because we have agreed to reject the hypothesis

$$H : \mu_U = \mu_V = \mu_W = \mu_X = \mu_Y$$

in favor of the alternative

$$A : \text{These true means are not all the same}$$

at level of significance $\alpha = .05$ if F turned out to be greater than 3.06, the result that $F = 7.95 > 3.06$ leads us to reject H. Therefore, at level $\alpha = .05$, we conclude that the costs of the five WP systems are not all the same.

The material dealing with the components of the variance that we have presented in this section provides the philosophical and mathematical basis of the group of statistical procedures known collectively as the analysis of variance. In the next two sections, we shall study in detail two of these procedures that are widely used in statistical analysis of applied problems. Before we proceed to those topics, however, let's take a look at the conditions under which the analysis of variance procedures are valid.

As you will perhaps recall from our discussion of tests for standard deviations in Sec. 5.E, it is proper to use the F distribution only when working with a set of normally distributed data. In particular, the valid application of the F test to the solution of analysis of variance problems requires that the populations (all m of them!) from which the random samples were selected be normally distributed with the same standard deviation. (The central limit theorem operates if the n's are larger than 5.)

It is *not* required, however, that the number of data points (what we have been representing by n) be the same for each sample. These conditions are reminiscent of those we encountered in our discussion of the two-sample t test in Sec. 5.C, where both sets of data were required to come from populations having normal distributions with the same standard deviations. It was not necessary that both sets have the same number of data points, and, in fact, we will see in the next section some analysis of variance examples where the samples do not all have the same number of data points.

What do we do in case the conditions (especially the one requiring equal standard deviations) are not fulfilled? The answer to this question is the same as it was in Chaps. 5 and 6: Use a nonparametric test. In Sec. 11.D, we will study the Kruskal-Wallis test, often called the "nonparametric analysis of variance" test.

SUPPLEMENT 10.1   THE COMPONENTS OF THE VARIANCE FORMULA

Our goal in this supplement is to provide the reader with the mathematical background underlying the "components of the variance formula,"

$$TV = nVB + VW$$

A typical set of data involved in an analysis of variance problem, for example the data of Table 10.1, is composed of m samples of n data points each. (In Table 10.1, $m = 4$ and $n = 5$.) A typical data point is labeled

$$x_{ij} = j\text{th data point of the } i\text{th sample}$$

If we set up the data in a table similar to Table 10.1, we obtain the following:

	Sample 1	Sample 2	. . .	Sample m	Entire Set of Data
	x_{11}	x_{21}	. . .	x_{m1}	
	x_{12}	x_{22}	. . .	x_{m2}	
	. . .	. . .	. . .	. . .	
	x_{1n}	x_{2n}	. . .	x_{mn}	
Sample means	$\bar{x}_1$	$\bar{x}_2$	. . .	$\bar{x}_m$	$\bar{x}$

The total variation of the data points from the grand mean $\bar{x}$ is given by the "double summation":

$$TV = \sum_{i=1}^{m} \sum_{j=1}^{n} (x_{ij} - \bar{x})^2$$

Here $\sum_{j=1}^{n} (x_{ij} - \bar{x}) = (x_{i1} - \bar{x}) + (x_{i2} - \bar{x}) + \cdots + (x_{in} - \bar{x}) = TV_i$

is the total variation of the ith sample from the grand mean $\bar{x}$. Then

$$TV = TV_1 + TV_2 + \cdots + TV_m = \sum_{i=1}^{m} TV_i$$

We use the following algebraic manipulations to break up TV into its components:

$$\sum_{i=1}^{m} \sum_{j=1}^{n} (x_{ij} - \bar{x})^2 = \sum_{i=1}^{m} \sum_{j=1}^{n} (x_{ij} - \bar{x}_i + \bar{x}_i - \bar{x})^2$$

$$= \sum_{i=1}^{m} \sum_{j=1}^{n} [(x_{ij} - \bar{x}_i)^2 + 2(x_{ij} - \bar{x}_i)(\bar{x}_i - \bar{x}) + (\bar{x}_i - \bar{x})^2]$$

$$= \sum_{i=1}^{m} \sum_{j=1}^{n} (x_{ij} - \bar{x}_i)^2 + 2 \sum_{i=1}^{m} \sum_{j=1}^{n} (x_{ij} - \bar{x}_i)(\bar{x}_i - \bar{x})$$

$$+ \sum_{i=1}^{m} \sum_{j=1}^{n} (\bar{x}_i - \bar{x})^2$$

$$= \sum_{j=1}^{n} \sum_{i=1}^{m} (\bar{x}_i - \bar{x})^2 + \sum_{i=1}^{m} \sum_{j=1}^{n} (x_{ij} - \bar{x}_i)^2$$

$$+ 2 \sum_{i=1}^{m} \sum_{j=1}^{n} (x_{ij} - \bar{x}_i)(\bar{x}_i - \bar{x})$$

$$\therefore TV = nVB + VW + 2 \sum_{i=1}^{m} \sum_{j=1}^{n} (x_{ij} - \bar{x}_i)(\bar{x}_i - \bar{x})$$

because $VB = \sum_{i=1}^{m} (\bar{x}_i - \bar{x})^2$ is the variation of the m sample means from the grand mean $\bar{x}$ and

$$VW = VW_1 + VW_2 + \cdots + VW_m$$

$$= \sum_{j=1}^{n} (x_{1j} - \bar{x}_1)^2 + \sum_{j=1}^{n} (x_{2j} - \bar{x}_2)^2 + \cdots + \sum_{j=1}^{n} (x_{mj} - \bar{x}_m)^2$$

$$= \sum_{i=1}^{m} \sum_{j=1}^{n} (x_{ij} - \bar{x}_i)^2$$

We now show that the third term in the expression for TV above is always zero no matter what the data points are. Namely, we show that

$$2 \sum_{i=1}^{m} \sum_{j=1}^{n} (x_{ij} - \bar{x}_i)(\bar{x}_i - \bar{x}) = 0$$

from which it will follow that $TV = nVB + VW$. We proceed as follows:

$$2 \sum_{i=1}^{m} \sum_{j=1}^{n} (x_{ij} - \bar{x}_i)(\bar{x}_i - \bar{x}) = 2 \sum_{i=1}^{m} \sum_{j=1}^{n} (x_{ij}\bar{x}_i - \bar{x}_i^2 - \bar{x}x_{ij} + \bar{x}_i\bar{x})$$

$$= 2 \sum_{i=1}^{m} \sum_{j=1}^{n} x_{ij}\bar{x}_i - 2 \sum_{i=1}^{m} \sum_{j=1}^{n} \bar{x}_i^2 - 2 \sum_{i=1}^{m} \sum_{j=1}^{n} \bar{x}x_{ij} + 2 \sum_{i=1}^{m} \sum_{j=1}^{n} \bar{x}_i\bar{x}$$

$$= 2 \sum_{i=1}^{m} \bar{x}_i \sum_{j=1}^{n} x_{ij} - 2 \sum_{i=1}^{m} n\bar{x}_i^2 - 2\bar{x} \sum_{i=1}^{m} \sum_{i=1}^{n} x_{ij} + 2\bar{x} \sum_{i=1}^{m} n\bar{x}_i$$

$$= 2 \sum_{i=1}^{m} \bar{x}_i(n\bar{x}_i) - 2n \sum_{i=1}^{m} \bar{x}_i^2 - 2\bar{x}nm\bar{x} + 2\bar{x}mn\bar{x}$$

$$= 2n \sum_{i=1}^{m} \bar{x}_i^2 - 2n \sum_{i=1}^{m} \bar{x}_i^2 - 2nm\bar{x}^2 + 2nm\bar{x}^2$$

$$= 0$$

It therefore follows that $TV = nVB + VW$, which is the "components of the variance formula."

1 An individual considering purchase of a 5000-dollar life insurance policy wants to find out whether there are substantial differences between insurance premiums of the various types of policies available. In particular, he investigates the prices for the following related types of policies: individual renewable term, individual convertible term, group renewable term, and government group insurance. For each of these four types of policies, he obtains price quotations from six different competing underwriters. These price quotations are as follows:

Price Quotations for Monthly Premium, dollars			
Individual Renewable Term	Individual Convertible Term	Group Renewable Term	Government Group Insurance
2.50	2.00	2.00	1.00
3.00	2.20	1.50	1.20
2.20	1.80	1.80	1.00
2.80	2.50	2.00	1.10
2.50	3.00	2.00	1.50
3.00	2.00	1.80	1.40

a Calculate the components of the variance, and write down the numbers appearing in the components of the variance formula $TV = nVB + VW$.

b Decide at level of significance $\alpha = .05$ whether or not there are significant differences among the average monthly premiums of the four types of policies under study.

2 A stock brokerage firm, members of the New York Stock Exchange and other major exchanges, thinks it has a psychological test which can pick out those individuals who will turn out to be successful stockbrokers. As the firm usually recruits its employees from outside the securities industry, it would like to know in which occupations it can expect to find those with a high level of aptitude as determined by the psychological test. Before the current recruiting drive begins, the firm therefore decides to give its psychological test to six randomly selected individuals in each of six occupational classifications. The aptitude scores follow:

Aptitude Scores for Persons in Six Occupational Classifications					
Commission Salesperson	Office Clerk	Teacher	Civil Servant	Career Military	Professional Athlete
80	70	80	40	50	70
70	80	60	90	10	60
80	60	90	50	50	30
90	30	70	40	60	50
60	50	80	30	30	80
40	70	10	20	40	70

a Calculate the components of the variance.

b At level $\alpha = .01$, can the stock brokerage firm conclude that there is a significant different in aptitude among persons in the six occupational classifications?

3 An ichthyologist interested in the development of coastal nuclear generating facilities wants to find out whether temperature changes in the ocean's water will have a significant effect on the growth of fish indigenous to the region. He sets up an experiment involving four groups of seven recently hatched specimens each of the same species of fish. Each group is placed in a simulated ocean environment in which all factors are controlled and identical, with the exception of the temperature of the water. Six months later, the 28 specimens are weighed, and their weights are recorded in the following table:

Weights of Specimens from Each of Four Water-Temperature Groups, ounces			
40°F	42°F	44°F	46°F
20	18	24	16
18	25	17	17
23	16	14	26
16	20	22	19
15	24	25	14
25	16	27	20
16	21	18	21

a Calculate the components of the variance.

b At level $\alpha = .05$, do the results of the experiment indicate that the temperature differences tested have a significant effect on the average weight of fish in the region?

4 As part of a study of the manner in which pork-barrel projects are funded, a political scientist gathered the following data from eight randomly selected federal budgets over the past 40 years. The data points are the total appropriation in the budgets for pork-barrel projects in each of the three socio-politico-economic subdivisions of the nation: urban, suburban, and rural.

Appropriation for Each Socio-Politico-Economic Subdivision, millions of dollars		
Urban	Suburban	Rural
8	4	11
10	8	10
12	9	8
15	10	9
18	12	10
24	17	12
30	25	16
27	35	20

a Calculate the components of the variance.

b At level $\alpha = .01$, can we conclude from the data that the average amounts of federal money dispensed to each of the three subdivisions are approximately the same?

SECTION 10.B TESTING FOR DIFFERENCES AMONG SEVERAL MEANS

We now proceed to the development of the techniques of "one-way analysis of variance," a method of testing for differences among the true means of several groups. The theory underlying one-way analysis of variance has already been dealt with in the previous section, and even a few simple examples have been presented. In this section, we will concentrate on constructing a more efficient way of organizing the data and computations needed in carrying out the F test for the equality of several means.

Let's begin by taking another look at Example 10.1 and the data of Table 10.4. In order to test the hypothesis

$$H : \mu_U = \mu_V = \mu_W = \mu_X = \mu_Y$$

against the alternative

$$A : \text{These true means are not all the same}$$

it was necessary to compute the number

$$F = \frac{nVB/(m - 1)}{VW/m(n - 1)}$$

From the data of Table 10.4, it was easy to see that $n = 4$ and $m = 5$, and it was not really difficult to calculate $VB = 88$. The calculation of VW, however, was another story. That required us to do six separate calculations, one each of VW_U, VW_V, VW_W, VW_X, and VW_Y, and one summing all of the latter quantities. You must admit that the work required to calculate VW was somewhat out of line with that needed for the remainder of the problem and that a shortcut method of computing VW would be much appreciated by all those working in the field (even by those having access to expensive calculators!). The level of general appreciation would increase as the number of samples involved grew, for the work required to compute VW by the method of Sec. 10.A would grow correspondingly.

As it turns out, there is a very simple way of getting the value of VW. It is based on the components of the variance formula studied in the previous section, which asserts that

$$TV = nVB + VW$$

The shortcut formula is an immediate algebraic consequence of the above, namely, the assertion that

$$VW = TV - nVB$$

As we have already pointed out, n and VB are relatively easy to compute, and so if we can find a quick way of getting TV, we will have done the job.

Fortunately, there is a quick way of calculating TV which is well suited to both machine and hand computation. The explicit formula for TV is similar to the shortcut formula (4.3) for the standard deviation that we first discussed way back in Chap. 2 (Sec. 2.C, to be sure) and later made extensive use of in Chaps. 4 and 5. In particular, TV is given by the formula

$$TV = \Sigma x^2 - \frac{(\Sigma x)^2}{N}$$

where
N = total number of data points in all the samples put together
Σx^2 = sum of the squares of all N data points
Σx = sum of all N data points

In Table 10.5, we set up the data of Example 10.1 (Table 10.4) for the computation of TV. (A somewhat related construction in Table 10.3 gave a slightly longer calculation of TV for a different set of data.) The computations in Table 10.5 show that

$$N = 20$$
$$\Sigma x = 240$$
$$\Sigma x^2 = 3398$$

for the WP systems data, from which it follows that

$$TV = \Sigma x^2 - \frac{(\Sigma x)^2}{N} = 3398 - \frac{(240)^2}{20} = 3398 - \frac{57,600}{20} = 3398 - 2880 = 518$$

As long as we are on the subject of shortcut formulas, let's complete the story by presenting and showing how to use a shortcut formula for computing the term nVB. Our calculational task will be immediately shortened by one step simply because we will be getting the term nVB in its entirety rather than merely VB alone. The shortcut formula for nVB is

$$nVB = \sum_{i=1}^{m} \left(\frac{S_i}{n}\right)^2 - \frac{(\Sigma x)^2}{N}$$

where S_i is the sum of the n data points of sample i. Referring to Table 10.4, we can see that

$$S_U = 40$$
$$S_V = 32$$
$$S_W = 32$$
$$S_X = 64$$
$$S_Y = 72$$

TABLE 10.5
Calculation of the Total Variation *TV*
WP Systems Data

	x	x²
	12	144
	13	169
	7	49
	8	64
	12	144
	10	100
	6	36
	4	16
	3	9
	6	36
	10	100
	13	169
	14	196
	18	324
	18	324
	14	196
	16	256
	21	441
	15	225
	20	400
Sums	240	3398

$$N = 20$$
$$\Sigma x = 240$$
$$\Sigma x^2 = 3398$$
$$TV = \Sigma x^2 - \frac{(\Sigma x)^2}{N} = 3398 - \frac{(240)^2}{20} = 518$$

Because $\Sigma x = 240$, $N = 20$, and $n = 4$ as before, the shortcut formula for nVB yields that

$$nVB = \left(\frac{40^2}{4} + \frac{32^2}{4} + \frac{32^2}{4} + \frac{64^2}{4} + \frac{72^2}{4}\right) - \frac{(240)^2}{20}$$

$$= \left(\frac{1600}{4} + \frac{1024}{4} + \frac{1024}{4} + \frac{4096}{4} + \frac{5184}{4}\right) - \frac{57,600}{20}$$

$$= (400 + 256 + 256 + 1024 + 1296) - 2880$$

$$= 3232 - 2880$$

$$= 352$$

TABLE 10.6
One-Way Analysis of Variance Calculations
WP Systems Data

x^2	x	S	S^2	n	S^2/n
144	12				
169	13	40	1600	4	400
49	7				
64	8				
144	12				
100	10	32	1024	4	256
36	6				
16	4				
9	3				
36	6	32	1024	4	256
100	10				
169	13				
196	14				
324	18	64	4096	4	1024
324	18				
196	14				
256	16				
441	21	72	5184	4	1296
225	15				
400	20				
3398	240			20	3232

$$TV = \Sigma x^2 - \frac{(\Sigma x)^2}{N} = 3398 - \frac{(240)^2}{20} = 518$$

$$nVB = \Sigma\left(\frac{S^2}{n}\right) - \frac{(\Sigma x)^2}{N} = 3232 - \frac{(240)^2}{20} = 352$$

$$VW = TV - nVB = 518 - 352 = 166$$

As we have just observed, $TV = 518$ and $nVB = 352$. It therefore follows that

$$VW = TV - nVB = 518 - 352 = 166$$

which is, of course, the same result we obtained in the last section for the WP systems data.

In Table 10.6, we illustrate how to combine the calculations made in Tables 10.4 and 10.5 into a single table and thereby to obtain a more direct computation of TV, nVB, and ultimately VW.

The algebraic details of justifying the shortcut formulas for TV and nVB are presented in Supplement 10.2.

Now that we have introduced the shortcut methods that are commonly used in applied work, we proceed to carry out the entire one-way analysis of variance of the following example dealing with the difference of means.

Example 10.2   Gasoline Efficiency   A motorist would like to find out whether or not he gets the same gasoline efficiency (in miles per gallon) regardless of the brand of fuel he uses. He chooses five brands of gasoline available in his locality, and he records the number of miles per gallon he obtained on several tankfuls of each brand. In Table 10.7, his recorded mileage (mpg) for each tankful is listed.

SOLUTION   The items of primary interest are the five true means:

$$\mu_T = \text{true mean number of miles per gallon of Thrifty gasoline}$$
$$\mu_B = \text{true mean number of miles per gallon of Broadway gasoline}$$
$$\mu_F = \text{true mean number of miles per gallon of Federated gasoline}$$
$$\mu_G = \text{true mean number of miles per gallon of Gibraltar gasoline}$$
$$\mu_H = \text{true mean number of miles per gallon of Holiday gasoline}$$

We want to test the hypothesis

$$H : \mu_T = \mu_B = \mu_F = \mu_G = \mu_H$$

against the alternative

$$A : \text{These true means are not all the same}$$

In one-way analysis of variance, we compute[1]

TABLE 10.7
Efficiency of Different Brands of Gasoline
Data in Miles per Gallon

Thrifty	Broadway	Federated	Gibraltar	Holiday
26	29	29	31	28
25	27	29	28	23
28	26	26	26	24
24	30	24	28	28
29	25	25	26	25
24	25	23	27	22
26		26	28	25
			30	
		n		
7	6	7	8	7

[1]If all n's are equal, then $N = mn$, so that $N - m = mn - m = m(n - 1)$, the number that appeared in formula (10.2).

$$F = \frac{nVB/(m-1)}{VW/(N-m)} \qquad (10.3)$$

and we reject H in favor of A at level α if

$$F > F_\alpha(m-1, N-m) \qquad (10.4)$$

In this gasoline efficiency example,

$$m = \text{number of samples} = 5$$

and

$$N = \text{total number of data points}$$
$$= n_T + n_B + n_F + n_G + n_H$$
$$= 7 + 6 + 7 + 8 + 7 = 35$$

Therefore the degrees of freedom for the appropriate percentage point of the F distribution are

$$m - 1 = 5 - 1 = 4 \text{ (for the numerator)}$$
$$N - m = 35 - 5 = 30 \text{ (for the denominator)}$$

If we consider a level of significance of $\alpha = .01$ to be appropriate for this problem, then we agree to reject H in favor of A if

$$F > F_{.01}(4, 30) = 4.02$$

It therefore remains only to calculate the numerical value of F and to observe whether or not it exceeds 4.02. Inserting the numbers $n = 5$ and $N = 35$, we have

$$F = \frac{nVB/(m-1)}{VW/(N-m)} = \frac{nVB/4}{VW/30} = \frac{30nVB}{4VW}$$

where

$$nVB = \Sigma\left(\frac{S^2}{n}\right) - \frac{(\Sigma x)^2}{N}$$

$$TV = \Sigma x^2 - \frac{(\Sigma x)^2}{N}$$

so that

$$VW = TV - nVB$$
$$= \left[\Sigma x^2 - \frac{(\Sigma x)^2}{N}\right] - \left[\Sigma\left(\frac{S^2}{n}\right) - \frac{(\Sigma x)^2}{N}\right]$$
$$= \Sigma x^2 - \Sigma\left(\frac{S^2}{n}\right)$$

In Table 10.8 we present the calculations required to obtain the value of F, especially the calculation of

$$nVB = \Sigma\left(\frac{S^2}{n}\right) - \frac{(\Sigma x)^2}{N} \qquad (10.5)$$

Sums of x^2	x^2	x	S	S^2	n	S^2/n
	676	26				
	625	25				
	784	28				
	576	24	182	33,124	7	4732
	841	29				
	576	24				
4754	676	26				
	841	29				
	729	27				
	676	26	162	26,244	6	4374
	900	30				
	625	25				
4396	625	25				
	841	29				
	841	29				
	676	26				
	576	24	182	33,124	7	4732
	625	25				
	529	23				
4764	676	26				
	961	31				
	784	28				
	676	26				
	784	28	224	50,176	8	6272
	676	26				
	729	27				
	784	28				
6294	900	30				
	784	28				
	529	23				
	576	24				
	784	28	175	30,625	7	4375
	625	25				
	484	22				
4407	625	25				
Sums	24,615	925			35	24,485

$$nVB = \Sigma\left(\frac{S^2}{n}\right) - \frac{(\Sigma x)^2}{N} = 24,485 - \frac{(925)^2}{35} = 38.57$$

$$VW = \Sigma x^2 - \Sigma\left(\frac{S^2}{n}\right) = 24,615 - 24,485 = 130$$

and

$$VW = \Sigma x^2 - \Sigma\left(\frac{S^2}{n}\right) \tag{10.6}$$

We see from the results of Table 10.8 that

$$nVB = 38.57$$

and

$$VW = 130$$

from which it follows that

$$F = \frac{30nVB}{4VW} = \frac{(30)(38.57)}{(4)(130)} = 2.23$$

As we have agreed to reject the hypothesis

$$H : \mu_T = \mu_B = \mu_F = \mu_G = \mu_H$$

in favor of the alternative

$$A : \text{These true means are not all the same}$$

at level $\alpha = .01$ if F turned out to be larger than 4.02, the result that

$$F = 2.23 < 4.02$$

indicates that we should not reject H. At significance level $\alpha = .01$, then, we conclude that the data do not reveal significant differences in mileage per gallon among the five brands of gasoline tested.

With this example, we complete our formal discussion of the one-way analysis of variance problem. Before we consider two-way analysis of variance in the next section, it would be useful to explain what we have done in terms of the traditional language and vocabulary of analysis of variance. Referring again to the basic gasoline efficiency data of Table 10.7, we can view the five different brands of gasoline as "treatments" affecting the car's mileage per gallon.

Under the analysis of variance restrictions discussed at the conclusion of Sec. 10.A, we operate under the assumption that each treatment yields a set of data whose underlying population is normally distributed, and all such populations have the same standard deviation σ. The hypothesis

$$H : \mu_T = \mu_B = \mu_F = \mu_G = \mu_H$$

asserts that all such populations also have the same mean μ (which is, of course, the common numerical value of μ_T, μ_B, μ_F, μ_G, and μ_H), and therefore that *all* the data points come from a *single* normally distributed population having mean μ and standard deviation σ.

Considered from this point of view, the quantity

$$nVB = \sum_{i=1}^{m} n_i(\bar{x}_i - \bar{x})^2 = \Sigma\left(\frac{S^2}{n}\right) - \frac{(\Sigma x)^2}{N}$$

which we have been referring to as the variation between the samples can be considered as a measure of variation between the treatments.

Traditionally, then, the quantity nVB has been called the "treatment sum of squares" and has often been denoted by the symbols SST. What about the quantity VW? This number, which we have been calling the variation within the samples, represents the standard error (as in Sec. 6.E) made in using the mean of each sample as an estimate of the true mean of the corresponding population. Historically, then,

$$VW = \sum_{i=1}^{m} \sum_{j=1}^{n_i} (x_{ij} - \bar{x}_i)^2 = \Sigma x^2 - \Sigma\left(\frac{S^2}{n}\right)$$

has been called the "error sum of squares" and has often been denoted by the symbol SSE. In the language of treatment sum of squares and error sum of squares, the F statistic used to test the hypothesis in question would be

$$F = \frac{SST/(m-1)}{SSE/(N-m)}$$

where
$$m = \text{number of samples of data}$$
$$N = \text{total number of data points}$$
$$SST = \Sigma\left(\frac{S^2}{n}\right) - \frac{(\Sigma x)^2}{N}$$
$$SSE = \Sigma x^2 - \Sigma\left(\frac{S^2}{n}\right)$$

We then reject H, concluding that the "treatment effects" differ among themselves, at level α if $F > F_\alpha(m - 1, N - m)$, i.e., if the treatment sum of squares is substantially larger than the error sum of squares.

SUPPLEMENT 10.2   SHORTCUT FORMULAS FOR ONE-WAY ANALYSIS OF VARIANCE

The general structure of data in the one-way analysis of variance problem provides for m samples consisting of $n_1, n_2, \ldots, n_m$ data points, respectively. We list these in the following table, which has the same format as Table 10.6:

x^2	x	S	S^2	n	S^2/n
x_{11}^2 x_{12}^2 $\ldots$ $x_{1n_1}^2$	x_{11} x_{12} $\ldots$ x_{1n_1}	S_1	S_1^2	n_1	S_1^2/n_1
x_{21}^2 x_{22}^2 $\ldots$ $x_{2n_2}^2$	x_{21} x_{22} $\ldots$ x_{2n_2}	S_2	S_2^2	n_2	S_2^2/n_2

(Continued)

x^2	x	S	S^2	n	S^2/n
$\cdots$	$\cdots$	$\cdots$	$\cdots$	$\cdots$	$\cdots$
$x_{m1}^{\,2}$ $x_{m2}^{\,2}$ $\cdots$ $x_{mn_m}^{\;2}$	x_{m1} x_{m2} $\cdots$ x_{mn_m}	S_m	$S_m^{\,2}$	n_m	$S_m^{\,2}/n_m$
Σx^2	Σx			N	$\Sigma(S^2/n)$

In the above table,

$$N = \sum_{i=1}^{m} n_i \quad \text{and} \quad S_i = \sum_{j=1}^{n} x_{ij}$$

for each $i = 1, 2, \ldots, m$. We are now in a position to study the formulas for TV and nVB, using as our point of departure the expressions for TV and VB given in Supplement 10.1 in connection with the components of the variance formula.

We apply algebraic techniques similar to those in the supplements of Chaps. 2 and 6 during our study of the mean, the standard deviation, regression, and correlation. We have that

$$TV = \sum_{i=1}^{m} \sum_{j=1}^{n_i} (x_{ij} - \bar{x})^2 = \sum_{i=1}^{m} \sum_{j=1}^{n_i} (x_{ij}^2 - 2\bar{x}x_{ij} + \bar{x}^2)$$

$$= \sum_{i=1}^{m} \sum_{j=1}^{n_i} x_{ij}^2 - 2\bar{x} \sum_{i=1}^{m} \sum_{j=1}^{n_i} x_{ij} + N\bar{x}^2$$

$$= \sum_{i=1}^{m} \sum_{j=1}^{n_i} x_{ij}^2 - 2\bar{x}(N\bar{x}) + N\bar{x}^2$$

$$= \Sigma x^2 - 2N \left(\frac{\Sigma x}{N}\right)^2 + N \left(\frac{\Sigma x}{N}\right)^2$$

$$= \Sigma x^2 - 2 \frac{(\Sigma x)^2}{N} + \frac{(\Sigma x)^2}{N}$$

$$= \Sigma x^2 - \frac{(\Sigma x)^2}{N}$$

which is the shortcut formula for the total variation. Insofar as the term nVB is concerned, we must make some changes in view of the fact that different samples may have different numbers of data points, and therefore there is no unique number n that we can multiply by VB. In particular, instead of nVB being $n \sum_{i=1}^{m} (\bar{x}_i - \bar{x})^2$, we must write

$$nVB = \sum_{i=1}^{m} n_i(\bar{x}_i - \bar{x})^2$$

as each sample possesses its own corresponding value of n_i. It follows then that

$$nVB = \sum_{i=1}^{m} n_i(\bar{x}_i - \bar{x})^2 = \sum_{i=1}^{m} n_i \left(\frac{S_i}{n_i} - \bar{x}\right)^2$$

$$= \sum_{i=1}^{m} n_i \left(\frac{S_i^2}{n_i^2} - 2\bar{x}\, \frac{S_i}{n_i} + \bar{x}^2 \right)$$

$$= \sum_{i=1}^{m} \frac{S_i^2}{n_i} - 2\bar{x} \sum_{i=1}^{m} S_i + \bar{x}^2 \sum_{i=1}^{m} n_i$$

$$= \Sigma \left(\frac{S^2}{n} \right) - 2\bar{x} \sum_{i=1}^{m} \sum_{j=1}^{n_i} x_{ij} + N\bar{x}^2$$

$$= \Sigma \left(\frac{S^2}{n} \right) - 2\bar{x}(\Sigma x) + N \left(\frac{\Sigma x}{N} \right)^2$$

$$= \Sigma \left(\frac{S^2}{n} \right) - 2\, \frac{(\Sigma x)^2}{N} + \frac{(\Sigma x)^2}{N}$$

$$= \Sigma \left(\frac{S^2}{n} \right) - \frac{(\Sigma x)^2}{N}$$

This completes the algebraic justification of the shortcut formulas for one-way analysis of variance, namely,

$$TV = \Sigma x^2 - \frac{(\Sigma x)^2}{N}$$

and

$$nVB = \Sigma \left(\frac{S^2}{n} \right) - \frac{(\Sigma x)^2}{N}$$

EXERCISES 10.B

1 In the situation of Exercise 10.A.1, use the procedures of one-way analysis of variance to decide at level $\alpha = .05$ whether or not there are significant differences among the monthly premiums of the four competing plans.

2 Carry out a one-way analysis of variance based on the data of Exercise 10.A.2 in order to decide at level $\alpha = .01$ whether or not there are significant differences in aptitude among persons in the six occupational classifications.

3 Apply the one-way analysis of variance technique to the ichthyological data of Exercise 10.A.3 to find out at level $\alpha = .05$ whether temperature changes affect the weights of fish.

4 Using the data of Exercise 10.A.4, run a one-way analysis of variance to answer the question whether the average appropriations are basically the same to each of the three subdivisions. Use a level $\alpha = .01$.

5 A New Jersey tomato farmer wants to find out if his land is more suitable for production of one kind of tomato or if all kinds of tomatoes grow equally well on his land. He conducts an experiment planting beefsteak tomatoes on 9 plots, cherry tomatoes on 7 plots, and pear tomatoes on 6 plots. The resulting yields, in pounds of tomatoes, follow:

Tomato Yield, pounds		
Beefsteak	Cherry	Pear
120	160	110
110	140	90
140	130	100
100	140	120
120	150	100
110	140	80
130	140	
140		
130		

At level $\alpha = .05$, do the results of the experiment indicate the existence of significant differences among the yields of the three kinds of tomatoes?

6 A psychologist's research assistant wants to find out whether porpoises, monkeys, and rats learn at the same rate of speed. She places one animal at a time in an appropriate T-maze, the left exit of which contains a punishment such as an electric shock and the right exit, a reward such as food. The assistant runs 6 porpoises, 9 monkeys, and 11 rats through the maze, and she records the number of times through that it takes each animal to learn that the right fork is the one to take. The learning times for each animal are as follows:

Porpoises	3	7	9	15	6	2					
Monkeys	18	14	7	5	11	9	3	11	12		
Rats	4	21	11	16	19	23	7	10	17	6	8

At level $\alpha = .05$, can she conclude that all three species of animals have the same average learning times?

SECTION 10.C TWO-WAY ANALYSIS OF VARIANCE

During the course of its work on a cure for the common cold, a pharmaceutical company has discovered a new ingredient which it believes will substantially relieve the symptoms of the illness. The company must find out whether or not the new ingredient, MM-75, is truly effective in the relief of cold symptoms. For this purpose, it sets up an experiment in which the ingredient, both alone and in combination with other standard preparations, is tested for its ability to treat five of the more prevalent cold symptoms, namely, nasal congestion, cough, fever, earache, and upset stomach. Those individuals conducting the experiment make observations of the length of time it takes for the treatments to yield noticeable relief of the symptoms. The resulting data appear in Table 10.9.

We define the following notation:

μ_P = true mean relief time when treated by MM-75 plain

μ_{AS} = true mean relief time when treated by aspirin with MM-75

μ_{AH} = true mean relief time when treated by antihistamine with MM-75

μ_{CS} = true mean relief time when treated by cough syrup with MM-75

μ_{ND} = true mean relief time when treated by nose drops with MM-75

μ_{NT} = true mean relief time when not treated

We can test the hypothesis, as we did in our discussion of one-way analysis of variance in the previous section, that the six treatments do not differ significantly in mean relief time, namely,

$$H_T : \mu_P = \mu_{AS} = \mu_{AH} = \mu_{CS} = \mu_{ND} = \mu_{NT}$$

against the alternative

$$A_T : \text{These true mean relief times are not all the same}$$

We use the symbols H_T and A_T to indicate that these are the "treatment hypothesis" and the "treatment alternative," respectively. The *failure* to reject H in favor of A would be evidence supporting an assertion that MM-75 is *not* effective, for it would mean that all the remedies, even with MM-75 added, are just as effective as no treatment at all, the last of the six treatments considered in the study.

In addition to the hypothesis of equal "treatment effects," the two-way structure of Table 10.9 provides us with enough information to deal with another question: whether or not the five symptoms differ in duration of time until relief begins. For this question, we test the hypothesis

$$H_B : \mu_N = \mu_C = \mu_F = \mu_E = \mu_U$$

against the alternative

$$A_B : \text{These true mean relief times are not all the same}$$

where

μ_N = true mean time until relief of nasal congestion

μ_C = true mean time until relief of cough

μ_F = true mean time until relief of fever

μ_E = true mean time until relief of earache

μ_U = true mean time until relief of upset stomach

Here we have used the symbols H_B and A_B to indicate that these are the "block hypothesis" and the "block alternative," respectively. In two-way analysis of variance, we refer to the categories along the top of the table as the "treatments" and the categories along the side as the "blocks." Then we view the data points in the body of the table (Table 10.9, for example) as depending, to various extents, on treatment effects and block effects. In this language, we can reformulate the treatment hypothesis as

TABLE 10.9
Effectiveness of New Ingredient MM-75 in the Treatment of Cold Symptoms
Data in Hours until Occurrence of Noticeable Relief

Treatment / Symptom	MM-75	Aspirin with MM-75	Antihistamine with MM-75	Cough Syrup with MM-75	Nose Drops with MM-75	No Treatment
Nasal congestion	35	30	25	41	37	42
Cough	38	28	23	29	30	38
Fever	41	32	24	32	28	47
Earache	29	35	24	36	25	37
Upset stomach	37	25	29	42	30	47

$$H_T : \text{All treatments have equal effects on relief time}$$

and the treatment alternative as

$$A_T : \text{The treatment effects are not all the same}$$

Similarly, we can express the block hypothesis as

$$H_B : \text{All blocks have equal effects on relief time}$$

and the block alternative as

$$A_B : \text{The block effects are not all the same}$$

Now that we have formulated the questions to be answered on the basis of the data of Table 10.9, we proceed to develop the two-way analysis of variance techniques used toward their solution. As in one-way analysis of variance, we use an F test to test each of the hypotheses H_T and H_B. Here, as in the previous section, the F test for H_T involves a treatment sum of squares SST and an error sum of squares SSE. By analogy, the F test for H_B involves a block sum of squares SSB and the error sum of squares SSE. In particular, if we set

$$m = \text{number of treatments}$$

and

$$n = \text{number of blocks}$$

then we reject H_T in favor of A_T at level α if

● $$F_T > F_\alpha(m - 1, (n - 1)(m - 1)) \tag{10.7}$$

where

● $$F_T = \frac{SST/(m - 1)}{SSE/(n - 1)(m - 1)} = \frac{(n - 1)SST}{SSE} \tag{10.8}$$

and we reject H_B in favor of A_B at level α if

● $$F_B > F_\alpha(n - 1, (n - 1)(m - 1)) \tag{10.9}$$

where

● $$F_B = \frac{SSB/(n - 1)}{SSE/(n - 1)(m - 1)} = \frac{(m - 1)SSB}{SSE} \tag{10.10}$$

If we denote by N the total number of data points ($N = mn$), then we have the following formulas which are useful in computing F_T and F_B:

● $$SST = \frac{\Sigma S_T{}^2}{n} - \frac{(\Sigma x)^2}{N} \tag{10.11}$$

● $$SSB = \frac{\Sigma S_B{}^2}{m} - \frac{(\Sigma x)^2}{N} \tag{10.12}$$

and

$$SSE = \Sigma x^2 - \frac{\Sigma S_T^2}{n} - \frac{\Sigma S_B^2}{m} + \frac{(\Sigma x)^2}{N} \tag{10.13}$$

In the above computing formulas, S_T stands for the column (treatment) sums and S_B stands for the row (block) sums. In Table 10.10, we illustrate the form a typical two-way analysis of variance table will take.

We now proceed to the solution of the two-way analysis of variance problem based on the data of Table 10.9. Using a level of significance of $\alpha = .05$, we agree to reject

H_T : All treatments have equal effects on relief time
(so that MM-75 cannot be considered effective)

in favor of

A_T : The treatment effects are not all the same

if

$$F_T > F_{.05}(5, 20) = 2.71$$

because $m = 6$ (number of treatments) and $n = 5$ (number of blocks). And, also at level $\alpha = .05$, we agree to reject

H_B : All blocks have equal effects on relief time
(so that all symptoms have the same duration)

in favor of

A_B : The block effects are not all the same
(so that some symptoms last longer than others)

if

$$F_B > F_{.05}(4, 20) = 2.87$$

TABLE 10.10
The Form of a Two-Way Analysis of Variance Table
For Use in Calculations

Treatments / Blocks	Treatment 1	Treatment 2	...	Treatment m	Block Sums	Squares of Block Sums
Block 1	x_{11}	x_{12}	...	x_{1m}	S_{B_1}	$S_{B_1}^2$
Block 2	x_{21}	x_{22}	...	x_{2m}	S_{B_2}	$S_{B_2}^2$
Block n	x_{n1}	x_{n2}	...	x_{nm}	S_{B_n}	$S_{B_n}^2$
Treatment sums	S_{T_1}	S_{T_2}	...	S_{T_m}	Σx	ΣS_B^2
Squares of treatment sums	$S_{T_1}^2$	$S_{T_2}^2$	...	$S_{T_m}^2$	ΣS_T^2	

It remains therefore only to compute the numerical values of F_T and F_B and to determine whether or not they exceed 2.71 and 2.87, respectively. The calculations leading to the computation of these numbers require a table having the form of Table 10.10. They are presented in Table 10.11, using the data of Table 10.9. The table at the bottom of Table 10.11 contains the square of each data point in Table 10.9. Its only objective is the numerical value of Σx^2 (the sum of squares of all the data points) which appears in the formula for SSE.

Using the results of the calculations appearing at the bottom of Table 10.11, we see that

$$SST = \frac{\Sigma S_T^2}{n} - \frac{(\Sigma x)^2}{N} = \frac{169{,}946}{5} - \frac{(996)^2}{30}$$

$$= 33{,}989.2 - \frac{992{,}016}{30} = 33{,}989.2 - 33{,}067.2 = 922$$

$$SSB = \frac{\Sigma S_B^2}{m} - \frac{(\Sigma x)^2}{N} = \frac{199{,}008}{6} - 33{,}067.2$$

$$= 33{,}168 - 33{,}067.2 = 100.8$$

and

$$SSE = \Sigma x^2 - \frac{\Sigma S_T^2}{n} - \frac{\Sigma S_B^2}{m} + \frac{(\Sigma x)^2}{N}$$

$$= 34{,}444 - 33{,}989.2 - 33{,}168 + 33{,}067.2$$

$$= 354$$

From these calculations, it then follows that

$$F_T = \frac{(n-1)SST}{SSE} = \frac{(4)(922)}{354} = 10.42$$

and

$$F_B = \frac{(m-1)SSB}{SSE} = \frac{(5)(100.8)}{354} = 1.42$$

We are now ready to draw our conclusions. We agreed to reject H_T (all treatments have equal effects) in favor of A_T (the treatment effects are not all the same) if F_T turned out to be greater than 2.71. And $F_T = 10.42 > 2.71$, and so we do reject H_T. This means that at level $\alpha = .05$ we conclude from the data of Table 10.9 that the treatment effects are not all the same. We therefore seem to have some statistical evidence that the various medications studied do have a detectable effect on the waiting time until noticeable relief occurs, at least when compared with no treatment at all.

In regard to the blocks, we have agreed to reject H_B (all symptoms have the same duration) in favor of A_B (some symptoms last longer than others) at level $\alpha = .05$ if $F_T > 2.87$. As it turned out, however, $F_T = 1.42 < 2.87$, and so we cannot reject H. We therefore assert at level $\alpha = .05$ that the data of Table 10.9 indicate that all the symptoms studied have the same duration, in the sense that the mean times until relief of the symptoms do not differ significantly among themselves.

TABLE 10.11
(a) Calculations for Two-Way Analysis of Variance
New Ingredient for Treating Cold Symptoms Data

Block \ Treatment	P	AS	AH	CS	ND	NT	Block Sums	Squares of Block Sums
N	35	30	25	41	37	42	210	44,100
C	38	28	23	29	30	38	186	34,596
F	41	32	24	32	28	47	204	41,616
E	29	35	24	36	25	37	186	34,596
U	37	25	29	42	30	47	210	44,100
							996	199,008
Treatment sums	180	150	125	180	150	211	996	
Squares of treatment sums	32,400	22,500	15,625	32,400	22,500	44,521	169,946	

(b) Calculation of Σx^2, the Sum of Squares of All Data Points

Block \ Treatment	P	AS	AH	CS	ND	NT	Σ
N	1225	900	625	1681	1369	1764	7564
C	1444	784	529	841	900	1444	5942
F	1681	1024	576	1024	784	2209	7298
E	841	1225	576	1296	625	1369	5932
U	1369	625	841	1764	900	2209	7708
Σ	6560	4558	3147	6606	4578	8995	34,444

$\Sigma x = 996$ $\Sigma S_T^2 = 169,946$
$\Sigma x^2 = 34,444$ $\Sigma S_B^2 = 199,008$
$m = 6$
$n = 5$ $N = 30$

1 Accident figures accumulated by the National Safety Council over a three-day holiday weekend for five consecutive years show the following number of fatalities in each of three cities of comparable size:

	Number of Fatalities		
Year	Baltimore	Detroit	San Diego
1972	20	22	16
1973	18	21	14
1974	25	28	15
1975	28	30	20
1976	29	29	25

a At level $\alpha = .05$, do the data indicate that the number of fatalities differs significantly among the three cities?

b At level $\alpha = .01$, do the data indicate that the number of fatalities differs significantly from year to year?

2 An economist for HEW is conducting a study of vegetable prices in various metropolitan areas for the purpose of estimating the cost to an individual of maintaining a diet containing all the recommended daily amounts of nutrients. The following data give the retail prices per pound of four basic vegetables in seven major metropolitan areas:

	Price per Pound			
City	Lettuce	Tomatoes	Carrots	Onions
New York	30	28	34	26
Los Angeles	18	25	17	17
Chicago	28	21	19	31
Philadelphia	26	30	32	29
Detroit	20	29	30	19
Houston	25	16	27	20
Miami	26	31	28	31

a At level $\alpha = .05$, do the results of the survey show that vegetable prices differ significantly among the metropolitan areas studied?

b At level $\alpha = .05$, does the study indicate that there are substantial price differences among the various vegetables listed?

3 As part of a study of the role of the coffee break in the proper functioning of the American economy, a corporation psychologist selected 20 employees of a major company classified according to position and number of cups of coffee drunk per day and rated each on a job performance scale. The ratings are:

Cups of Coffee Drunk per Day	Position			
	Secretary	Production Supervisor	Accountant	Division Manager
No coffee	15	10	9	11
1–3 cups	8	9	11	14
4–6 cups	12	13	8	15
7–9 cups	18	12	17	17
10 or more cups	12	12	15	18

a At level of significance $\alpha = .05$, do the various positions seem to have differing ratings?

b At level $\alpha = .01$, do differences in the number of cups of coffee consumed lead to significant differences in job performance ratings?

4 As part of an effort to increase the world's food supply, an agricultural research organization conducted a test of the effects of various kinds of fertilizer-soil combinations on the production of lettuce. In all, four varieties of lettuce were grown intensively in six different fertilizer-soil combinations, and the following yields (in pounds) were obtained from the 24 plots of land:

Fertilizer-Soil Combinations	Yield, pounds			
	Iceberg	Simpson	Red Leaf	Butter
Organic-loam	150	160	140	160
Organic-clay	130	140	130	150
Organic-sand	80	90	100	140
Chemical-loam	170	170	160	180
Chemical-clay	120	150	120	160
Chemical-sand	90	100	110	140

a At level $\alpha = .01$, do the data reveal significant differences in yield among the four varieties of lettuce?

b At level $\alpha = .01$, do the data point up significant differences in yield among the six fertilizer-soil combinations?

5 Some economists believe that in a period of simultaneous inflation and recession the real growth of the GNP, the overall price increase, and the unemployment rate will be nearly identical (when each is expressed as a percentage). The following data[1] list the economic forecasts for 1976 by 24 leading economists:

[1]*Business Week*, Dec. 29, 1975, p. 44.

	Forecasts for 1976		
Economist	Real GNP Growth	Price Increase	Unemployment Rate
R.D.	6.7	6.7	6.9
A.T.S.	6.9	6.1	8.0
K.B.S.	7.1	5.9	7.8
R.O.	6.7	6.0	7.6
D.A.H.	5.9	7.3	7.7
D.S.A.	6.6	6.5	7.8
R.J.	6.0	6.7	7.6
P.J.M.	7.1	5.1	7.8
R.G.D.	6.0	6.0	7.6
F.H.S.	6.0	6.2	7.8
J.R.F.	5.5	6.5	7.8
B.A.G.	5.9	6.0	7.7
P.L.B.	6.4	5.5	7.5
A.G.H.	5.6	5.8	7.8
H.E.N.	6.0	5.5	8.0
A.G.M.	5.4	5.9	7.7
M.C.	5.9	5.4	7.8
D.R.C.	5.4	5.6	7.9
W.C.F.	5.0	5.5	7.6
R.E.	5.4	5.3	8.0
R.H.P.	4.9	7.7	7.8
G.W.M.	5.7	4.8	8.1
J.J.O.	5.1	4.9	8.4
A.G.S.	4.9	4.1	7.5

a At level $\alpha = .01$, do the forecasts of the various economic quantities seem to differ significantly on the average?

b At level $\alpha = .05$, do the data indicate that the 24 economists have significantly differing forecasts?

SUMMARY AND DISCUSSION

In the present chapter, we have extended the Chap. 5 discussion of two-sample tests for the difference of means to the case of more than two samples. Using one-way analysis of variance, we were able to consider several samples simultaneously testing the hypothesis that all their population means are equal against the alternative that significant differences exist among the means.

The statistical analysis was based on the relationship of the variation between the sample means and the variation of the data points within each sample. The ratio between these two variations is mathematically very similar to the ratio between two squares of standard deviations, and therefore has the F distribution. It followed, then, that an F test of the sort used in Chap. 5 to test the difference of standard deviations can be applied in analysis of variance to test the difference of several means.

Finally, we have discussed the process of two-way analysis of variance, which allows us to test the effects of two types of influences (called treatments and blocks) on the underlying populations. The mathematical techniques of two-way analysis of variance were not new but were merely minor revisions of those used in one-way analysis of variance.

BIBLIOGRAPHY

Detailed Treatment of Analysis of Variance
Guenther, W. C.: "Analysis of Variance," Prentice-Hall, Englewood Cliffs, N.J., 1964.

Special Topics
D'Agostino, R. B.: Relation between the Chi-Squared and ANOVA Tests for Testing the Equality of k Independent Dichotomous Populations, *The American Statistician*, June 1972, pp. 30–32.

Harter, H. L.: Multiple Comparison Procedures for Interactions, *The American Statistician*, December 1970, pp. 30–32.

Sirotnik, K.: On the Meaning of the Mean in ANOVA (or the Case of the Missing Degree of Freedom), *The American Statistician*, October 1971, pp. 36–37.

SUPPLEMENTARY EXERCISES

1   In a study of whether or not various kinds of pollutants in the air inhibit the growth of mice, 28 mice born at the same time were fed exactly the same diet in order to prepare them for the experiment. All other living conditions were identical, except that the mice were divided into four groups according to the type of pollution in their atmosphere. A biologist participating in the study collected the following data on weight gains of the mice over a period of time:

Weight Gains of Mice, grams			
Atmosphere with Cigarette Smoke	Atmosphere with Auto Exhaust	Atmosphere with Industrial Smoke	Clean Air
6	9	10	11
7	7	9	8
6	9	8	10
10	8	5	9
8	5	10	8
5	10	8	6
7	8	6	11

a Calculate the components of the variance.

b Using the components of the variance, decide at level $\alpha = .05$ whether or not there are significant differences in weight gains of mice in the four types of atmosphere.

c Carry out a one-way analysis of variance to answer the question of part b.

2 It is the job of a tea tester for an English tea-importing company to decide which tea leaves are most suitable for English Breakfast tea. The tea tester must decide among three lots of leaves: an inexpensive lot, a moderately priced lot, and an expensive lot. The tea tester and his laboratory staff discover the following levels of impurities in eight randomly selected leaves of each lot:

Inexpensive lot	3	4	25	9	11	4	12	20
Moderately priced lot	4	9	16	9	12	6	4	8
Expensive lot	6	5	7	10	6	8	5	4

a Calculate the components of the variance.

b Use the components of the variance to test at level $\alpha = .01$ whether or not all three lots of leaves have the same average impurity levels.

c Apply the one-way analysis of variance technique to test the same hypothesis as in part b.

3 Peanuts are known to be an excellent source of protein, and they are therefore a good substitute for meat and fish when prices are high. To find out whether storage of peanuts for a length of time tends to change their protein content, a nutritionist selected several bags of peanuts from lots that had been in storage for various periods of time and then she measured the protein content of each bag. The data, in grams of protein per bag, follow:

Fresh Peanuts	Peanuts Stored 6 Months	Peanuts Stored 12 Months	Peanuts Stored 18 Months	Peanuts Stored 24 Months
10	8	5	10	7
8	6	8	6	8
6	5	9	9	7
10	10	8	10	11
8	7	8	6	5
10	9	6	8	9
6	6	10	7	6
8		5	7	8
6		7		10
10		5		
4				

At level $\alpha = .05$, do the data indicate that storage of peanuts tends to alter their protein content?

4 A factory, in the process of assessing its energy costs, conducted a study of how much it costs to run each of four types of machines that it uses with each of five different energy sources. The monthly costs, in hundreds of dollars, are listed below:

Type of Machine	Energy Source				
	Electricity	Natural Gas	Oil	Coal	Manual
Drill press	6	5	8	4	5
Metal stamping	4	3	6	2	4
Sorter	3	3	4	2	2
Conveyor	2	3	3	3	6

a At level $\alpha = .01$, do the data provide evidence sufficient to assert that different energy sources result in different operating costs, on the average, for the types of machines under study?

b At level $\alpha = .01$, can we conclude from the data that the four types of machines differ significantly among themselves in energy costs?

5 The following data[1] concern 1958 agricultural production (in thousands of tons of various crops) in the People's Republic of China. The data consist of early estimates of the 1958 production, a final total of the production, and a revised total presumably obtained from a thorough compilation of the agricultural records.

Crop	Jan. 1959 Estimate	Apr. 1959 Final Total	Sept. 1959 Revised Total
Ginned cotton	3,350	3,319	2,100
Soy beans	12,500	12,500	10,500
Ground nuts	6,300	4,000	2,800
Jute and hemp	375	325	320
Rapeseed	1,385	1,100	1,100
Sugar cane	20,000	13,525	13,525
Sugar beets	3,000	2,900	2,900

a Calculate the components of the variance.

b Use the components of the variance to decide at level $\alpha = .05$ whether or not there exist significant differences in the average agricultural production measured at the three different times.

c Apply the one-way analysis of variance procedure in testing the same hypothesis as in part b.

6 An agricultural analysis of several minority ethnic-group Soviet republics yielded the following data[1] on the total crop area (in thousands of hectares) for

[1]Choh-Ming Li, "The Statistical System of Communist China," p. 90, University of California Press, Berkeley, 1962.

several years in the past:

Republic	1913	1940	1950	1954	1955
Uzbek SSR	2166	3014	2773	2887	2939
Kazakh SSR	4194	6831	7885	11552	20662
Georgian SSR	748	907	914	944	951
Azerbajian SSR	962	1124	1057	1126	1272
Lithuanian SSR	1890	2497	2294	2242	2055
Moldavian SSR	2072	2057	1895	1904	1982
Latvian SSR	1396	1964	1413	1620	1447
Kirghiz SSR	640	1056	1061	1135	1207
Tajik SSR	494	807	837	802	811
Estonian SSR	697	918	813	854	782

a At level $\alpha = .05$, do the data indicate that the average crop area in these Soviet republics did not change over the years?
b Can we conclude at level $\alpha = .05$ that the various republics differ in average yearly crop area?

[1]Central Statistical Board of the U.S.S.R. Council of Ministers, "National Economy of the U.S.S.R.: Statistical Returns," p. 102, Foreign Languages Publishing House, Moscow, 1957.

NONPARAMETRIC STATISTICS

In most of the applied situations studied in the previous chapters using the format of testing statistical hypotheses, there were somewhat stringent requirements on the nature of the data involved. Many of the small-sample ($n < 30$) tests, for example, could not be applied unless it was reasonable to assume that the underlying populations from which the random samples were drawn were normally distributed. The restriction to normally distributed data was needed because an integral part of the testing procedure made use of the table of the normal distribution or some other table derived from it, such as that of the t distribution. Prime among the statistical tests falling into this category were the t tests of means in Chap. 5 and the correlation test for linearity of Chap. 6. Our objective in this chapter is to explain how to handle applied situations in which the prerequisites for using these tests are not met.

All the statistical tests studied in Chaps. 5, 6, and 10 dealt with parameters of populations. The concept of a parameter, as you can perhaps recall, was discussed in some detail in Chap. 3 in connection with the binomial and normal distributions. It was pointed out there that both these distributions possess certain characteristic numbers, called parameters, from which the complete behavior pattern of the distribution can be determined. For the binomial distribution, the parameters were denoted by n and p, while for the normal distribution they were called μ and σ. The tests of Chaps. 5 and 10 all involved the study of one or another of these parameters: we had one- and two-sample tests for population means μ (of normal distributions), one- and two-sample tests for population standard deviations σ (of normal distributions), and one- and two-sample tests for population proportions p (of binomial distributions). Because these tests all involve parameters, they could be referred to as "parametric" tests.

If the underlying populations from which our random samples were selected do not have distributions that behave in accordance with the table of the normal distribution, then these populations cannot be characterized by parameters μ and σ. (A statistical test of whether or not a population could be considered as having the normal distribution was presented in Sec. 9.C.) It logically follows that applied problems involving means and standard deviations of nonnormal populations cannot be solved[1] using the usual parametric tests of μ and σ, because such populations do not possess parameters having the same statistical properties as μ and σ. However, because small-sample problems involving nonnormal populations often arise out of applied contexts, it is necessary to develop new statistical techniques for analyzing and solving them. For reasons that should now be obvious, these new techniques are called "nonparametric."

The organized study of nonparametric statistics originated in the search for substitute methods for the two-sample t test for the difference of means when the populations involved are nonnormal. Because of the nonparametric nature of the problem, we can no longer test for the difference of means, but only for the difference of other types of "averages." The tests we study in this chapter, namely, the "sign test" and the "Mann-Whitney test," are really tests of the difference of medians, rather than the difference of means. As a rule, nonparametric tests yield less information than parametric tests because of their inability to measure the mean, which is a mathematically more precise measure of average than the median. For this reason, they should never be used when we know we are working with normally distributed data.

In place of the parametric F test for the difference of standard deviations, we introduce the nonparametric "Mann-Whitney test for equal dispersions," the notion of dispersion of a set of data away from its average being less precise than the concept of the standard deviation of a set of data from its mean. In the same spirit, we present the "Kruskal-Wallis nonparametric analysis of variance" procedure as a substitute for the normality-based analysis of variance of Chap. 10. The correlation test (appearing in Chap. 6) of linearity between two sets of data, the x's and the y's, is also, you may be surprised to know, a parametric test. The parameter involved is the "true correlation" ρ relating the two normally distributed populations of x's and y's. The correlation ρ is a parameter of the "bivariate normal distribution," a two-dimensional probability distribution formed by the intertwining of two normal populations. In the nonnormal case, we use a nonparametric replacement, the "Spearman test of rank correlation." (The word "rank" is not used here in a derogatory sense, but merely refers to the mathemati-

---

[1] Unless enough data points are available to satisfy the criteria of the central limit theorem.

TABLE 11.1
Parametric Tests and Their Nonparametric Replacements

Section	Parametric Test	Nonparametric Replacement	Section
5.B	One-sample t test	Sign test	11.A
5.C	Two-sample t test	Mann-Whitney test	11.B
5.D	Paired-sample t test	Sign test	11.A
5.E	Two-sample F test	Mann-Whitney test	11.C
6.D	Correlation test	Spearman rank correlation test	11.F
10.B	One-way analysis of variance	Kruskal-Wallis test	11.D

cal methods used.) These parametric tests and their nonparametric substitutes are listed formally in Table 11.1.

Because the nonparametric tests discussed above are freed from the restrictive conditions which tied the earlier tests to the normal distribution, they are often referred to as "distribution-free" tests. More precisely, distribution-free tests are tests that can be applied to many types of data regardless of the underlying distribution of the population.

In addition to the nonparametric tests mentioned above, there are some other kinds of distribution-free tests which are not nonparametric substitutes for parametric tests. Such tests, which for lack of a standard term we can refer to as "aparametric," are not involved in any way with parameters or substitutes for parameters, such as averages or dispersions. We have encountered an example of such an aparametric test already, namely, the chi-square test of independence in Sec. 9.A. That test has as its objective to decide whether or not two different ways of classifying members of a population are independent or dependent. Obviously nothing even remotely resembling a parameter is involved here; independence is not a numerical characteristic, but rather a qualitative one. In the present chapter, we discuss a second example of an aparametric distribution-free test, the "Wald-Wolfowitz test of randomness," the objective of which is to judge whether a sequence of data indicates random behavior in a population or whether it instead reveals the presence of "trends" or "cyclical" behavior.

Having discussed the logical background for the development of nonparametric statistics, we are now ready to proceed to the sign test, the simplest of all the nonparametric tests.

SECTION 11.A THE SIGN TEST

When we would like to use the one-sample t test or the paired-sample t test on an applied problem but are prevented from doing so by the fact that the population from which the data points were selected is probably not normally distributed, the sign test provides a valid way of analyzing the situation. Although the sign test has the valuable advantage that it can be used in cases involving nonnormal data, it suffers from the defect common to all nonparametric tests—a lack of precision regarding exactly what is being tested. For example, as we shall soon see, the one-sample sign test pays no attention at all to the actual numerical values of the data points, but merely notes whether each is below or above the number that plays the role of the hypothesized mean. This means that the one-sample sign test is really a test of whether or not the "true median" is equal to the "hypothesized median," rather than a test of whether the true mean is equal to the hypothesized mean. In fact, the sign test can be viewed as a particular type of test for the binomial proportion $p = \frac{1}{2}$, where the two classifications for the binomial distribution are the categories above and below the hypothesized median. The two-sample sign test, used as a replacement for the paired-sample t test, bears a similar relationship to the binomial distribution. Small-sample versions of the sign test base their rejection of the hypothesis on the table of binomial distribution, while large-sample versions use the normal approximation to the binomial.

It is instructive to illustrate the sign test by using it to analyze the same situations we faced in the comparable sections of Chap. 5. This method points up very clearly the distinctions between the parallel parametric and nonparametric tests. We begin with the situation of Example 5.2, the setup of which we repeat below:

Example 11.1 Body Temperatures A medical doctor who is also an amateur anthropologist is interested in finding out whether the average body temperature of Alaskan Eskimos is significantly lower than the usual American average, which is 98.6°F. Eight Eskimos selected at random from the State of Alaska census lists had the following recorded body temperatures in degrees Fahrenheit:

$$98.5 \quad 98.1 \quad 98.6 \quad 98.7 \quad 98.4 \quad 98.9 \quad 98.0 \quad 98.4$$

At significance level $\alpha = .05$, do the results of the study support the assertion that Eskimos really have lower body temperatures than Americans who are natives of warmer climates?

SOLUTION   For purposes of illustration, we will apply the one-sample sign test to this problem. If we denote by μ^* the true median body temperature of Alaskan Eskimos, the sign test is capable of testing the hypothesis

$$H : \mu^* = 98.6 \text{ (Eskimos have the same average body}$$
$$\text{temperature as other Americans)}$$

against the alternative

$$A : \mu^* < 98.6 \text{ (Eskimos have lower body temperatures)}$$

The number 98.6 here is called the hypothesized median, by analogy with the hypothesized mean appearing in the one-sample tests of Chap. 5, and we use the symbol

$$\mu_0^* = \text{hypothesized median}$$

The preliminary computational setup for the one-sample sign test appears in Table 11.2. Major components of the actual test for the rejection of H in favor of A are the quantities

θ = number of data points exceeding the hypothesized median μ_0^*

= number of plus (+) signs in the list of signs

n = number of data points not equaling μ_0^*

= number of pluses plus number of minuses

$p = \frac{1}{2}$

The numbers n and p are used in the small-sample case as the parameters of the appropriate binomial distribution. For a test such as the sign test, which involves the median, we always use $p = \frac{1}{2}$ because, if the hypothesis H is really true, we should ideally have half pluses and half minuses in the list of signs.

From the calculations of Table 11.2, we find that we must use the binomial distribution with parameters $n = 7$ and $p = \frac{1}{2}$. The portion of Table A.2 which is reproduced in Table 11.3 indicates that if we use a significance level of $\alpha = .0625$ instead of $\alpha = .05$, we can reject

$$H : \mu^* = 98.6$$

in favor of

$$A : \mu^* < 98.6$$

if

$$\theta < 2$$

We derive this rejection rule from the fact that

$$P(\theta < 2) = P(\theta \leq 1) = .0625$$

according to Table 11.3, if θ is a binomial random variable with parameters $n = 7$ and $p = \frac{1}{2}$.

TABLE 11.2
Calculations for the
One-Sample Sign Test
Body Temperature Data

x	μ_0^*	Sign of $x - \mu_0^*$
98.5	98.6	−
98.1	98.6	−
98.6	98.6	0
98.7	98.6	+
98.4	98.6	−
98.9	98.6	+
98.0	98.6	−
98.4	98.6	−

Total Number of +'s = 2

$\theta = 2$

$n = 8 - 1 = 7$

$p = \frac{1}{2}$

TABLE 11.3
A Small Portion of
Table A.2

n	k	p = .50
7	0	.0078
	1	.0625
	2	.2266

Therefore, our rejection rule for H against A requires us to reject H if $\theta < 2$. As it has turned out, however, we have obtained from the calculations in Table 11.2 that $\theta = 2$. And if $\theta = 2$, then it is not true that $\theta < 2$. Therefore, we cannot reject H at level $\alpha = .0625$. Because .05 is between .0078 and .0625, this decision also implies that we cannot reject H at level $\alpha = .05$. Our conclusion, then, at level $\alpha = .05$ is that the data do not support the assertion that Eskimos have lower body temperatures on the average than Americans of warmer climates.

Our decision, in regard to the question of Example 11.1, to support the assertion that Eskimos do not have lower body temperatures than Americans of warmer climates was the same decision to which we came in Example 5.2, using the one-sample t test. Although it is reassuring that this happened, it is frequently true that parametric and nonparametric tests applied to the same set of data will recommend "opposite" decisions. In the sign test versus the t test, this disparity could occur because the true median, for example, is larger than the hypothesized number, while the true mean is not. The decisions are not really opposite because both tests do not really test exactly the same thing, but the appearance of opposite is given if we think about the problem in terms of average instead of the more precise terms mean and median.

Which test is to be believed in the case of opposite decisions? Well, if the data come from a normally distributed population, the parametric test takes precedence because it makes use of the data directly instead of "shadow" properties of the data, such as signs or ranks. On the other hand, if the normality or other conditions for the parametric test are not met, then the nonparametric test dominates because the parametric test simply does not apply.

One more comment on the one-sample sign test: If $n > 20$, then the binomial table of the Appendix gives way to the normal approximation to the binomial distribution, by analogy with the large-sample test of proportion discussed in Sec. 5.F. We then base our decision to reject or accept H on a z test, where

●
$$z = \frac{\theta - n/2}{\sqrt{n/4}} \qquad (11.1)$$

This formula is a consequence of formula (5.10) of the large-sample test of proportion where $\hat{p} = \theta/n$ and $p_o = \frac{1}{2}$.

The above large-sample formula can also be used for the two-sample sign test. To

TABLE 11.4
Nebraska Precipitation Data from
Cloud-Seeding Experiment

| Month | Inches of Precipitation | |
	Norfolk (Unseeded)	Fairbury (Seeded)
Jan.	1.5	1.4
Feb.	1.4	1.4
Mar.	2.2	2.6
Apr.	2.6	2.5
May	3.9	4.8
June	4.5	4.3
July	3.7	4.0
Aug.	3.4	3.5
Sept.	4.0	3.9
Oct.	2.5	2.6
Nov.	1.9	1.7
Dec.	1.5	1.4

illustrate the small-sample techniques used in the two-sample sign test, we use the setup from the cloud-seeding experiment of Example 5.7.

Example 11.2   Cloud Seeding   A meteorologist participating in a project to determine to what extent, if any, mankind can influence local weather conditions has set up an experiment to test the effectiveness of present methods of cloud seeding in the artificial production of rainfall. Two farming areas in Nebraska with similar past meteorological records lying 150 miles apart in a north–south direction were selected for the experiment. The Fairbury area is regularly seeded throughout the year, while the Norfolk area is left unseeded. Their monthly precipitations are recorded in Table 11.4 (the same as Table 5.14). At level $\alpha = .01$, do the results of the experiment indicate that cloud seeding significantly increases monthly precipitation?

SOLUTION   We apply the two-sample sign test as a nonparametric substitute for the paired-sample t test used in Example 5.7. If we use the notation

$\mu_d^* = $ true median of the monthly differences in precipitation

then we want to test the hypothesis

$H : \mu_d^* = 0$ (cloud seeding does not increase precipitation)

against the alternative

$A : \mu_d^* < 0$ (cloud seeding increases precipitation)

Table 11.5 contains the calculations necessary to carry out the statistical analysis.
    The results of the calculations done in Table 11.5 show that we should base our

TABLE 11.5
Calculations for the Two-Sample Sign Test
Cloud-Seeding Data

Month	Inches of Precipitation		Sign of $d = x_u - x_s$
	Unseeded Area x_u	Seeded Area x_s	
Jan.	1.5	1.4	+
Feb.	1.4	1.4	0
Mar.	2.2	2.6	−
Apr.	2.6	2.5	+
May	3.9	4.8	−
June	4.5	4.3	+
July	3.7	4.0	−
Aug.	3.4	3.5	−
Sept.	4.0	3.9	+
Oct.	2.5	2.6	−
Nov.	1.9	1.7	+
Dec.	1.5	1.4	+
		Total Number of +'s = 6	

$\theta = 6$
$n = 12 - 1 = 11$
$p = \frac{1}{2}$

statistical test on the binomial distribution having parameters $n = 11$ and $p = \frac{1}{2}$. From the portion of Table A.2 that is reproduced in Table 11.6, we then see that we should reject

$$H : \mu_d^* = 0$$

in favor of

$$A : \mu_d^* < 0$$

at level $\alpha = .0059$ if $\theta \leq 1$ or $\theta < 2$ and at level $\alpha = .0329$ if $\theta \leq 2$ or $\theta < 3$. Because $\theta = 6$ and $.0059 < .01 < .0329$, we therefore cannot reject H at level $\alpha = .01$. Our conclusion at level $\alpha = .01$ is that cloud seeding does not seem to significantly increase rainfall, according to the given data.

TABLE 11.6
A Small Portion of
Table A.2

n	k	$p = .50$
11	1	.0059
	2	.0327

Although the decision based on the sign test agreed with our earlier decision in Chap. 5 based on the t test, a comparison of the computations in Tables 5.15 and 11.5 reveals a serious disparity between the two methods. From Table 5.15, we see that the total difference Σd is negative ($= -1.0$), indicating that the mean difference in rainfall is negative, so that on the average (in terms of the mean) more rain fell in the seeded area. From Table 11.5, on the other hand, we see that the number $\theta = 6$, indicating that 6 out of the 11 nonzero signs and so the median difference are positive; therefore, on the average (in terms of the median) more months have greater precipitation in the unseeded area.

The reason for this disparity lies in the major distinction between the mean and the median introduced back in Sec. 2.A: The mean takes into account extraordinary large values, while the median does not. The month of May, for example, in Table 5.15, has a difference d of -0.9, overcoming the total differences of the months January, April, June, September, November, and December. In Table 11.5, however, May counts as only one minus sign, while those six months contribute a total of six plus signs. In a case of an extreme difference between the mean and the median, it should not be surprising if the t test and the sign test recommend different decisions.

EXERCISES 11.A

1 A psychologist, conducting a study of the average person's ability to judge distances, sets up a test of depth perception in which randomly selected individuals attempt to estimate the distance between two markers. The markers were actually 2.5 feet apart, but the 10 participants in the study gave the following estimates:

$$2.1 \quad 1.8 \quad 2.3 \quad 2.3 \quad 2.6 \quad 2.5 \quad 2.3 \quad 2.5 \quad 2.1 \quad 2.5$$

Use the sign test at level $\alpha = .05$ to find whether or not the results of the study indicate that the average person has difficulty in estimating the correct distance of 2.5 feet.

2 In Supplementary Exercise 5.3 of Chap. 5, a company that markets canned carrots advertises that their 16-ounce cans actually weigh 16.1 ounces on the average. From the random sample of 11 weights listed in that exercise, can we conclude at level $\alpha = .05$ that
a The true median contents are below the 16.1 ounces claimed by the company?
b The true median contents are below the 16.0 ounces printed on the label?

3 It is the policy of one school district to hire credentialed reading specialists whenever the true median reading score of the district's sixth-graders falls below 40 as measured by a particular standard test. Of a random sample of 25 pupils, 18 had scores below 40 and the other 7 had scores above 40. Is the hiring of reading specialists justified:
a At significance level $\alpha = .05$?
b At significance level $\alpha = .005$?

4 From the data of Exercise 5.D.1, does the sign test support at level $\alpha = .05$ the theory that employees have less absenteeism after they stop smoking?

5 From the data of Exercise 5.D.6, does the sign test support at level $\alpha = .05$ the assertion that the kindergarten experience seems to help first-graders?

6 Ten sets of identical twins were administered drugs intended to increase the pulse rate, in an experiment designed to determine whether identical twins have similar reactions to medication. In four of the ten sets, the older twin had a larger increase in pulse rate, but in the remaining six sets, the younger twin had the larger increase. At level $\alpha = .10$, does the experiment tend to support the hypothesis that there is no general rule concerning which twin, the older or the younger, has the larger increase?

7 Using the data of Exercise 5.D.7, does the sign test support at level $\alpha = .05$ the assertion that exports of member nations of the West African Customs Union increased significantly in the period between 1961 and 1966?

SECTION 11.B THE MANN-WHITNEY TEST
FOR EQUAL AVERAGES

When a lack of normally distributed data or other conditions required for valid operation of the two-sample t test (conditions listed at the end of Sec. 5.C) prevents us from using that test, the Mann-Whitney test is sometimes an appropriate nonparametric substitute. Just as the sign test did not directly use the actual numerical values of the data points, neither does the Mann-Whitney test. However, whereas the sign test used only facts about which of two numbers is the larger, the Mann-Whitney test employs a more detailed method of comparing the data points—ranking all of them in order. By this ranking method, additional information about the data points, short of their actual values, is taken into consideration.

Because the Mann-Whitney test's ranking method manages to squeeze more information out of the data than the paired comparisons of the sign test, the Mann-Whitney test generally yields more accurate results than the sign test. By "more accurate" in this context, we mean that, for the same probability α of Type I error, the Mann-Whitney test will have a lower probability of Type II error than the sign test. In this chapter, we will study only the Mann-Whitney substitute for the two-sample t test. There exists also a test based on ranks that can be applied to cases involving paired samples, but we will only list references to this test in the bibliography at the end of the chapter. In Supplementary Exercise 11.1, you will see an example that illustrates why the Mann-Whitney test we describe here cannot be used for paired samples.

Following the procedure used in the previous section, we shall reevaluate an example from Sec. 5.C (where we discussed the two-sample t test) from the point of view of the Mann-Whitney test. In particular, we shall look at the situation of Example 5.4 through the eyes of the Mann-Whitney test.

Example 11.3 Pest Control A state department of public health, faced with a serious infestation of houseflies, decides to conduct a pilot test of effectiveness between two methods of insect control: the organic method and the chemical method. The organic method, which consists of saturating a community with nonpoisonous spiders, is applied in one neighborhood, while the chemical method, using a combination of poisonous sprays and bait, is applied in a second neighborhood comparable to the first.

Useful data are collected on six houses in the first neighborhood and on nine in the second during a 48-hour test period. The data are obtained by counting the number of houseflies visually observed making nuisances of themselves during the test period. In the first neighborhood, the one treated by the organic method, the six houses checked had, respectively, 41, 20, 19, 36, 38, and 26 houseflies roaming free. In the second neighborhood, the one treated by the chemical method, the nine houses checked had, respectively, 9, 26, 16, 10, 31, 28, 35, 15, and 10 houseflies roaming free. At significance level $\alpha = .01$, do the test results indicate that the chemical method is more effective than the organic method in reducing the number of houseflies roaming free throughout the neighborhood?

SOLUTION   To apply the Mann-Whitney test for equal averages, we first introduce the notation:

$$n_o = \text{number of data points related to the organic method}$$
$$n_c = \text{number of data points related to the chemical method}$$

Because

Set of data related to organic method = {41, 20, 19, 36, 38, 26}

and   Set of data related to chemical method = {9, 26, 16, 10, 31, 28, 35, 15, 10}

we have that

$$n_o = 6 \quad \text{and} \quad n_c = 9$$

We next rank all $n_o + n_c = 6 + 9 = 15$ data points together in order from smallest to largest, as illustrated in Table 11.7. We carry out the Mann-Whitney analysis to test the hypothesis

$$H : A_o = A_c$$

against the alternative

$$A : A_o > A_c$$

$$A_o = \text{true average number of houseflies roaming}$$
$$\text{free in neighborhood treated by organic method}$$

TABLE 11.7
Ranking of Combined Sets of Data
Pest Control Data

Rank position	1	2	3	4	5	6	7	8	9	10	11	12	13	14	15
Data points	9	10	10	15	16	19	20	26	26	28	31	35	36	38	41
Organic (O) or chemical (C)	C	C	C	C	C	O	O	O	C	C	C	C	O	O	O

$$A_c = \text{true average number of houseflies roaming}$$
free in neighborhood treated by chemical method

We next assign each of the 15 data points a rank. The "assigned rank" of a point is usually its rank position as shown in Table 11.7, but in the case of ties it may be slightly different. Consider, for example, the two 26s, which are tied for the rank positions 8 and 9. What we do is give each of them an assigned rank of 8.5. [Suppose, in some other problem, that we have three data points tied for rank positions 10, 11, and 12. Then we would average the rank positions in question and give each of the three data points an assigned rank of $(10 + 11 + 12)/3 = 11$.]

After giving each of the 15 data points its assigned rank, we then decide which of the two samples has the smaller number of data points. (If both samples have the same number of data points, either can be used.) Here $n_o < n_c$ because $n_o = y$ while $n_c = 9$, so that the organic sample is the smaller, and we compute the number

$$T = \text{sum of assigned ranks of smaller sample}$$

As worked out in Table 11.8, we obtain that

$$T = 63.5$$

We then refer to Table A.8 of the Appendix in order to determine the rejection rule at level $\alpha = .01$ for

$$H : A_o = A_c$$

TABLE 11.8
Calculations for the Mann-Whitney Test for Equal Averages
Pest Control Data

Rank Position	Data Point	Organic (O) or Chemical (C)	Assigned Rank	Of Smaller Sample (O)
1	9	C	1	
2	10	C	2.5	
3	10	C	2.5	
4	15	C	4	
5	16	C	5	
6	19	O	6	6
7	20	O	7	7
8	26	O	8.5	8.5
9	26	C	8.5	
10	28	C	10	
11	31	C	11	
12	35	C	12	
13	36	O	13	13
14	38	O	14	14
15	41	O	15	15
				Sum $T = 63.5$

TABLE 11.9
A Small Portion of
Table A.8

T_α	$T_{1-\alpha}$	α
	(6, 9)	
28	68	.009
29	67	.013

versus

$$A : A_0 > A_c$$

If it is really true that $A_0 > A_c$, we would expect the organic method data to be larger than the chemical method data. Therefore, the organic method data should occupy the higher ranks. It follows that we should reject H in favor of A if T exceeds the appropriate comparison value listed in the table. We reproduce the relevant portion of Table A.8 in Table 11.9. Because .01 is between .009 and .013, the relevant rejection rules are as follows: We should reject $H : A_0 = A_c$ in favor of $A : A_0 > A_c$ at level $\alpha = .009$ if $T \geq 68$ and at level $\alpha = .013$ if $T \geq 67$.

It turns out, however, that $T = 63.5$, and so we cannot reject H at either of these levels. It follows that we cannot reject H at level $\alpha = .01$ either. Therefore, at level $\alpha = .01$, we must conclude that the data do not support the assertion that the chemical method is more effective than the organic method in reducing the number of houseflies.

As long as the number of ties is not too large, the procedure used in the above example works satisfactorily. However, if there is an excessive number of ties, an additional correction factor enters into the calculations. The bibliography at the end of the chapter gives a reference where further details can be found in such a situation.

Example 11.4   Tea Testing   It is the job of a tea tester for an English tea-importing company to decide which tea leaves are most suitable for English Breakfast tea. The tea tester must decide between two lots of leaves: an inexpensive lot and an expensive lot. The tea tester and his laboratory staff find the following levels of impurities in eight randomly selected leaves of each lot:

Inexpensive lot	3	4	25	9	11	4	12	20
Expensive lot	6	5	7	10	6	8	5	4

At significance level $\alpha = .05$, do the data indicate that both lots of leaves have approximately the same average impurity level?

SOLUTION   We will use the Mann-Whitney test to test the hypothesis

$$H : A_I = A_E$$

TABLE 11.10
A Small Portion of
Table A.8

T_α	$T_{1-\alpha}$	α
	(8, 8)	
49	87	.025

versus

$$A : A_I \neq A_E$$

where the symbolism should be self-explanatory. Because of the format of the alternative A, we would reject H in favor of A if T turned out to be either too large or too small. Because $\alpha = .05$, we would therefore compare T with percentage points at .025 level in both directions. The numbers of data points in each sample are

$$n_I = 8 \quad \text{and} \quad n_E = 8$$

and the appropriate portion of Table A.8 is excerpted in Table 11.10. We should therefore reject $H : A_I = A_E$ in favor of $A : A_I \neq A_E$ if either $T \leq 49$ or $T \geq 87$. It now remains only to compute the numerical value of T, and we do this in Table 11.11. (Since both sets contain the same number of data points, we can use either sample as the smaller.)

TABLE 11.11
Calculations for the Mann-Whitney Test for Equal Averages
Tea Testing Data

Rank Position	Data Point	Inexpensive (I) or Expensive (E)	Assigned Rank	Of Smaller Sample (I)
1	3	I	1	1
2	4	I	3	3
3	4	I	3	3
4	4	E	3	
5	5	E	5.5	
6	5	E	5.5	
7	6	E	7.5	
8	6	E	7.5	
9	7	E	9	
10	8	E	10	
11	9	I	11	11
12	10	E	12	
13	11	I	13	13
14	12	I	14	14
15	20	I	15	15
16	25	I	16	16
				Sum $T = 76$

The calculations in Table 11.11 result in the fact that $T = 76$. As we have agreed to reject H if either $T \leq 49$ or $T \geq 87$, we can see that a numerical value of $T = 76$ does not permit us to reject H. At level $\alpha = .05$, we therefore conclude that the two lots of tea do not differ significantly in impurity levels.

Like the sign test, the large-sample version of the Mann-Whitney test is based on the percentage points of the normal distribution. It is a z test using the statistic

●
$$z = \frac{T - \mu_T}{\sigma_T} \tag{11.2}$$

where

●
$$\mu_T = \frac{n_1(n_1 + n_2 + 1)}{2} \tag{11.3}$$

●
$$\sigma_T = \sqrt{\frac{n_1 n_2(n_1 + n_2 + 1)}{12}} \tag{11.4}$$

Here, n_1 is the number of data points in the smaller sample, and n_2 is the number of data points in the larger sample. The hypothesis must always be of the form $H : A_1 = A_2$, while the alternative must be either $A_1 < A_2$, $A_1 > A_2$, or $A_1 \neq A_2$. The normal approximation takes effect when n_1 and n_2 are both greater than 10, and the rejection rules are given by Table 5.7.

EXERCISES 11.B

1   The following data are the prices (in cents) of eight-ounce cans of a local brand of tomato sauce at 10 randomly selected supermarkets in New York and another 10 across the river in suburban northern New Jersey:

At New York City Supermarkets	At New Jersey Supermarkets
11	11
10	7
7	8
14	9
10	11
12	7
16	7
9	13
10	8
11	9

At significance level $\alpha = .05$, does the information collected support the assertion that this brand of tomato sauce costs more in New York City than it does in northern New Jersey?

2   Using the data of Exercise 5.C.3, apply the Mann-Whitney test to decide at level $\alpha = .05$ whether or not cigarette smoke significantly inhibits the growth of mice.

3 Using the data of Exercise 5.C.4, apply the Mann-Whitney test to decide at level $\alpha = .01$ whether or not the roasting process reduces the protein content of peanuts. (Use the z test.)

4 Using the data of Exercise 5.C.5, apply the Mann-Whitney test to decide at level $\alpha = .10$ whether or not English cheese sells at lower prices in Canada.

5 Using the data of Supplementary Exercise 5.8 of Chap. 5, decide at level $\alpha = .10$ whether or not the new technique produces stronger string.

SECTION 11.C THE MANN-WHITNEY TEST FOR EQUAL DISPERSIONS

Everything we have learned about making the calculations for the Mann-Whitney test for equal averages holds true for the Mann-Whitney test for equal dispersions as well, except the method of determining the assigned ranks. To test for equal "dispersions," we give the lowest assigned ranks to data points having the extreme, high and low, rank positions. This way a more dispersed sample has a relatively lower sum of ranks than a less dispersed sample, and we are then able to decide whether or not the two samples differ in dispersion. By the dispersion of a set of data, we mean the extent to which the data vary from their "average." The standard deviation, the mean absolute deviation, and the range, which we discussed in Sec. 2.C, are all measures of dispersion, but in the nonparametric framework, we refer to the more general concept of dispersion rather than any specific way to measure it.

Let's take another look, for example, at the tea testing data of Example 11.4 of the previous section. Suppose we want to determine whether or not the two lots, inexpensive and expensive, differed in degree of internal variability of impurity level. In other words, we want to find out whether the inexpensive tea leaves have the same degree of uniformity as the expensive tea leaves. In particular, if we set

$$D_I = \text{lack of uniformity of impurity level in inexpensive lot}$$

and $\qquad D_E = \text{lack of uniformity of impurity level in expensive lot}$

to be the true dispersions of the two samples, we want to test the hypothesis

$$H : D_I = D_E$$

against the alternative

$$A : D_I \neq D_E$$

We still have $n_I = 8$ and $n_E = 8$, and if we use the level of significance $\alpha = .05$, Table 11.10 shows that we again should reject H in favor of A if $T \leq 49$ or $T \geq 87$. The only place where something different occurs is in the calculation of T. We compute T for the test of equal dispersions in Table 11.12.

There is an important difference in the treatment of ties between Tables 11.11 and 11.12. In Table 11.11, the data points equal to 4 were tied for the assigned ranks 2, 3, and 4 and were each assigned a rank of $(2 + 3 + 4)/3 = 3$; in Table 11.12, however,

TABLE 11.12
Calculations for the Mann-Whitney Test for Equal Dispersions
Tea Testing Data

Rank Position	Data Point	Inexpensive (I) or Expensive (E)	Adjusted Rank Position	Assigned Rank	Of Smaller Sample (I)
1	3	I	1	1	1
2	4	I	3	5	5
3	4	I	5	5	5
4	4	E	7	5	
5	5	E	9	10	
6	5	E	11	10	
7	6	E	13	14	
8	6	E	15	14	
9	7	E	16	16	
10	8	E	14	14	
11	9	I	12	12	12
12	10	E	10	10	
13	11	I	8	8	8
14	12	I	6	6	6
15	20	I	4	4	4
16	25	I	2	2	2
				Sum T = 43	

these three data points are tied for the assigned ranks 3, 5, and 7 and are therefore each assigned a rank of $(3 + 5 + 7)/3 = 5$.

From the computations in Table 11.12, we see that $T = 43$. As we agreed to reject H at level $\alpha = .05$ if either $T \leq 49$ or $T \geq 87$, and, in fact, $T = 43 \leq 49$, we must reject H. At level $\alpha = .05$, we conclude on the basis of the given data that the two lots of tea have different dispersions and therefore do not have the same internal variability.

Our decision to the effect that the two lots of tea have different dispersions most likely implies that the standard deviations of the two underlying populations are different. As it is one of the requirements for the validity of the two-sample t test that the underlying populations have the same standard deviation, it would be inappropriate for the tea tester to use the t test to test his tea.

One additional point, if the medians of the two sets of data are very far apart, the Mann-Whitney test may not be able to discern a difference which really exists between the dispersions. This is due to the fact that the low (nonadjusted) rank positions will all be occupied by the members of one sample, while the high positions will be filled by those of the other, regardless of the degree of internal dispersion within the separate samples. One way to counteract this effect would be to subtract off the median of each sample from each data point in that sample. This operation would have no effect at all on the dispersions, but it would give us two sets of numbers having the same median. It would then be reasonably valid in some situations to run the Mann-Whitney test of equal dispersions. For an example of such a problem, see Supplementary Exercise 11.2.

EXERCISES 11.C

1   Using the data of Exercise 11.B.1, apply the Mann-Whitney test to decide at level $\alpha = .10$ whether or not the tomato sauce prices are equally dispersed in the two marketing regions.

2   On the basis of the data of Exercise 5.E.4, can we say at level $\alpha = .05$ that the index of industrial production is significantly more variable than the wholesale price index? Use the Mann-Whitney test.

3   To the data of Exercise 5.C.5, apply the Mann-Whitney test in an effort to determine at level $\alpha = .10$ whether or not prices of English cheese vary to the same extent in the United States and Canada.

4   Using the data of Supplementary Exercise 5.8 of Chap. 5, decide at level $\alpha = .10$ whether or not the breaking points of both groups of string have approximately the same degree of variation.

5   Some supermarkets give trading stamps with purchases, while others do not. One consumer organization believes that prices are more variable among supermarkets not offering trading stamps, because of their supposed tendency to offer "loss-leaders" or other price-adjustment gimmicks. To test out its guess, the organization checks banana prices (per pound) at four supermarkets that give trading stamps and six that do not. The results of the survey follow:

At supermarkets that give stamps	12	18	14	16		
At supermarkets that give no stamps	10	16	16	12	8	10

At level $\alpha = .05$, do the data support the allegation that prices are more variable at supermarkets not giving trading stamps?

SECTION 11.D KRUSKAL-WALLIS NONPARAMETRIC ANALYSIS OF VARIANCE

As we emphasized in Chap. 10, the F test on which analysis of variance is based can be used with confidence only when the data points involved come from normally distributed populations, each having the same standard deviation. When it is unreasonable to operate under this assumption, the analysis of variance techniques studied in Chap. 10 do not apply. The nonparametric substitute generally used is the Kruskal-Wallis test, also called "nonparametric analysis of variance" or "analysis of variance by ranks."

Rejection rules for the Kruskal-Wallis test are based on the tables of the chi-square distribution, which we have had occasion to refer to in several places in this text, most

FIGURE 11.1
A T maze.

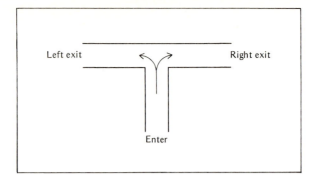

recently in Chap. 9. There is one basic requirement that must be met in order that the use of the chi-square distribution be valid, namely, that each of the samples involved must contain at least 5 data points. (This requirement is similar to one we encountered in connection with the chi-square test of independence in Sec. 9.A to the effect that each cell of a contingency table must have an expected frequency of at least 5.)

In this section, we will restrict ourselves to a nonparametric replacement for the one-way analysis of variance which we studied in Sec. 10.B. The following example deals with a psychological experiment in learning theory.

Example 11.5   Learning Times   A psychologist's research assistant wants to find out whether porpoises, monkeys, and rats learn at the same rate of speed. She places one animal at a time in an appropriate T maze of the sort illustrated in Fig. 11.1, where the left exit contains a punishment such as an electric shock and the right exit contains a reward such as food. The assistant runs 6 porpoises, 9 monkeys, and 11 rats through the maze, and she records the number of times through that it takes each animal to learn that the right fork is the one to take. The learning times for each animal are as follows:

Learning Times

Porpoises	3, 7, 9, 15, 6, 2
Monkeys	18, 14, 7, 5, 11, 9, 3, 11, 12
Rats	4, 21, 11, 16, 19, 23, 7, 10, 17, 6, 8

At level $\alpha = .05$, can she conclude that all three species of animals have the same average learning time?

SOLUTION   We'll carry out the Kruskal-Wallis analysis of variance by ranks. We use the following notation:

$$m = \text{number of samples}$$

$$n_P = \text{number of data points in the porpoise sample}$$
$$n_M = \text{number of data points in the monkey sample}$$
$$n_R = \text{number of data points in the rat sample}$$
$$n = n_P + n_M + n_R = \text{total number of data points}$$

In the example under study, we see that

$$m = 3$$
$$n_P = 6$$
$$n_M = 9$$
$$n_R = 11$$
$$n = 6 + 9 + 11 = 26$$

The Kruskal-Wallis test is a test of the hypothesis

$$H : A_P = A_M = A_R \text{ (all groups have the same average)}$$

against the alternative

$$A : \text{These averages are not all the same}$$

The rejection rule for the Kruskal-Wallis test is as follows:

We compute

● $$H = \frac{12}{n(n+1)} \Sigma \frac{R_i^2}{n_i} - 3(n+1) \qquad (11.5)$$

and we reject H at level α if

$$H > \chi_\alpha^2(m-1)$$

where $m - 1$ is the degrees of freedom for the chi-square statistic. In the situation at hand, we are using $\alpha = .05$ and df $= m - 1 = 3 - 1 = 2$. It follows that we should reject H in favor of A at level $\alpha = .05$ if

$$H > \chi_{.05}^2(2) = 5.991$$

It now remains only to compute H. We rank all the $n = 26$ data points in order (as though we were running a Mann-Whitney test). Then we calculate

$$R_P = \text{sum of assigned ranks of porpoise sample}$$
$$R_M = \text{sum of assigned ranks of monkey sample}$$
$$R_R = \text{sum of assigned ranks of rat sample}$$

The Kruskal-Wallis statistic is then

$$H = \frac{12}{n(n+1)} \left(\frac{R_P}{n_P} + \frac{R_M}{n_M} + \frac{R_R}{n_R} \right) - 3(n+1)$$

The preliminary calculations appear in Table 11.13.

Upon inserting the results at the bottom of Table 11.13 into the formula for H, we obtain that

TABLE 11.13
Preliminary Calculations for Kruskal-Wallis
Nonparametric Analysis of Variance
Learning Times Data

Rank Position	Data Point	Sample (P, M, or R)	Assigned Rank	Of Porpoise Sample (P)	Of Monkey Sample (M)	Of Rat Sample (R)
1	2	P	1	1		
2	3	P	2.5	2.5		
3	3	M	2.5		2.5	
4	4	R	4			4
5	5	M	5		5	
6	6	P	6.5	6.5		
7	6	R	6.5			6.5
8	7	P	9	9		
9	7	M	9		9	
10	7	R	9			9
11	8	R	11			11
12	9	P	12.5	12.5		
13	9	M	12.5		12.5	
14	10	R	14			14
15	11	M	16		16	
16	11	M	16		16	
17	11	R	16			16
18	12	M	18		18	
19	14	M	19		19	
20	15	P	20	20		
21	16	R	21			21
22	17	R	22			22
23	18	M	23		23	
24	19	R	24			24
25	21	R	25			25
26	23	R	26			26

Rank Sums $R_P = 51.5$ $R_M = 121$ $R_R = 178.5$

Number of Data Points $n_P = 6$ $n_M = 9$ $n_R = 11$

$$H = \frac{12}{26(26 + 1)}\left[\frac{(51.5)^2}{6} + \frac{(121)^2}{9} + \frac{(178.5)^2}{11}\right] - 3(26 + 1)$$

$$= \frac{12}{(26)(27)}\left(\frac{2652.25}{6} + \frac{14641}{9} + \frac{31862.25}{11}\right) - 3(27)$$

$$= \frac{12}{702}(442.04 + 1626.78 + 2896.57) - 81 = \frac{12}{702}(4965.39) - 81$$

$$= 84.88 - 81$$

$$= 3.88$$

We have agreed to reject the hypothesis if $H > 5.991$, but H turned out to be $3.88 < 5.991$. It follows that we cannot reject the hypothesis that all groups have the same average learning time. At level $\alpha = .05$, we therefore assert that on the basis of the data porpoises, monkeys, and rats take the same amount of time on the average to learn the workings of a T maze.

EXERCISES 11.D

1 From the data of Exercise 10.A.1, carry out the Kruskal-Wallis nonparametric analysis of variance to test at level of significance $\alpha = .05$ whether or not there are real differences between the average monthly premiums of the four types of life insurance policies.

2 Using the ichthyological data of Exercise 10.A.3, apply the Kruskal-Wallis test at level $\alpha = .05$ to decide whether or not temperature differences in the range tested have a significant effect on the average weight of fish in the geographical region.

3 On the basis of the Kruskal-Wallis test applied to the pork-barrel data of Exercise 10.A.4, can we conclude at level $\alpha = .01$ that the average amounts of money dispensed to each of the three socio-politico-economic subdivisions are approximately the same?

4 Apply the Kruskal-Wallis test to the tomato yield data of Exercise 10.B.5 with the aim of deciding at level $\alpha = .05$ whether or not there are significant differences among the yields of the three kinds of tomatoes.

SECTION 11.E THE WALD-WOLFOWITZ TEST OF RANDOMNESS

In many problems of applied interest, the questions do not involve numerical characteristics of a population, such as averages, dispersions, or proportions, but instead are concerned with more qualitative behavior of the population. We saw in Sec. 9.A one example of this type of question, where the chi-square test of independence was used to determine whether or not two different ways of classifying elements of a population were independent of each other. In this section, our objective will be to develop a method of testing whether or not a given sequence of data has been produced by some sort of random process or whether the elements of the sequence are following some identifiable pattern. Let's take a look at the following example.

Example 11.6   London Gold Prices   Each day on the London gold exchange the price of gold either rises or falls. Although some observers consider the price of gold to fluctuate randomly, others believe that the prices rise and fall in trends, more precisely that there are several consecutive days of falling prices followed by several consecutive days of rising prices, and so on. To check on the opposing theories, one financial analyst of a major international banking concern recorded the behavior of the price of gold over a recent 40-day period as follows:

FRRRRRFFRRRRRRRFRFRRRFFFRRRRRRRFFFRRRRRRRR

where a fall in price from the preceding day's close is indicated by F and a rise is indicated by R. At significance level $\alpha = .05$, do the data indicate that gold prices rise and fall in trends, as opposed to random fluctuation?

SOLUTION   According to mathematical principles on which the Wald-Wolfowitz test of randomness is based, the mean number of "runs" occurring in a string of F's and R's is

●
$$\mu_W = \frac{2n_F n_R}{n_F + n_R} + 1 \qquad (11.6)$$

where
$$n_F = \text{number of F's in the sequence}$$
$$n_R = \text{number of R's in the sequence}$$

By a run, we mean a string of consecutive letters. For example, the sequence of data on the behavior of London gold prices is composed of the following runs:

1   A run of one F
2   A run of five R's
3   A run of two F's
4   A run of six R's
5   A run of one F
6   A run of one R
7   A run of one F
8   A run of three R's
9   A run of three F's
10   A run of six R's
11   A run of three F's
12   A run of eight R's

We denote the number of runs by the letter W, and for this example, we have

$$W = 12$$

It is convenient to mark off the runs directly on the sequence of data, as done in Fig. 11.2. It can be clearly seen from the figure that $W = 12$.

Now, how can the number of runs be used to test for trends in the data? Basically the idea is this: If the F's and R's were generated randomly instead of according to any noticeable pattern, the actual number of runs would be somewhat close to the mean number μ_W calculated by formula (11.6). However, if there are noticeable trends in the data, there would be fewer runs than expected because trend behavior will tend to produce longer runs than usual. The longer runs occur because trend behavior means that several consecutive days of falling prices are followed by several consecutive days of rising prices, etc. If the runs are longer, there would have to be fewer of them in the sequence of 40 data points. Therefore, a numerical value of W significantly *below* μ_W would be evidence in favor of trends, as opposed to random behavior. In particular, we test the hypothesis

FIGURE 11.2
Marking off the runs (London gold data).

2 4 6 8 10 12

FRRRRRFFRRRRRRFRFRRRFFFRRRRRRFFFRRRRRRRR

1 3 5 7 9 11

$W = 12, n_F = 11, n_R = 29$

$$H: \text{The data sequence exhibits random behavior}$$

against the alternative

$$A: \text{The data sequence reveals trends}$$

If both n_F and n_R exceed 10, a version of the central limit theorem shows that the numbers

- $$z = \frac{W - \mu_W}{\sigma_W} \tag{11.7}$$

have the standard normal distribution, where

- $$\sigma_W = \sqrt{\frac{2n_F n_R(2n_F n_R - n_F - n_R)}{(n_F + n_R)^2(n_F + n_R - 1)}} \tag{11.8}$$

Although even the simplest method of justifying the formulas for μ_W and σ_W require mathematical techniques beyond the prerequisites for this text, a reasonably elementary algebraic derivation of them can be found in the article by C. W. Marshall listed in the bibliography at the end of this chapter.

Because $W < \mu_W$ tends to indicate the presence of trends, it makes sense to reject H in favor of A at level α if

$$z < -z_\alpha$$

because the farther W is below μ_W, the more negative z will be. Therefore, at level $\alpha = .05$, we will agree to reject H in favor of A if

$$z < -z_{.05} = -1.64$$

It remains only to compute z. We already know that $W = 12$, and so all we have to do is to calculate μ_W and σ_W. But

$$n_F = 11$$

and

$$n_R = 29$$

as we can discover by counting the F's and R's in the data sequence appearing in Fig.

11.2. (Both of these exceed 10, and so the use of the normal distribution is valid.) Therefore,

$$\mu_W = \frac{2n_F n_R}{n_F + n_R} + 1 = \frac{2(11)(29)}{11 + 29} + 1 = \frac{638}{40} + 1 = 15.95 + 1 = 16.95$$

and

$$\sigma_W = \sqrt{\frac{2n_F n_R (2n_F n_R - n_F - n_R)}{(n_F + n_R)^2 (n_F + n_R - 1)}} = \sqrt{\frac{2(11)(29)[2(11)(29) - 11 - 29]}{(11 + 29)^2 (11 + 29 - 1)}}$$

$$= \sqrt{\frac{638(638 - 11 - 29)}{(40)^2(39)}} = \sqrt{\frac{(638)(598)}{(1600)(39)}} = \sqrt{6.114} = 2.473$$

Therefore,

$$z = \frac{W - \mu_W}{\sigma_W} = \frac{12 - 16.95}{2.473} = \frac{-4.95}{2.473} - -2.00$$

Now $z = -2.00$, and we have agreed to reject H if $z < -1.64$. But, in fact, $z = -2.00 < -1.64$, and so we reject H. At level $\alpha = .05$, we therefore conclude that the 40-day sequence of data indicates that London gold prices rise and fall according to trends rather than randomly.

In the above example, we have explained how a fewer than average number of runs indicates the presence of trends in the data sequence. Suppose, instead, that there is "cyclic" behavior in the data sequence. We would then expect more than the average number of runs, for cyclic behavior would seem to indicate that the data sequence fluctuates quite often between the two letters. The large number of runs present in cyclic behavior can be represented pictorially to some extent by the graph of Fig. 1.6. The previous graph, that of Fig. 1.5, would provide an analogy to the case of trends, having fewer runs. On the basis of this discussion, we can construct a table of rejection rules for the Wald-Wolfowitz test of randomness, a table analogous to Tables 5.7, 5.16, and 5.19. This table appears as Table 11.14. The next example provides a situation where it is appropriate to test for cyclic behavior.

Example 11.7 Employee Absenteeism A major corporation has hired a psychological consultant to analyze the phenomenon of employee absenteeism. By its investigation, the corporation hopes to be able to cut down on absenteeism and, better yet, its tendency to reduce productivity. As part of his analysis, the psychologist carefully checks the records of one particular employee who has an unusually large number of absences. On 50 consecutive working days, this employee had the following attendance record, where A indicates "absent" and P indicates "present."

PPPPAPPPAPPPPAPAPPPAPPPAPAPPPPAAPPPPAPPPAPPPAAPPPA

The psychologist decides to test the employee's attendance record for cyclic behavior, which may indicate that the employee's job is so boring, tiring, or aggravating that he must take a day off after every few days at work. Randomness, on the other hand, would

TABLE 11.14
Rejection Rules in the Presence of
Various Alternatives for the
Wald-Wolfowitz Test of Randomness
H : The Data Sequence Exhibits Random Behavior

Alternative	Reject H in Favor of A at Significance Level α if:
A : Cyclic behavior	$z > z_\alpha$
A : Trends	$z < -z_\alpha$
A : Nonrandom behavior	$\|z\| > z_{\alpha/2}$

NOTE: The use of z is valid only in the large-sample case when both components of the data sequence are represented by more than 10 data points. For the small-sample case, special tables must be used.

indicate that the employee merely takes a day off once in a while for no apparent reason. At level $\alpha = .05$, can the psychologist conclude from the employee's attendance record that the absences occur in cycles?

SOLUTION   We test the hypothesis

$$H : \text{The absences occur randomly}$$

against the alternative

$$A : \text{The absences occur in cycles}$$

at level $\alpha = .05$. In the case of cycles, we would expect more than the average number of runs. Therefore, W, and consequently z, would need to be large for us to agree to reject H. Referring to Table 11.14, the table of rejection rules, we see that we are to compute

$$z = \frac{W - \mu_W}{\sigma_W}$$

FIGURE 11.3
Marking off the runs (employee absenteeism data).

```
        2    4    6  8   10   12 14    16    18  20   22   24
 PPPPAPPPAPPPPAPAPPPAPPPAPAPPPPAAPPPPAPPPAPPPAAPPPA
   1    3    5   7  9    11 13 15     17   19   21    23
           W = 24, n_p = 36, n_A = 14
```

and to reject H in favor of A if

$$z > z_{.05} = 1.64$$

We proceed to the computation of z. We see from Fig. 11.3 that $W = 24$, $n_P = 36$, and $n_A = 14$. It follows that

$$\mu_W = \frac{2n_P n_A}{n_P + n_A} + 1 = \frac{2(36)(14)}{36 + 14} + 1 = \frac{1008}{50} + 1 = 20.16 + 1 = 21.16$$

and

$$\sigma_W = \sqrt{\frac{2n_P n_A(2n_P n_A - n_P - n_A)}{(n_P + n_A)^2(n_P + n_A - 1)}} = \sqrt{\frac{2(36)(14)[2(36)(14) - 36 - 14]}{(36 + 14)^2(36 + 14 - 1)}}$$

$$= \sqrt{\frac{1008(1008 - 36 - 14)}{(50)^2(49)}} = \sqrt{\frac{(1008)(958)}{(2500)(49)}} = \sqrt{7.883} = 2.808$$

Therefore,

$$z = \frac{W - \mu_W}{\sigma_W} = \frac{24 - 21.16}{2.808} = \frac{2.84}{2.808} = 1.012$$

Since we have agreed to reject H if $z > 1.64$ but $z = 1.012 < 1.64$, we cannot reject H. At level $\alpha = .05$, therefore, the psychologist concludes that the employee's record does not show evidence of a cyclic pattern of absenteeism.

In each of the two examples presented in this section, there have been more than 10 data points of each symbol; i.e., $n_F > 10$ and $n_R > 10$ in Example 11.6, and $n_P > 10$ and $n_A > 10$ in Example 11.7. Situations in which this occurs are said to be large-sample problems, and we can use the z test to solve them. When this criterion is not satisfied, we are dealing with a small-sample problem, and, just as in Chaps. 4 and 5, use of the central limit theorem is no longer valid. In such instances we simply compute W and compare the number we get with entries in a special table for the small-sample Wald-Wolfowitz test. This table and instructions for its use are available in several journals and books, one of which is the book by Dixon and Massey listed in the bibliography at the end of Chap. 6.

EXERCISES 11.E

1   A stockbroker observes the behavior of the Dow-Jones industrial average on the New York Stock Exchange for a period of 30 days. When the average goes up, he marks a U for that day, and when the average goes down, he marks a D. The 30-day record of market behavior is as follows:

UUUUUDDUUUUDDDUUUDDDDUUUUUDDUU

Can the ups and downs of the stock market during the 30-day period be considered random at a level of significance of $\alpha = .01$?

2   A quality control supervisor wants to know whether or not defective items come off

his assembly line randomly or according to some identifiable pattern. He sampled 35 consecutive items and recorded the defective ones as D's and the acceptable ones as A's. The data follow:

DDAAAAADAAAAAAADDAAAAAADDDAAADAAADA

At level $\alpha = .05$, do the data indicate that the defectives come off the assembly line randomly or nonrandomly?

3 One psychologist has a theory to the effect that children whose last names begin with letters toward the end of the alphabet (namely, the "Z" end of the alphabet) do not do as well in elementary school as other children. He attributes this situation to various factors, including the facts that such children are often called upon less frequently and they often are assigned seats in the back of the room. In any case, he conducted a test of his theory by listing all the children in a fifth-grade class in alphabetical order and then recording next to each name an "A" if the child scored *above* the median on a standardized achievement test and a "B" if the child scored *below* the median. The 32 data points turned out as follows:

AABBBABAABAAABBBAAABBBABBABAAABB

At level $\alpha = .10$, do the results of the experiment support the theory that scores above and below the median occur in some nonrandom pattern?

4 A biologist who wants to know whether or not ant colonies are established in randomly selected locations divides a hundred-yard strip of meadow into 40 consecutive pieces and observes which of the 40 pieces contain resident ant colonies. Those that do are recorded with an "A" (for "ants") and those that don't are listed as "N" (for "no ants"). The data follow:

NNANNNNNANANNNNNNNAAAAANNNNNAANNNNNNNNN

At level $\alpha = .05$, can she assert with confidence that ant colonies are located according to some nonrandom distribution?

SECTION 11.F THE SPEARMAN TEST OF RANK CORRELATION

We used the (Pearson) coefficient of linear correlation in Sec. 6.D to test for the existence of a linear relationship between two sets of corresponding data points. We pointed out at that time the requirement that both sets of data be normally distributed in order for the t test of linearity to be valid. In this section, we will explain what to do when it is unreasonable to believe that we are working with normally distributed data. In brief, we can say that the technique, called the Spearman test of "rank correlation," requires us to rank each set of data and then compute the correlation coefficient of the ranks.

As we have discovered during the course of our work with nonparametric statistics in this chapter, we lose some control over the problem when we convert the original data points into ranks. The unpleasant consequences of the conversion can be best seen in the fact that we cannot use the Mann-Whitney test to check on the difference of means and standard deviations but only on the differences of averages and dispersions.

A similar situation occurs in connection with the Spearman test of rank correlation. It turns out that we cannot use the Spearman test to check on the existence of a linear relationship between two sets of data, but only to check on the existence of a "monotone" relationship. By a monotone relationship between x and y, we mean a relationship in which y increases as x increases, or y decreases as x increases. Examples of monotone increasing relationships can be seen graphically in Figs. 6.3, 6.6, 6.9b, and 6.12. Monotone decreasing relationships are illustrated in Figs. 6.1, 6.8, and 6.9c. All linear relationships are monotone, but, as Figs. 6.9b and 6.9c show, not all monotone relationships are linear. On the other hand, the parabolic data of Fig. 6.9a express a relationship that is not monotone, because, as x increases, y starts to decrease, but after a while y turns upward and begins to increase.

Because the Spearman test uses only the ranks of the data points and not their original numerical values, it can only judge whether or not two sets of data are related in a monotone way. It is not sharp enough to be able to judge the degree of linearity. On the other hand, the Spearman test has two distinct advantages: (1) Its usage is valid in the presence of nonnormal data, and (2) it works excellently in situations where at least one of the two data sets consists entirely of rankings rather than measurements. One widely known example of data that consist entirely of rankings is the weekly (during the football season) rankings of the top 10 college football teams. One interesting rank correlation problem here might be to test the correlation between the national rankings of the top 10 teams (ranked data) and the total weight of their seven first-string linemen (measurement data). It makes no sense to apply the Pearson test of linearity to this question, but the Spearman test fits in quite well.

As our first example of the use of the Spearman test of rank correlation, we present a large-sample problem dealing with a situation in medicine.

Example 11.8   Control of Fever by Aspirin Substitute   Because a significant fraction of the general population is known to be allergic to aspirin, the need for an effective substitute is evident. In the course of the laboratory development of a possible substitute, DAT-1, one seriously ill patient is given a dose every 4 hours for 48 hours, and his body temperature is recorded 2 hours after each dosage. His temperature, as time passes, is recorded in Table 11.15. At level $\alpha = .01$, can we assert with confidence that one's body temperature decreases as we continue to administer the drug DAT-1 every 4 hours?

SOLUTION   We are asking whether or not temperature decreases as hour of dosage increases. It is therefore appropriate to test whether or not temperature and hour of dosage are related to a monotone decreasing manner, and so the coefficient of rank correlation will be useful to us. For illustrative purposes, we present the scattergram of the aspirin-substitute–temperature data in Fig. 11.4.

In particular, we are testing the hypothesis

H : The relationship between temperature and hour
of dosage is not monotone

FIGURE 11.4
Scattergram of aspirin-substitute-temperature data.

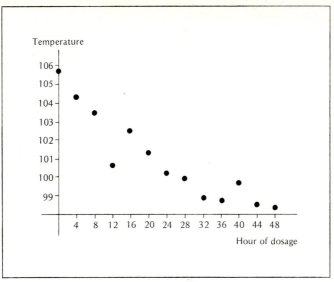

against the alternative

> A : The relationship is a monotone decreasing one

To test H against A, we first determine

$$n = \text{number of pairs of data points}$$

TABLE 11.15
Periodic Body Temperatures of
Patient Undergoing Aspirin
Substitute Treatment

Hour of Dosage	Temperature 2 Hours After Dosage
0	105.7
4	104.2
8	103.4
12	100.6
16	102.4
20	101.2
24	100.1
28	99.7
32	98.9
36	98.8
40	99.5
44	98.7
48	98.6

TABLE 11.16
Rejection Rules in the Presence of Various Alternatives
for the Spearman Test of Rank Correlation
H : The Relationship Is Not Monotone

Alternative	Reject H in Favor of A at Significance Level α if:
A : Monotone increasing	$z > z_\alpha$
A : Monotone decreasing	$z < -z_\alpha$
A : Monotone	$\lvert z \rvert > z_{\alpha/2}$

NOTE: The use of z is valid only in the large-sample case when there are more than 10 pairs of data points. For the small-sample case, special tables must be used.

If $n > 10$, then a modified version of the central limit theorem allows us to use a z test of H against A, where

● $$z = r_s \sqrt{n - 1} \qquad (11.9)$$

Here r_s = Spearman coefficient of rank correlation

The rejection rule is as follows, selected appropriately from Table 11.16: We reject H in favor of A at level $\alpha = .01$ if

$$z < -z_{.01} = -2.33$$

All that now remains is the computation of z, and most of this computation is concerned with the calculation of r_s. We perform the calculation of r_s in Table 11.17, using the formula

● $$r_s = 1 - \frac{6\Sigma d^2}{n(n^2 - 1)} \qquad (11.10)$$

The algebraic origin of this formula is discussed in Supplement 11.1. The operative term in the formula for r_s is the quantity Σd^2, the sum of the squared differences of the *ranks* of the data points in each pair. As shown in Table 11.17, we rank the data points in each column separately, lowest to highest; we compute the pairwise differences d between the *ranks* of x and y; and we then come up with Σd^2. From the calculations in Table 11.17, we find that

$$r_s = 1 - \frac{6\Sigma d^2}{n(n^2 - 1)} = 1 - \frac{6(716)}{13(13^2 - 1)} = 1 - \frac{4296}{(13)(169 - 1)}$$

$$= 1 - \frac{4296}{(13)(168)} = 1 - 1.967 = -0.967$$

It follows at once that

$$z = r_s \sqrt{n - 1} = (-0.967)\sqrt{13 - 1} = (-0.967)\sqrt{12} = (-0.967)(3.464)$$
$$= -3.35$$

TABLE 11.17
Calculations Leading to the Spearman Coefficient
of Rank Correlation
Aspirin-Substitute–Temperature Data

Hour of Dosage x	Rank r_x	Temperature 2 Hours after Dosage y	Rank r_y	$d = r_x - r_y$	d^2
0	1	105.7	13	−12	144
4	2	104.2	12	−10	100
8	3	103.4	11	−8	64
12	4	100.6	8	−4	16
16	5	102.4	10	−5	25
20	6	101.2	9	−3	9
24	7	100.1	7	0	0
28	8	99.7	6	2	4
32	9	98.9	4	5	25
36	10	98.8	3	7	49
40	11	99.5	5	6	36
44	12	98.7	2	10	100
48	13	98.6	1	12	144
Sums				0	716

$n = 13$

$$r_s = 1 - \frac{6\Sigma d^2}{n(n^2 - 1)} = 1 - \frac{6(716)}{13(13^2 - 1)} = 1 - \frac{4296}{(13)(169 - 1)} = 1 - \frac{4296}{(13)(168)}$$
$$= 1 - 1.967 = -.967$$

We have agreed to reject H in favor of A if $z < -2.33$. Since it turned out that $z = -3.35 < -2.33$, we do, in fact, reject H. At level $\alpha = .01$, we therefore conclude that temperature is monotone decreasing as the hours of dosage with the aspirin substitute DAT-1 pass. The data do contain evidence for the assertion that the periodic use of DAT-1 tends to reduce body temperature.

When fewer than 11 pairs of data points are available, it is not valid to use the percentage points of the normal distribution for the Spearman test of rank correlation. We must refer instead to a special table, appearing as Table A.9 of the Appendix, giving the levels of significance for small-sample tests of rank correlation. The following example, based on the parabolic data of Fig. 6.9a and Table 6.4a, presents an illustration of the small-sample procedure.

Example 11.9 Psychological Learning Theory A psychologist wants to find out whether she can teach color differentiation to monkeys by using bananas as rewards. She sets up a row of several balls of different colors, hands each monkey a blue ball and several bananas, and asks the monkey to choose the ball of the same color from the row. She records the number of bananas given to and the number of mistakes made by each

TABLE 11.18
Psychological Learning Data

Monkey	Number of Bananas	Number of Mistakes
A	8	9
B	1	16
C	6	1
D	3	4
E	5	0
F	2	10

of the monkeys. The resulting data are presented in Table 11.18. At level $\alpha = .05$, do the results of the experiment indicate that the number of mistakes made by a monkey bears a monotone relationship with the number of bananas it received?

SOLUTION We apply the Spearman test of rank correlation to test the hypothesis

$$H : \text{The relationship is not monotone}$$

against the alternative

$$A : \text{The relationship is monotone}$$

According to the information on rejection rules presented in Table 11.16 (adjusted mentally to the small-sample case), we compute r_s and we reject H at level α if

$$|r_s| > r_s(n, \alpha/2)$$

Here $n = 6$, and Table A.9 contains the information that

$$r_s(6, .029) = 0.83$$

As we have often found when working with small-sample tests, especially those involving the binomial distribution, we cannot always conduct a test at the exact significance level desired. However, we can usually do it at a very close level. In this case, we want to conduct the test at level $\alpha = .05$; however, the nature of r_s allows us to have $\alpha/2 = .029$ instead of .025, and so we must use a level $\alpha = (.029)(2) = .058$. At level $\alpha = .058$, then, we would reject H in favor of A if

$$|r_s| > r_s(6, .029) = 0.83$$

It remains now only to compute r_s for the data of Table 11.18. We do the computations in Table 11.19, and they show that

$$r_s = 1 - \frac{6\Sigma d^2}{n(n^2 - 1)} = 1 - \frac{6(56)}{6(6^2 - 1)} = 1 - \frac{336}{(6)(36 - 1)} = 1 - \frac{336}{(6)(35)}$$
$$= 1 - 1.6 = -0.6$$

Therefore $|r_s| = .6$, which does not exceed our rejection level of $r_s(6, .029) = .83$. Consequently, we cannot reject H at level $\alpha = .058$. Our conclusion is that the relation-

TABLE 11.19

Calculations Leading to the Spearman Coefficient of Rank Correlation
Psychological Learning Data

Monkey	Number of Bananas x	Rank r_x	Number of Mistakes y	Rank r_y	$d = r_x - r_y$	d^2
A	8	6	9	4	2	4
B	1	1	16	6	−5	25
C	6	5	1	2	3	9
D	3	3	4	3	0	0
E	5	4	0	1	3	9
F	2	2	10	5	−3	9
Sums					0	56

$n = 6$

$$r_s = 1 - \frac{6\Sigma d^2}{n(n^2 - 1)} = 1 - \frac{6(56)}{(6)(6^2 - 1)} = 1 - \frac{336}{(6)(36 - 1)} = 1 - \frac{336}{(6)(35)} = 1 - 1.6 = -0.6$$

ship between number of mistakes made by the monkeys and number of bananas given to them is not monotone. There seems to be no trend in either direction, monotone increasing or monotone decreasing.

One final remark, if there are two or more data points that have the same numerical value and therefore have tied ranks, we divide up the rank positions equally in the same manner as we did for the Mann-Whitney test. For example, if the 3rd, 4th, and 5th ranking data points were all equal, they would each be assigned a rank of

$$\frac{3 + 4 + 5}{3} = \frac{12}{3} = 4$$

Unless there are an unusually large number of ties, no adjustment of the formula for r_s is necessary.

SUPPLEMENT 11.1 ALGEBRAIC JUSTIFICATION OF THE FORMULA FOR THE COEFFICIENT OF RANK CORRELATION

In Supplement 6.3 we developed the formula for r, the coefficient of linear correlation. The formula for r_s, the coefficient of rank correlation, has its origin in the formula

$$r = \frac{n\sum_{k=1}^{n} x_k y_k - \sum_{k=1}^{n} x_k \sum_{k=1}^{n} y_k}{\sqrt{n\sum_{k=1}^{n} x_k^2 - \left(\sum_{k=1}^{n} x_k\right)^2}\sqrt{n\sum_{k=1}^{n} y_k^2 - \left(\sum_{k=1}^{n} y_k\right)^2}}$$

because r_s is merely the coefficient of linear correlation between the *ranks* of the original data points, instead of between the original data points themselves. To obtain formula (11.10) for r_s, we can consider x_k and y_k to be the *ranks* of the *k*th data point of each column. Then

$$d_k = x_k - y_k$$

is the difference between the ranks of the *k*th pair.

Now, if there are *n* data points, then both the *x* column and the *y* column will have the ranks 1, 2, 3, . . . , *n*. Therefore the sum of *x* ranks, denoted by

$$\sum_{k=1}^{n} x_k$$

and the sum of *y* ranks, denoted by

$$\sum_{k=1}^{n} y_k$$

will both be equal to

$$1 + 2 + 3 + \cdots + n = \frac{n(n + 1)}{2}$$

The above statement is true for the following reason: If *n* is even, then

$$1 + 2 + 3 + \cdots + n = (1 + n) + (2 + n - 1) + \cdots + \left[\frac{n}{2} + \left(\frac{n}{2} + 1 \right) \right]$$

$$= (1 + n) + (1 + n) + \cdots + (1 + n)$$

$$= \frac{n}{2}(n + 1)$$

$$= \frac{n(n + 1)}{2}$$

because the sum consists of $n/2$ terms, each of which is equal to $n + 1$. If *n* is odd, then $n - 1$ is even, and so

$$1 + 2 + 3 + \cdots + n = 1 + 2 + 3 + \cdots + (n - 1) + n$$

$$= \frac{(n - 1)(n)}{2} + n$$

$$= \frac{(n - 1)(n) + 2n}{2}$$

$$= \frac{(n - 1 + 2)n}{2}$$

$$= \frac{n(n + 1)}{2}$$

Therefore,

$$\sum_{k=1}^{n} x_k = \frac{n(n + 1)}{2}$$

and

$$\sum_{k=1}^{n} y_k = \frac{n(n + 1)}{2}$$

Furthermore,

$$\sum_{k=1}^{n} x_k^2 \quad \text{and} \quad \sum_{k=1}^{n} y_k^2$$

are sums of squares of the numbers $1, 2, 3, \ldots, n$. An algebraic result related to the one just discussed above reveals that

$$1^2 + 2^2 + 3^2 + \cdots + n^2 = \frac{n(n + 1)(2n + 1)}{6}$$

and so

$$\sum_{k=1}^{n} x_k{}^2 = \frac{n(n + 1)(2n + 1)}{6}$$

and

$$\sum_{k=1}^{n} y_k{}^2 = \frac{n(n + 1)(2n + 1)}{6}$$

Therefore,

$$n \sum_{k=1}^{n} x_k{}^2 - \left(\sum_{k=1}^{n} x_k\right)^2 = n \left[\frac{n(n + 1)(2n + 1)}{6}\right] - \left[\frac{n(n + 1)}{2}\right]^2$$

$$= \frac{n^2(n + 1)(2n + 1)}{6} - \frac{n^2(n + 1)^2}{4}$$

$$= n^2(n + 1)\left[\frac{2n + 1}{6} - \frac{n + 1}{4}\right]$$

$$= n^2(n + 1)\left[\frac{2(2n + 1) - 3(n + 1)}{12}\right]$$

$$= n^2(n + 1)\left[\frac{n - 1}{12}\right] = \frac{n^2(n^2 - 1)}{12}$$

But

$$n \sum_{k=1}^{n} y_k{}^2 - \left(\sum_{k=1}^{n} y_k\right)^2$$

equals the same thing, so that upon inserting these items into the formula for r, we obtain that

$$r_s = \frac{n \sum_{k=1}^{n} x_k y_k - \frac{n^2(n + 1)^2}{4}}{\frac{n^2(n^2 - 1)}{12}}$$

What about $\sum_{k=1}^{n} x_k y_k$? Well,

$$\sum_{k=1}^{n} d_k{}^2 = \sum_{k=1}^{n} (x_k - y_k)^2 = \sum_{k=1}^{n} x_k{}^2 - 2 \sum_{k=1}^{n} x_k y_k + \sum_{k=1}^{n} y_k{}^2$$

$$= \frac{n(n + 1)(2n + 1)}{6} - 2 \sum_{k=1}^{n} x_k y_k + \frac{n(n + 1)(2n + 1)}{6}$$

$$= \frac{n(n + 1)(2n + 1)}{3} - 2 \sum_{k=1}^{n} x_k y_k$$

so that

$$\sum_{k=1}^{n} x_k y_k = \frac{\frac{n(n + 1)(2n + 1)}{3} - \sum_{k=1}^{n} d_k{}^2}{2}$$

$$= \frac{n(n + 1)(2n + 1) - 3 \sum_{k=1}^{n} d_k{}^2}{6}$$

Therefore,

$$n \sum_{k=1}^{n} x_k y_k = \frac{n^2(n + 1)(2n + 1) - 3n \sum_{k=1}^{n} d_k^2}{6}$$

Inserting this value back into the formula for r_s that we have developed so far, we have that

$$r_s = \frac{\dfrac{n^2(n + 1)(2n + 1) - 3n \sum_{k=1}^{n} d_k^2}{6} - \dfrac{n^2(n + 1)^2}{4}}{\dfrac{n^2(n^2 - 1)}{12}}$$

$$= \frac{2n^2(n + 1)(2n + 1) - 6n \sum_{k=1}^{n} d_k^2 - 3n^2(n + 1)^2}{n^2(n^2 - 1)}$$

$$= \frac{2n^2(n + 1)(2n + 1) - 3n^2(n + 1)^2}{n^2(n^2 - 1)} - \frac{6n \sum_{k=1}^{n} d_k^2}{n^2(n^2 - 1)}$$

$$= \frac{n^2(n + 1)[2(2n + 1) - 3(n + 1)]}{n^2(n + 1)(n - 1)} - \frac{6 \sum_{k=1}^{n} d_k^2}{n(n^2 - 1)}$$

$$= \frac{2(2n + 1) - 3(n + 1)}{n - 1} - \frac{6 \sum_{k=1}^{n} d_k^2}{n(n^2 - 1)}$$

$$= 1 - \frac{6 \sum_{k=1}^{n} d_k^2}{n(n^2 - 1)}$$

which is the formula for the Spearman coefficient of rank correlation.

EXERCISES 11.F

1 Referring to the data of Exercise 6.C.7,
   a At level $\alpha = .05$, does one's job performance rating seem to be related in a monotone way with one's math aptitude score?
   b Draw a scattergram of the data.

2 Based on the data of Exercise 6.C.5,
   a Does the total cost of production seem to increase monotonically at level $\alpha = .01$ as the quantity produced increases?
   b Draw a scattergram of the data.

3 A geographer would like to discover whether or not the amount of cropland in a region bears a monotone relationship to the amount of lowland in that region. The following data compare the percentage of lowland in each of 20 randomly selected counties across the United States with the percentage of cropland in the county:

County	Percentage of Lowland	Percentage of Cropland
A	48	28
B	62	45
C	35	22
D	49	35
E	100	75
F	88	62
G	0	3
H	41	24
I	40	21
J	40	21
K	61	25
L	68	26
M	33	17
N	0	4
O	100	70
P	37	29
Q	44	36
R	90	73
S	100	82
T	100	84

a At level $\alpha = .05$, do the data indicate that percentage of cropland is related in a monotonically increasing manner with the percentage of lowland?

b Draw a scattergram of the data.

4 Twelve persons selected at random are asked the following two questions by a psychologist as part of a job satisfaction study: (1) What is your hourly pay? (2) How would you rate your job satisfaction on a scale of 0 (low) to 10 (high)? The results of the survey are presented below:

Respondent	A	B	C	D	E	F	G	H	I	J	K	L
Hourly pay, dollars	8	2	6	4	4	20	10	6	6	4	11	15
Job satisfaction	6	4	5	4	3	8	7	4	1	2	9	5

a At level $\alpha = .05$, does there seem to be a monotone relationship between job satisfaction and hourly pay?

b Draw a scattergram of the data.

5 As part of a study on the question of whether or not you should judge a book by its cover, an advertising executive hired two groups of consultants to judge each of 10 books according to artistry of the cover and internal literary content. Each book was given a ranking, from 1 to 10, for its cover and again for its content. The results follow:

Book	#1	#2	#3	#4	#5	#6	#7	#8	#9	#10
Cover rank	6	4	10	9	7	8	3	2	5	1
Content rank	5	7	1	2	4	3	8	9	6	10

Do the results of the study indicate at level $\alpha = .01$ that you should judge a book by its cover?

6 An analysis of the relation between the net incomes of 16 major banking companies and the total return on investment to their stockholders yielded the following information:

Banking Company	Rank According to Net Income	Rank According to Return to Stockholders
Bank of America	2	9
First National City	1	12
Chase Manhattan	3	15
J.P. Morgan	4	2
Manufacturers Hanover	5	6
Chemical New York	8	3
Bankers Trust	10	7
Continental Illinois	7	14
First Chicago	6	16
United California	9	11
Security Pacific	12	10
Wells Fargo	13	13
Marine Midland	14	8
Irving Trust	15	4
Crocker National	16	5
Mellon National	11	1

On the basis of the above data (which described the situation for one year in the recent past), could a financial analyst have concluded at level $\alpha = .10$ that the relationship between net income and return to stockholders was monotone?

7 In a historical study of the changing patterns of employment, the following data[1] were gathered on the percentage of the working population employed in various sectors of the economy in nineteenth century England and Wales. The objective was to determine whether or not the leading sectors of the economy in 1881 were the same ones as in 1851, namely, whether or not the relationship between the employment percentages of the two years was a monotone increasing one.

Sector	Employment Percentage	
	1851	1881
Agriculture	20.9	11.5
Fishing	0.2	0.3
Mining	4.0	4.8
Building	5.5	6.8
Manufacture	32.7	30.7

[1]E. A. Wrigley (ed.), "Nineteenth-Century Society: Essays in the Use of Quantitative Methods for the Study of Social Data," p. 229, Cambridge University Press, New York, 1972.

Sector	Employment Percentage	
	1851	1881
Transport	4.1	5.6
Dealing	6.5	7.8
Industrial service	4.5	6.7
Public service and professional	4.6	5.6
Domestic service	13.3	15.7
Other	3.7	4.5

At level $\alpha = .05$, does the relationship between employment percentages of the years studied seem to be a monotone increasing one?

SUMMARY AND DISCUSSION

In this concluding chapter, we have tied together a number of loose ends which were left hanging in earlier chapters. Many of the small-sample tests of statistical hypotheses that we studied in Chaps. 5, 6, and 10 were based on tables of the t and F distributions, and this meant that they could be validly applied only to sets of data satisfying stringent conditions. In the present chapter, we have freed ourselves of some of these restrictions by introducing statistical tests based on shadow properties, such as ranks, runs, and signs, of a set of data, instead of on the numerical values of the data points themselves.

Although, on the one hand, we can apply these tests to almost all data regardless of the distribution, on the other hand, we get less precise results because we are forced to work with these shadow properties instead of with the actual data points. Working with a nonparametric test is somewhat analogous to studying the shadow of an object in order to draw conclusions about the object itself. We can see the general picture, but we lose sight of the sharp distinctions. However, when the data do not satisfy the required conditions of the parametric tests, we have no other choice but to use nonparametric tests because the parametric tests simply are invalid in such contexts. The moral of the story is this: When a parametric test is validly applicable, use it; when it is not, make the best of a nonparametric test that is available.

BIBLIOGRAPHY

Detailed Treatments including Corrections for Ties
Mosteller, F., and R. E. K. Rourke: "Sturdy Statistics," Addison-Wesley, Reading, Mass., 1973.

Noether, G. E.: "Introduction to Statistics: A Nonparametric Approach," 2d ed., Houghton Mifflin, Boston, 1976.

Special Topics
Marshall, C. W.: A Simple Derivation of the Mean and Variance of the Number of Runs in an Ordered Sample, *The American Statistician*, October 1970, pp. 27–28.

Noether, G. E.: Distribution-Free Confidence Intervals, *The American Statistician*, February 1972, pp. 39–41.

Zar, J. H.: Significance Testing of the Spearman Rank Correlation Coefficient, *Journal of the American Statistical Association*, September 1972, pp. 578–580.

SUPPLEMENTARY EXERCISES

1 A person who feels the need to lose some weight is considering going on one of the two famous reducing diets, the Drink-More diet and the Eat-Less diet, but only if statistical analysis shows that one or both of these diets are effective in inducing weight loss. As it turns out, five acquaintances have tried the Drink-More diet, and another five have tried the Eat-Less diet. The weights of the 10 acquaintances, both before beginning and after completing their diet programs, are recorded below:

	Drink-More Diet			Eat-Less Diet	
Acquain-tance	Weight Before	Weight After	Acquain-tance	Weight Before	Weight After
A	150	150	F	150	141
B	160	160	G	160	150
C	170	170	H	170	160
D	180	179	I	180	170
E	190	141	J	190	179

a Apply the Mann-Whitney test for equal averages to find at level $\alpha = .05$ whether or not the Drink-More diet is effective.
b Apply the Mann-Whitney test for equal averages to find at level $\alpha = .05$ whether or not the Eat-Less diet is effective.
c Compare the results of parts a and b above, and explain why the Mann-Whitney test does not have the ability to distinguish between the two diets.
d Apply the paired-sample sign test to determine whether or not the hypothesis that the Drink-More diet is not effective can be rejected at level $\alpha = .05$. (At what exact level of significance does the paired-sample sign test reject the hypothesis that the Drink-More diet is not effective?)
e Apply the paired-sample sign test to determine whether or not the hypothesis that the Eat-Less diet is not effective can be rejected at level $\alpha = .05$.
f At what exact level of significance does the paired-sample sign test reject the hypothesis that the Eat-Less diet is not effective?
g Explain how the answers to parts d and f show that the paired-sample sign test has the ability to distinguish between the two diets.

2 As part of a sociological comparison of employment patterns in the central city and the suburbs, a researcher accumulates the following data on the monthly incomes of seven randomly selected residents of the central city and five randomly selected residents of the suburbs:

Central city incomes	550	429	401	528	659	896	764
Suburban incomes	1804	811	960	3124	1204		

a Use the F test of Sec. 5.E to test at level $\alpha = .05$ whether or not suburban incomes are more variable than incomes in the central city.

b Use the Mann-Whitney test of equal dispersion at level $\alpha = .05$ to answer the same question as in part a on the basis of the above data.

c Compare the calculations involved in parts a and b above, and explain why the Mann-Whitney test does not have the ability to distinguish between the two dispersions.

d Upon observing that the medians of the two sets of data are 550 and 1204, respectively, we can subtract from each data point the median of its own set. We then obtain the following "adjusted data points":

Central city incomes (minus median)	0	−121	−149	−22	109	346	214
Suburban incomes (minus median)	600	−393	−244	1920	0		

Apply the Mann-Whitney test of equal dispersion to the above adjusted data points in order to decide, at level $\alpha = .05$, whether or not suburban incomes are more variable than incomes in the central city.

e Compare the different results obtained in parts b and d, and explain why the method of part d provides a truer measure of the difference between the two dispersions.

3 The value of the dollar on international currency markets goes up (U) some days and down (D) other days. For a recent 40-day period, the ups and downs went as follows:

DDDDUDDDDDDDDDUUDDUUUUUUDDDDDDDDUDDDDDDDDD

At level $\alpha = .05$, can we consider the price fluctuations of the dollar to be random?

4 As part of a study of unemployment and unemployment compensation, an economist noted the following data on the situation in seven industrialized countries:

Country	Percentage of Work Force Unemployed	Percentage of Work Force Covered by Unemployment Compensation
United States	8.4	95
Canada	7.2	99
Great Britain	4.7	80
West Germany	4.5	93
France	3.8	61
Italy	3.4	51
Japan	1.7	45

Does the data support the conclusion that a country which ranks high in percentage of work force covered by unemployment compensation also ranks high in percentage of work force unemployed? Use a level of significance of $\alpha = .01$.

5 A social worker is assigned the job of finding whether or not a particular 6-month vocational training program is effective in increasing the income of its participants. Fourteen participants are selected at random from a group that is just about to begin training, and their monthly incomes are recorded. Eighteen months later (one year after completion of the training program) their monthly incomes are again noted. The data (in hundreds of dollars) follow:

Participant	Income at Start Of Program	Income 18 Months Later
#1	2.2	3.8
#2	0.0	4.1
#3	3.9	4.8
#4	6.0	6.0
#5	1.3	4.1
#6	3.9	5.8
#7	2.3	0.0
#8	0.0	0.0
#9	1.5	0.0
#10	4.2	4.7
#11	1.6	4.0
#12	2.4	4.2
#13	3.6	4.1
#14	3.5	0.0

At level $\alpha = .05$, decide whether the program yields a significant increase in the monthly income of its participants.

6 It is the job of a tea tester for an English tea-importing company to decide which tea leaves are most suitable for English Breakfast tea. The tea tester must decide among three lots of leaves, an inexpensive lot, a moderately priced lot, and an expensive lot. The tea tester and his laboratory staff discover the following levels of impurities in eight randomly selected leaves of each lot:

Inexpensive lot	3	4	25	9	11	4	12	20
Moderately priced lot	4	9	16	9	12	6	4	8
Expensive lot	6	5	7	10	6	8	5	4

Do the data indicate at level $\alpha = .01$ that the three lots of leaves manifest significant differences in impurity content?

7   Applicants for the position of public school security guard are routinely given a psychological test of their ability to remain cool under extremely trying conditions. The following set of data classifies some recent test scores according to the age group of the applicant:

21–25	26–30	31–35	36–40	41–45
80	70	41	68	72
69	64	73	84	90
92	80	65	29	44
83	76	88	63	88
94	91	67	87	66
75	59	78	56	34
85		77	86	98
82		95	48	87
77			96	

At level $\alpha = .05$, do the data provide sufficient evidence to assert that there are significant differences between the various age groups in ability to remain cool?

8   The following data[1] concern the rate of unemployment in ten countries belonging to the Organization for Economic Co-operation and Development (OECD) early in the 1970s:

Country	Second-Quarter Unemployment Rate	
	1971	1972
U.K.	2.9	3.5
U.S.	6.0	5.8
France	2.2	2.4
Germany	0.7	1.0
Italy	3.2	3.7
Belgium	1.8	2.2
Canada	6.4	6.1
Japan	1.0	1.1
Australia	1.2	1.8
Sweden	2.1	2.0

[1]The OECD Observer, December 1972, p. 23.

At significance level $\alpha = .02$, can we say that unemployment increased in OECD countries during the one-year period?

9   At level $\alpha = .05$, do the data of Exercise 9.B.7 point up the existence of a significantly monotone relationship between the numbers of workers in various industries covered and not covered by collective bargaining agreements which mention pension plans?

GLOSSARY OF FORMULAS

sample mean (Sec. 4.A)

$$\bar{x} = \frac{\Sigma x}{n} \tag{4.1}$$

sample standard deviation, defining formula (Sec. 4.B)

$$s = \sqrt{\frac{\Sigma(x - \bar{x})^2}{n - 1}} \tag{4.2}$$

sample standard deviation, short-cut formula (Sec. 4.B)

$$s = \sqrt{\frac{n\Sigma x^2 - (\Sigma x)^2}{n(n - 1)}} \tag{4.3}$$

$(1 - \alpha)$ 100% **large-sample confidence interval for mean** (Sec. 4.B)

$$\bar{x} - z_{\alpha/2}\frac{s}{\sqrt{n}} < \mu < \bar{x} + z_{\alpha/2}\frac{s}{\sqrt{n}} \tag{4.4}$$

$(1 - \alpha)$ 100% **error of estimation** (Sec. 4.B)

$$E = z_{\alpha/2}\frac{s}{\sqrt{n}} \tag{4.5}$$

$(1 - \alpha)$ 100% **small-sample confidence interval for mean** (Sec. 4.C)

$$\bar{x} - t_{\alpha/2}(n - 1)\frac{s}{\sqrt{n}} < \mu < \bar{x} + t_{\alpha/2}(n - 1)\frac{s}{\sqrt{n}} \tag{4.6}$$

$(1 - \alpha)$ 100% **small-sample confidence interval for standard deviation** (Sec. 4.D)

$$\sqrt{\frac{(n - 1)s^2}{\chi_{\alpha/2}^2(n - 1)}} < \sigma < \sqrt{\frac{(n - 1)s^2}{\chi_{1-\alpha/2}^2(n - 1)}} \tag{4.7}$$

$(1 - \alpha)$ 100% **large-sample confidence interval for standard deviation** (Sec. 4.D)

$$\frac{s}{1 + \dfrac{z_{\alpha/2}}{\sqrt{2n}}} < \sigma < \frac{s}{1 - \dfrac{z_{\alpha/2}}{\sqrt{2n}}} \tag{4.8}$$

$(1 - \alpha)$ 100% **large-sample confidence interval for proportion** (Sec. 4.E)

$$\hat{p} - z_{\alpha/2}\sqrt{\frac{\hat{p}(1 - \hat{p})}{n}} < p < \hat{p} + z_{\alpha/2}\sqrt{\frac{\hat{p}(1 - \hat{p})}{n}} \tag{4.9}$$

one-sample z statistic (Sec. 5.B)

$$z = \frac{\bar{x} - \mu_0}{s/\sqrt{n}} \tag{5.1}$$

one-sample t statistic (Sec. 5.B)

$$t = \frac{\bar{x} - \mu_0}{s/\sqrt{n}} \tag{5.2}$$

$$df = n - 1$$

two-sample z statistic (Sec. 5.C)

$$z = \frac{\bar{x}_A - \bar{x}_B}{\sqrt{\dfrac{s_A^2}{n_A} + \dfrac{s_B^2}{n_B}}} \tag{5.3}$$

two-sample t statistic (Sec. 5.C)

$$t = \frac{\bar{x}_A - \bar{x}_B}{\sqrt{\dfrac{\Sigma(x_A - \bar{x}_A)^2 + \Sigma(x_B - \bar{x}_B)^2}{n_A + n_B - 2}\left(\dfrac{1}{n_A} + \dfrac{1}{n_B}\right)}} \tag{5.4}$$

$$df = n_A + n_B - 2$$

paired-sample t statistic (Sec. 5.D)

$$t = \frac{\bar{d}\sqrt{n}}{s_d} \tag{5.6}$$

$$df = n - 1$$

one-sample chi-square statistic (Sec. 5.E)

$$\chi^2 = \frac{(n - 1)s^2}{\sigma_0^2} = \frac{\Sigma(x - \bar{x})^2}{\sigma_0^2} \tag{5.7}$$

$$df = n - 1$$

one-sample z statistic for standard deviation (Sec. 5.E)

$$z = \sqrt{2\chi^2} - \sqrt{2(n-1)} \qquad (5.8)$$

two-sample F statistic (Sec. 5.E)

$$F = \frac{s_L^2}{s_S^2} \qquad (5.9)$$

$$\text{dfn} = n_L - 1 \qquad \text{dfd} = n_S - 1$$

one-sample z statistic for proportion (Sec. 5.F)

$$z = \frac{\hat{p} - p_0}{\sqrt{\dfrac{p_0(1 - p_0)}{n}}} \qquad (5.10)$$

two-sample z statistic for proportions (Sec. 5.F)

$$z = \frac{\hat{p}_A - \hat{p}_B}{\sqrt{\hat{p}(1 - \hat{p})\left(\dfrac{1}{n_A} + \dfrac{1}{n_B}\right)}} \qquad (5.11)$$

$$\hat{p} = \frac{n_A\hat{p}_A + n_B\hat{p}_B}{n_A + n_B}$$

linear regression equation (Sec. 6.B)

$$y = a + bx$$

$$b = \frac{n\Sigma xy - (\Sigma x)(\Sigma y)}{n\Sigma x^2 - (\Sigma x)^2} \qquad (6.1)$$

$$a = \frac{\Sigma y - b\Sigma x}{n} \qquad (6.2)$$

coefficient of linear correlation (Sec. 6.C)

$$r = \frac{n\Sigma xy - (\Sigma x)(\Sigma y)}{\sqrt{n\Sigma x^2 - (\Sigma x)^2}\sqrt{n\Sigma y^2 - (\Sigma y)^2}} \qquad (6.3)$$

coefficient of linear determination (Sec. 6.C)

$$r^2$$

t statistic for linearity (Sec. 6.D)

$$t = \frac{r\sqrt{n-2}}{\sqrt{1 - r^2}} \qquad (6.4)$$

$$\text{df} = n - 2$$

standard error of the estimate (Sec. 6.E)

$$s_e = \sqrt{\frac{\Sigma y^2 - a\Sigma y - b\Sigma xy}{n - 2}} \qquad (6.5)$$

(1 − α) 100% confidence interval for prediction (Sec. 6.E)

$$a + bx_0 - E_0 \le y_0 \le a + bx_0 + E_0 \qquad (6.6)$$

$$E_0 = t_{\alpha/2}(n - 2)s_e\sqrt{1 + \frac{1}{n} + \frac{n(x_0 - \bar{x})^2}{n\Sigma x^2 - (\Sigma x)^2}} \qquad (6.7)$$

logarithmic regression equation (Sec. 7.A)

$$y = a + b \log x \qquad (7.1)$$

simple exponential regression equation (Sec. 7.A)

$$y = a \cdot b^x \qquad (7.2)$$

power regression equation (Sec. 7.A)

$$y = a \cdot x^b \qquad (7.3)$$

parabolic regression equation (Sec. 7.B)

$$y = a + bx + cx^2 \qquad (7.4)$$

$$\begin{aligned} an + b\Sigma x + c\Sigma x^2 &= y \\ a\Sigma x + b\Sigma x^2 + c\Sigma x^3 &= \Sigma xy \\ a\Sigma x^2 + b\Sigma x^3 + c\Sigma x^4 &= \Sigma x^2 y \end{aligned} \qquad (7.5)$$

multiple linear regression equation (Sec. 7.C)

$$y = a + bx + cw \qquad (7.6)$$

$$\begin{aligned} an + b\Sigma x + c\Sigma w &= \Sigma y \\ a\Sigma x + b\Sigma x^2 + c\Sigma xw &= \Sigma xy \\ a\Sigma w + b\Sigma xw + c\Sigma w^2 &= \Sigma wy \end{aligned} \qquad (7.11)$$

coefficients of partial correlation (Sec. 7.C)

$$r_{wy:x} = \frac{r_{wy} - r_{xw}r_{xy}}{\sqrt{(1 - r_{xw}^2)(1 - r_{xy}^2)}} \qquad (7.9)$$

$$r_{xy:w} = \frac{r_{xy} - r_{xw}r_{wy}}{\sqrt{(1 - r_{xw}^2)(1 - r_{wy}^2)}} \qquad (7.10)$$

coefficient of multiple determination (Sec. 7.C)

$$R^2 = r_{xy}^2 - (1 - r_{xy}^2)r_{wy:x}^2 \qquad (7.7)$$

$$R^2 = r_{wy}^2 - (1 - r_{wy}^2)r_{xy:w}^2 \qquad (7.8)$$

arithmetic of events (Sec. 8.A)

$$E = (E \cap F) \cup (E \cap F^c) \qquad (8.1)$$

$$E = (E \cap F_1) \cup (E \cap F_2) \cup \cdots \cup (E \cap F_n) \qquad (8.2)$$

probability of union of events (Sec. 8.B)

$$P(C \cup D) = P(C) + P(D) - P(C \cap D) \qquad (8.3)$$

probability of an event (Sec. 8.B)

$$P(E) = P(E \cap F) + P(E \cap F^c) \qquad (8.4)$$

$$P(E) = P(E \cap F_1) + P(E \cap F_2) + \cdots + P(E \cap F_n) \qquad (8.5)$$

conditional probability (Sec. 8.C)

$$P(C \mid D) = \frac{P(C \cap D)}{P(D)} \qquad (8.6)$$

probability of intersection of events (Sec. 8.C)

$$P(C \cap D) = P(C \mid D)P(D) \qquad (8.7)$$

probability of intersection of independent events (Sec. 8.D)

$$P(C \cap D) = P(C)P(D) \tag{8.8}$$

$$P(C_1 \cap C_2 \cap \cdots \cap C_n) = P(C_1)P(C_2) \cdots P(C_n) \tag{8.9}$$

permutations (Sec. 8.E)

$$P(n, k) = n \times (n - 1) \times \cdots \times (n - k + 1) \tag{8.10}$$

$$P(n, k) = \frac{n!}{(n - k)!} \tag{8.11}$$

combinations (Sec. 8.E)

$$C(n, k) = \frac{n!}{k!(n - k)!} \tag{8.12}$$

binomial probabilities (Sec. 8.E)

$$P(X = j) = C(n, j)p^j(1 - p)^{n-j} \tag{8.13}$$

$$P(X \leq k) = \sum_{j=0}^{k} C(n, j)p^j(1 - p)^{n-j} \tag{8.14}$$

Bayes' theorem (Sec. 8.F)

$$P(B \mid A) = \frac{P(A \mid B)P(B)}{P(A \mid B)P(B) + P(A \mid B^c)P(B^c)} \tag{8.15}$$

$$P(B_j \mid A) = \frac{P(A \mid B_j)P(B_j)}{\sum_{k=1}^{n} P(A \mid B_k)P(B_k)} \tag{8.16}$$

chi-square test of independence (Sec. 9.A)

$$\chi^2 = \sum_{k=1}^{n} \frac{(f_k - e_k)^2}{e_k} \tag{9.2}$$

$$df = (r - 1)(c - 1)$$

Goodman-Kruskal index of predictive association (Sec. 9.B)

$$\lambda = \frac{\Sigma(LR) - L(\Sigma C)}{T - L(\Sigma C)} \tag{9.3}$$

chi-square test of goodness of fit (Sec. 9.C)

$$\chi^2 = \sum_{k=1}^{m} \frac{(f_k - e_k)^2}{e_k}$$

$$df = m - 1 \text{ (uniform, binomial)}$$

$$df = m - 3 \text{ (normal)}$$

components of the variance (Sec. 10.A)

$$TV = nVB + VW \tag{10.1}$$

F statistic for analysis of variance (Sec. 10.A)

$$F = \frac{nVB/(m - 1)}{VW/m(n - 1)} \tag{10.2}$$

$$m = \text{number of samples}$$

$$n = \text{number of data points per sample}$$

$$\text{dfn} = m - 1$$

$$\text{dfd} = m(n - 1)$$

F statistic for one-way analysis of variance (Sec. 10.B)

$$F = \frac{nVB/(m - 1)}{VW/(N - m)} \tag{10.3}$$

$$m = \text{number of samples}$$

$$N = \text{total number of data points}$$

$$\begin{aligned}\text{dfn} &= m - 1 \\ \text{dfd} &= N - m\end{aligned} \tag{10.4}$$

$$nVB = \Sigma\left(\frac{S^2}{n}\right) - \frac{(\Sigma x)^2}{N} \tag{10.5}$$

$$VW = \Sigma x^2 - \Sigma\left(\frac{S^2}{n}\right) \tag{10.6}$$

F statistic for treatment effects in two-way analysis of variance (Sec. 10.C)

$$F_T = \frac{(n - 1)SST}{SSE} \tag{10.8}$$

$$n = \text{number of blocks}$$

$$\begin{aligned}\text{dfn} &= m - 1 \\ \text{dfd} &= (n - 1)(m - 1)\end{aligned} \tag{10.7}$$

$$m = \text{number of treatments}$$

$$SST = \frac{\Sigma S_T^2}{n} - \frac{(\Sigma x)^2}{N} \tag{10.11}$$

$$SSE = \Sigma x^2 - \frac{\Sigma S_T^2}{n} - \frac{\Sigma S_B^2}{m} + \frac{(\Sigma x)^2}{N} \tag{10.13}$$

F statistic for block effects in two-way analysis of variance (Sec. 10.C)

$$F_B = \frac{(m - 1)SSB}{SSE} \tag{10.10}$$

$$\begin{aligned}\text{dfn} &= n - 1 \\ \text{dfd} &= (n - 1)(m - 1)\end{aligned} \tag{10.9}$$

$$SSB = \frac{\Sigma S_B^2}{m} - \frac{(\Sigma x)^2}{N} \tag{10.12}$$

z statistic for large-sample sign test (Sec. 11.A)

$$z = \frac{\theta - n/2}{\sqrt{n/4}}$$

(11.1)

$$\theta = \text{number of pluses}$$

z statistic for large-sample Mann-Whitney test (Sec. 11.B)

$$z = \frac{T - \mu_T}{\sigma_T}$$

(11.2)

$$\mu_T = \frac{n_1(n_1 + n_2 + 1)}{2}$$

(11.3)

$$\sigma_T = \sqrt{\frac{n_1 n_2 (n_1 + n_2 + 1)}{12}}$$

(11.4)

chi-square statistic for Kruskal-Wallis test (Sec. 11.D)

$$H = \frac{12}{n(n + 1)} \Sigma \frac{R_i^2}{n_i} - 3(n + 1)$$

(11.5)

$$df = m - 1$$

z statistic for Wald-Wolfowitz test (Sec. 11.E)

$$z = \frac{W - \mu_W}{\sigma_W}$$

(11.7)

$$\mu_W = \frac{2 n_F n_R}{n_F + n_R} + 1$$

(11.6)

$$\sigma_W = \sqrt{\frac{2 n_F n_R (2 n_F n_R - n_F - n_R)}{(n_F + n_R)^2 (n_F + n_R - 1)}}$$

(11.8)

Spearman rank correlation statistic (Sec. 11.F)

$$r_S = 1 - \frac{6 \Sigma d^2}{n(n^2 - 1)}$$

(11.10)

z statistic for rank correlation test (Sec. 11.F)

$$z = r_S \sqrt{n - 1}$$

(11.9)

ANSWERS TO SELECTED ODD-NUMBERED EXERCISES

Exercises 1.A

1	(a)			(b)			(c)		
		1–4	10		0.5–4.5	10		2.5	10
		5–8	20		4.5–8.5	20		6.5	20
		9–12	10		8.5–12.5	10		10.5	10
		13–16	5		12.5–16.5	5		14.5	5
		17–20	5		16.5–20.5	5		18.5	5

3			5	(a)			(b)		
	1500.00	3			1–5	39		3	39
	4500.00	5			6–10	31		8	31
	7500.00	18			11–15	10		13	10
	10,500.00	15			16–20	10		18	10
	13,500.00	6			21–25	5		23	5
	16,500.00	0			26–30	3		28	3
	19,500.00	3			31–35	2		33	2

Exercises 1.C

1 **(c)** 59%   **3** **(b)** 64%
   (d) 94.9%   **(c)** 12.5%

Supplementary Exercises, Chapter 1

1 **(a)**

1–7	22	**(b)** 4	22
8–14	21	11	21
15–21	9	18	9
22–28	4	25	4
29–35	4	32	4

7 **(c)** 35%
   (d) 22.9%

Exercises 2.A

1 **(a)** 107   **3** **(a)** 444   **7** **(a)** 7
   (b) 107.5   **(b)** 460   **(b)** 4
   (c) 110   **(c)** No   **(c)** 4
   (d) 110   **(d)** 525   **(d)** 8

Exercises 2.B

5 **(a)** -6 **7** -240 **9** 39   **11** 8.5   **13** 139.73
   (b) -36
   (c) -36

Exercises 2.C

1 **(a)** 11.1   **3** **(a)** 230.1   **5** 4.82   **7** 2.12
   (b) 8.83   **(b)** 230.1
   (c) 40   **(c)** 197.1
   (d) 11.1   **(d)** 810

Exercises 2.D

1 Yes   **3** 16%   **5** 27   **7** 97.22%   **9** 40

Exercises 2.E

1 **(d)**

A	2.03
B	2.07
C	-2.61
D	2.24
E	-2.32
F	-1.41

Applicant D gets the job.

3 #1 (1.00), #3 (0.60), #4 (-0.33), #2 (-0.50)

1 (a) 17
  (b) 17
  (c) 20
  (d) 19.5
  (e) 7.15
  (f) 6
  (g) 25

3 (a) 14.35
  (c) 14
  (d) 6.36

5 93.75%

7 888

9 44,444   **11** Yes   **13** Yes   **15** (a) 0.14
                                                   (b) 1.31

Exercises 3.A

1

1–4	.20
5–8	.40
9–12	.20
13–16	.10
17–20	.10

3

0.0–19.5	.30
19.5–39.5	.15
39.5–59.5	.05
59.5–79.5	.10
79.5–100.0	.40

Exercises 3.B

1 (a) $n = 4, p = \frac{1}{2}$
  (b) 0, 1, 2, 3, 4
  (c)

k	$P(H = k)$
0	.0625
1	.2500
2	.3750
3	.2500
4	.0625

3 (a) $n = 12, p = .2$
  (b) 0, 1, 2, . , . , 12
  (c) 2.4
  (d) 1.39
  (e)

k	$P(A = k)$
0	.0687
1	.2062
2	.2834
3	.2363
4	.1328
5	.0532
6	.0155
7	.0033
8	.0005
9	.0001
10	.0000
11	.0000
12	.0000

5 (a) $n = 20, p = .05$
  (b) .3585
  (c) .3773
  (d) .0003
  (e) 1.0000
  (f) .0003

Exercises 3.C

1 70.16

3 (a) 3.44%
  (b) 91.08%

5 (a) 87.9%
  (b) 59.34%
  (c) 0.38%

7 .043   **9** 8.93   **11** 1136

Exercises 3.D

1 (a) 62.5%
(b) 56.31%
(c) 51.99%
(d) 50.8%

3 (a) 7.55%
(b) 0.03%
(c) 0.00%
(d) 0.00%

5 (a) 16.42%
(b) 1.62%
(c) 0.00%
(d) 0.00%

Exercises 3.E

1 (a) 0.52
(b) 1.04
(c) 1.44
(d) 1.88
(e) 2.17
(f) 2.41
(g) 2.88
(h) 3.09

3 .0793

5 .0286

Supplementary Exercises, Chapter 3

1 (a) 15.03%
(b) 1.1%

3 0.01%

5 (a) 15.13%
(b) 1.39%

7 (a) 92.8
(b) 96.4
(c) 103.3

9 86.4

11 5.86

13 (a) 18.7%
(b) 1.04%

15
0–39.0	F
39.1–42.5	D
42.6–64.8	C
64.9–82.8	B
82.9–100.0	A

17 (a) .012
(b) .091

19 (a) .006
(b) .202
(c) .475

Exercises 4.B

1 16,182.70–16,266.75

3 (a) 1.97–2.03
(b) 1.97–2.04
(c) 1.96–2.04

5 99.13%

7 246

Exercises 4.C

1 16,140.64–16,308.80

3 (a) 1.91–2.09
(b) 1.89–2.11
(c) 1.84–2.16

5 2393.05–4025.58

Exercises 4.D

1   (a) 4.53–9.40   3   (a) 5.32–6.88   5   1.78–2.68
    (b) 4.29–10.29   (b) 5.15–7.18
    (c) 3.89–12.46   (c) 5.02–7.46

Exercises 4.E

1   (a) .70–.83   3   .52–.67   5   (a) .37–.41
    (b) 2401   (b) .37–.41
       (c) .36–.42
       (d) 4269

Supplementary Exercises, Chapter 4

1   952   3   (a) 2.29–2.71   5   2.19–2.55
      (b) .24–.58
      (c) 67
      (d) 92.98%

7   (a) 5.65–7.64   9   .12–.18   11   (a) .023–.037
    (b) 5.36–8.26   (b) 27,225

Exercises 5.B

1   $z = -1.8$, yes   3 $z = 2.95$, yes   5 $t = 2.67$, yes   7 $t = .583$, yes

Exercises 5.C

1   $z = 6.53$, yes   3 $t = -2.32$, yes   5 $t = .86$, no   7 $t = .87$, no

Exercises 5.D

1   $t = 1.90$, yes   3 $t = 1.68$, no   5 $t = .62$, no   7 $t = -1.934$, no

Exercises 5.E

1   $z = 1.98$, yes   3 $\chi^2 = 38$, yes   5 $F = 2.70$, yes

Exercises 5.F

1   (a) Yes   3   Yes   5   (a) $z = -2.42$, yes   7 $z = .32$, no
    (b) $z = 2$, no   (b) $z = 2.42$, yes

Supplementary Exercises, Chapter 5

1 (a) $z = -4.54$, yes   **3** (a) $t = -2.06$, yes   **5** $t = 2.059$, yes
    (b) $z = 4.89$, yes   (b) $t = 0$, no
                         (c) $\chi^2 = 26$, yes

7 $F = 17.16$, no   **9** $t = 3.54$, yes   **11** $z = -3.51$, yes

13 $z = 3.82$, yes   **15** $t = 1.67$, no

Exercises 6.A

1 (a) $y = 1.75 + .25x$ **3** (a) (0, 6) and (4, 14)
    (b) $y = -2.0 + 1.5x$ (b) (0, 8) and (3, −1)
    (c) $y = 10 - x$ (c) (0, −3) and (1, 11)

5 $y = 1.47 + .10x$, (0, 1.47)

Exercises 6.B

1 (b) $y = 3.33 + .67x$ **3** (b) $y = -3.60 + 2.80x$
    (c) (2, 4.67) and (8, 8.67)   (c) (4, 7.6) and (12, 30)

Exercises 6.C

1 92.3%   **3** 99.5%   **5** (b) .723   **7** (b) 0   **9** 54.7%
                                               (c) 0%

Exercises 6.D

1 $t = 6.93$, yes   **3** $t = 27.93$, yes   **5** $t = 3.94$, yes   **7** $t = 0$, no

Exercises 6.E

1 6.00–8.68   **3** 16.16–21.28   **5** 2.19–5.52   **7** 2.58–3.00

Exercises 6.F

1 (a) 95.6%   **3** (a) 69.8%   **5** (a) 57.9%
    (c) $t = -8.07$, yes   (c) $t = -3.04$, yes   (c) $t = 2.62$, yes
    (d) $y = 3.23 - .91x$ (d) $y = 28.8 - 1.22x$ (d) yes
    (e) 1.27–2.48   (e) 1.55–31.6   (e) $y = 22.9 + 6.0x$
                                                      (f) 35,340–106,350
                                                      (g) 4740–77,010

Supplementary Exercises, Chapter 6

1 **(a)** 88.8%
   (c) $t = 6.90$, yes
   (d) $y = 1.06 + 5.72x$
   (e) 20.57–32.77

3 **(a)** 93.46%
   (c) $t = -6.55$, yes
   (d) $y = 2148 - 800x$
   (e) 282.08–493.92

5 **(a)** 96.57%
   (c) $t = 10.61$, yes
   (d) $y = 2.5 + .65x$
   (e) 6.35–7.75

7 **(a)** 95.29%
   (c) $t = -7.79$, yes
   (d) $y = 112.35 - 5.29x$
   (e) 22.40–43.48

9 **(a)** 90.34%
   (c) $t = 6.12$, yes
   (d) $y = -.65 + 8.70x$
   (e) 43.25–77.18

11 **(a)** 71.3%
   (c) $t = 5.46$, yes
   (d) $y = 12.51 + .78x$
   (e) 20.47–35.75

Exercises 7.A

1 **(a)** logarithmic
   (b) 99.78%
   (c) $y = 5.09 + 1.95 \log x$
   (d) 10.35

3 **(b)** logarithmic: 38.6%
       power: 31.4%
       simple exponential: 17.5%
   (c) $y = 85.0 - 3.79 \log x$ or $y = 86.2x^{-.303}$
   (d) logarithmic: 29.0
       power: 30.8

Exercises 7.B

1 **(a)** parabolic
   (b) $y = .88 + 5.29x - .88x^2$
   (c) 6.87

3 **(a)** parabolic
   (b) $y = 18.30 - 6.25x + .98x^2$
   (c) 8.36

Exercises 7.C

1 **(a)** 80.2%
   (b) 23.5%
   (c) 87.8%
   (d) $y = 3.92 + 6.17x + 1.09w$
   (e) 67.47

5 $F = 25.19$, significant

3 **(a)** x: 56.7%
       w: 79.4%
   (b) 95%
   (c) $y = 38.90 - .20x + .27w$, 38.76

7 $F = 28.5$, not significant

Supplementary Exercises, Chapter 7

1 **(a)** logarithmic
   (b) linear: 84.3%
       logarithmic: 99.85%
   (c) $y = .20 + .31 \log x$
   (d) .70

3 **(a)** exponential or linear
   (b) linear: 79.7%
       power: 63.15%
       simple exponential: 77.8%
   (c) $y = -11.67 + 5.69x$
   (d) 11,075

5 **(a)** 93%
  (b) 93%
  (c) $F = 19.92$, significant
  (d) $y = -3.56 + .75x + .41w$
  (e) 16.64

Exercises 8.A

1 **(c)** $W \cup L = \{7, 11, 2, 3, 12\}$
    $W \cap L = \phi$
    $W^c = \{2, 3, 4, 5, 6, 8, 9, 10, 12\}$
    $L^c = \{4, 5, 6, 7, 8, 9, 10, 11\}$
    $W^c \cup L^c = S$
    $W^c \cap L^c = \{4, 5, 6, 8, 9, 10\}$
    $E^c = D$
    $E^c \cap D = D$
    $E^c \cup D = D$

3 **(a)** $S = \{3, 4, 5, \ldots, 17, 18\}$
  (c) $L \cup E = \{3, 4, 6, 8, 10, 13, 14, 16, 18\}$
    $(L \cup E)^c = \{5, 7, 9, 11, 15, 17\}$
    $L^c = \{5, 6, 7, 8, 9, 10, 11, 14, 15, 16, 17\}$
    $E^c = D$
    $L^c \cap E^c = \{5, 7, 9, 11, 15, 17\}$

Exercises 8.B

1 **(a)** 1/6   **3** **(a)** 1/2
  (b) 1/18   **(b)** 5/8
  (c) 2/9   **(c)** 1/8
  (d) 1/9   **(d)** 3/8
  (e) 1/2
  (f) 1/2
  (g) 1/18
  (h) 5/9
  (i) 0
  (j) 1/3
  (k) 0
  (l) 1

Exercises 8.C

1 **(a)** 4/9   **3** **(a)** 3/10   **5** **(a)** .96
  (b) 1/9   **(b)** 2/5   **(b)** .9728
  (c) 1   **(c)** 1/10
  (d) 1/2   **(d)** 7/10
  (e) 0
  (f) 1/9

Exercises 8.D

1 (a) .015625
  (b) .09375
  (c) .234375
  (d) .3125
  (e) .234375
  (f) .09375
  (g) .015625

3 (a) .45
  (b) .45
  (c) .05

5 28.98%

Exercises 8.E

1 (a) $C(10, 5) = 252$
  (b) $P(10, 5) = 30{,}240$

3 (a) $P(26, 4) = 358{,}800$
  (c) $26^4 = 456{,}976$

5 $C(1000, 10) = (1000 \times 999 \times \cdots \times 991)/10! = 2.634 \times 10^{23}$

Exercises 8.F

1 94.9%   **3** 70.3%   **5** 72.7%   **7** 13.6%

Supplementary Exercises, Chapter 8

1 (a) True
  (b) True
  (c) False
  (d) False

3 (a) 66.5%
  (b) 61.9%

5 37.5%

7 (a) 3/4
  (b) 1
  (c) 1/3
  (d) 0

9 (a) 70%
  (b) 42.9%

11 7,893,600

13 54.8%

15 22.9%

17 63.6%

19 64% (heat-seeking)
  36% (laser-guided)

21 (a) 6.07%
  (b) 16%

Exercises 9.A

1 Yes   **3** Yes   **5** Yes   **7** No   **9** Yes

Exercises 9.B

1 0%   **3** 0%   **5** 15%   **7** 13.7%

Exercises 9.C

1 Yes   **3** No   **5** Yes   **7** No

Supplementary Exercises, Chapter 9

1 **(a)** Yes   **3** **(a)** Yes   **5** Yes   **7** **(a)** Dependent
   (b) 36%   **(b)** 11.1%   **(b)** .0004 = 4/100 of 1%
                                                      (c) 0%

Exercises 10.A

1 **(a)** 8.90 = 7.02 + 1.88   **3** **(a)** 413.25 = 19.25 + 394
   (b) $F = 24.83$, yes   **(b)** $F = .39$, no

Exercises 10.B

1 $F = 24.83$, yes   **3** $F = .39$, no   **5** $F = 18.21$, yes

Exercises 10.C

1 **(a)** $F_T = 23.63$, yes   **3** **(a)** $F_T = 2.18$, no   **5** **(a)** $F_T = 80.19$, yes
   (b) $F_B = 14.77$, yes   **(b)** $F_B = 3.35$, no   **(b)** $F_B = 1.07$, no

Supplementary Exercises, Chapter 10

1 **(a)** 88 = 14 + 74   **3** $F = .36$, no   **5** **(a)** $nVB = 13,890,480$
   (b), (c) $F = 1.51$, no $VW = 634, 324, 716.5$
                                                       (b), (c) $F = .20$, no

Exercises 11.A

1 They do not   **3** **(a)** $z = -2.2$, yes   **5** No   **7** No
                        (b) $z = -2.2$, no

Exercises 11.B

1 $T = 81.5$, yes   **3** $z = -1.16$, it does not   **5** $T = 31$, it does not

Exercises 11.C

1 $T = 94$, they are   **3** $T = 42$, they do   **5** $T = 26$, no

Exercises 11.D

1 $H = 18.33$, yes   **3** $H = 2.00$, yes

Exercises 11.E

1 $z = -2.38$, yes   **3** $z = -0.36$, no

Exercises 11.F

1 **(a)** $r_s = .027$, no   **3** **(a)** $r_s = .91$, yes   **5** $r_s = -1$, yes
7 $r_s = .975$, yes

Supplementary Exercises, Chapter 11

1 **(a)** $T = 22.5$, it is not   **3** $z = -3.02$, no   **5** It does
    (b) $T = 22.5$, it is not
    (d) $\alpha = .25$
    (e) It can
    (f) $\alpha = .0312$

7 $H = 2.40$, no   **9** $r_s = .18$, no

APPENDIX

Square Roots

N	$\sqrt{N}$	$\sqrt{10N}$	N	$\sqrt{N}$	$\sqrt{10N}$	N	$\sqrt{N}$	$\sqrt{10N}$
1.00	1.000	3.162	1.50	1.225	3,873	2.00	1.414	4.472
1.01	1.005	3.178	1.51	1.229	3.886	2.01	1.418	4.483
1.02	1.010	3.194	1.52	1.233	3.899	2.02	1.421	4.494
1.03	1.015	3.209	1.53	1.237	3.912	2.03	1.425	4.506
1.04	1.020	3.225	1.54	1.241	3.924	2.04	1.428	4.517
1.05	1.025	3.240	1.55	1.245	3.937	2.05	1.432	4.528
1.06	1.030	3.256	1.56	1.249	3.950	2.06	1.435	4.539
1.07	1.034	3.271	1.57	1.253	3.962	2.07	1.439	4.550
1.08	1.039	3.286	1.58	1.257	3.975	2.08	1.442	4.571
1.09	1.044	3.302	1.59	1.261	3.987	2.09	1.446	4.572
1.10	1.049	3.317	1.60	1.265	4.000	2.10	1.449	4.583
1.11	1.054	3.332	1.61	1.269	4.012	2.11	1.453	4.593
1.12	1.058	3.347	1.62	1.273	4.025	2.12	1.456	4.604
1.13	1.063	3.362	1.63	1.277	4.037	2.13	1.459	4.615
1.14	1.068	3.376	1.64	1.281	4.050	2.14	1.463	4.626
1.15	1.072	3.391	1.65	1.285	4.062	2.15	1.466	4.637
1.16	1.077	3.406	1.66	1.288	4.074	2.16	1.470	4.648
1.17	1.082	3.421	1.67	1.292	4.087	2.17	1.473	4.658
1.18	1.086	3.435	1.68	1.296	4.099	2.18	1.476	4.669
1.19	1.091	3.450	1.69	1.300	4.111	2.19	1.480	4.680
1.20	1.095	3.464	1.70	1.304	4.123	2.20	1.483	4.690
1.21	1.100	3.479	1.71	1.308	4.135	2.21	1.487	4.701
1.22	1.105	3.493	1.72	1.311	4.147	2.22	1.490	4.712
1.23	1.109	3.507	1.73	1.315	4.159	2.23	1.493	4.722
1.24	1.114	3.521	1.74	1.319	4.171	2.24	1.497	4.733
1.25	1.118	3.536	1.75	1.323	4.183	2.25	1.500	4.743
1.26	1.122	3.550	1.76	1.327	4.195	2.26	1.503	4.754
1.27	1.127	3.564	1.77	1.330	4.207	2.27	1.507	4.764
1.28	1.131	3.578	1.78	1.334	4.219	2.28	1.510	4.775
1.29	1.136	3.592	1.79	1.338	4.231	2.29	1.513	4.785
1.30	1.140	3.606	1.80	1.342	4.243	2.30	1.517	4.796
1.31	1.145	3.619	1.81	1.345	4.254	2.31	1.520	4.806
1.32	1.149	3.633	1.82	1.349	4.266	2.32	1.523	4,817
1.33	1.153	3.647	1.83	1.353	4.278	2.33	1.526	4,827
1.34	1.158	3.661	1.84	1.356	4.290	2.34	1.530	4,837
1.35	1.162	3.674	1.85	1.360	4.301	2.35	1.533	4.848
1.36	1.166	3.688	1.86	1.364	4.313	2.36	1.536	4.858
1.37	1.170	3.701	1.87	1.367	4.324	2.37	1.539	4.868
1.38	1.175	3.715	1.88	1.371	4.336	2.38	1.543	4.879
1.39	1.179	3.728	1.89	1.375	4.347	2.39	1.546	4.889
1.40	1.183	3.742	1.90	1.378	4.359	2.40	1.549	4.899
1.41	1.187	3.755	1.91	1.382	4.370	2.41	1.552	4.909
1.42	1.192	3.768	1.92	1.386	4.382	2.42	1.556	4.919
1.43	1.196	3.782	1.93	1.389	4.393	2.43	1.559	4.930
1.44	1.200	3.795	1.94	1.393	4.405	2.44	1.562	4.940
1.45	1.204	3.808	1.95	1.396	4.416	2.45	1.565	4.950
1.46	1.208	3.821	1.96	1.400	4.427	2.46	1.568	4.960
1.47	1.212	3.834	1.97	1.404	4.438	2.47	1.572	4.970
1.48	1.217	3.847	1.98	1.407	4.450	2.48	1.575	4.980
1.49	1.221	3.860	1.99	1.411	4.461	2.49	1.578	4.990

N	$\sqrt{N}$	$\sqrt{10N}$	N	$\sqrt{N}$	$\sqrt{10N}$	N	$\sqrt{N}$	$\sqrt{10N}$
2.50	1.581	5.000	3.00	1.732	5.477	3.50	1.871	5.916
2.51	1.584	5.010	3.01	1.735	5.486	3.51	1.873	5.925
2.52	1.587	5.020	3.02	1.738	5.495	3.52	1.876	5.933
2.53	1.591	5.030	3.03	1.741	5.505	3.53	1.879	5.941
2.54	1.594	5.040	3.04	1.744	5.514	3.54	1.881	5.950
2.55	1.597	5.050	3.05	1.746	5.523	3.55	1.884	5.958
2.56	1.600	5.060	3.06	1.749	5.532	3.56	1.887	5.967
2.57	1.603	5.070	3.07	1.752	5.541	3.57	1.889	5.975
2.58	1.606	5.079	3.08	1.755	5.550	3.58	1.892	5.983
2.59	1.609	5.089	3.09	1.758	5.559	3.59	1.895	5.992
2.60	1.612	5.099	3.10	1.761	5.568	3.60	1.897	6.000
2.61	1.616	5.109	3.11	1.764	5.577	3.61	1.900	6.008
2.62	1.619	5.119	3.12	1.766	5.586	3.62	1.903	6.017
2.63	1.622	5.128	3.13	1.769	5.595	3.63	1.905	6.025
2.64	1.625	5.138	3.14	1.772	5.604	3.64	1.908	6.033
2.65	1.628	5.148	3.15	1.775	5.612	3.65	1.910	6.042
2.66	1.631	5.158	3.16	1.778	5.621	3.66	1.913	6.050
2.67	1.634	5.167	3.17	1.780	5.630	3.67	1.916	6.058
2.68	1.637	5.177	3.18	1.783	5.639	3.68	1.918	6.066
2.69	1.640	5.187	3.19	1.786	5.648	3.69	1.921	6.075
2.70	1.643	5.196	3.20	1.789	5.657	3.70	1.924	6.083
2.71	1.646	5.206	3.21	1.792	5.666	3.71	1.926	6.091
2.72	1.649	5.215	3.22	1.794	5.675	3.72	1.929	6.099
2.73	1.652	5.225	3.23	1.797	5.683	3.73	1.931	6.107
2.74	1.655	5.234	3.24	1.800	5.692	3.74	1.934	6.116
2.75	1.658	5.244	3.25	1.803	5.701	3.75	1.936	6.124
2.76	1.661	5.254	3.26	1.806	5.710	3.76	1.939	6.132
2.77	1.664	5.263	3.27	1.808	5.718	3.77	1.942	6.140
2.78	1.667	5.273	3.28	1.811	5.727	3.78	1.944	6.148
2.79	1.670	5.282	3.29	1.814	5.736	3.79	1.947	6.156
2.80	1.673	5.292	3.30	1.817	5.745	3.80	1.949	6.164
2.81	1.676	5.301	3.31	1.819	5.753	3.81	1.852	6.173
2.82	1.679	5.310	3.32	1.822	5.762	3.82	1.954	6.181
2.83	1.682	5.320	3.33	1.825	5.771	3.83	1.957	6.189
2.84	1.685	5.329	3.34	1.828	5.779	3.84	1.960	6.197
2.85	1.688	5.339	3.35	1.830	5.788	3.85	1.962	6.205
2.86	1.691	5.348	3.36	1.833	5.797	3.86	1.965	6.213
2.87	1.694	5.357	3.37	1.836	5.805	3.87	1.967	6.221
2.88	1.697	5.367	3.38	1.838	5.814	3.88	1.970	6.229
2.89	1.700	5.376	3.39	1.841	5.822	3.89	1.972	6.237
2.90	1.703	5.385	3.40	1.844	5.831	3.90	1.975	6.245
2.91	1.706	5.394	3.41	1.847	5.840	3.91	1.977	6.253
2.92	1.709	5.404	3.42	1.849	5.848	3.92	1.980	6.261
2.93	1.712	5.413	3.43	1.852	5.857	3.93	1.982	6.269
2.94	1.715	5.422	3.44	1.855	5.865	3.94	1.985	6.277
2.95	1.718	5.431	3.45	1.857	5.874	3.95	1.987	6.285
2.96	1.720	5.441	3.46	1.860	5.882	3.96	1.990	6.293
2.97	1.723	5.450	3.47	1.863	5.891	3.97	1.992	6.301
2.98	1.726	5.459	3.48	1.865	5.899	3.98	1.995	6.309
2.99	1.729	5.468	3.49	1.868	5.908	3.99	1.997	6.317

N	$\sqrt{N}$	$\sqrt{10N}$	N	$\sqrt{N}$	$\sqrt{10N}$	N	$\sqrt{N}$	$\sqrt{10N}$
4.00	2.000	6.325	4.50	2.121	6.708	5.00	2.236	7.071
4.01	2.002	6.332	4.51	2.124	6.716	5.01	2.238	7.078
4.02	2.005	6.340	4.52	2.126	6.723	5.02	2.241	7.085
4.03	2.007	6.348	4.53	2.128	6.731	5.03	2.243	7.092
4.04	2.010	6.356	4.54	2.131	6.738	5.04	2.245	7.099
4.05	2.012	6.364	4.55	2.133	6.745	5.05	2.247	7.106
4.06	2.015	6.372	4.56	2.135	6.753	5.06	2.249	7.113
4.07	2.017	6.380	4.57	2.138	6.760	5.07	2.252	7.120
4.08	2.020	6.387	4.58	2.140	6.768	5.08	2.254	7.127
4.09	2.022	6.395	4.59	2.142	6.775	5.09	2.256	7.134
4.10	2.025	6.403	4.60	2.145	6.782	5.10	2.258	7.141
4.11	2.027	6.411	4.61	2.147	6.790	5.11	2.261	7.148
4.12	2.030	6.419	4.62	2.149	6.797	5.12	2.263	7.155
4.13	2.032	6.427	4.63	2.152	6.804	5.13	2.265	7.162
4.14	2.035	6.434	4.64	2.154	6.812	5.14	2.267	7.169
4.15	2.037	6.442	4.65	2.156	6.819	5.15	2.269	7.176
4.16	2.040	6.450	4.66	2.159	6.826	5.16	2.272	7.183
4.17	2.042	6.458	4.67	2.161	6.834	5.17	2.274	7.190
4.18	2.045	6.465	4.68	2.163	6.841	5.18	2.276	7.197
4.19	2.047	6.473	4.69	2.166	6.848	5.19	2.278	7.204
4.20	2.049	6.481	4.70	2.168	6.856	5.20	2.280	7.211
4.21	2.052	6.488	4.71	2.170	6.863	5.21	2.280	7.218
4.22	2.054	6.496	4.72	2.173	6.870	5.22	2.285	7.225
4.23	2.057	6.504	4.73	2.175	6.877	5.23	2.287	7.232
4.24	2.059	6.512	4.74	2.177	6.885	5.24	2.289	7.239
4.25	2.062	6.519	4.75	2.179	6.892	5.25	2.291	7.246
4.26	2.064	6.527	4.76	2.182	6.899	5.26	2.293	7.253
4.27	2.066	6.535	4.77	2.184	6.907	5.27	2.296	7.259
4.28	2.069	6.542	4.78	2.186	6.914	5.28	2.298	7.266
4.29	2.071	6.550	4.79	2.189	6.921	5.29	2.300	7.273
4.30	2.074	6.557	4.80	2.191	6.928	5.30	2.302	7.280
4.31	2.076	6.565	4.81	2.193	6.935	5.31	2.304	7.287
4.32	2.078	6.573	4.82	2.195	6.943	5.32	2.307	7.294
4.33	2.081	6.580	4.83	2.198	6.950	5.33	2.309	7.301
4.34	2.083	6.588	4.84	2.200	6.957	5.34	2.311	7.308
4.35	2.086	6.595	4.85	2.202	6.964	5.35	2.313	7.314
4.36	2.088	6.603	4.86	2.205	6.971	5.36	2.315	7.321
4.37	2.090	6.611	4.87	2.207	6.979	5.37	2.317	7.328
4.38	2.093	6.618	4.88	2.209	6.986	5.38	2.319	7.335
4.39	2.095	6.626	4.89	2.211	6.993	5.39	2.322	7.342
4.40	2.098	6.633	4.90	2.214	7.000	5.40	2.324	7.348
4.41	2.100	6.641	4.91	2.216	7.007	5.41	2.326	7.355
4.42	2.102	6.648	4.92	2.218	7.014	5.42	2.328	7.632
4.43	2.105	6.656	4.93	2.220	7.021	5.43	2.330	7.369
4.44	2.107	6.663	4.94	2.223	7.029	5.44	2.332	7.376
4.45	2.110	6.671	4.95	2.225	7.036	5.45	2.335	7.382
4.46	2.112	6.678	4.96	2.227	7.043	5.46	2.337	7.389
4.47	2.114	6.686	4.97	2.229	7.050	5.47	2.339	7.396
4.48	2.117	6.693	4.98	2.232	7.057	5.48	2.341	7.403
4.49	2.119	6.701	4.99	2.234	7.064	5.49	2.343	7.409

N	$\sqrt{N}$	$\sqrt{10N}$	N	$\sqrt{N}$	$\sqrt{10N}$	N	$\sqrt{N}$	$\sqrt{10N}$
5.50	2.345	7.416	6.00	2.449	7.746	6.50	2.550	8.062
5.51	2.347	7.423	6.01	2.452	7.752	6.51	2.551	8.068
5.52	2.349	7.430	6.02	2.454	7.759	6.52	2.553	8.075
5.53	2.352	7.436	6.03	2.456	7.765	6.53	2.555	8.081
5.54	2.354	7.443	6.04	2.458	7.772	6.54	2.557	8.087
5.55	2.356	7.450	6.05	2.460	7.778	6.55	2.559	8.093
5.56	2.358	7.457	6.06	2.462	7.785	6.56	2.561	8.099
5.57	2.360	7.463	6.07	2.464	7.791	6.57	2.563	8.106
5.58	2.362	7.470	6.08	2.466	7.797	6.58	2.565	8.112
5.59	2.364	7.477	6.09	2.468	7.804	6.59	2.567	8.118
5.60	2.366	7.483	6.10	2.470	7.810	6.60	2.569	8.124
5.61	2.369	7.490	6.11	2.472	7.817	6.61	2.571	8.130
5.62	2.371	7.497	6.12	2.474	7.823	6.62	2.573	8.136
5.63	2.373	7.503	6.13	2.476	7.829	6.63	2.575	8.142
5.64	2.375	7.510	6.14	2.478	7.836	6.64	2.577	8.149
5.65	2.377	7.517	6.15	2.480	7.842	6.65	2.579	8.155
5.66	2.379	7.523	6.16	2.482	7.849	6.66	2.581	8.161
5.67	2.381	7.530	6.17	2.484	7.855	6.67	2.583	8.167
5.68	2.383	7.537	6.18	2.486	7.861	6.68	2.585	8.173
5.69	2.385	7.543	6.19	2.488	7.868	6.69	2.587	8.179
5.70	2.387	7.550	6.20	2.490	7.874	6.70	2.588	8.185
5.71	2.390	7.556	6.21	2.492	7.880	6.71	2.590	8.191
5.72	2.392	7.563	6.22	2.494	7.887	6.72	2.592	8.198
5.73	2.394	7.570	6.23	2.496	7.893	6.73	2.594	8.204
5.74	2.396	7.576	6.24	2.498	7.899	6.74	2.596	8.210
5.75	2.398	7.583	6.25	2.500	7.906	6.75	2.598	8.216
5.76	2.400	7.589	6.26	2.502	7.912	6.76	2.600	8.222
5.77	2.402	7.596	6.27	2.504	7.918	6.77	2.602	8.228
5.78	2.404	7.603	6.28	2.506	7.925	6.78	2.604	8.234
5.79	2.406	7.609	6.29	2.508	7.931	6.79	2.606	8.240
5.80	2.408	7.616	6.30	2.510	7.937	6.80	2.608	8.246
5.81	2.410	7.622	6.31	2.512	7.944	6.81	2.610	8.252
5.82	2.412	7.629	6.32	2.514	7.950	6.82	2.612	8.258
5.83	2.415	7.635	6.33	2.516	7.956	6.83	2.613	8.264
5.84	2.417	7.642	6.34	2.518	7.962	6.84	2.615	8.270
5.85	2.419	7.649	6.35	2.520	7.969	6.85	2.617	8.276
5.86	2.421	7.655	6.36	2.522	7.975	6.86	2.619	8.283
5.87	2.423	7.662	6.37	2.524	7.981	6.87	2.621	8.289
5.88	2.425	7.668	6.38	2.526	7.987	6.88	2.623	8.295
5.89	2.427	7.675	6.39	2.528	7.994	6.89	2.625	8.301
5.90	2.429	7.681	6.40	2.530	8.000	6.90	2.627	8.307
5.91	2.431	7.688	6.41	2.532	8.006	6.91	2.629	8.313
5.92	2.433	7.694	6.42	2.534	8.012	6.92	2.631	8.319
5.93	2.435	7.701	6.43	2.536	8.019	6.93	2.632	8.325
5.94	2.437	7.707	6.44	2.538	8.025	6.94	2.634	8.331
5.95	2.439	7.714	6.45	2.540	8.031	6.95	2.636	8.337
5.96	2.441	7.720	6.46	2.542	8.037	6.96	2.638	8.343
5.97	2.443	7.727	6.47	2.544	8.044	6.97	2.640	8.349
5.98	2.445	7.733	6.48	2.546	8.50	6.98	2.642	8.355
5.99	2.447	7.740	6.49	2.548	8.056	6.99	2.644	8.361

N	$\sqrt{N}$	$\sqrt{10N}$	N	$\sqrt{N}$	$\sqrt{10N}$	N	$\sqrt{N}$	$\sqrt{10N}$
7.00	2.646	8.367	7.50	2.739	8.660	8.00	2.828	8.944
7.01	2.648	8.373	7.51	2.740	8.666	8.01	2.830	8.950
7.02	2.650	8.379	7.52	2.742	8.672	8.02	2.832	8.955
7.03	2.651	8.385	7.53	2.744	8.678	8.03	2.834	8.961
7.04	2.653	8.390	7.54	2.746	8.683	8.04	2.835	8.967
7.05	2.655	8.396	7.55	2.748	8.689	8.05	2.837	8.972
7.06	2.657	8.402	7.56	2.750	8.695	8.06	2.839	8.978
7.07	2.659	8.408	7.57	2.751	8.701	8.07	2.841	8.983
7.08	2.661	8.414	7.58	2.753	8.706	8.08	2.843	8.989
7.09	2.663	8.420	7.59	2.755	8.712	8.09	2.844	8.994
7.10	2.665	8.426	7.60	2.757	8.718	8.10	2.846	9.000
7.11	2.666	8.432	7.61	2.759	8.724	8.11	2.848	9.006
7.12	2.668	8.438	7.62	2.760	8.729	8.12	2.850	9.011
7.13	2.670	8.444	7.63	2.762	8.735	8.13	2.851	9.017
7.14	2.672	8.450	7.64	2.764	8.741	8.14	2.853	9.022
7.15	2.674	8.456	7.65	2.766	8.746	8.15	2.855	9.028
7.16	2.676	8.462	7.66	2.768	8.752	8.16	2.857	9.033
7.17	2.678	8.468	7.67	2.769	8.758	8.17	2.858	9.039
7.18	2.680	8.473	7.68	2.771	8.764	8.18	2.860	9.044
7.19	2.681	8.479	7.69	2.773	8.769	8.19	2.862	9.050
7.20	2.683	8.485	7.70	2.775	8.775	8.20	2.864	9.055
7.21	2.685	8.491	7.71	2.777	8.781	8.21	2.865	9.061
7.22	2.687	8.497	7.72	2.778	8.786	8.22	2.867	9.066
7.23	2.689	8.503	7.73	2.780	8.792	8.23	2.869	9.072
7.24	2.691	8.509	7.74	2.782	8.798	8.24	2.871	9.077
7.25	2.693	8.515	7.75	2.784	8.803	8.25	2.872	9.083
7.26	2.694	8.521	7.76	2.786	8.809	8.26	2.874	9.088
7.27	2.696	8.526	7.77	2.787	8.815	8.27	2.876	9.094
7.28	2.698	8.532	7.78	2.789	8.820	8.28	2.877	9.099
7.29	2.700	8.538	7.79	2.791	8.826	8.29	2.879	9.105
7.30	2.702	8.544	7.80	2.793	8.832	8.30	2.881	9.110
7.31	2.704	8.550	7.81	2.795	8.837	8.31	2.883	9.116
7.32	2.706	8.556	7.82	2.796	8.843	8.32	2.884	9.121
7.33	2.707	8.562	7.83	2.798	8.849	8.33	2.886	9.127
7.34	2.709	8.567	7.84	2.800	8.854	8.34	2.888	9.132
7.35	2.711	8.573	7.85	2.802	8.860	8.35	2.890	9.138
7.36	2.713	8.579	7.86	2.804	8.866	8.36	2.891	9.143
7.37	2.715	8.585	7.87	2.805	8.871	8.37	2.893	9.149
7.38	2.717	8.591	7.88	2.807	8.877	8.38	2.895	9.154
7.39	2.718	8.597	7.89	2.809	8.883	8.39	2.897	9.160
7.40	2.720	8.602	7.90	2.811	8.888	8.40	2.898	9.165
7.41	2.722	8.608	7.91	2.812	8.894	8.41	2.900	9.171
7.42	2.724	8.614	7.92	2.814	8.899	8.42	2.902	9.176
7.43	2.726	8.620	7.93	2.816	8.905	8.43	2.903	9.182
7.44	2.728	8.626	7.94	2.818	8.911	8.44	2.905	9.187
7.45	2.729	8.631	7.95	2.820	8.916	8.45	2.907	9.192
7.46	2.731	8.637	7.96	2.821	8.922	8.46	2.909	9.198
7.47	2.733	8.643	7.97	2.823	8.927	8.47	2.910	9.203
7.48	2.735	8.649	7.98	2.825	8.933	8.48	2.912	9.209
7.49	2.737	8.654	7.99	2.827	8.939	8.49	2.914	9.214

N	$\sqrt{N}$	$\sqrt{10N}$	N	$\sqrt{N}$	$\sqrt{10N}$	N	$\sqrt{N}$	$\sqrt{10N}$
8.50	2.915	9.220	9.00	3.000	9.480	9.50	3.082	9.747
8.51	2.917	9.225	9.01	3.002	9.492	9.51	3.084	9.752
8.52	2.919	9.230	9.02	3.003	9.497	9.52	3.085	9.757
8.53	2.921	9.236	9.03	3.005	9.503	9.53	3.087	9.762
8.54	2.922	9.241	9.04	3.007	9.508	9.54	3.089	9.767
8.55	2.924	9.247	9.05	3.008	9.513	9.55	3.090	9.772
8.56	2.926	9.252	9.06	3.010	9.518	9.56	3.092	9.778
8.57	2.927	9.257	9.07	3.012	9.524	9.57	3.094	9.783
8.58	2.929	9.263	9.08	3.013	9.529	9.58	3.095	9.788
8.59	2.931	9.268	9.09	3.015	9.534	9.59	3.097	9.793
8.60	2.933	9.274	9.10	3.017	9.539	9.60	3.098	9.798
8.61	2.934	9.279	9.11	3.017	9.545	9.61	3.100	9.803
8.62	2.936	9.284	9.12	3.020	9.550	9.62	3.102	9.808
8.63	2.938	9.290	9.13	3.022	9.555	9.63	3.103	9.813
8.64	2.939	9.295	9.14	3.023	9.560	9.64	3.105	9.818
8.65	2.941	9.301	9.15	3.025	9.566	9.65	3.106	9.823
8.66	2.943	9.306	9.16	3.027	9.571	9.66	3.108	9.829
8.67	2.944	9.311	9.17	3.028	9.576	9.67	3.110	9.834
8.68	2.946	9.317	9.18	3.030	9.581	9.68	3.111	9.839
8.69	2.948	9.322	9.19	3.031	9.586	9.69	3.113	9.844
8.70	2.950	9.327	9.20	3.033	9.592	9.70	3.114	9.849
8.71	2.951	9.333	9.21	3.035	9.597	9.71	3.116	9.854
8.72	2.953	9.338	9.22	3.036	9.602	9.72	3.118	9.859
8.73	2.955	9.343	9.23	3.038	9.607	9.73	3.119	9.864
8.74	2.956	9.349	9.24	3.040	9.612	9.74	3.121	9.869
8.75	2.958	9.354	9.25	3.041	9.618	9.75	3.122	9.874
8.76	2.960	9.359	9.26	3.043	9.623	9.76	3.124	9.879
8.77	2.961	9.365	9.27	3.045	9.628	9.77	3.126	9.884
8.78	2.963	9.370	9.28	3.046	9.633	9.78	3.127	9.889
8.79	2.965	9.375	9.29	3.048	9.638	9.79	3.129	9.894
8.80	2.966	9.381	9.30	3.050	9.644	9.80	3.130	9.899
8.81	2.968	9.386	9.31	3.051	9.649	9.81	3.132	9.905
8.82	2.970	9.391	9.32	3.053	9.654	9.82	3.134	9.910
8.83	2.972	9.397	9.33	3.055	9.659	9.83	3.135	9.915
8.84	2.973	9.402	9.34	3.056	9.664	9.84	3.137	9.920
8.85	2.975	9.407	9.35	3.058	9.670	9.85	3.138	9.925
8.86	2.977	9.413	9.36	3.059	9.675	9.86	3.140	9.930
8.87	2.978	9.418	9.37	3.061	9.680	9.87	3.142	9.935
8.88	2.980	9.423	9.38	3.063	9.685	9.88	3.143	9.940
8.89	2.982	9.429	9.39	3.064	9.690	9.89	3.145	9.945
8.90	2.983	9.434	9.40	3.066	9.695	9.90	3.146	9.950
8.91	2.985	9.439	9.41	3.068	9.701	9.91	3.148	9.955
8.92	2.987	9.445	9.42	3.069	9.706	9.92	3.150	9.960
8.93	2.988	9.450	9.43	3.071	9.711	9.93	3.151	9.965
8.94	2.990	9.455	9.44	3.072	9.716	9.94	3.153	9.970
8.95	2.992	9.460	9.45	3.074	9.721	9.95	3.154	9.975
8.96	2.993	9.466	9.46	3.076	9.726	9.96	3.156	9.980
8.97	2.995	9.471	9.47	3.077	9.731	9.97	3.158	9.985
8.98	2.997	9.476	9.48	3.079	9.737	9.98	3.158	9.990
8.99	2.998	9.482	9.49	3.081	9.742	9.99	3.161	9.995

SOURCE: M. Orkin and R. Drogin, "Vital Statistics," pp. 332–337, McGraw-Hill, New York, 1975.

Table A.2 gives cumulative binomial probabilities $P(S \leq k)$ for a variable S having a binomial distribution with parameters n and p. For example, if $n = 10$ and $p = .40$, then $P(S \leq 3) = .3823$, or if $n = 13$ and $p = .85$, then $P(S \leq 10) = .2704$. Other probabilities of interest may be obtained according to the formulas

$P(S > k) = 1 - P(S \leq k)$
$P(S = k) = P(S \leq k) - P(S \leq k - 1)$

TABLE A.2
Binomial Distribution

n	k	$p = .05$	.10	.15	.20	.25	.30	.35	.40	.45
1	0	.9500	.9000	.8500	.8000	.7500	.7000	.6500	.6000	.5500
	1	1.0000	1.0000	1.0000	1.0000	1.0000	1.0000	1.0000	1.0000	1.0000
2	0	.9025	.8100	.7225	.6400	.5625	.4900	.4225	.3600	.3025
	1	.9975	.9900	.9775	.9600	.9375	.9100	.8775	.8400	.7975
	2	1.0000	1.0000	1.0000	1.0000	1.0000	1.0000	1.0000	1.0000	1.0000
3	0	.8574	.7290	.6141	.5129	.4219	.3439	.2746	.2160	.1664
	1	.9928	.9720	.9392	.8960	.8438	.7840	.7182	.6480	.5748
	2	.9999	.9990	.9966	.9929	.9844	.9730	.9571	.9360	.9089
	3	1.0000	1.0000	1.0000	1.0000	1.0000	1.0000	1.0000	1.0000	1.0000
4	0	.8145	.6561	.5200	.4096	.3164	.2401	.1785	.1296	.0915
	1	.9860	.9477	.8905	.8192	.7383	.6517	.5630	.4752	.3910
	2	.9995	.9963	.9880	.9728	.9492	.9163	.8735	.8208	.7585
	3	1.0000	.9999	.9995	.9984	.9961	.9919	.9850	.9743	.9590
	4	1.0000	1.0000	1.0000	1.0000	1.0000	1.0000	1.0000	1.0000	1.0000
5	0	.7738	.5905	.4437	.3277	.2373	.1681	.1160	.0778	.0503
	1	.9774	.9185	.8352	.7373	.6328	.5282	.4284	.3370	.2562
	2	.9988	.9924	.9734	.9421	.8965	.8369	.7648	.6826	.5931
	3	1.0000	.9995	.9978	.9933	.9844	.9692	.9460	.9130	.8688
	4	1.0000	1.0000	.9999	.9997	.9990	.9976	.9947	.9898	.9815
	5	1.0000	1.0000	1.0000	1.0000	1.0000	1.0000	1.0000	1.0000	1.0000
6	0	.7351	.5314	.3771	.2621	.1780	.1176	.0754	.0467	.0277
	1	.9672	.8857	.7765	.6534	.5339	.4202	.3191	.2333	.1636
	2	.9978	.9842	.9527	.9011	.8306	.7443	.6471	.5443	.4415
	3	.9999	.9987	.9941	.9830	.9624	.9295	.8826	.9208	.7447
	4	1.0000	.9999	.9996	.9984	.9954	.9891	.9777	.9590	.9308
	5	1.0000	1.0000	1.0000	.9999	.9998	.9993	.9982	.9959	.9917
	6	1.0000	1.0000	1.0000	1.0000	1.0000	1.0000	1.0000	1.0000	1.0000
7	0	.6983	.4783	.3206	.2097	.1335	.0824	.0490	.0280	.0152
	1	.9556	.8503	.7166	.5767	.4449	.3294	.2338	.1586	.1024
	2	.9962	.9743	.9262	.8520	.7564	.6471	.5323	.4199	.3164
	3	.9998	.9973	.9879	.9667	.9294	.8740	.8002	.7102	.6083
	4	1.0000	.9998	.9988	.9953	.9871	.9812	.9444	.9037	.8643
	5	1.0000	1.0000	.9999	.9996	.9987	.9962	.9910	.9812	.9643
	6	1.0000	1.0000	1.0000	1.0000	.9999	.9998	.9994	.9984	.0063
	7	1.0000	1.0000	1.0000	1.0000	1.0000	1.0000	1.0000	1.0000	1.0000

$p = .50$	.55	.60	.65	.70	.75	.80	.85	.90	.95
.5000	.4500	.4000	.3500	.3000	.2500	.2000	.1500	.1000	.0500
1.0000	1.0000	1.0000	1.0000	1.0000	1.0000	1.0000	1.0000	1.0000	1.0000
.2500	.2025	.1600	.1225	.0900	.0625	.0400	.0225	.0100	.0025
.7500	.6975	.6400	.5775	.5100	.4373	.3600	.2775	.1900	.0975
1.0000	1.0000	1.0000	1.0000	1.0000	1.0000	1.0000	1.0000	1.0000	1.0000
.1250	.0911	.0640	.0429	.0270	.0156	.0080	.0034	.0010	.0001
.5000	.4252	.3520	.2818	.2160	.1562	.1040	.0608	.0280	.0072
.8750	.8336	.7840	.7254	.6570	.5781	.4880	.3959	.2710	.1426
1.0000	1.0000	1.0000	1.0000	1.0000	1.0000	1.0000	1.0000	1.0000	1.0000
.0625	.0410	.0256	.0150	.0081	.0039	.0016	.0005	.0001	.0000
.3125	.2415	.1792	.1265	.0837	.0508	.0272	.0120	.0037	.0005
.6875	.6090	.5248	.4370	.3483	.2617	.1808	.1095	.0523	.0140
.9375	.9085	.8704	.8215	.7599	.6836	.5904	.4780	.3439	.1855
1.0000	1.0000	1.0000	1.0000	1.0000	1.0000	1.0000	1.0000	1.0000	1.0000
.0312	.0185	.0102	.0053	.0024	.0010	.0003	.0001	.0000	.0000
.1875	1312	.0870	.0540	.0308	.0156	.0067	.0022	.0005	.0000
.5000	.4069	.3174	.2352	.1631	.1035	.0579	.0266	.0086	.0012
.8125	.7438	.6630	.5716	.4718	.3672	.2627	.1648	.0815	.0226
.9688	.9497	.9222	.8840	.8319	.7627	.6723	.5563	.4095	.2262
1.0000	1.0000	1.0000	1.0000	1.0000	1.0000	1.0000	1.0000	1.0000	1.0000
.0156	.0083	.0041	.0018	.0007	.0002	.0001	.0000	.0000	.0000
.1094	.0692	.0410	.0023	.0109	.0046	.0016	.0004	.0001	.0000
.3438	.2553	.1792	.1174	.0705	.0376	.0170	.0059	.0013	.0001
.6562	.5585	.4557	.3529	.2557	.1694	.0989	.0473	.0158	.0022
.8906	.8364	.7667	.6809	.5798	.4661	.3446	.2235	.1143	.0328
.9844	.9723	.9533	.9246	.8824	.8220	.7379	.6229	.4686	.2649
1.0000	1.0000	1.0000	1.0000	1.0000	1.0000	1.0000	1.0000	1.0000	1.0000
.0078	.0037	.0016	.0006	.0002	.0001	.0000	.0000	.0000	.0000
.0625	.0357	.0188	.0090	.0038	.0013	.0004	.0001	.0000	.0000
.2266	.1529	.0963	.0556	.0288	.0129	.0047	.0012	.0002	.0000
.5000	.3917	.2898	.1998	.1260	.0706	.0333	.0121	.0027	.0002
.7734	.6836	.5801	.4677	.3529	.2436	.1480	.0738	.0257	.0038
.9375	.8976	.8414	.7662	.6706	.5551	.4233	.2834	.1497	.0444
.9922	.9848	.9720	.9510	.9176	.8665	.7903	.6794	.5217	.3917
1.0000	1.0000	1.0000	1.0000	1.0000	1.0000	1.0000	1.0000	1.0000	1.0000

Binomial Distribution

n	k	p = .05	.10	.15	.20	.25	.30	.35	.40	.45
8	0	.6634	.4305	.2725	.1678	.1001	.0576	.0319	.0168	.0084
	1	.9428	.8131	.6572	.5033	.3671	.2553	.1691	.1064	.0632
	2	.9942	.9619	.8948	.7969	.6785	.5518	.4278	.3154	.2201
	3	.9996	.9950	.9786	.9437	.8862	.8059	.7064	.5941	.4770
	4	1.0000	.9996	.9971	.9896	.9727	.9420	.8939	.8263	.7396
	5	1.0000	1.0000	.9998	.9988	.9958	.9887	.9747	.9502	.9115
	6	1.0000	1.0000	1.0000	.9999	.9996	.9987	.9964	.9915	.9819
	7	1.0000	1.0000	1.0000	1.0000	1.0000	.9999	.9988	.9993	.9983
	8	1.0000	1.0000	1.0000	1.0000	1.0000	1.0000	1.0000	1.0000	1.0000
9	0	.6302	.3874	.2316	.1342	.0751	.0404	.0207	.0101	.0046
	1	.9288	.7748	.5995	.4362	.3003	.1960	.1211	.0705	.0385
	2	.9916	.9470	.8591	.7382	.6007	.4628	.3373	.2318	.1495
	3	.9994	.9917	.9661	.9144	.8343	.7297	.6089	.4826	.3614
	4	1.0000	.9991	.9944	.9804	.9511	.9012	.8283	.7334	.6214
	5	1.0000	.9999	.9994	.9969	.9900	.9747	.9464	.9006	.9342
	6	1.0000	1.0000	1.0000	.9997	.9987	.9957	.9888	.9750	.9502
	7	1.0000	1.0000	1.0000	1.0000	.9999	.9996	.9986	.9962	.9909
	8	1.0000	1.0000	1.0000	1.0000	1.0000	1.0000	.9999	.9997	.9992
	9	1.0000	1.0000	1.0000	1.0000	1.0000	1.0000	1.0000	1.0000	1.0000
10	0	.5987	.3487	.1969	.1074	.0563	.0282	.0135	.0060	.0025
	1	.9139	.7361	.5443	.3758	.2440	.1493	.0860	.0464	.0233
	2	.9885	.9298	.8202	.6778	.5256	.3828	.2616	.1673	.0996
	3	.9990	.9872	.9500	.8791	.7759	.6496	.5138	.3823	.2660
	4	.9999	.9984	.9901	.9672	.9219	.8497	.7515	.6331	.5044
	5	1.0000	.9999	.9986	.9936	.9803	.9527	.9051	.8338	.7384
	6	1.0000	1.0000	.9999	.9991	.9965	.9894	.9740	.9452	.8980
	7	1.0000	1.0000	1.0000	.9999	.9996	.9984	.9952	.9877	.9726
	8	1.0000	1.0000	1.0000	1.0000	1.0000	.9999	.9995	.9983	.9955
	9	1.0000	1.0000	1.0000	1.0000	1.0000	1.0000	1.0000	.9999	.9997
	10	1.0000	1.0000	1.0000	1.0000	1.0000	1.0000	1.0000	1.0000	1.0000
11	0	.5688	.3138	.1673	.0859	.0422	.0198	.0088	.0036	.0014
	1	.8981	.6974	.4922	.3221	.1971	.1130	.0606	.0302	.0139
	2	.9848	.9104	.7788	.6174	.4552	.3127	.2001	.1189	.0652
	3	.9984	.9815	.9306	.8389	.7133	.5696	.4256	.2963	.1911
	4	.9999	.9972	.9841	.9496	.8854	.7897	.6683	.5328	.3971
	5	1.0000	.9997	.9973	.9883	.9657	.9218	.8513	.7535	.6331
	6	1.0000	1.0000	.9997	.9980	.9924	.9784	.9499	.9006	.8262
	7	1.0000	1.0000	1.0000	.9998	.9988	.9957	.9878	.9707	.9390
	8	1.0000	1.0000	1.0000	1.0000	.9999	.9994	.9980	.9941	.9852
	9	1.0000	1.0000	1.0000	1.0000	1.0000	1.0000	.9998	.9993	.9978
	10	1.0000	1.0000	1.0000	1.0000	1.0000	1.0000	1.0000	1.0000	.9998
	11	1.0000	1.0000	1.0000	1.0000	1.0000	1.0000	1.0000	1.0000	1.0000

p = .50	.55	.60	.65	.70	.75	.80	.85	.90	.95
.0030	.0017	.0007	.9992	.0001	.0000	.0000	.0000	.0000	.0000
.0352	.0181	.0085	.0086	.0013	.0004	.0001	.0000	.0000	.0000
.1445	.0885	.0498	.0253	.0113	.0042	.0012	.0002	.0000	.0000
.3633	.2604	.1737	.1061	.0580	.0273	.0104	.0029	.0004	.0000
.6367	.5230	.4059	.2936	.1941	.1138	.0563	.0214	.0050	.0004
.8555	.7799	.6846	.5722	.4482	.3215	.2031	.1052	.0381	.0058
.9648	.9368	.8936	.8309	.7447	.6329	.4967	.3428	.1869	.0572
.9961	.9916	.9832	.9681	.9424	.8999	.8322	.7275	.5695	.3366
1.0000	1.0000	1.0000	1.0000	1.0000	1.0000	1.0000	1.0000	1.0000	1.0000
.0020	.0008	.0003	.0001	.0000	.0000	.0000	.0000	.0000	.0000
.0195	.0091	.0038	.0014	.0004	.0001	.0000	.0000	.0000	.0000
.0898	.0498	.0250	.0112	.0043	.0013	.0003	.0000	.0000	.0000
.2539	.1658	.0994	.0536	.0253	.0100	.0031	.0006	.0001	.0000
.5000	.3786	.2666	.1717	.0988	.0489	.0196	.0056	.0009	.0000
.7461	.6386	.5174	.3911	.2703	.1657	.0856	.0339	.0083	.0006
.9102	.8505	.7682	.6627	.5372	.3993	.2618	.1409	.0530	.0084
.9805	.9615	.9295	.8789	.8040	.6997	.5638	.4005	.2252	.0712
.9980	.9954	.9899	.9793	.9596	.9249	.8658	.7684	.6126	.3698
1.0000	1.0000	1.0000	1.0000	1.0000	1.0000	1.0000	1.0000	1.0000	1.0000
.0010	.0003	.0001	.0000	.0000	.0000	.0000	.0000	.0000	.0000
.0107	.0045	.0017	.0005	.0001	.0000	.0000	.0000	.0000	.0000
.0547	.0274	.0123	.0048	.0016	.0004	.0001	.0000	.0000	.0000
.1719	.1020	.0548	.0260	.0106	.0035	.0009	.0001	.0000	.0000
.3770	.2616	.1662	.0949	.0473	.0197	.0064	.0014	.0001	.0000
.6230	.4956	.3669	.2485	.1503	.0781	.0328	.0099	.0016	.0001
.8281	.7340	.6177	.4862	.3504	.2241	.1209	.0500	.0128	.0010
.9453	.9004	.8327	.7184	.6172	.4744	.3222	.1798	.0702	.0115
.9893	.9767	.9536	.9140	.8507	.7560	.6242	.4557	.2639	.0861
.9990	.9975	.9940	.9865	.9718	.9437	.8926	.8031	.6513	.4013
1.0000	1.0000	1.0000	1.0000	1.0000	1.0000	1.0000	1.0000	1.0000	1.0000
.0005	.0002	.0000	.0000	.0000	.0000	.0000	.0000	.0000	.0000
.0059	.0022	.0007	.0002	.0000	.0000	.0000	.0000	.0000	.0000
.0327	.0148	.0059	.0020	.0006	.0001	.0000	.0000	.0000	.0000
.1133	.0610	.0293	.0122	.0043	.0012	.0002	.0000	.0000	.0000
.2744	.1738	.0994	.0501	.0216	.0076	.0020	.0003	.0000	.0000
.5000	.3669	.2465	.1487	.0782	.0343	.0117	.0027	.0003	.0000
.7256	.6029	.4672	.3317	.2103	.1146	.0504	.0159	.0028	.0001
.8867	.8089	.7037	.5744	.4304	.2867	.1611	.0694	.0185	.0016
.9673	.9348	.8811	.7999	.6873	.5448	.3826	.2212	.0896	.0152
.9941	.9861	.9698	.9394	.8870	.8029	.6779	.5078	.3026	.1019
.9995	.9986	.9964	.9912	.9802	.0578	.9141	.8327	.6862	.4312
1.0000	1.0000	1.0000	1.0000	1.0000	1.0000	1.0000	1.0000	1.0000	1.0000

n	k	p = .05	.10	.15	.20	.25	.30	.35	.40	.45
12	0	.5404	.2824	.1422	.0687	.0317	.0138	.0057	.0022	.0008
	1	.8816	.6590	.4435	.2749	.1584	.0850	.0424	.0424	.0083
	2	.9804	.8891	.7358	.5583	.3907	.2528	.1513	.0834	.0421
	3	.9978	.9744	.9078	.7946	.6488	.4925	.3467	.2253	.1345
	4	.9998	.9956	.9761	.9274	.8424	.7237	.5833	.4382	.3044
	5	1.0000	.9995	.9954	.9806	.9456	.8822	.7873	.6652	.5269
	6	1.0000	.9999	.9993	.9961	.9857	.9614	.9154	.8418	.7393
	7	1.0000	1.0000	.9999	.9994	.9972	.9905	.9745	.9427	.8883
	8	1.0000	1.0000	1.0000	.9999	.9996	.9983	.9944	.9847	.9644
	9	1.0000	1.0000	1.0000	1.0000	1.0000	.9998	.9992	.9972	.9921
	10	1.0000	1.0000	1.0000	1.0000	1.0000	1.0000	.9999	.9997	.9989
	11	1.0000	1.0000	1.0000	1.0000	1.0000	1.0000	1.0000	1.0000	.9999
	12	1.0000	1.0000	1.0000	1.0000	1.0000	1.0000	1.0000	1.0000	1.0000
13	0	.5133	.2542	.1209	.0550	.0238	.0097	.0037	.0013	.0004
	1	.8646	.6213	.3983	.2336	.1267	.0637	.9296	.0126	.0049
	2	.9755	.8661	.7296	.5017	.3326	.2025	.1132	.0579	.0269
	3	.9969	.9658	.9033	.7473	.5843	.4206	.2783	.1686	.0929
	4	.9997	.9935	.9740	.9009	.7940	.6543	.5005	.3530	.2279
	5	1.0000	.9991	.9947	.9700	.9198	.8346	.7159	.5744	.4268
	6	1.0000	.9999	.9987	.9930	.9757	.9376	.8705	.7712	.6437
	7	1.0000	1.0000	.9998	.9988	.9944	.9818	.9538	.9023	.8212
	8	1.0000	1.0000	1.0000	.9998	.9990	.9960	.9874	.9679	.9302
	9	1.0000	1.0000	1.0000	1.0000	.9999	.9993	.9975	.9922	.9797
	10	1.0000	1.0000	1.0000	1.0000	1.0000	.9999	.9997	.9987	.9959
	11	1.0000	1.0000	1.0000	1.0000	1.0000	10000	1.0000	.9999	.9995
	12	1.0000	1.0000	1.0000	1.0000	1.0000	1.0000	1.0000	1.0000	1.0000
	13	1.0000	1.0000	1.0000	1.0000	1.0000	1.0000	1.0000	1.0000	1.0000
14	0	.4877	.2288	.1028	.0440	.0178	.0068	.0024	.0008	.0002
	1	.8470	.5846	.3567	.1979	.1010	.0475	.0205	.0081	.0029
	2	.9699	.8416	.6479	.4481	.2811	.1608	.0839	.0398	.0170
	3	.9958	.9559	.8535	.6982	.5213	.3552	.2205	.1243	.0632
	4	.9996	.9908	.9533	.8702	.7415	.5842	.4227	.2793	.1672
	5	1.0000	.9985	.9885	.9561	.8883	.7805	.6405	.4859	.3373
	6	1.0000	.9998	.9978	.9884	.9617	.9067	.8164	.6925	.5461
	7	1.0000	1.0000	.9997	.9976	.9897	.9685	.9247	.8499	.7414
	8	1.0000	1.0000	1.0000	.9996	.9978	.9917	.9757	.9417	.8811
	9	1.0000	1.0000	1.0000	1.0000	.9997	.9983	.9940	.9825	.9574
	10	1.0000	1.0000	1.0000	1.0000	1.0000	.9998	.9989	.9961	.9886
	11	1.0000	1.0000	1.0000	1.0000	1.0000	1.0000	.9999	.9994	.9978
	12	1.0000	1.0000	1.0000	1.0000	1.0000	1.0000	1.0000	.9999	.9997
	13	1.0000	1.0000	1.0000	1.0000	1.0000	1.0000	1.0000	1.0000	1.0000
	14	1.0000	1.0000	1.0000	1.0000	1.0000	1.0000	1.0000	1.0000	1.0000

p = .50	.55	.60	.65	.70	.75	.80	.85	.90	.95
.0002	.0001	.0000	.0000	.0000	.0000	.0000	.0000	.0000	.0000
.0032	.0011	.0003	.0001	.0000	.0000	.0000	.0000	.0000	.0000
.0193	.0079	.0028	.0008	.0002	.0000	.0000	.0000	.0000	.0000
.0730	.0356	.0153	.0056	.0017	.0004	.0001	.0000	.0000	.0000
.1938	.1117	.0573	.0255	.0095	.0028	.0006	.0001	.0000	.0000
.3872	.2607	.1582	.0846	.0386	.0143	.0039	.0007	.0001	.0000
.6128	.4731	.3348	.2127	.1178	.0544	.0194	.0046	.0005	.0000
.8062	.6956	.5618	.4167	.2763	.1576	.0726	.0239	.0043	.0002
.9270	.8655	.7747	.6533	.5075	.3512	.2054	.0922	.0256	.0022
.9807	.9579	.9166	.8487	.7472	.6093	.4417	.2642	.1109	.0196
.9968	.9917	.9804	.9576	.9150	.8416	.7251	.5565	.3410	.1184
.9998	.9992	.9978	.9943	.9862	.9683	.9313	.8578	.7176	.4596
1.0000	1.0000	1.0000	1.0000	1.0000	1.0000	1.0000	1.0000	1.0000	1.0000
.0001	.0000	.0000	.0000	.0000	.0000	.0000	.0000	.0000	.0000
.0017	.0005	.0001	.0000	.0000	.0000	.0000	.0000	.0000	.0000
.0112	.0041	.0013	.0003	.0001	.0000	.0000	.0000	.0000	.0000
.0461	.0203	.0078	.0025	.0007	.0001	.0000	.0000	.0000	.0000
.1334	.0698	.0321	.0126	.0040	.0010	.0002	.0000	.0000	.0000
.2905	.1788	.0977	.0462	.0182	.0056	.0012	.0002	.0000	.0000
.5000	.3563	.2288	.1295	.0624	.0243	.0070	.0013	.0001	.0000
.7095	.5732	.4256	.2841	.1654	.0802	.0300	.0053	.0009	.0000
.8666	.7721	.6470	.4995	.3457	.2060	.0991	.0260	.0065	.0003
.9539	.9071	.8314	.7217	.5794	.4157	.2527	.0967	.0342	.0031
.9888	.9731	.9421	.8868	.7975	.6674	.4983	.2704	.1339	.0245
.9983	.9951	.9874	.9704	.9363	.8733	.7664	.6017	.3787	.1354
.9999	.9996	.9987	.9963	.9903	.9762	.9450	.8791	.7458	.4867
1.0000	1.0000	1.0000	1.0000	1.0000	1.0000	1.0000	1.0000	1.0000	1.0000
.0000	.0000	.0000	.0000	.0000	.0000	.0000	.0000	.0000	.0000
.0009	.0003	.0001	.0000	.0000	.0000	.0000	.0000	.0000	.0000
.0065	.0022	.0006	.0001	.0000	.0000	.0000	.0000	.0000	.0000
.0287	.0114	.0039	.0011	.0002	.0000	.0000	.0000	.0000	.0000
.0898	.0462	.0175	.0060	.0017	.0003	.0000	.0000	.0000	.0000
.2120	.1189	.0583	.0243	.0083	.0022	.0004	.0000	.0000	.0000
.3953	.2586	.1501	.0753	.0315	.0108	.0024	.0003	.0000	.0000
.6047	.4539	.3075	.1836	.0933	.0383	.0116	.0022	.0002	.0000
.7880	.6627	.5141	.3595	.2195	.1117	.0439	.0115	.0015	.0000
.9102	.8328	.7207	.5773	.4158	.2585	.1298	.0467	.0092	.0004
.9713	.9368	.8757	.7795	.6448	.4787	.3018	.1465	.0441	.0042
.9935	.9830	.9602	.9161	.8392	.7189	.5519	.3521	.1584	.0301
.9991	.9971	.9919	.9795	.9525	.8990	.8021	.6433	.4154	.1530
.9999	.9998	.9992	.9976	.9932	.9822	.9560	.8972	.7712	.5123
1.0000	1.0000	1.0000	1.0000	1.0000	1.0000	1.0000	1.0000	1.0000	1.0000

n	k	p = .05	.10	.15	.20	.25	.30	.35	.40	.45
15	0	.4633	.2059	.0874	.0352	.0134	.0047	.0016	.0005	.0001
	1	.8290	.5490	.3186	.1671	.0802	.0353	.0142	.0052	.0017
	2	.9638	.8159	.6042	.3980	.2361	.1268	.0617	.0271	.0107
	3	.9945	.9444	.8227	.6482	.4613	.2969	.1727	.0905	.1424
	4	.9994	.9873	.9383	.8358	.6865	.5155	.3519	.2173	.1204
	5	.9999	.9978	.9832	.9389	.8516	.7216	.5643	.4032	.2608
	6	1.0000	.9997	.9964	.9819	.9434	.8689	.7548	.6098	.4522
	7	1.0000	1.0000	.9994	.9958	.9827	.9500	.8868	.7869	.6535
	8	1.0000	1.0000	.9999	.9992	.9958	.9848	.9578	.9050	.8121
	9	1.0000	1.0000	1.0000	.9999	.9992	.9963	.9876	.9662	.9231
	10	1.0000	1.0000	1.0000	1.0000	.9999	.9993	.9972	.9907	.9745
	11	1.0000	1.0000	1.0000	1.0000	1.0000	.9999	.9995	.9981	.9937
	12	1.0000	1.0000	1.0000	1.0000	1.0000	1.0000	.9999	.9997	.9989
	13	1.0000	1.0000	1.0000	1.0000	1.0000	1.0000	1.0000	1.0000	.9999
	14	1.0000	1.0000	1.0000	1.0000	1.0000	1.0000	1.0000	1.0000	1.0000
	15	1.0000	1.0000	1.0000	1.0000	1.0000	1.0000	1.0000	1.0000	1.0000
16	0	.4401	.1853	.0743	.0281	.0100	.0033	.0010	.0003	.0001
	1	.8108	.5147	.2839	.1407	.0635	.0261	.0098	.0088	.0010
	2	.9571	.7892	.5614	.3518	.1971	.0904	.0451	.0183	.0066
	3	.9930	.9316	.7899	.5981	.4050	.2459	.1339	.0651	.0281
	4	.9991	.9830	.9209	.7982	.6302	.4499	.2892	.1666	.0853
	5	.9999	.9967	.9765	.9183	.8103	.6598	.4900	.3288	.1970
	6	1.0000	.9995	.9944	.9733	.9204	.8247	.6881	.5272	.3660
	7	1.0000	.9999	.9989	.9930	.9729	.9256	.8406	.7161	.5629
	8	1.0000	1.0000	.9998	.9985	.9925	.9743	.9329	.8577	.7441
	9	1.0000	1.0000	1.0000	.9998	.9984	.9938	.9809	.9514	.8759
	10	1.0000	1.0000	1.0000	1.0000	.9997	.9984	.9938	.9809	.9514
	11	1.0000	1.0000	1.0000	1.0000	1.0000	.9997	.9987	.9951	.9851
	12	1.0000	1.0000	1.0000	1.0000	1.0000	1.0000	.9998	.9991	.9965
	13	1.0000	1.0000	1.0000	1.0000	1.0000	1.0000	1.0000	.9999	.9965
	14	1.0000	1.0000	1.0000	1.0000	1.0000	1.0000	1.0000	1.0000	.9999
	15	1.0000	1.0000	1.0000	1.0000	1.0000	1.0000	1.0000	1.0000	1.0000
	16	1.0000	1.0000	1.0000	1.0000	1.0000	1.0000	1.0000	1.0000	1.0000

p = .50	.55	.60	.65	.70	.75	.80	.85	.90	.95
.0000	.0000	.0000	.0000	.0000	.0000	.0000	.0000	.0000	.0000
.0005	.0001	.0000	.0000	.0000	.0000	.0000	.0000	.0000	.0000
.0037	.0011	.0003	.0001	.0000	.0000	.0000	.0000	.0000	.0000
.0176	.0063	.0019	.0005	.0001	.0000	.0000	.0000	.0000	.0000
.0592	.0255	.0093	.0028	.0007	.0001	.0000	.0000	.0000	.0000
.1509	.1769	.0228	.0124	.0037	.0008	.0001	.0000	.0000	.0000
.3036	.1818	.0950	.0422	.0152	.0042	.0008	.0001	.0000	.0000
.5000	.3465	.2131	.1132	.0500	.0173	.0042	.0006	.0000	.0000
.6964	.5478	.3902	.2452	.1311	.0566	.0181	.0036	.0003	.0000
.8491	.7392	.5968	.4357	.2784	.1484	.0611	.0168	.0022	.0001
.9408	.8796	.7827	.6481	.4845	.3135	.1642	.0617	.0127	.0006
.9824	.9576	.9095	.8273	.7031	.5387	.3518	.1773	.0556	.0055
.9963	.9893	.9729	.9383	.8732	.7639	.6020	.3958	.1841	.0362
.9995	.9983	.9948	.9858	.9647	.9198	.8329	.6814	.4510	.1710
1.0000	.9999	.9995	.9984	.9953	.9866	.9648	.9126	.7941	.5367
1.0000	1.0000	1.0000	1.0000	1.0000	1.0000	1.0000	1.0000	1.0000	1.0000
.0000	.0000	.0000	.0000	.0000	.0000	.0000	.0000	.0000	.0000
.0003	.0001	.0000	.0000	.0000	.0000	.0000	.0000	.0000	.0000
.0021	.0006	.0001	.0000	.0000	.0000	.0000	.0000	.0000	.0000
.0106	.0035	.0009	.0002	.0000	.0000	.0000	.0000	.0000	.0000
.0384	.0149	.0049	.0013	.0003	.0000	.0000	.0000	.0000	.0000
.1051	.0486	.0191	.0062	.0016	.0003	.0000	.0000	.0000	.0000
.2272	.1241	.0583	.0229	.0071	.0016	.0002	.0000	.0000	.0000
.4018	.2559	.1423	.0671	.0257	.0075	.0015	.0002	.0000	.0000
.5982	.4371	.2839	.1594	.0744	.0271	.0070	.0011	.0001	.0000
.7228	.6340	.4728	.3119	.1753	.0796	.0267	.0056	.0005	.0000
.8949	.8024	.6712	.5100	.3402	.1897	.0817	.0235	.0033	.0001
.9616	.9147	.8334	.7108	.5501	.3698	.2018	.0791	.0170	.0009
.9894	9719	.9349	.8661	.7541	.5950	.4019	.2101	.0684	.0070
.9979	.9934	.9817	.9549	.9006	.8729	.6482	.4386	.2108	.0429
.9997	.9990	.9967	.9902	.9739	.9365	.8593	.7176	.4853	.1892
1.0000	.9999	.9997	.9990	.9967	.9900	.9719	.9257	.8147	.5599
1.0000	1.0000	1.0000	1.0000	1.0000	1.0000	1.0000	1.0000	1.0000	1.0000

n	k	p = .05	.10	.15	.20	.25	.30	.35	.40	.45
17	0	.4181	.1668	.0631	.0225	.0075	.0023	.0007	.0002	.0000
	1	.7922	.4818	.2525	.1182	.0501	.0193	.0067	.0021	.0006
	2	.9497	.7618	.5198	.3096	.1637	.0774	.0327	.0123	.0041
	3	.9912	.9174	.7556	.5489	.3530	.2019	.1028	.0464	.0184
	4	.9988	.9779	.9013	.7582	.5739	.3887	.2348	.1260	.0596
	5	.9999	.9953	.9681	.8943	.7653	.5968	.4197	.2639	.1471
	6	1.0000	.9992	.9917	.9623	.8929	.7752	.6188	.4478	.2902
	7	1.0000	.9999	.9983	.9891	.9598	.8954	.7872	.6405	.4743
	8	1.0000	1.0000	.9997	.9974	.9876	.9597	.9006	.8011	.6626
	9	1.0000	1.0000	1.0000	.9995	.9969	.9873	.9611	.9081	.8166
	10	1.0000	1.0000	1.0000	.9999	.9994	.9968	.9880	.9652	.9174
	11	1.0000	1.0000	1.0000	1.0000	.9999	.9993	.9970	.9894	.9699
	12	1.0000	1.0000	1.0000	1.0000	1.0000	.9999	.9994	.9975	.9914
	13	1.0000	1.0000	1.0000	1.0000	1.0000	1.0000	.9999	.9995	.9981
	14	1.0000	1.0000	1.0000	1.0000	1.0000	1.0000	1.0000	.9999	.9997
	15	1.0000	1.0000	1.0000	1.0000	1.0000	1.0000	1.0000	1.0000	1.0000
	16	1.0000	1.0000	1.0000	1.0000	1.0000	1.0000	1.0000	1.0000	1.0000
	17	1.0000	1.0000	1.0000	1.0000	1.0000	1.0000	1.0000	1.0000	1.0000
18	0	3972	.1501	.0536	.0180	.0056	.0016	.0004	.0001	.0000
	1	.7735	.4503	.2241	.0991	.0395	.0142	.0046	.0013	.0003
	2	.9419	.7338	.4797	.2713	.1353	.0600	.0236	.0082	.0025
	3	.9891	.9018	.7202	.5010	.3057	.1646	.0783	.0328	.0120
	4	.9985	.9718	.8794	.7164	.5187	.3327	.1886	.0942	.0411
	5	.9998	.9936	.9581	.8671	.7175	.5344	.3550	.2088	.1077
	6	1.0000	.9988	.9973	.9837	.9431	.8593	.7283	.5634	.3915
	7	1.0000	.9998	.9973	.9837	.9431	.8593	.7283	.5634	.3915
	8	1.0000	1.0000	.9995	.9957	.9807	.9404	.8609	.7368	.5778
	9	1.0000	1.0000	.9999	.9991	.9946	.9790	.9403	.8653	.7473
	10	1.0000	1.0000	1.0000	.9998	.9988	.9939	.9788	.9424	.8720
	11	1.0000	1.0000	1.0000	1.0000	.9998	.9986	.9938	.9797	.9463
	12	1.0000	1.0000	1.0000	1.0000	1.0000	.9997	.9986	.9942	.9817
	13	1.0000	1.0000	1.0000	1.0000	1.0000	1.0000	.9997	.9987	.9951
	14	1.0000	1.0000	1.0000	1.0000	1.0000	1.0000	1.0000	.9998	.9990
	15	1.0000	1.0000	1.0000	1.0000	1.0000	1.0000	1.0000	1.0000	.9999
	16	1.0000	1.0000	1.0000	1.0000	1.0000	1.0000	1.0000	1.0000	1.0000
	17	1.0000	1.0000	1.0000	1.0000	1.0000	1.0000	1.0000	1.0000	1.0000
	18	1.0000	1.0000	1.0000	1.0000	1.0000	1.0000	1.0000	1.0000	1.0000

p = .50	.55	.60	.65	.70	.75	.80	.85	.90	.95
.0000	.0000	.0000	.0000	.0000	.0000	.0000	.0000	.0000	.0000
.0001	.0000	.0000	.0000	.0000	.0000	.0000	.0000	.0000	.0000
.0012	.0003	.0001	.0000	.0000	.0000	.0000	.0000	.0000	.0000
.0064	.0019	.0005	.0001	.0000	.0000	.0000	.0000	.0000	.0000
.0245	.0086	.0025	.0006	.0001	.0000	.0000	.0000	.0000	.0000
.0717	.0301	.0106	.0030	.0007	.0001	.0000	.0000	.0000	.0000
.1662	.0826	.0348	.0120	.0032	.0006	.0001	.0000	.0000	.0000
.3145	.1834	.0919	.0383	.0127	.0031	.0005	.0000	.0000	.0000
.5000	.3374	.1989	.0994	.0403	.0124	.0026	.0003	.0000	.0000
.6855	.5257	.3595	.2128	.1046	.0402	.0109	.0017	.0001	.0000
.8338	.7098	.5522	.3812	.2248	.1071	.0377	.0083	.0008	.0000
.9283	.8529	.7361	.5803	.4032	.2347	.1057	.0319	.0047	.0001
.9755	.9404	.8740	.7652	.6113	.4261	.2418	.0987	.0221	.0012
.9936	.9816	.9536	.8972	.7981	.6470	.4511	.2444	.0826	.0088
.9988	.9959	.9877	.9673	.9226	.8363	.6904	.4802	.2382	.0503
.9999	.9994	.9979	.9933	.9807	.9499	.8818	.7475	.5182	.2078
1.0000	1.0000	.9998	.0003	.0077	.9925	.9775	.9369	.8332	.5819
1.0000	1.0000	1.0000	1.0000	1.0000	1.0000	1.0000	1.0000	1.0000	1.0000
.0000	.0000	.0000	.0000	.0000	.0000	.0000	.0000	.0000	.0000
.0001	.0000	.0000	.0000	.0000	.0000	.0000	.0000	.0000	.0000
.0007	.0001	.0000	.0000	.0000	.0000	.0000	.0000	.0000	.0000
.0038	.0010	.0002	.0000	.0000	.0000	.0000	.0000	.0000	.0000
.0154	.0049	.0013	.0003	.0000	.0000	.0000	.0000	.0000	.0000
.0481	.0183	.0058	.0014	.0003	.0000	.0000	.0000	.0000	.0000
.1189	.0537	.0203	.0062	.0014	.0002	.0000	.0000	.0000	.0000
.2403	.1280	.0576	.0212	.0061	.0012	.0002	.0000	.0000	.0000
.4073	.2527	.1347	.0597	.0210	.0054	.0009	.0001	.0000	.0000
.5927	.4222	.2632	.2717	.1407	.0569	.0163	.0027	.0002	.0000
.7597	.6085	.4366	.2717	.1407	.0569	.0163	.0027	.0002	.0000
.8811	.7742	.6457	.4509	.2783	.1390	.0513	.0118	.0012	.0000
.9519	.8923	.7912	.6450	.4656	.2825	.1329	.0419	.0064	.0002
.9846	.9589	.9058	.8114	.6673	.4813	.2836	.1206	.0282	.0015
.9962	.9880	.9672	.9217	.8354	.6943	.4990	.2798	.0982	.0109
.9993	.9975	.9918	.9764	.9400	.8647	.7287	.5203	.2662	.0581
.9999	.9997	.9987	.9954	.9858	.9605	.9009	.7759	.5497	.2265
1.0000	1.0000	.9999	.9996	.9984	.9944	.9820	.9464	.8499	.6028
1.0000	1.0000	1.0000	1.0000	1.0000	1.0000	1.0000	1.0000	1.0000	1.0000

n	k	p = .05	.10	.15	.20	.25	.30	.35	.40	.45
19	0	.3774	.1351	.0456	.0144	.0042	.0011	.0003	.0001	.0000
	1	.7547	.4203	.1985	.0829	.0310	.0008	.0002	.0008	.0002
	2	.9335	.7054	.4413	.2369	.1113	.0462	.0170	.0055	.0015
	3	.9869	.8850	.6841	.4551	.2631	.1332	.0591	.0230	.0077
	4	.9869	.8850	.6841	.4551	.2631	.1332	.0591	.0230	.0077
	5	.9998	.9914	.9463	.8369	.6678	.4739	.2968	.1629	.0777
	6	1.0000	.9983	.9837	.9324	.8251	.6655	.4912	.3081	.1727
	7	1.0000	.9997	.9959	.9767	.9225	.8180	.6656	.4878	.3169
	8	1.0000	1.0000	.9992	.9933	.9713	.9161	.8145	.6675	.4940
	9	1.0000	1.0000	.9999	.9984	.9911	.9674	.9125	.8139	.6710
	10	1.0000	1.0000	1.0000	.9997	.9977	.9895	.9653	.9115	.8159
	11	1.0000	1.0000	1.0000	1.0000	.9995	.9972	.9886	.9648	.9129
	12	1.0000	1.0000	1.0000	1.0000	.9999	.9994	.9969	.9884	.9658
	13	1.0000	1.0000	1.0000	1.0000	1.0000	.9999	.9993	.9969	.9891
	14	1.0000	1.0000	1.0000	1.0000	1.0000	1.0000	.9999	.9994	.9972
	15	1.0000	1.0000	1.0000	1.0000	1.0000	1.0000	1.0000	.9999	.9995
	16	1.0000	1.0000	1.0000	1.0000	1.0000	1.0000	1.0000	1.0000	.9999
	17	1.0000	1.0000	1.0000	1.0000	1.0000	1.0000	1.0000	1.0000	1.0000
	18	1.0000	1.0000	1.0000	1.0000	1.0000	1.0000	1.0000	1.0000	1.0000
	19	1.0000	1.0000	1.0000	1.0000	1.0000	1.0000	1.0000	1.0000	1.0000
20	0	.3585	.1261	.0388	.0115	.0032	.0008	.0000	.0000	.0000
	1	.7358	.3917	.1756	.0692	.0243	.0076	.0021	.0005	.0001
	2	.9245	.6769	.4049	.2061	.0913	.0355	.0121	.0036	.0009
	3	.9841	.8670	.6477	.4114	.2252	.1071	.0444	.0160	.0049
	4	.9974	.9568	.8298	.6296	.4148	.2375	.1182	.0510	.0189
	5	.9997	.9887	.9327	.8042	.6172	.4164	.2454	.1256	.0553
	6	1.0000	.9976	.9781	.9133	.7858	.6080	.4166	.2500	.1299
	7	1.0000	.9996	.9941	.9679	.8982	.7723	.6010	.4159	.2520
	8	1.0000	.9999	.9987	.9900	.9591	.8867	.7624	.5956	.4143
	9	1.0000	1.0000	.9998	.9974	.9861	.9520	.8782	.7553	.5914
	10	1.0000	1.0000	1.0000	.9994	.9961	.9829	.9468	.8725	.7507
	11	1.0000	1.0000	1.0000	.9999	.9991	.9949	.9804	.9435	.8692
	12	1.0000	1.0000	1.0000	1.0000	.9998	.9987	.9940	.9790	.9420
	13	1.0000	1.0000	1.0000	1.0000	1.0000	.9997	.9985	.9935	.9786
	14	1.0000	1.0000	1.0000	1.0000	1.0000	1.0000	.9997	.9984	.9936
	15	1.0000	1.0000	1.0000	1.0000	1.0000	1.0000	1.0000	.9997	.9985
	16	1.0000	1.0000	1.0000	1.0000	1.0000	1.0000	1.0000	1.0000	.9997
	17	1.0000	1.0000	1.0000	1.0000	1.0000	1.0000	1.0000	1.0000	1.0000
	18	1.0000	1.0000	1.0000	1.0000	1.0000	1.0000	1.0000	1.0000	1.0000
	19	1.0000	1.0000	1.0000	1.0000	1.0000	1.0000	1.0000	1.0000	1.0000
	20	1.0000	1.0000	1.0000	1.0000	1.0000	1.0000	1.0000	1.0000	1.0000

p = .50	.55	.60	.65	.70	.75	.80	.85	.90	.95
.0000	.0000	.0000	.0000	.0000	.0000	.0000	.0000	.0000	.0000
.0000	.0000	.0000	.0000	.0000	.0000	.0000	.0000	.0000	.0000
.0004	.0001	.0000	.0000	.0000	.0000	.0000	.0000	.0000	.0000
.0022	.0005	.0001	.0000	.0000	.0000	.0000	.0000	.0000	.0000
.0096	.0028	.0006	.0001	.0000	.0000	.0000	.0000	.0000	.0000
.0318	.0109	.0031	.0007	.0001	.0000	.0000	.0000	.0000	.0000
.0835	.0342	.0116	.0031	.0006	.0001	.0000	.0000	.0000	.0000
.1796	.0871	.0352	.0114	.0028	.0005	.0000	.0000	.0000	.0000
.3238	.1841	.0885	.0347	.0105	.0023	.0003	.0000	.0000	.0000
.5000	.3290	.1861	.0875	.0287	.0067	.0008	.0000	.0000	.0000
.6762	.5060	.3325	.1855	.0839	.0287	.0067	.0008	.0000	.0000
.8204	.6831	.5122	.3344	.1820	.0775	.0233	.0041	.0008	.0000
.9165	.8273	.6919	.5188	.3345	.1749	.0676	.0163	.0017	.0000
.9682	.9223	.8371	.7032	.5261	.3322	.1631	.0537	.0086	.0002
.9904	.9720	.9304	.8500	.7178	.5346	.3267	.1444	.0352	.0020
.9978	.9923	.9770	.9409	.8668	.7369	.5449	.3159	.1150	.0132
.9996	.9985	.9945	.9830	.9538	.8887	.7631	.5587	.2946	.0665
1.0000	.9998	.9992	.9969	.9896	.9690	.9171	.8015	.5797	.2453
1.0000	1.0000	.9999	.9997	.9989	.9958	.9856	.9544	.8649	.6226
1.0000	1.0000	1.0000	1.0000	1.0000	1.0000	1.0000	1.0000	1.0000	1.0000
.0000	.0000	.0000	.0000	.0000	.0000	.0000	.0000	.0000	.0000
.0000	.0000	.0000	.0000	.0000	.0000	.0000	.0000	.0000	.0000
.0002	.0000	.0000	.0000	.0000	.0000	.0000	.0000	.0000	.0000
.0013	.0003	.0000	.0000	.0000	.0000	.0000	.0000	.0000	.0000
.0059	.0015	.0003	.0000	.0000	.0000	.0000	.0000	.0000	.0000
.0207	.0064	.0016	.0003	.0000	.0000	.0000	.0000	.0000	.0000
.0577	.0214	.0065	.0015	.0003	.0000	.0000	.0000	.0000	.0000
.1316	.0580	.0210	.0060	.0013	.0002	.0000	.0000	.0000	.0000
.2517	.1308	.0565	.0196	.0051	.0009	.0001	.0000	.0000	.0000
.4119	.2493	.1275	.0532	.0171	.0039	.0006	.0000	.0000	.0000
.5881	.4086	.2447	.1218	.0480	.0139	.0026	.0002	.0000	.0000
.7483	.5857	.4044	.2376	.1133	.0409	.0100	.0013	.0001	.0000
.8684	.7480	.5841	.3990	.2277	.1018	.0321	.0059	.0004	.0000
.9423	.8701	.7500	.5834	.3920	.2142	.0867	.0219	.0024	.0000
.9793	.9447	.8744	.7546	.5836	.3828	.1958	.0673	.0113	.0003
.9941	.9811	.9490	.8818	.7625	.5852	.3704	.1702	.0432	.0026
.9987	.9951	.9840	.9556	.8929	.7748	.5886	.3523	.1330	.0159
.9998	.9991	.9964	.9879	.9645	.9087	.7939	.5951	.3231	.0755
1.0000	.9999	.9995	.9979	.9924	.9757	.9308	.8244	.6083	.2642
1.0000	1.0000	1.0000	.9998	.9992	.9968	.9885	.9612	.8784	.6415
1.0000	1.0000	1.0000	1.0000	1.0000	1.0000	1.0000	1.0000	1.0000	1.0000

SOURCE: M. Orkin and R. Drogin, "Vital Statistics," pp. 338–349, McGraw-Hill, New York, 1975.

Table A.3 gives the area to the left of z under a standard normal curve, for various values of z.

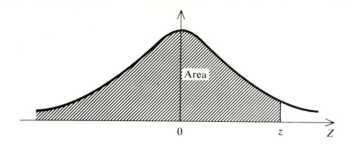

To obtain probabilities use the fact that

$$\text{Area to the left of } z = P(Z < z)$$

where Z is a variable with a standard normal distribution.

z	Area	z	Area	z	Area
−4	.00003	−2.74	.0031	−2.29	.0110
−3.9	.00005	−2.73	.0032	−2.28	.0113
−3.8	.0001	−2.72	.0033	−2.27	.0116
−3.7	.0001	−2.71	.0034	−2.26	.0119
−3.6	.0002	−2.70	.0035	−2.25	.0122
−3.5	.0002	−2.69	.0036	−2.24	.0125
−3.4	.0003	−2.68	.0037	−2.23	.0129
−3.3	.0005	−2.67	.0038	−2.22	.0132
−3.2	.0007	−2.66	.0039	−2.21	.0136
−3.1	.0010	−2.65	.0040	−2.20	.0139
−3.09	.0010	−2.64	.0041	−2.19	.0143
−3.08	.0010	−2.63	.0043	−2.18	.0146
−3.07	.0011	−2.62	.0044	−2.17	.0150
−3.06	.0011	−2.61	.0045	−2.16	.0154
−3.05	.0011	−2.60	.0047	−2.15	.0158
−3.04	.0012	−2.59	.0048	−2.14	.0162
−3.03	.0012	−2.58	.0049	−2.13	.0166
−3.02	.0013	−2.57	.0051	−2.12	.0170
−3.01	.0013	−2.56	.0052	−2.11	.0174
−3.00	.0013	−2.55	.0054	−2.10	.0179
−2.99	.0014	−2.54	.0055	−2.09	.0183
−2.98	.0014	−2.53	.0057	−2.08	.0188
−2.97	.0015	−2.52	.0059	−2.07	.0182
−2.96	.0015	−2.51	.0060	−2.06	.0197
−2.95	.0016	−2.50	.0062	−2.05	.0202
−2.94	.0016	−2.49	.0064	−2.04	.0207
−2.93	.0017	−2.48	.0066	−2.03	.0212
−2.92	.0017	−2.47	.0068	−2.02	.0217
−2.91	.0018	−2.46	.0069	−2.01	.0222
−2.90	.0019	−2.45	.0071	−2.00	.0228
−2.89	.0019	−2.44	.0073	−1.99	.0233
−2.88	.0020	−2.43	.0075	−1.98	.0239
−2.87	.0021	−2.42	.0078	−1.97	.0244
−2.86	.0021	−2.41	.0080	−1.96	.0250
−2.85	.0022	−2.40	.0082	−1.95	.0256
−2.84	.0023	−2.39	.0084	−1.94	0262
−2.83	.0023	−2.38	.0087	−1.93	.0268
−2.82	.0024	−2.37	.0089	1.92	.0274
−2.81	.0025	−2.36	.0091	−1.91	.0281
−2.80	.0026	−2.35	.0094	−1.90	.0287
−2.79	.0026	−2.34	.0096	−1.89	.0294
−2.78	.0027	−2.33	.0099	−1.88	.030
−2.77	.0028	−2.32	.0102	−1.87	.0307
−2.76	.0029	−2.31	.0104	−1.86	.0314
−2.75	.0030	−2.30	.0107	−1.85	.0322

z	Area		z	Area		z	Area
−1.84	.0329		−1.39	.0823		− .94	.1736
−1.83	.0336		−1.38	.0838		− .93	.1762
−1.82	.0344		−1.37	.0853		− .92	.1788
−1.81	.0352		−1.36	.0869		− .91	.1814
−1.80	.0359		−1.35	.0885		− .90	.1841
−1.79	.0367		−1.34	.0901		− .89	.1867
−1.78	.0375		−1.33	.0918		− .88	.1894
−1.77	.0384		−1.32	.0934		− .87	.1922
−1.76	.0392		−1.31	.0951		− .86	.1949
−1.75	.0401		−1.30	.0968		− .85	.1977
−1.74	.0409		−1.29	.0985		− .84	.2005
−1.73	.0418		−1.28	.1003		− .83	.2033
−1.72	.0427		−1.27	.1020		− .82	.2061
−1.71	.0436		−1.26	.1038		− .81	.2090
−1.70	.0446		−1.25	.1056		− .80	.2119
−1.69	.0455		−1.24	.1075		− .79	.2148
−1.68	.0465		−1.23	.1093		− .78	.2177
−1.67	.0475		−1.22	.1112		− .77	.2296
−1.66	.0485		−1.21	.1131		− .76	.2236
−1.65	.0495		−1.20	.1151		− .75	.2266
−1.64	.0505		−1.19	.1170		− .74	.2296
−1.63	.0516		−1.18	.1190		− .73	.2327
−1.62	.0526		−1.17	.1210		− .72	.2358
−1.61	.0537		−1.16	.1230		− .71	.2389
−1.60	.0548		−1.15	.1251		− .70	.2420
−1.59	.0559		−1.14	.1271		− .69	.2451
−1.58	.0571		−1.13	.1292		− .68	.2483
−1.57	.0582		−1.12	.1314		− .67	.2514
−1.56	.0594		−1.11	.1335		− .66	.2546
−1.55	.0606		−1.10	.1357		− .65	.2578
−1.54	.0618		−1.09	.1379		− .64	.2611
−1.53	.0630		−1.08	.1401		− .63	.2643
−1.52	.0643		−1.07	.1423		− .62	.2676
−1.51	.0655		−1.06	.1446		− .61	.2709
−1.50	.0668		−1.05	.1469		− .60	.2743
−1.49	.0681		−1.04	.1492		− .59	.2776
−1.48	.0694		−1.03	.1515		− .58	.2810
−1.47	.0708		−1.02	.1539		− .57	.2843
−1.46	.0722		−1.01	.1562		− .56	.2877
−1.45	.0735		−1.00	.1587		− .55	.2912
−1.44	.0749		− .99	.1611		− .54	.2946
−1.43	.0764		− .98	.1635		− .53	.2981
−1.42	.0778		− .97	.1660		− .52	.3015
−1.41	.0793		− .96	.1685		− .51	.3050
−1.40	.0808		− .95	.1711		− .50	.3085

z	Area	z	Area	z	Area
− .49	.3121	− .04	.4840	.41	.6591
− .48	.3156	− .03	.4880	.42	.6628
− .47	.3192	− .02	.4920	.43	.6664
− .46	.3228	− .01	.4960	.44	.6700
− .45	.3264	.00	.5000	.45	.6736
− .44	.3300	.01	.5040	.46	.6772
− .43	.3336	.02	.5080	.47	.6808
− .42	.3372	.03	.5120	.48	.6844
− .41	.3409	.04	.5160	.49	.6849
− .40	.3446	.05	.5199	.50	.6915
− .39	.3483	.06	.5239	.51	.6950
− .38	.3520	.07	.5279	.52	.6985
− .37	.3557	.08	.5319	.53	.7019
− .36	.3594	.09	.5359	.54	.7054
− .35	.3632	.10	.5398	.55	.7088
− .34	.3669	.11	.5438	.56	.7123
− .33	.3707	.12	.5478	.57	.7157
− .32	.3745	.13	.5517	.58	.7190
− .31	.3783	.14	.5557	.59	.7224
− .30	.3821	.15	.5596	.60	.7257
− .29	.3859	.16	.5636	.61	.7291
− .28	.3897	.17	.5675	.62	.7324
− .27	.3936	.18	.5714	.63	.7357
− .26	.3974	.19	.5753	.64	.7389
− .25	.4013	.20	.5793	.65	.7422
− .24	.4052	.21	.5832	.66	.7454
− .23	.4090	.22	.5871	.67	.7486
− .22	.4129	.23	.5910	.68	.7517
− .21	.4168	.24	.5948	.69	.7549
− .20	.4207	.25	.5987	.70	.7580
− .19	.4247	.26	.6026	.71	.7611
− .18	.4286	.27	.6064	.72	.7642
− .17	.4325	.28	.6103	.73	.7673
− .16	.4364	.29	.6141	.74	.7704
− .15	.4404	.30	.6179	.75	.7734
− .14	.4443	.31	.6217	.76	.7764
− .13	.4483	.32	.6255	.77	.7794
− .12	.4522	.33	.6293	.78	.7823
− .11	.4562	.34	.6331	.79	.7852
− .10	.4602	.35	.6368	.80	.7881
− .09	.4641	.36	.6406	.81	.7910
− .08	.4681	.37	.6443	.82	.7939
− .07	.4721	.38	.6480	.83	.7967
− .06	.4761	.39	.6517	.84	.7995
− .05	.4801	.40	.6554	.85	.8023

z	Area	z	Area	z	Area
.86	.8051	1.31	.9049	1.76	.9608
.87	.8078	1.32	.9066	1.77	.9616
.88	.8106	1.33	.9082	1.78	.9625
.89	.8133	1.34	.9099	1.79	.9633
.90	.8159	1.35	.9115	1.80	.9641
.91	.8186	1.36	.9131	1.81	.9649
.92	.8212	1.37	.9147	1.82	.9656
.93	.8238	1.38	.9162	1.83	.9664
.94	.8264	1.39	.9177	1.84	.9671
.95	.8289	1.40	.9192	1.85	.9678
.96	.8315	1.41	.9207	1.86	.9686
.97	.8340	1.42	.9222	1.87	.9693
.98	.8365	1.43	.9236	1.88	.9699
.99	.8389	1.44	.9251	1.89	.9706
1.00	.8413	1.45	.9265	1.90	.9713
1.01	.8438	1.46	.9278	1.91	.9719
1.02	.8461	1.47	.9292	1.92	.9726
1.03	.8485	1.48	.9306	1.93	.9732
1.04	.8508	1.49	.9319	1.94	.9738
1.05	.8531	1.50	.9332	1.95	.9744
1.06	.8554	1.51	.9345	1.96	.9750
1.07	.8577	1.52	.9357	1.97	.9756
1.08	.8599	1.53	.9370	1.98	.9761
1.09	.8621	1.54	.9382	1.99	.9767
1.10	.8643	1.55	.9394	2.00	.9772
1.11	.8665	1.56	.9406	2.01	.9778
1.12	.8686	1.57	.9418	2.02	.9783
1.13	.8708	1.58	.9429	2.03	.9788
1.14	.8729	1.59	.9441	2.04	.9793
1.15	.8749	1.60	.9452	2.05	.9798
1.16	.8770	1.61	.9463	2.06	.9803
1.17	.8790	1.62	.9474	2.07	.9808
1.18	.8810	1.63	.9484	2.08	.9812
1.19	.8830	1.64	.9495	2.09	.9817
1.20	.8849	1.65	.9505	2.10	.9821
1.21	.8869	1.66	.9515	2.11	.9826
1.22	.8888	1.67	.9525	2.12	.9830
1.23	.8907	1.68	.9535	2.13	.9834
1.24	.8925	1.69	.9545	2.14	.9838
1.25	.8944	1.70	.9554	2.15	.9842
1.26	.8962	1.71	.9564	2.16	.9846
1.27	.8980	1.72	.9573	2.17	.9850
1.28	.8997	1.73	.9582	2.18	.9854
1.29	.9015	1.74	.9591	2.19	.9857
1.30	.9032	1.75	.9599	2.20	.9861

z	Area	z	Area	z	Area
2.21	.9864	2.66	.9961	3.2	.9993
2.22	.9868	2.67	.9962	3.3	.9995
2.23	.9871	2.68	.9963	3.4	.9997
2.24	.9875	2.69	.9964	3.5	.9998
2.25	.9878	2.70	.9965	3.6	.9998
				3.7	.9999
2.26	.9881	2.71	.9966	3.8	.9999
2.27	.9884	2.72	.9967		
2.28	.9887	2.73	.9968	3.9	.99995
2.29	.9890	2.74	.9969	4.0	.99997
2.30	.9893	2.75	.9970		
2.31	.9896	2.76	.9971		
2.32	.9898	2.77	.9972		
2.33	.9901	2.78	.9973		
2.34	.9904	2.79	.9974		
2.35	.9906	2.80	.9974		
2.36	.9909	2.81	.9975		
2.37	.9911	2.82	.9976		
2.38	.9913	2.83	.9977		
2.39	.9916	2.84	.9977		
2.40	.9918	2.85	.9978		
2.41	.9920	2.86	.9979		
2.42	.9922	2.87	.9979		
2.43	.9925	2.88	.9980		
2.44	.9927	2.89	.9981		
2.45	.9929	2.90	.9981		
2.46	.9931	2.91	.9982		
2.47	.9932	2.92	.9982		
2.48	.9934	2.93	.9983		
2.49	.9936	2.94	.9984		
2.50	.9938	2.95	.9984		
2.51	.9940	2.96	.9985		
2.52	.9941	2.97	.9985		
2.53	.9943	2.98	.9986		
2.54	.9945	2.99	.9986		
2.55	.9946	3.00	.9987		
2.56	.9948	3.01	.9987		
2.57	.9949	3.02	.9987		
2.58	.9951	3.03	.9988		
2.59	.9952	3.04	.9988		
2.60	.9953	3.05	.9989		
2.61	.9955	3.06	.9989		
2.62	.9956	3.07	.9989		
2.63	.9957	3.08	.9990		
2.64	.9959	3.09	.9990		
2.65	.9960	3.10	.9990		

SOURCE: M. Orkin and R. Drogin, "Vital Statistics," pp. 350–355, McGraw-Hill, New York, 1975.

TABLE A.4
The t Distribution (Values of t_α)*

df	$t_{.100}$	$t_{.050}$	$t_{.025}$	$t_{.010}$	$t_{.005}$	df
1	3.078	6.314	12.706	31.821	63.657	1
2	1.886	2.920	4.303	6.965	9.925	2
3	1.638	2.353	3.182	4.541	5.841	3
4	1.533	2.132	2.776	3.747	4.604	4
5	1.476	2.015	2.571	3.365	4.032	5
6	1.440	1.943	2.447	3.143	3.707	6
7	1.415	1.895	2.365	2.998	3.499	7
8	1.397	1.860	2.306	2.896	3.355	8
9	1.383	1.833	2.262	2.821	3.250	9
10	1.372	1.812	2.228	2.764	3.169	10
11	1.363	1.796	2.201	2.718	3.106	11
12	1.356	1.782	2.179	2.681	3.055	12
13	1.350	1.771	2.160	2.650	3.012	13
14	1.345	1.761	2.145	2.624	2.977	14
15	1.341	1.753	2.131	2.602	2.947	15
16	1.337	1.746	2.120	2.583	2.921	16
17	1.333	1.740	2.110	2.567	2.898	17
18	1.330	1.734	2.101	2.552	2.878	18
19	1.328	1.729	2.093	2.539	2.861	19
20	1.325	1.725	2.086	2.528	2.845	20
21	1.323	1.721	2.080	2.518	2.831	21
22	1.321	1.717	2.074	2.508	2.819	22
23	1.319	1.714	2.069	2.500	2.807	23
24	1.318	1.711	2.064	2.492	2.797	24
25	1.316	1.708	2.060	2.485	2.787	25
26	1.315	1.706	2.056	2.479	2.779	26
27	1.314	1.703	2.052	2.473	2.771	27
28	1.313	1.701	2.048	2.467	2.763	28
29	1.311	1.699	2.045	2.462	2.756	29
inf.	1.282	1.645	1.960	2.326	2.576	inf.

*This table is abridged from Table IV of R. A. Fisher, "Statistical Methods for Research Workers," copyright 1972 by Hafner Press, and is reproduced with permission of the Hafner Press.

DEGREES OF FREEDOM FOR NUMERATOR

df	1	2	3	4	5	6	7	8	9	10	12	15	20	24	30	40	60	120	∞
1	161	200	216	225	230	234	237	239	241	242	244	246	248	249	250	251	252	253	254
2	18.5	19.0	19.2	19.2	19.3	19.3	19.4	19.4	19.4	19.4	19.4	19.4	19.4	19.5	19.5	19.5	19.5	19.5	19.5
3	10.1	9.55	9.28	9.12	9.01	8.94	8.89	8.85	8.81	8.79	8.74	8.70	8.66	8.64	8.62	8.59	8.57	8.55	8.53
4	7.71	6.94	6.59	6.39	6.26	6.16	6.09	6.04	6.00	5.96	5.91	5.86	5.80	5.77	5.75	5.72	5.69	5.66	5.63
5	6.61	5.79	5.41	5.19	5.05	4.95	4.88	4.82	4.77	4.74	4.68	4.62	4.56	4.53	4.50	4.46	4.43	4.40	4.37
6	5.99	5.14	4.76	4.53	4.39	4.28	4.21	4.15	4.10	4.06	4.00	3.94	3.87	3.84	3.81	3.77	3.74	3.70	3.67
7	5.59	4.74	4.35	4.12	3.97	3.87	3.79	3.73	3.68	3.64	3.57	3.51	3.44	3.41	3.38	3.34	3.30	3.27	3.23
8	5.32	4.46	4.07	3.84	3.69	3.58	3.50	3.44	3.39	3.35	3.28	3.22	3.15	3.12	3.08	3.04	3.01	2.97	2.93
9	5.12	4.26	3.86	3.63	3.48	3.37	3.29	3.23	3.18	3.14	3.07	3.01	2.94	2.90	2.86	2.83	2.79	2.75	2.71
10	4.96	4.10	3.71	3.48	3.33	3.22	3.14	3.07	3.02	2.98	2.91	2.85	2.77	2.74	2.70	2.66	2.62	2.58	2.54
11	4.84	3.98	3.59	3.36	3.20	3.09	3.01	2.95	2.90	2.85	2.79	2.72	2.65	2.61	2.57	2.53	2.49	2.45	2.40
12	4.75	3.89	3.49	3.26	3.11	3.00	2.91	2.85	2.80	2.75	2.69	2.62	2.54	2.51	2.47	2.43	2.38	2.34	2.30
13	4.67	3.81	3.41	3.18	3.03	2.92	2.83	2.77	2.71	2.67	2.60	2.53	2.46	2.42	2.38	2.34	2.30	2.25	2.21
14	4.60	3.74	3.34	3.11	2.96	2.85	2.76	2.70	2.65	2.60	2.53	2.46	2.39	2.35	2.31	2.27	2.22	2.18	2.13
15	4.54	3.68	3.29	3.06	2.90	2.79	2.71	2.64	2.59	2.54	2.48	2.40	2.33	2.29	2.25	2.20	2.16	2.11	2.07
16	4.49	3.63	3.24	3.01	2.85	2.74	2.66	2.59	2.54	2.49	2.42	2.35	2.28	2.24	2.19	2.15	2.11	2.06	2.01
17	4.45	3.59	3.20	2.96	2.81	2.70	2.61	2.55	2.49	2.45	2.38	2.31	2.23	2.19	2.15	2.10	2.06	2.01	1.96
18	4.41	3.55	3.16	2.93	2.77	2.66	2.58	2.51	2.46	2.41	2.34	2.27	2.19	2.15	2.11	2.06	2.02	1.97	1.92
19	4.38	3.52	3.13	2.90	2.74	2.63	2.54	2.48	2.42	2.38	2.31	2.23	2.16	2.11	2.07	2.03	1.98	1.93	1.88
20	4.35	3.49	3.10	2.87	2.71	2.60	2.51	2.45	2.39	2.35	2.28	2.20	2.12	2.08	2.04	1.99	1.95	1.90	1.84
21	4.32	3.47	3.07	2.84	2.68	2.57	2.49	2.42	2.37	2.32	2.25	2.18	2.10	2.05	2.01	1.96	1.92	1.87	1.81
22	4.30	3.44	3.05	2.82	2.66	2.55	2.46	2.40	2.34	2.30	2.23	2.15	2.07	2.03	1.98	1.94	1.89	1.84	1.78
23	4.28	3.42	3.03	2.80	2.64	2.53	2.44	2.37	2.32	2.27	2.20	2.13	2.05	2.01	1.96	1.91	1.86	1.81	1.76
24	4.26	3.40	3.01	2.78	2.62	2.51	2.42	2.36	2.30	2.25	2.18	2.11	2.03	1.98	1.94	1.89	1.84	1.79	1.73
25	4.24	3.39	2.99	2.76	2.60	2.49	2.40	2.34	2.28	2.24	2.16	2.09	2.01	1.96	1.92	1.87	1.82	1.77	1.71
30	4.17	3.32	2.92	2.69	2.53	2.42	2.33	2.27	2.21	2.16	2.09	2.01	1.93	1.89	1.84	1.79	1.74	1.68	1.62
40	4.08	3.23	2.84	2.61	2.45	2.34	2.25	2.18	2.12	2.08	2.00	1.92	1.84	1.79	1.74	1.69	1.64	1.58	1.51
60	4.00	3.15	2.76	2.53	2.37	2.25	2.17	2.10	2.04	1.99	1.92	1.84	1.75	1.70	1.65	1.59	1.53	1.47	1.39
120	3.92	3.07	2.68	2.45	2.29	2.18	2.09	2.02	1.96	1.91	1.83	1.75	1.66	1.61	1.55	1.50	1.43	1.35	1.25
∞	3.84	3.00	2.60	2.37	2.21	2.10	2.01	1.94	1.88	1.83	1.75	1.67	1.57	1.52	1.46	1.39	1.32	1.22	1.00

DEGREES OF FREEDOM FOR DENOMINATOR

TABLE A.5 (Continued)
The F Distribution (Values of $F_{.05}$)*

	DEGREES OF FREEDOM FOR NUMERATOR																		
	1	2	3	4	5	6	7	8	9	10	12	15	20	24	30	40	60	120	∞
1	4,052	5,000	5,403	5,625	5,764	5,859	5,928	5,982	6,023	6,056	6,106	6,157	6,209	6,235	6,261	6,287	6,313	6,339	6,366
2	98.5	99.0	99.2	99.2	99.3	99.3	99.4	99.4	99.4	99.4	99.4	99.4	99.4	99.5	99.5	99.5	99.5	99.5	99.5
3	34.1	30.8	29.5	28.7	28.2	27.9	27.7	27.5	27.3	27.2	27.1	26.9	26.7	26.6	26.5	26.4	26.3	26.2	26.1
4	21.2	18.0	16.7	16.0	15.5	15.2	15.0	14.8	14.7	14.5	14.4	14.2	14.0	13.9	13.8	13.7	13.7	13.6	13.5
5	16.3	13.3	12.1	11.4	11.0	10.7	10.5	10.3	10.2	10.1	9.89	9.72	9.55	9.47	9.38	9.29	9.20	9.11	9.02
6	13.7	10.9	9.78	9.15	8.75	8.47	8.26	8.10	7.98	7.87	7.72	7.56	7.40	7.31	7.23	7.14	7.06	6.97	6.88
7	12.2	9.55	8.45	7.85	7.46	7.19	6.99	6.84	6.72	6.62	6.47	6.31	6.16	6.07	5.99	5.91	5.82	5.74	5.65
8	11.3	8.65	7.59	7.01	6.63	6.37	6.18	6.03	5.91	5.81	5.67	5.52	5.36	5.28	5.20	5.12	5.03	4.95	4.86
9	10.6	8.02	6.99	6.42	6.06	5.80	5.61	5.47	5.35	5.26	5.11	4.96	4.81	4.73	4.65	4.57	4.48	4.40	4.31
10	10.0	7.56	6.55	5.99	5.64	5.39	5.20	5.06	4.94	4.85	4.71	4.56	4.41	4.33	4.25	4.17	4.08	4.00	3.91
11	9.65	7.21	6.22	5.67	5.32	5.07	4.89	4.74	4.63	4.54	4.40	4.25	4.10	4.02	3.94	3.86	3.78	3.69	3.60
12	9.33	6.93	5.95	5.41	5.06	4.82	4.64	4.50	4.39	4.30	4.16	4.01	3.86	3.78	3.70	3.62	3.54	3.45	3.36
13	9.07	6.70	5.74	5.21	4.86	4.62	4.44	4.30	4.19	4.10	3.96	3.82	3.66	3.59	3.51	3.43	3.34	3.25	3.17
14	8.86	6.51	5.56	5.04	4.70	4.46	4.28	4.14	4.03	3.94	3.80	3.66	3.51	3.43	3.35	3.27	3.18	3.09	3.00
15	8.68	6.36	5.42	4.89	4.56	4.32	4.14	4.00	3.89	3.80	3.67	3.52	3.37	3.29	3.21	3.13	3.05	2.96	2.87
16	8.53	6.23	5.29	4.77	4.44	4.20	4.03	3.89	3.78	3.69	3.55	3.41	3.26	3.18	3.10	3.02	2.93	2.84	2.75
17	8.40	6.11	5.19	4.67	4.34	4.10	3.93	3.79	3.68	3.59	3.46	3.31	3.16	3.08	3.00	2.92	2.83	2.75	2.65
18	8.29	6.01	5.09	4.58	4.25	4.01	3.84	3.71	3.60	3.51	3.37	3.23	3.08	3.00	2.92	2.84	2.75	2.66	2.57
19	8.19	5.93	5.01	4.50	4.17	3.94	3.77	3.63	3.52	3.43	3.30	3.15	3.00	2.92	2.84	2.76	2.67	2.58	2.49
20	8.10	5.85	4.94	4.43	4.10	3.87	3.70	3.56	3.46	3.37	3.23	3.09	2.94	2.86	2.78	2.69	2.61	2.52	2.42
21	8.02	5.78	4.87	4.37	4.04	3.81	3.64	3.51	3.40	3.31	3.17	3.03	2.88	2.80	2.72	2.64	2.55	2.46	2.36
22	7.95	5.72	4.82	4.31	3.99	3.76	3.59	3.45	3.35	3.26	3.12	2.98	2.83	2.75	2.67	2.58	2.50	2.40	2.31
23	7.88	5.66	4.76	4.26	3.94	3.71	3.54	3.41	3.30	3.21	3.07	2.93	2.78	2.70	2.62	2.54	2.45	2.35	2.26
24	7.82	5.61	4.72	4.22	3.90	3.67	3.50	3.36	3.26	3.17	3.03	2.89	2.74	2.66	2.58	2.49	2.40	2.31	2.21
25	7.77	5.57	4.68	4.18	3.86	3.63	3.46	3.32	3.22	3.13	2.99	2.85	2.70	2.62	2.53	2.45	2.36	2.27	2.17
30	7.56	5.39	4.51	4.02	3.70	3.47	3.30	3.17	3.07	2.98	2.84	2.70	2.55	2.47	2.39	2.30	2.21	2.11	2.01
40	7.31	5.18	4.31	3.83	3.51	3.29	3.12	2.99	2.89	2.80	2.66	2.52	2.37	2.29	2.20	2.11	2.02	1.92	1.80
60	7.08	4.98	4.13	3.65	3.34	3.12	2.95	2.82	2.72	2.63	2.50	2.35	2.20	2.12	2.03	1.94	1.84	1.73	1.60
120	6.85	4.79	3.95	3.48	3.17	2.96	2.79	2.66	2.56	2.47	2.34	2.19	2.03	1.95	1.86	1.76	1.66	1.53	1.38
∞	6.63	4.61	3.78	3.32	3.02	2.80	2.64	2.51	2.41	2.32	2.18	2.04	1.88	1.79	1.70	1.59	1.47	1.32	1.00

DEGREES OF FREEDOM FOR DENOMINATOR

SOURCE: This table is abridged from E. S. Pearson and H. O. Hartley (eds.), "Biometrika Tables for Statisticians," vol. 1, table 18, pp. 159 and 161, Cambridge University Press, 1962, and is reproduced with permission of the "Biometrika" trustees.

TABLE A.6
The Chi-Square Distribution (Values of χ_α^2)*

df	$\chi_{.995}^2$	$\chi_{.99}^2$	$\chi_{.975}^2$	$\chi_{.95}^2$	$\chi_{.05}^2$	$\chi_{.025}^2$	$\chi_{.01}^2$	$\chi_{.005}^2$	df
1	.0000393	.000157	.000982	.00393	3.841	5.024	6.635	7.879	1
2	.0100	.0201	.0506	.103	5.991	7.378	9.210	10.597	2
3	.0717	.115	.216	.352	7.815	9.348	11.345	12.838	3
4	.207	.297	.484	.711	9.488	11.143	13.277	14.860	4
5	.412	.554	.831	1.145	11.070	12.832	15.086	16.750	5
6	.676	.872	1.237	1.635	12.592	14.449	16.812	18.548	6
7	.989	1.239	1.690	2.167	14.067	16.013	18.475	20.278	7
8	1.344	1.646	2.180	2.733	15.507	17.535	20.090	21.955	8
9	1.735	2.088	2.700	3.325	16.919	19.023	21.666	23.589	9
10	2.156	2.558	3.247	3.940	18.307	20.483	23.209	25.188	10
11	2.603	3.053	3.816	4.575	19.675	21.920	24.725	26.757	11
12	3.074	3.571	4.404	5.226	21.026	23.337	26.217	28.300	12
13	3.565	4.107	5.009	5.892	22.362	24.736	27.688	29.819	13
14	4.075	4.660	5.629	6.571	23.685	26.119	29.141	31.319	14
15	4.601	5.229	6.262	7.261	24.996	27.488	30.578	32.801	15
16	5.142	5.812	6.908	7.962	26.296	28.845	32.000	34.267	16
17	5.697	6.408	7.564	8.672	27.587	30.191	33.409	35.718	17
18	6.265	7.015	8.231	9.390	28.869	31.526	34.805	37.156	18
19	6.844	7.633	8.907	10.117	30.144	32.852	36.191	38.582	19
20	7.434	8.260	9.591	10.851	31.410	34.170	37.566	39.997	20
21	8.034	8.897	10.283	11.591	32.671	35.479	38.932	41.401	21
22	8.643	9.542	10.982	12.338	33.924	36.781	40.289	42.796	22
23	9.260	10.196	11.689	13.091	35.172	38.076	41.638	44.181	23
24	9.886	10.856	12.401	13.848	36.415	39.364	42.980	45.558	24
25	10.520	11.524	13.120	14.611	37.652	40.646	44.314	46.928	25
26	11.160	12.198	13.844	15.379	38.885	41.923	45.642	48.290	26
27	11.808	12.879	14.573	16.151	40.113	43.194	46.963	49.645	27
28	12.461	13.565	15.308	16.928	41.337	44.461	48.278	50.993	28
29	13.121	14.256	16.047	17.708	42.557	45.722	49.588	52.336	29
30	13.787	14.953	16.791	18.493	43.773	46.979	50.892	53.672	30

*This table is abridged from table 8, pages 130 and 131, of "Biometrika Tables for Statisticians," vol. 1, edited by E. S. Pearson and H. O. Hartley, published by Cambridge University Press, 1962, and is reproduced with permission of the "Biometrika" trustees.

TABLE A.7
Common Logarithms (Base 10)*

N	0	1	2	3	4	5	6	7	8	9
10	0000	0043	0086	0128	0170	0212	0253	0294	0334	0374
11	0414	0453	0492	0531	0569	0607	0645	0682	0179	0755
12	0792	0828	0864	0899	0934	0969	1004	1038	1072	1106
13	1139	1173	1206	1239	1271	1303	1335	1367	1399	1430
14	1461	1492	1523	1553	1584	1614	1644	1673	1703	1732
15	1761	1790	1818	1847	1875	1903	1931	1959	1987	2014
16	2041	2068	2095	2122	2148	2175	2201	2227	2253	2279
17	2304	2330	2355	2380	2405	2430	2455	2480	2504	2529
18	2553	2577	2601	2625	2648	2672	2695	2718	2742	2765
19	2788	2810	2833	2856	2878	2900	2923	2945	2967	2989
20	3010	3032	3054	3075	3096	3118	3139	3160	3181	3201
21	3222	3243	3263	3284	3304	3324	3345	3365	3385	3404
22	3424	3444	3464	3483	3502	3522	3541	3560	3579	3598
23	3617	3636	3655	3674	3692	3711	3729	3747	3766	3784
24	3802	3820	3838	3856	3874	3892	3909	3927	3945	3962
25	3979	3997	4014	4031	4048	4065	4082	4099	4116	4133
26	4150	4166	4183	4200	4216	4322	4249	4265	4281	4298
27	4314	4330	4346	4362	4378	4393	4409	4425	4440	4456
28	4472	4487	4502	4518	4533	4548	4564	4579	4594	4609
29	4624	4639	4654	4669	4683	4698	4713	4728	4742	4757
30	4771	4786	4800	4814	4829	4843	4857	4871	4886	4900
31	4914	4928	4942	4955	4969	4983	4997	5011	5024	5038
32	5051	5065	5079	5092	5105	5119	5132	5145	5159	5172
33	5185	5198	5211	5224	5237	5250	5263	5276	5289	5302
34	5315	5328	5340	5353	5366	5378	5391	5403	5416	5428
35	5441	5453	5465	5478	5490	5502	5514	5527	5539	5551
36	5563	5575	5587	5599	5611	5623	5635	5647	5658	5670
37	5682	5694	5705	5717	5729	5740	5752	5763	5775	5786
38	5798	5809	5821	5832	5843	5855	5866	5877	5888	5899
39	5911	5922	5933	5944	5955	5966	5977	5988	5999	6010
40	6021	6031	6042	6053	6064	6075	6085	6096	6107	6117
41	6128	6138	6149	6160	6170	6180	6191	6201	6212	6222
42	6232	6243	6253	6263	6274	6284	6294	6304	6314	6325
43	6335	6345	6355	6365	6375	6385	6395	6405	6415	6425
44	6435	6444	6454	6464	6474	6484	6493	6503	6513	6522
45	6532	6542	6551	6561	6571	6580	6590	6599	6609	6618
46	6628	6637	6646	6656	6665	6675	6684	6693	6702	6712
47	6721	6730	6739	6749	6758	6767	6776	6785	6794	6803
48	6812	6821	6830	6839	6848	6857	6866	6875	6884	6893
49	6902	6911	6920	6928	6937	6946	6955	6964	6972	6981
50	6990	6998	7007	7016	7024	7033	7042	7050	7059	7067
51	7076	7084	7093	7101	7110	7118	7126	7135	7143	7152
52	7160	7168	7177	7185	7193	7202	7210	7218	7226	7235
53	7243	7251	7259	7267	7275	7284	7292	7300	7308	7316
54	7324	7332	7340	7348	7356	7364	7372	7380	7388	7396
N	0	1	2	3	4	5	6	7	8	9

N	0	1	2	3	4	5	6	7	8	9
55	7404	7412	7419	7427	7435	7443	7451	7459	7466	7474
56	7482	7490	7497	7505	7513	7520	7528	7536	7543	7551
57	7559	7566	7574	7582	7589	7597	7604	7612	7619	7627
58	7634	7642	7649	7657	7664	7672	7679	7686	7694	7701
59	7709	7716	7723	7731	7738	7745	7752	7760	7767	7774
60	7782	7789	7796	7803	7810	7818	7825	7832	7839	7846
61	7853	7860	7868	7875	7882	7889	7896	7903	7910	7917
62	7924	7931	7938	7945	7952	7959	7966	7973	7980	7987
63	7993	8000	8007	8014	8021	8028	8035	8041	8048	8055
64	8062	8069	8075	8082	8089	8096	8102	8109	8116	8122
65	8129	8136	8142	8149	8156	8162	8169	8176	8182	8189
66	8195	8202	8209	8215	8222	8228	8235	8241	8248	8254
67	8261	8267	8274	8280	8287	8293	8299	8306	8312	8319
68	8325	8331	8338	8344	8351	8357	8363	8370	8376	8382
69	8388	8395	8401	8407	8414	8420	8426	8432	8439	8445
70	8451	8457	8463	8470	8476	8482	8488	8494	8500	8506
71	8513	8519	8525	8531	8537	8543	8549	8555	8561	8567
72	8573	8579	8585	8591	8597	8603	8609	8615	8621	8627
73	8633	8639	8645	8651	8657	8663	8669	8675	8681	8686
74	8692	8698	8704	8710	8716	8722	8727	8733	8739	8745
75	8751	8756	8762	8768	8774	8779	8785	8791	8797	8802
76	8808	8814	8820	8825	8831	8837	8842	8848	8854	8859
77	8865	8871	8876	8882	8887	8893	8899	8904	8910	8915
78	8921	8927	8932	8938	8943	8949	8954	8960	8965	8971
79	8976	8982	8987	8993	8998	9004	9009	9015	9020	9025
80	9031	9036	9042	9047	9053	9058	9063	9069	9074	9079
81	9085	9090	9096	9101	9106	9112	9117	9122	9128	9133
82	9138	9143	9149	9154	9159	9165	9170	9175	9180	9186
83	9191	9196	9201	9206	9212	9217	9222	9227	9232	9238
84	9243	9248	9253	9258	9263	9269	9274	9279	9284	9289
85	9294	9299	9304	9309	9315	9320	9325	9330	9335	9340
86	9345	9350	9355	9360	9365	9370	9375	9380	9385	9390
87	9395	9400	9405	9410	9415	9420	9425	9430	9435	9440
88	9445	9450	9455	9460	9465	9469	9474	9479	9484	9489
89	9494	9499	9504	9509	9513	9518	9523	9528	9533	9538
90	9542	9547	9552	9557	9562	9566	9571	9576	9581	9586
91	9590	9595	9600	9605	9609	9614	9619	9624	9628	9633
92	9638	9643	9647	9652	9657	9661	9666	9671	9675	9680
93	9685	9689	9694	9699	9703	9708	9713	9717	9722	9727
94	9731	9736	9741	9745	9750	9754	9759	9763	9768	9773
95	9777	9782	9786	9791	9795	9800	9805	9809	9814	9818
96	9823	9827	9832	9836	9841	9845	9850	9854	9859	9863
97	9868	9872	9877	9881	9886	9890	9894	9899	9903	9908
98	9912	9917	9921	9926	9930	9934	9939	9943	9948	9952
99	9956	9961	9965	9969	9974	9978	9983	9987	9991	9996
N	0	1	2	3	4	5	6	7	8	9

*This table lists the logarithm to four decimal places to the base 10 of numbers in the range from 1 to 9.99. For instance, log 3.57 = 0.5527. (First find 35 under N.) Logarithms of numbers outside this range can be calculated as shown in these two examples:

$\log_{10} 357 = \log_{10} (10^2)(3.57) =$
$$\log_{10} 10^2 + \log_{10} 3.57 = 2 + 0.5527 = 2.5527$$

$\log_{10} 0.357 = \log_{10} (10^{-1})(3.57) =$
$$\log_{10} 10^{-1} + \log_{10} 3.57 = (-1) + 0.5527 = -0.4473$$

SOURCE: S. K. Stein, "Calculus and Analytic Geometry," pp. 1014, McGraw-Hill, New York, 1973.

The values of T_α, $T_{1-\alpha}$, and α are such that if the N_1 and N_2 observations are chosen at random from the same population, the chance that the rank sum T of the N_1 observations in the smaller sample is equal to or less than T_α is α and the chance that T is equal to or greater than $T_{1-\alpha}$ is α. The sample sizes are shown in parentheses (N_1, N_2).

TABLE A.8
The Mann-Whitney Test

T_α	$T_{1-\alpha}$	α	T_α	$T_{1-\alpha}$	α	T_α	$T_{1-\alpha}$	α	T_α	$T_{1-\alpha}$	α
	(1,1)			(2,2)			(2,8) (Cont.)			(3,5) (Cont.)	
1	2	.500	3	7	.167	8	14	.267	8	19	.071
	(1,2)		4	6	.333	9	13	.356	9	18	.125
1	3	.333	5	5	.667	10	12	.444	10	17	.196
2	2	.667		(2,3)		11	11	.556	11	16	.286
	(1,3)		3	9	.100		(2,9)		12	15	.393
1	4	.250	4	8	.200	3	21	.018	13	14	.500
2	3	.500	5	7	.400	4	20	.036			
	(1,4)		6	6	.600	5	19	.073		(3,6)	
1	5	.200		(2,4)		6	18	.109	6	24	.012
2	4	.400	3	11	.067	7	17	.164	7	23	.024
3	3	.600	4	10	.133	8	16	.218	8	22	.048
	(1,5)		5	9	.267	9	15	.291	9	21	.083
1	6	.167	6	8	.400	10	14	.364	10	20	.131
2	5	.333	7	7	.600	11	13	.455	11	19	.190
3	4	.500		(2,5)		12	12	.545	12	18	.274
	(1,6)		3	13	.047		(2,10)		13	17	.357
1	7	.143	4	12	.095	3	23	.015	14	16	.452
2	6	.286	5	11	.190	4	22	.030	15	15	.548
3	5	.428	6	10	.286	5	21	.061			
4	4	.571	7	9	.429	6	20	.091		(3,7)	
	(1,7)		8	8	.571	7	19	.136	6	27	.008
1	8	.125		(2,6)		8	18	.182	7	26	.017
2	7	.250	3	15	.036	9	17	.242	8	25	.033
3	6	.375	4	14	.071	10	16	.303	9	24	.058
4	5	.500	5	13	.143	11	15	.379	10	23	.092
	(1,8)		6	12	.214	12	14	.455	11	22	.133
1	9	.111	7	11	.321	13	13	.545	12	21	.192
2	8	.222	8	10	.429		(3,3)		13	20	.258
3	7	.333	9	9	.571	6	15	.050	14	19	.333
4	6	.444		(2,7)		7	14	.100	15	18	.417
5	5	.556	3	17	.028	8	13	.200	16	17	.500
	(1,9)		4	16	.056	9	12	.350		(3,8)	
1	10	.100	5	15	.111	10	11	.500	6	30	.006
2	9	.200	6	14	.167		(3,4)		7	29	.012
3	8	.300	7	13	.250	6	18	.028	8	28	.024
4	7	.400	8	12	.333	7	17	.057	9	27	.042
5	6	.500	9	11	.444	8	16	.114	10	26	.067
	(1,10)		10	10	.556	9	15	.200	11	25	.097
1	11	.091		(2,8)		10	14	.314	12	24	.139
2	10	.182	3	19	.022	11	13	.429	13	23	.188
3	9	.273	4	18	.044	12	12	.571	14	22	.248
4	8	.364	5	17	.089		(3,5)		15	21	.315
5	7	.455	6	16	.133	6	21	.018	16	20	.387
6	6	.545	7	15	.200	7	20	.036	17	19	.461
									18	18	.539

T_α	$T_{1-\alpha}$	α	T_α	$T_{1-\alpha}$	α	T_α	$T_{1-\alpha}$	α	T_α	$T_{1-\alpha}$	α
	(3,9)			(4,5) (*Cont.*)			(4,8) (*Cont.*)			(5,5) (*Cont.*)	
6	33	.005	17	23	.278	24	28	.404	18	37	.028
7	32	.009	18	22	.365	25	27	.467	19	36	.048
8	31	.018	19	21	.452	26	26	.533	20	35	.075
9	30	.032	20	20	.548		(4,9)		21	34	.111
10	29	.050		(4,6)		10	46	.001	22	33	.155
11	28	.073	10	34	.005	11	45	.003	23	32	.210
12	27	.105	11	33	.010	12	44	.006	24	31	.274
13	26	.141	12	32	.019	13	43	.010	25	30	.345
14	25	.186	13	31	.033	14	42	.017	26	29	.421
15	24	.241	14	30	.057	15	41	.025	27	28	.500
16	23	.300	15	29	.086	16	40	.038		(5,6)	
17	22	.363	16	28	.129	17	39	.053	15	45	.002
18	2i	.432	17	27	.176	18	38	.074	16	44	.004
19	20	.500	18	26	.238	19	37	.099	17	43	.009
	(3,10)		19	25	.305	20	36	.130	18	42	.015
6	36	.003	20	24	.381	21	35	.165	19	41	.026
7	35	.007	21	23	.457	22	34	.207	20	40	.041
8	34	.014	22	22	.545	23	33	.252	21	39	.063
9	33	.024		(4,7)		24	32	.302	22	38	.089
10	32	.038	10	38	.003	25	31	.355	23	37	.123
11	31	.056	11	37	.006	26	30	.413	24	36	.165
12	30	.080	12	36	.012	27	29	.470	25	35	.214
13	29	.108	13	35	.021	28	28	.530	26	34	.268
14	28	.143	14	34	.036		(4,10)		27	33	.331
15	27	.185	15	33	.055	10	50	.001	28	32	.396
16	26	.234	16	32	.082	11	49	.002	29	31	.465
17	25	.287	17	31	.115	12	48	.004	30	30	.535
18	24	.346	18	30	.158	13	47	.007		(5,7)	
19	23	.406	19	29	.206	14	46	.012	15	50	.001
20	22	.469	20	28	.264	15	45	.018	16	49	.003
21	21	.531	21	27	.324	16	44	.026	17	48	.005
	(4,4)		22	26	.394	17	43	.038	18	47	.009
10	26	.014	23	25	.464	18	42	.053	19	46	.015
11	25	.029	24	24	.538	19	41	.071	20	45	.024
12	24	.057		(4,8)		20	40	.094	21	44	.037
13	23	.100	10	42	.002	21	39	.120	22	43	.053
14	22	.171	11	41	.004	22	38	.152	23	42	.074
15	21	.243	12	40	.008	23	37	.187	24	41	.101
16	20	.343	13	39	.014	24	36	.227	25	40	.134
17	19	.443	14	38	.024	25	35	.270	26	39	.172
18	18	.557	15	37	.036	26	34	.318	27	38	.216
	(4,5)		16	36	.055	27	33	.367	28	37	.265
10	30	.008	17	35	.077	28	32	.420	29	36	.319
11	29	.016	18	34	.107	29	31	.473	30	35	.378
12	28	.032	19	33	.141	30	30	.527	31	34	.438
13	27	.056	20	32	.184		(5,5)		32	33	.500
14	26	.095	21	31	.230	15	40	.004		(5,8)	
15	25	.143	22	30	.285	16	39	.008	15	55	.001
16	24	.206	23	29	.341	17	38	.016	16	54	.002

TABLE A.8 (Continued)
The Mann-Whitney Test

T_α	$T_{1-\alpha}$	α	T_α	$T_{1-\alpha}$	α	T_α	$T_{1-\alpha}$	α	T_α	$T_{1-\alpha}$	α
(5,8) (Cont.)			(5,10) (Cont.)			(6,7) (Cont.)			(6,9) (Cont.)		
17	53	.003	20	60	.006	28	56	.026	28	68	.009
18	52	.005	21	59	.010	29	55	.037	29	67	.013
19	51	.009	22	58	.014	30	54	.051	30	66	.018
20	50	.015	23	57	.020	31	53	.069	31	65	.025
21	49	.023	24	56	.028	32	52	.090	32	64	.033
22	48	.033	25	55	.038	33	51	.117	33	63	.044
23	47	.047	26	54	.050	34	50	.147	34	62	.057
24	46	.064	27	53	.065	35	49	.183	35	61	.072
25	45	.085	28	52	.082	36	48	.223	36	60	.091
26	44	.111	29	51	.103	37	47	.267	37	59	.112
27	43	.142	30	50	.127	38	46	.314	38	58	.136
28	42	.177	31	49	.155	39	45	.365	39	57	.164
29	41	.217	32	48	.185	40	44	.418	40	56	.194
30	40	.262	33	47	.220	41	43	.473	41	55	.228
31	39	.311	34	46	.257	42	42	.527	42	54	.264
32	38	.362	35	45	.297	(6,8)			43	53	.303
33	37	.416	36	44	.339	21	69	.000	44	52	.344
34	36	.472	37	43	.384	22	68	.001	45	51	.388
35	35	.528	38	42	.430	23	67	.001	46	50	.432
(5,9)			39	41	.477	24	66	.002	47	49	.477
15	60	.000	40	40	.523	25	65	.004	48	48	.523
16	59	.001	(6,6)			26	64	.006	(6,10)		
17	58	.002	21	57	.001	27	63	.010	21	81	.000
18	57	.003	22	56	.002	28	62	.015	22	80	.000
19	56	.006	23	55	.004	29	61	.021	23	79	.000
20	55	.009	24	54	.008	30	60	.030	24	78	.001
21	54	.014	25	53	.013	31	59	.041	25	77	.001
22	53	.021	26	52	.021	32	58	.054	26	76	.002
23	52	.030	27	51	.032	33	57	.071	27	75	.004
24	51	.041	28	50	.047	34	56	.091	28	74	.005
25	50	.056	29	49	.066	35	55	.114	29	73	.008
26	49	.073	30	48	.090	36	54	.141	30	72	.011
27	48	.095	31	47	.120	37	53	.172	31	71	.016
28	47	.120	32	46	.155	38	52	.207	32	70	.021
29	46	.149	33	45	.197	39	51	.245	33	69	.028
30	45	.182	34	44	.242	40	50	.286	34	68	.036
31	44	.219	35	43	.294	41	49	.331	35	67	.047
32	43	.259	36	42	.350	42	48	.377	36	66	.059
33	42	.303	37	41	.409	43	47	.426	37	65	.074
34	41	.350	38	40	.469	44	46	.475	38	64	.090
35	40	.399	39	39	.531	45	45	.525	39	63	.110
36	39	.449	(6,7)			(6,9)			40	62	.132
37	38	.500	21	63	.001	21	75	.000	41	61	.157
(5,10)			22	62	.001	22	74	.000	42	60	.184
15	65	.000	23	61	.002	23	73	.001	43	59	.214
16	64	.001	24	60	.004	24	72	.001	44	58	.246
17	63	.001	25	59	.007	25	71	.002	45	57	.281
18	62	.002	26	58	.011	26	70	.004	46	56	.318
19	61	.004	27	57	.017	27	69	.006	47	55	.356

TABLE A.8 (Continued)
The Mann-Whitney Test

T_α	$T_{1-\alpha}$	α	T_α	$T_{1-\alpha}$	α	T_α	$T_{1-\alpha}$	α	T_α	$T_{1-\alpha}$	α
(6,10) *(Cont.)*			(7,8) *(Cont.)*			(7,10) *(Cont.)*			(8,8) *(Cont.)*		
48	54	.396	46	66	.140	32	94	.001	52	84	.052
49	53	.437	47	65	.168	33	93	.001	53	83	.065
50	52	.479	48	64	.198	34	92	.001	54	82	.080
51	51	.521	49	63	.232	35	91	.002	55	81	.097
(7,7)			50	62	.268	36	90	.003	56	80	.117
28	77	.000	51	61	.306	37	89	.005	57	79	.139
29	76	.001	52	60	.347	38	88	.007	58	78	.164
30	75	.001	53	59	.389	39	87	.009	59	77	.191
31	74	.002	54	58	.433	40	86	.012	60	76	.221
32	73	.003	55	57	.478	41	85	.017	61	75	.253
33	72	.006	56	56	.522	42	84	.022	62	74	.287
34	71	.009	(7,9)			43	83	.028	63	73	.323
35	70	.013	28	91	.000	44	82	.035	64	72	.360
36	69	.019	29	90	.000	45	81	.044	65	71	.399
37	68	.027	30	89	.000	46	80	.054	66	70	.439
38	67	.036	31	88	.001	47	79	.067	67	69	.480
39	66	.049	32	87	.001	48	78	.081	68	68	.520
40	65	.064	33	86	.002	49	77	.097	(8,9)		
41	64	.082	34	85	.003	50	76	.115	36	108	.000
42	63	.104	35	84	.004	51	75	.135	40	104	.000
43	62	.130	36	83	.006	52	74	.157	41	103	.001
44	61	.159	37	82	.008	53	73	.182	42	102	.001
45	60	.191	38	81	.011	54	72	.209	43	101	.002
46	59	.228	39	80	.016	55	71	.237	44	100	.003
47	58	.267	40	79	.021	56	70	.268	45	99	.004
48	57	.310	41	78	.027	57	69	.300	46	98	.006
49	56	.355	42	77	.036	58	68	.335	47	97	.008
50	55	.402	43	76	.045	59	67	.370	48	96	.010
51	54	.451	44	75	.057	60	66	.406	49	95	.014
52	53	.500	45	74	.071	61	65	.443	50	94	.018
(7,8)			46	73	.087	62	64	.481	51	93	.023
28	84	.000	47	72	.105	63	63	.519	52	92	.030
29	83	.000	48	71	.126	(8,8)			53	91	.037
30	82	.001	49	70	.150	36	100	.000	54	90	.046
31	81	.001	50	69	.175	37	99	.000	55	89	.057
32	80	.002	51	68	.204	38	98	.000	56	88	.069
33	79	.003	52	67	.235	39	97	.001	57	87	.084
34	78	.005	53	66	.268	40	96	.001	58	86	.100
35	77	.007	54	65	.303	41	95	.001	59	85	.118
36	76	.010	55	64	.340	42	94	.002	60	84	.138
37	75	.014	56	63	.379	43	93	.003	61	83	.161
38	74	.020	57	62	.419	44	92	.005	62	82	.185
39	73	.027	58	61	.459	45	91	.007	63	81	.212
40	72	.036	59	60	.500	46	90	.010	64	80	.240
41	71	.047	(7,10)			47	89	.014	65	79	.271
42	70	.060	28	98	.000	48	88	.019	66	78	.303
43	69	.076	29	97	.000	49	87	.025	67	77	.336
44	68	.095	30	96	.000	50	86	.032	68	76	.371
45	67	.116	31	95	.000	51	85	.041	69	75	.407

T_α	$T_{1-\alpha}$	α	T_α	$T_{1-\alpha}$	α	T_α	$T_{1-\alpha}$	α	T_α	$T_{1-\alpha}$	α
(8,9) (*Cont.*)			(9,9)			(9,10) (*Cont.*)			(10,10) (*Cont.*)		
70	74	.444	45	126	.000	54	126	.001	65	145	.001
71	73	.481	50	121	.000	55	125	.001	66	144	.001
72	72	.519	51	120	.001	56	124	.002	67	143	.001
(8,10)			52	119	.001	57	123	.003	68	142	.002
36	116	.000	53	118	.001	58	122	.004	69	141	.003
41	111	.000	54	117	.002	59	121	.005	70	140	.003
42	110	.001	55	116	.003	60	120	.007	71	139	.004
43	109	.001	56	115	.004	61	119	.009	72	138	.006
44	108	.002	57	114	.005	62	118	.011	73	137	.007
45	107	.002	58	113	.007	63	117	.014	74	136	.009
46	106	.003	59	112	.009	64	116	.017	75	135	.012
47	105	.004	60	111	.012	65	115	.022	76	134	.014
48	104	.006	61	110	.016	66	114	.027	77	133	.018
49	103	.008	62	109	.020	67	113	.033	78	132	.022
50	102	.010	63	108	.025	68	112	.039	79	131	.026
51	101	.013	64	107	.031	69	111	.047	80	130	.032
52	100	.017	65	106	.039	70	110	.056	81	129	.038
53	99	.022	66	105	.047	71	109	.067	82	128	.045
54	98	.027	67	104	.057	72	108	.078	83	127	.053
55	97	.034	68	103	.068	73	107	.091	84	126	.062
56	96	.042	69	102	.081	74	106	.106	85	125	.072
57	95	.051	70	101	.095	75	105	.121	86	124	.083
58	94	.061	71	100	.111	76	104	.139	87	123	.095
59	93	.073	72	99	.129	77	103	.158	88	122	.109
60	92	.086	73	98	.149	78	102	.178	89	121	.124
61	91	.102	74	97	.170	79	101	.200	90	120	.140
62	90	.118	75	96	.193	80	100	.223	91	119	.157
63	89	.137	76	95	.218	81	99	.248	92	118	.176
64	88	.158	77	94	.245	82	98	.274	93	117	.197
65	87	.180	78	93	.273	83	97	.302	94	116	.218
66	86	.204	79	92	.302	84	96	.330	95	115	.241
67	85	.230	80	91	.333	85	95	.360	96	114	.264
68	84	.257	81	90	.365	86	94	.390	97	113	.289
69	83	.286	82	89	.398	87	93	.421	98	112	.315
70	82	.317	83	88	.432	88	92	.452	99	111	.342
71	81	.348	84	87	.466	89	91	.484	100	110	.370
72	80	.381	85	86	.500	90	90	.516	101	109	.398
73	79	.414	(9,10)			(10,10)			102	108	.427
74	78	.448	45	135	.000	55	155	.000	103	107	.456
75	77	.483	52	128	.000	63	147	.000	104	106	.485
76	76	.517	53	127	.001	64	146	.001	105	105	.515

NOTE: For sample sizes greater than 10 the chance that the statistic T will be less than or equal to an integer k is given approximately by the area under the standard normal curve to the left of

$$z = \frac{k + \frac{1}{2} - N_1(N_1 + N_2 + 1)/2}{\sqrt{N_1 N_2 (N_1 + N_2 + 1)/12}}$$

SOURCE: W. J. Dixon and F. J. Massey, Jr., "Introduction to Statistical Analysis," 3d ed., pp. 545–549, McGraw-Hill, New York, 1969.

The value α is the probability that Σd_i^2 is less than or equal to the table value (or that r_s is greater than or equal to the table value). An observed value of Σd_i^2 is located in the table heading and the corresponding value of r_s and α are in the body of the table arranged by sample size N.

TABLE A.9
Rank Correlation Tables

N		VALUES OF Σd_i^2												
		0	2	4	6	8	10	12	14	16	18	20	22	24
2	α	.500	1.000											
	r_s	1.00	−1.00											
3	α	.167	.500	—	.833									
	r_s	1.00	.50	—	−.50									
4	α	.042	.167	.208	.375	.458	.542							
	r_s	1.00	.80	.60	.40	.20	.00							
5	α	.008	.042	.067	.117	.175	.225	.258	.342	.392	.475	.525		
	r_s	1.00	.90	.80	.70	.60	.50	.40	.30	.20	.10	.00		
6	α	.001	.008	.017	.029	.051	.068	.088	.121	.149	.178	.210	.249	.282
	r_s	1.00	.94	.89	.83	.77	.71	.66	.60	.54	.49	.43	.37	.31
7	α	.000	.001	.003	.006	.012	.017	.024	.033	.044	.055	.069	.083	.100
	r_s	1.00	.96	.93	.89	.86	.82	.79	.75	.71	.68	.64	.61	.57
8	α	.000	.000	.001	.001	.002	.004	.005	.008	.011	.014	.018	.023	.029
	r_s	1.00	.98	.95	.93	.90	.88	.86	.83	.81	.79	.76	.74	.71
9	α	.000	.000	.000	.000	.000	.001	.001	.002	.002	.003	.004	.005	.007
	r_s	1.00	.98	.97	.95	.93	.92	.90	.88	.87	.85	.83	.82	.80
10	α	.000	.000	.000	.000	.000	.000	.000	.000	.000	.001	.001	.001	.001
	r_s	1.00	.99	.98	.96	.95	.94	.93	.92	.90	.89	.88	.87	.86

TABLE A. 9 (*Continued*)
Rank Correlation Tables

N		26	28	30	32	34	36	38	40	42	44	46	48
5	α	.742	.775	.825	.883	.933	.958	.992	1.000				
	r_s	-.30	-.40	-.50	-.60	-.70	-.80	-.90	-1.00				
6	α	.329	.357	.401	.460	.500	.540	.599	.643	.671	.718	.751	.790
	r_s	.26	.20	.14	.09	.03	-.03	-.09	-.14	-.20	-.26	-.31	-.37
7	α	.118	.133	.151	.177	.200	.222	.249	.278	.297	.331	.357	.391
	r_s	.54	.50	.46	.43	.39	.36	.32	.29	.25	.21	.18	.14
8	α	.035	.042	.048	.057	.066	.076	.085	.098	.108	.122	.134	.150
	r_s	.69	.67	.64	.62	.60	.57	.55	.52	.50	.48	.45	.43
9	α	.009	.011	.013	.016	.018	.022	.025	.029	.033	.038	.043	.048
	r_s	.78	.77	.75	.73	.72	.70	.68	.67	.65	.63	.62	.60
10	α	.002	.002	.003	.004	.004	.005	.006	.007	.009	.010	.012	.013
	r_s	.84	.83	.82	.81	.79	.78	.77	.76	.75	.73	.72	.71

N		50	52	54	56	58	60	62	64	66	68	70	72	74
6	α	.822	.851	.879	.913	.932	.949	.971	.983	.992	.999	1.000		
	r_s	-.43	-.49	-.54	-.60	-.66	-.71	-.77	-.83	-.89	-.94	-1.00		
7	α	.420	.453	.482	.518	.547	.580	.609	.643	.669	.703	.723	.751	.778
	r_s	.11	.07	.04	.00	-.04	-.07	-.11	-.14	-.18	-.21	-.25	-.29	-.32
8	α	.163	.180	.195	.214	.231	.250	.268	.291	.310	.332	.352	.376	.397
	r_s	.40	.38	.36	.33	.31	.29	.26	.24	.21	.19	.17	.14	.12
9	α	.054	.060	.066	.074	.081	.089	.097	.106	.115	.125	.135	.146	.156
	r_s	.58	.57	.55	.53	.52	.50	.48	.47	.45	.43	.42	.40	.38
10	α	.015	.017	.019	.022	.024	.027	.030	.033	.037	.040	.044	.048	.052
	r_s	.70	.68	.67	.66	.65	.64	.62	.61	.60	.59	.58	.56	.55

Block 1

N		76	78	80	82	84	86	88	90	92	94	96	98	100
7	α	.802	.823	.849	.867	.882	.900	.917	.931	.945	.956	.967	.976	.983
	r_s	−.36	−.39	−.43	−.46	−.50	−.54	−.57	−.61	−.64	−.68	−.71	−.75	−.76
8	α	.420	.441	.467	.488	.512	.533	.559	.580	.603	.624	.648	.668	.690
	r_s	.10	.07	.05	.02	.00	−.02	−.05	−.07	−.10	−.12	−.14	−.17	−.19
9	α	.168	.179	.193	.205	.218	.231	.243	.260	.276	.290	.307	.322	.339
	r_s	.37	.35	.33	.32	.30	.28	.27	.25	.23	.22	.20	.18	.17
10	α	.057	.062	.067	.072	.077	.083	.089	.096	.102	.109	.116	.124	.132
	r_s	.54	.53	.52	.50	.49	.48	.47	.45	.44	.43	.42	.41	.39

Block 2

N		102	104	106	108	110	112	114	116	118	120	122	124	126
7	α	.988	.944	.997	.999	1.000	1.000							
	r_s	−.82	−.86	−.89	−.93	−.96	−1.00							
8	α	.709	.732	.750	.769	.786	.805	.820	.837	.850	.866	.878	.892	.902
	r_s	−.21	−.24	−.26	−.29	−.31	−.33	−.36	−.38	−.40	−.43	−.45	−.48	−.50
9	α	.354	.372	.388	.405	.422	.440	.456	.474	.491	.509	.526	.544	.560
	r_s	.15	.13	.12	.10	.08	.07	.05	.03	.02	.00	−.02	−.03	−.05
10	α	.139	.148	.156	.165	.174	.184	.193	.203	.214	.224	.235	.246	.257
	r_s	.38	.37	.36	.35	.33	.32	.31	.30	.29	.27	.26	.25	.24

Block 3

N		128	130	132	134	136	138	140	142	144	146	148	150	152
8	α	.915	.924	.934	.943	.952	.958	.965	.971	.977	.982	.986	.989	.992
	r_s	−.52	−.55	−.57	−.60	−.62	−.64	−.67	−.69	−.71	−.74	−.76	−.79	−.81
9	α	.578	.595	.612	.628	.646	.661	.678	.693	.710	.724	.740	.753	.769
	r_s	−.07	−.08	−.10	−.12	−.13	−.15	−.17	−.18	−.20	−.22	−.23	−.25	−.27
10	α	.268	.280	.292	.304	.316	.328	.341	.354	.367	.379	.393	.406	.419
	r_s	.22	.21	.20	.19	.18	.16	.15	.14	.13	.12	.10	.09	.08

TABLE A.9 (*Continued*)
Rank Correlation Tables

N	154	156	158	160	162	164	166	168	170	172	174	176	178
8 α	.995	.996	.998	.999	.999	1.000	1.000	1.000					
r_s	−.83	−.86	−.88	−.90	−.93	−.95	−.98	−1.00					
9 α	.782	.795	.807	.821	.832	.844	.854	.865	.875	.885	.894	.903	.911
r_s	−.28	−.30	−.32	−.33	−.35	−.37	−.38	−.40	−.42	−.43	−.45	−.47	−.48
10 α	.433	.446	.459	.473	.486	.500	.514	.527	.541	.554	.567	.581	.594
r_s	.07	.05	.04	.03	.02	.01	−.01	−.02	−.03	−.04	−.05	−.07	−.08

N	180	182	184	186	188	190	192	194	196	198	200	202	204
9 α	.919	.926	.934	.940	.946	.952	.957	.962	.967	.971	.975	.978	.982
r_s	−.50	−.52	−.53	−.55	−.57	−.58	−.60	−.61	−.63	−.65	−.67	−.68	−.70
10 α	.607	.621	.633	.646	.659	.672	.684	.696	.708	.720	.732	.743	.754
r_s	−.09	−.10	−.12	−.13	−.14	−.15	−.16	−.18	−.19	−.20	−.21	−.22	−.24

N	206	208	210	212	214	216	218	220	222	224	226	228	230
9 α	.984	.987	.989	.991	.993	.995	.996	.997	.998	.998	.999	.999	1.000
r_s	−.72	−.73	−.75	−.77	−.78	−.80	−.82	−.83	−.85	−.87	−.88	−.90	−.92
10 α	.765	.776	.786	.797	.807	.816	.826	.835	.844	.852	.861	.868	.876
r_s	−.25	−.26	−.27	−.28	−.30	−.31	−.32	−.33	−.35	−.36	−.37	−.38	−.39

N	232	234	236	238	240	242	244	246	248	250	252	254	256
9 α	1.000	1.000	1.000	1.000	1.000								
r_s	−.93	−.95	−.97	−.98	−1.00								
10 α	.884	.891	.898	.904	.911	.917	.923	.928	.933	.938	.943	.948	.952
r^*	−.41	−.42	−.43	−.44	−.45	−.47	−.48	−.49	−.50	−.52	−.53	−.54	−.55

N	258	260	262	264	266	268	270	272	274	276	278	280	282
10 α	.956	.960	.963	.967	.970	.973	.976	.978	.981	.983	.985	.987	.988
r_s	−.56	−.58	−.59	−.60	−.61	−.62	−.64	−.65	−.66	−.67	−.68	−.70	−.71

N	284	286	288	290	292	294	296	298	300	302	304	306	308
10 α	.990	.991	.993	.994	.995	.996	.996	.997	.998	.998	.999	.999	.999
r_s	−.72	−.73	−.75	−.76	−.77	−.78	−.79	−.81	−.82	−.83	−.84	−.85	−.87

N	310	312	314	316	318	320	322	324	326	328	330
10 α	.999	1.000	1.000	1.000	1.000	1.000	1.000	1.000	1.000	1.000	1.000
r_s	−.88	−.89	−.90	−.92	−.93	−.94	−.95	−.96	−.98	−.99	−1.00

SOURCE: W. J. Dixon and F. J. Massey, Jr., "Introduction to Statistical Analysis," 3d ed., pp. 570–574, McGraw-Hill, New York, 1969.

$P(X \leq N)$

λ \ N	0	1	2	3	4	5	6	7	8	9	10	11	12
.05	.951	.999	1.000										
.10	.905	.995	1.000										
.15	.861	.990	.999	1.000									
.20	.819	.982	.999	1.000									
.25	.779	.974	.998	1.000									
.30	.741	.963	.996	1.000									
.35	.705	.951	.994	1.000									
.40	.670	.938	.992	.999	1.000								
.45	.638	.925	.989	.999	1.000								
.50	.607	.910	.986	.998	1.000								
.55	.577	.894	.982	.998	1.000								
.60	.549	.878	.977	.997	1.000								
.65	.522	.861	.972	.996	.999	1.000							
.70	.497	.844	.966	.994	.999	1.000							
.75	.472	.827	.959	.993	.999	1.000							
.80	.449	.809	.953	.991	.999	1.000							
.85	.427	.791	.945	.989	.998	1.000							
.90	.407	.772	.937	.987	.998	1.000							
.95	.387	.754	.929	.984	.997	1.000							
1.00	.368	.736	.920	.981	.996	.999	1.000						
1.1	.333	.699	.900	.974	.995	.999	1.000						
1.2	.301	.663	.879	.966	.992	.998	1.000						
1.3	.273	.627	.857	.957	.989	.998	1.000						
1.4	.247	.592	.833	.946	.986	.997	.999	1.000					
1.5	.223	.558	.809	.934	.981	.996	.999	1.000					
1.6	.202	.525	.783	.921	.976	.994	.999	1.000					
1.7	.183	.493	.757	.907	.970	.992	.998	1.000					
1.8	.165	.463	.731	.891	.964	.990	.997	.999	1.000				
1.9	.150	.434	.704	.875	.956	.987	.997	.999	1.000				
2.0	.135	.406	.677	.857	.947	.983	.995	.999	1.000				
2.2	.111	.355	.623	.819	.928	.975	.993	.998	1.000				
2.4	.091	.308	.570	.779	.904	.964	.988	.997	.999	1.000			
2.6	.074	.267	.518	.736	.877	.951	.983	.995	.999	1.000			
2.8	.061	.231	.469	.692	.848	.935	.976	.992	.998	.999	1.000		
3.0	.050	.199	.423	.647	.815	.916	.966	.988	.996	.999	1.000		
3.2	.041	.171	.380	.603	.781	.895	.955	.983	.994	.998	1.000		
3.4	.033	.147	.340	.558	.744	.871	.942	.977	.992	.997	.999	1.000	
3.6	.027	.126	.303	.515	.706	.844	.927	.969	.988	.996	.999	1.000	
3.8	.022	.107	.269	.473	.668	.816	.909	.960	.984	.994	.998	.999	1.000
4.0	.018	.092	.238	.433	.629	.785	.889	.949	.979	.992	.997	.999	1.000

SOURCE: W. J. Dixon and F. J. Massey, Jr., "Introduction to Statistical Analysis," 3d ed., p. 533, McGraw-Hill, New York, 1969.

INDEX

Dependence:
  measure of, 337
  nature of, 337
  in probability theory, 298,
      323–325, 337, 345
  in regression, 208–209
Descriptive statistics, 96, 101
Determinant, 264
Determination:
  coefficient of, 220–223, 253, 268
  linear, 220
  multiple, 269
Deviation:
  from mean, 32
    (See also Mean, balancing
        property of)
  mean absolute (see Mean
        absolute deviation)
  squared, 210
  standard (see Standard deviation)
  from straight line, 209
  vertical, 209
Die, 279, 291
  fair, 284
Discrete, 86, 189
Disjoint, 280, 288
Dispersion, 401, 415
  Mann-Whitney test for, 415–417
Distribution-free, 402

Error:
  of compounding tests, 362–363
  of estimation in confidence
      intervals, 111–113
  of predicting from contingency
      tables, 338–344
    (See also Index of predictive
        association)
  round-off, 44
  standard, 233
  sum of squares, 383, 389
  Type I, 135–139, 362–363, 409
  Type II, 135–139, 409
  worse, 135–138

Event, 279
  arithmetic of, 279–280
  disjoint, 280, 288
  empty, 280
  favorable, 292
  independent, 293, 298–302,
      324–334
  probability of, 284, 290
  unfavorable, 292
Expected frequencies, 328, 334
Exponential data, 218, 223, 229–230
Exponential regression, 255–260
  power, 256–260
  simple, 256–258

F distribution, 179–182
  degrees of freedom of, 181–185,
      367–368, 380
  percentage points of, 181–186
  table of, 179, 182, 487–488
F test:
  in analysis of variance, 367–370,
      379–382, 389–391, 417
  for multiple linearity, 273
  rejection rules for, 182–185
  for standard deviations, 180–186,
      367, 401
Factorial, 306, 309
Freedom, degrees of (see Degrees
      of freedom)
Frequency distribution, 3–7
  construction of, 4
  graphs of, 5–6
  testing for normality, 352–353
Frequency polygon, 5–7, 80–81,
      123

Goodman-Kruskal index, 337
  (See also Index of predictive
      association)
Goodness-of-fit test, 347–355
Graph:
  bar (see Histogram)
  line (see Frequency polygon)